Retrovirology: Clinical Aspects

Retrovirology: Clinical Aspects

Editor: Adeline Foley

AMERICAN
MEDICAL PUBLISHERS
www.americanmedicalpublishers.com

AMERICAN
MEDICAL PUBLISHERS
www.americanmedicalpublishers.com

Cataloging-in-Publication Data

Retrovirology : clinical aspects / edited by Adeline Foley.
p. cm.
Includes bibliographical references and index.
ISBN 978-1-63927-264-8
1. Retroviruses. 2. Retrovirus infections--Treatment. 3. Oncogenic viruses. 4. RNA viruses.
5. Medical microbiology. I. Foley, Adeline.
QR414.5 .R48 2021
579.256 9--dc23

American Medical Publishers,
41 Flatbush Avenue,
1st Floor, New York,
NY 11217, USA

ISBN 978-1-63927-264-8 (Hardback)

Contents

Preface

The world is advancing at a fast pace like never before. Therefore, the need is to keep up with the latest developments. This book was an idea that came to fruition when the specialists in the area realized the need to coordinate together and document essential themes in the subject. That's when I was requested to be the editor. Editing this book has been an honour as it brings together diverse authors researching on different streams of the field. The book collates essential materials contributed by veterans in the area which can be utilized by students and researchers alike.

The study of retroviruses is known as retrovirology. Retrovirus is any group of viruses that belong to the family of retroviridae, these viruses carry their genetic blueprint in the form of ribonucleic acid. In most viruses, DNA is translated into RNA, and then RNA is translated into protein. However, retroviruses function differently, as their RNA is reversed into DNA. Retroviruses affect both humans and animals. In animals, it causes tumor growth and certain cancers that are linked with slow infections such as equine infectious anemia. Whereas in human beings, retroviruses such as HTLV-1 cause cancer called adult T-cell leukemia. HIV retrovirus in humans causes disease like AIDS. The treatment of infection by retroviruses includes antiretroviral drugs and medications. This book contains some path-breaking studies in the field of retrovirology. It presents researches and studies performed by experts across the globe. Students, researchers, experts and all associated with retrovirology will benefit alike from this book.

Each chapter is a sole-standing publication that reflects each author´s interpretation. Thus, the book displays a multi-facetted picture of our current understanding of application, resources and aspects of the field. I would like to thank the contributors of this book and my family for their endless support.

Editor

1

Modulation of chromatin structure by the FACT histone chaperone complex regulates HIV-1 integration

Julien Matysiak[1,8†], Paul Lesbats[1,8†], Eric Mauro[1,8†], Delphine Lapaillerie[1,8,9], Jean-William Dupuy[2], Angelica P. Lopez[3], Mohamed Salah Benleulmi[1,8,9], Christina Calmels[1,8,9], Marie-Line Andreola[1,8,9], Marc Ruff[4], Manuel Llano[3], Olivier Delelis[5,9], Marc Lavigne[6,7,9] and Vincent Parissi[1,8,9*]

Abstract

Background: Insertion of retroviral genome DNA occurs in the chromatin of the host cell. This step is modulated by chromatin structure as nucleosomes compaction was shown to prevent HIV-1 integration and chromatin remodeling has been reported to affect integration efficiency. LEDGF/p75-mediated targeting of the integration complex toward RNA polymerase II (polII) transcribed regions ensures optimal access to dynamic regions that are suitable for integration. Consequently, we have investigated the involvement of polII-associated factors in the regulation of HIV-1 integration.

Results: Using a pull down approach coupled with mass spectrometry, we have selected the FACT (FAcilitates Chromatin Transcription) complex as a new potential cofactor of HIV-1 integration. FACT is a histone chaperone complex associated with the polII transcription machinery and recently shown to bind LEDGF/p75. We report here that a tripartite complex can be formed between HIV-1 integrase, LEDGF/p75 and FACT in vitro and in cells. Biochemical analyzes show that FACT-dependent nucleosome disassembly promotes HIV-1 integration into chromatinized templates, and generates highly favored nucleosomal structures in vitro. This effect was found to be amplified by LEDGF/p75. Promotion of this FACT-mediated chromatin remodeling in cells both increases chromatin accessibility and stimulates HIV-1 infectivity and integration.

Conclusions: Altogether, our data indicate that FACT regulates HIV-1 integration by inducing local nucleosomes dissociation that modulates the functional association between the incoming intasome and the targeted nucleosome.

Keywords: Retroviral integration, HIV-1, Integrase, Chromatin, Nucleosome, FACT

Background

Integration of the retroviral genome into the host chromosomes, catalyzed by the integrase protein (IN), is a prerequisite for viral replication (for a recent review on integration see [1]). This process is regulated by cellular factors at several stages including nuclear import of the preintegration complex (PIC) and association with chromatin loci [2, 3]. Integration appears to occur preferentially into nucleosomal target DNA both in vitro and in infected cells but the chromatin structures and regions targeted by the integration complexes depend on the virus [3–8]. This preferential integration into nucleosomes was assumed to be due to the preference of IN for bent DNA, which can be found at the surface of the nucleosome [9]. This was confirmed by determining the complex formed between the foamy virus (PFV) intasome capture complex and the human nucleosome, showing a close association between the retroviral IN and highly bent target DNA at the nucleosome surface

*Correspondence: vincent.parissi@u-bordeaux.fr
† Julien Matysiak, Paul Lesbats and Eric Mauro are contributed equally to this work
[1] Fundamental Microbiology and Pathogenicity Laboratory, UMR 5234 CNRS, University of Bordeaux, SFR TransBioMed, 146 rue Léo Saignat, 33076 Bordeaux Cedex, France
Full list of author information is available at the end of the article

[7]. Recent works confirmed that nucleosomes and DNA bendability are major determinants of the HIV-1 IN selectivity [3, 9–11]. Additionally, the functional association between retroviral intasomes and chromatin has been shown to be modulated both by chromatin and intasome structures, thereby governing their preference for specific target DNA flexibility and nucleosome density both in vitro and in vivo depending on the retroviral genus [8]. Consequently, this leads to a specific and distinct requirement for chromatin structures depending on the viral integration machinery. Indeed, previous works showed that, while Avian Sarkoma Leukosis Virus (ASLV) and HIV-1 integration was preferred in regions of the chromatin with low nucleosomes density both in vitro and in vivo, PFV and Murine Leukemia Virus (MLV) integration accommodate more easily different chromatin structures with a significant preference for regions of high nucleosomes density [8].

In the cell, these suitable chromatin loci can be reached thanks to specific interactions between retroviral intasomes and cellular targeting cofactors such as LEDGF/p75, CPSF6 and BET proteins (for reviews about integration selectivity see [3, 12, 13]). While previous studies have shown that mononucleosomes are preferential substrates for retroviral integration in vitro, it has also been reported that the physiological full-site HIV-1 integration of both viral DNA ends into polynucleosomal compacted chromatin may require coupling with local additional remodeling activity [14, 15]. Interestingly, HIV-1 integration is promoted in the chromatin regions highly transcribed by the RNA polymerase II (PolII) machinery, where the nucleosomes are highly dynamic [8, 16]. Moreover, the structure of the complex formed between the PFV strand transfer complex and the human nucleosome indicates close integrase/histones interactions that allow the target DNA to reach a suitable degree of bending for integration [7]. Taken together these data suggest that cellular chromatin remodeling activities, especially those found in the vicinity of the integration sites, may control the efficiency of retroviral integration by modulating the number of intasome/nucleosome contacts. We have investigated this issue by selecting cellular cofactors of HIV-1 integration associated with chromatin. In this work, we report the identification and functional characterization of the FACT (FAcilitate Chromatin Transcription) complex as a modulator of HIV-1 integration.

FACT is a histone chaperone heterodimeric complex composed of human homolog of the suppressor of Ty16 (hSpt16) and the structure-specific recognition protein 1 (SSRP1) [17]. This chromatin remodeling complex is tightly associated with the PolII transcription machinery [17–19] and was previously shown to participate in the regulation of HIV-1 LTR-driven transcription [20]. Moreover, a recent study reported that the SSRP1 FACT component interacts with LEDGF/p75 via their HMG and PWWP domains [21] (a summary of the previously identified interactions is shown in Fig. 1a). The close proximity found between SSRP1 and LEDGF/p75 protein in cells strongly supports the enrichment of FACT in the chromatin regions targeted by HIV-1 integration complexes [21]. Consequently, in view of (i) the histone chaperone activity of FACT, (ii) the previously reported links between this complex and HIV-1 replication and (iii) the importance of chromatin remodeling in regulating HIV-1 integration we further analyzed its potential function in modulating the insertion of viral DNA into chromatin.

Results

Selection of cellular binding partners of HIV-1 IN•DNA complex

To identify new integration partners, we used a pull-down strategy depicted in Additional file 1: Figure S1. Fractions enriched in IN•U5 viral DNA end nucleocomplexes were generated under previously reported conditions [22] using recombinant IN and viral DNA fragments fused to a biotin in their 5′ end. IN•viral DNA complexes were checked by in vitro concerted integration activity (see Additional file 1: Figure S1). These fractions were then incubated with cellular protein extracts from HIV-1 permissive HeLa P4 cells previously counter-selected against beads coupled to DNA alone to limit the unspecific selection of potential DNA and avidin binders. Cellular partners were then sorted using magnetized streptavidin-coupled beads. The sorted proteins were analyzed on SDS-PAGE and compared to proteins found to be non-specifically associated with beads coupled to the viral DNA fragment without IN. As shown in Additional file 1: Figure S1, the experiments performed using the IN•U5 complex led to an apparent enrichment of cellular factors when compared to conditions using beads coupled to U5 DNA without IN. The cellular proteins selected under each condition were further digested *in gel* by trypsin and peptides analyzed by Liquid chromatography coupled to tandem mass spectrometry (LC–MS/MS). Experiments were conducted in triplicate and a set of about 75 proteins specifically and systematically found to be associated with the IN/DNA complex and not with control DNA alone were selected.

Among the selected proteins, most of the IN cofactors or proteins previously associated with integration were also selected under these conditions including LEDGF/p75, INI1, VBP1, FEN1, BAF, and RAD51 [12, 23–25], thereby validating our approach. In addition to these previously identified factors, several other proteins were also selected. These proteins were found to be mainly associated with nuclear import, DNA repair,

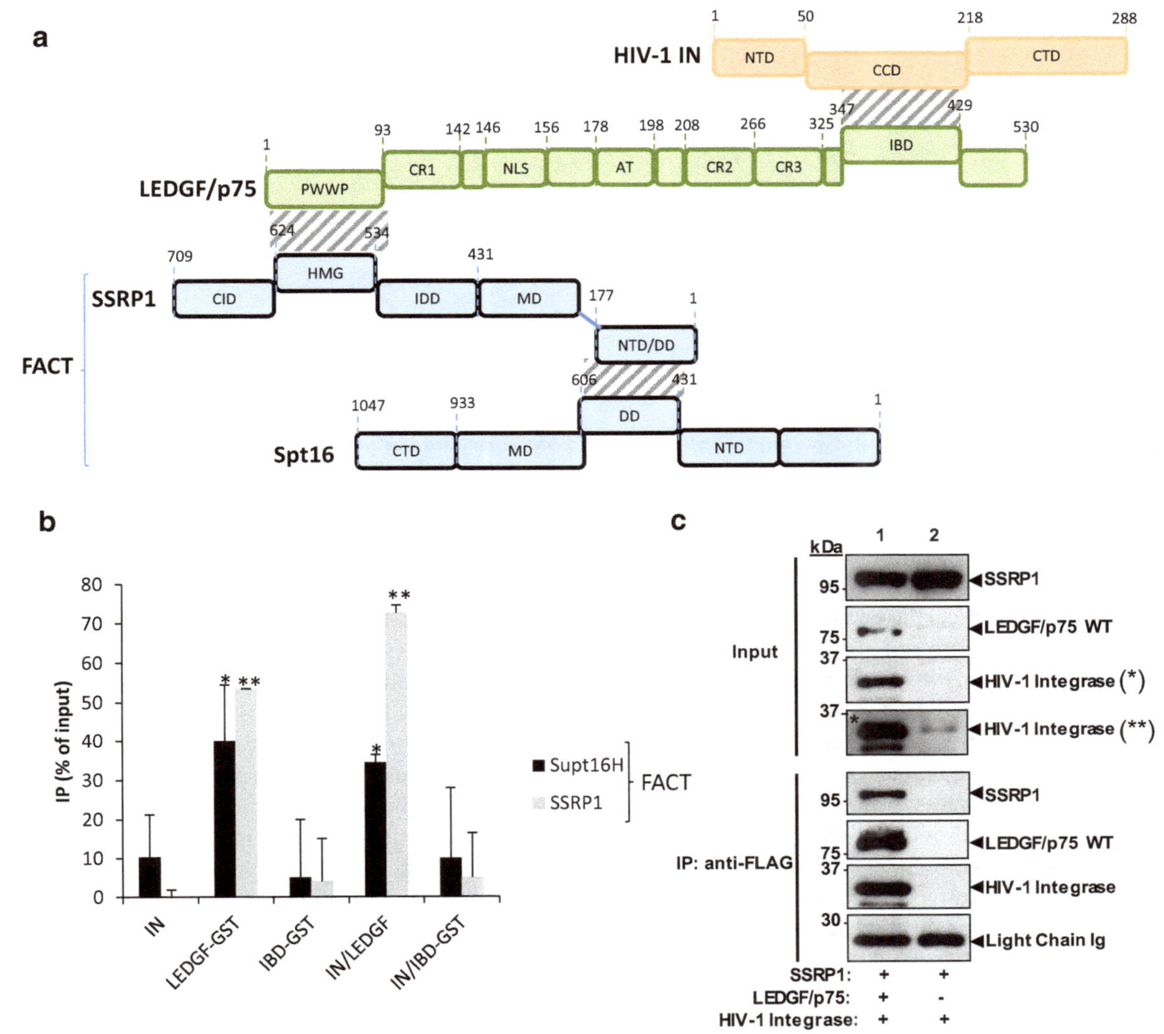

Fig. 1 Interaction between HIV-1 IN, LEDGF/p75 and FACT complex. A Schematic representation of the previously reported interactions between HIV-1 IN, LEDGF/p75 and FACT complex is shown in **a**. IN/FACT, IN/LEDGF, LEDGF/FACT, IN/LEDGF/FACT and IN/IBD/FACT interactions were analyzed by co-immunoprecipitation using recombinant cofactors and polyclonal anti-HIV-1 IN antibodies. IBD/FACT interactions were analyzed by GST-pull down using IBD-GST and FACT recombinant proteins. The interactions were monitored either direct gel staining using colloidal blue or western blot using the corresponding antibodies and quantified by Image J software (see quantification in **b** and representative experiments in Additional file 3: Figure S3). All values are shown as the mean ± standard deviation (*error bars*) of at least three independent sets of experiments. The p-values were calculated by Student's *t* test and are shown as *$p < 0.05$ and **$p < 0.005$ to represent the probability of obtaining significant differences compared with the data obtained with the negative background obtained with the beads alone. Cellular interaction between SSRP1, LEDGF/p75 and IN was checked by immunoprecipitation with an anti-FLAG antibodies in cells lysates obtained from LEDGF/p75-deficient cells (si1340/1428 cells) and transfected with plasmid expressing SSRP1-Myc and HIV-1 IN-Myc, and either FLAG-LEDGF/p75 (*lane 1*) or an empty plasmid (*lane 2*) (**c**). Then, immunoprecipitated proteins were evaluated for the presence of the expressed proteins by immunoblotting with tag-specific antibodies. (**) represents a longer exposure of (*). Detection of light chain Igs were used as loading control. The experiment was performed twice with identical results

protein degradation, and chromatin maintenance pathways (see Table 1 reporting the main selected proteins and complementary list Additional file 2: Table S2). In the present work, only the selected proteins associated with chromatin or transcription were considered in view of the impact of the chromatin structure on integration. Interestingly, most of these proteins were also found to be associated with the PolII transcription machinery. Due to (i) the possible modulation of HIV-1 integration by chromatin remodeling, (ii) the recently reported interaction between FACT and LEDGF/p75 IN cofactor and (iii) the enrichment of FACT in transcribed region of the chromatin [26], we investigated the potential role of this complex on HIV-1 integration.

Table 1 Selection of cellular interact ants of the IN•viral DNA complex

Protein name	Protein family	Protein complex	Accession number	Number of peptide
LEDGF/p75	Transcription	PolII	O75475	4
PFD3 (VBP1)	Protein chaperone	Prefoldin PFD	F5H2A7	12
SMARCB1 (INI1)	Transcription	SWI/SNF	B5MC5	3
BAF	DNA binding/chromatin	–	O75531	2
FEN1	DNA repair	–	P39748	15
RAD51	DNA repair	–	Q06609	7
SSRP1	Histone chaperone	FACT	E9PPZ7	10
SPT16	Histone chaperone	FACT	Q9Y5B9	31
IWS1	Transcription	PolII	E7EX51	2
SMARCC1 (BAF155)	Transcription	SWI/SNF	Q92922	2
SMARCD1 (BAF60)	Transcription	SWI/SNF	Q96GM5	2

Cellular extracts from HeLa P4 cells were incubated with streptavidin beads coupled to fraction enriched in active IN•viral DNA complexes as shown in SI. The elution of the interacting proteins was loaded on 1256 SDS-PAGE gel stained with silver nitrate and the bands were excised from gel and submitted to electroelution. The selected proteins were identified by MS–MS. A list of proteins found only associated to IN•viral DNA complexes but not to control DNA alone and linked to chromatin or transcription is provided here. The number of peptides identified per protein is also reported. Selected proteins previously reported as playing a role in retroviral integration are underlined

FACT complex forms a tri-partite complex with HIV-1 IN and LEDGF/p75 both in vitro and in cells

We first checked by in vitro co-immunoprecipitation whether the selection of FACT was due to a direct or indirect interaction between the complex and the retroviral IN. As reported in Fig. 1b (see representative experiments in Additional file 3: Figure S3), no direct interaction was observed between the purified recombinant FACT complex and IN. Since SSRP1 has been shown to bind LEDGF/75, we tested whether FACT could interact with the recombinant purified IN•LEDGF/p75 complex. Unlike IN alone, a reproducible association between the IN•LEDGF complex and FACT was detected, indicating that the the three proteins can form a complex altogether in vitro. This finding is supported by the interaction between FACT and LEDGF/p75-GST detected in GST-pull down experiments (Fig. 1b and Additional file 3: Figure S3). To determine whether the IN/LEDGF/FACT interaction occurs via the LEDGF/SSRP1 interaction we used a 326–471 amino-acid LEDGF/p75-GST construct lacking the PWWP SSRP1 interacting domain (as determined previously, [21]) but carrying the integrase binding domain (IBD). We first confirmed that the IBD interacts with HIV-1 IN but not FACT by GST pull down (Additional file 3: Figure S3). As shown in Fig. 1b, no interaction was detected between IN/IBD/FACT by co-immunoprecipitation suggesting that the formation of the IN/LEDGF/FACT complex requires the previously reported physical association between FACT and LEDGF/p75 mediated by SSRP1 and PWWP domains [21].

To define whether LEDGF/p75 also forms a complex with HIV-1 integrase and components of the FACT complex in cells, lysate from LEDGF/p75-deficient HEK293T cells transiently transfected with plasmids expressing Myc-tagged SSRP1 and HIV-1 IN, and either FLAG-tagged LEDGF/p75 or an empty plasmid were subjected to immunoprecipitation with anti-FLAG antibodies. The presence of these proteins in the immunoprecipitated samples was evaluated by immunoblot with tag-specific antibodies. In support to our in vitro findings, data in Fig. 1c indicate that SSRP1 and IN strongly associated with LEDGF/p75 in cells. The FACT subunit SSRP1 was sufficient for the formation of this complex, indicating its Spt16-independence. Similarly, SSRP1 was also reported to be sufficient to associate the FACT complex to LEDGF/p75 [21]. Low levels of HIV IN were observed in the control cells lacking LEDGF/p75 (lane 2, Fig. 1c) as expected considering the inhibitory effect of LEDGF/p75 on the proteasome-mediated HIV-1 IN degradation previously reported [27].

Based on these identified interactions we next investigated the effect of FACT on the integration process both in vitro and in cells.

FACT-dependent nucleosome disassembly promotes HIV-1 integration into chromatin in vitro

While HIV-1 IN integrates efficiently in mononucleosomes [4, 5] the concerted full site integration was found to be inhibited by nucleosome compaction in vitro [8, 14]. We, thus, tested the effect of FACT-dependent nucleosome disassembly on in vitro integration performed in reconstituted chromatin templates. Recombinant purified FACT complex was added in a typical concerted integration assay using polynucleosomal substrates (PN) and HIV-1 IN. As reported in Fig. 2a, b,

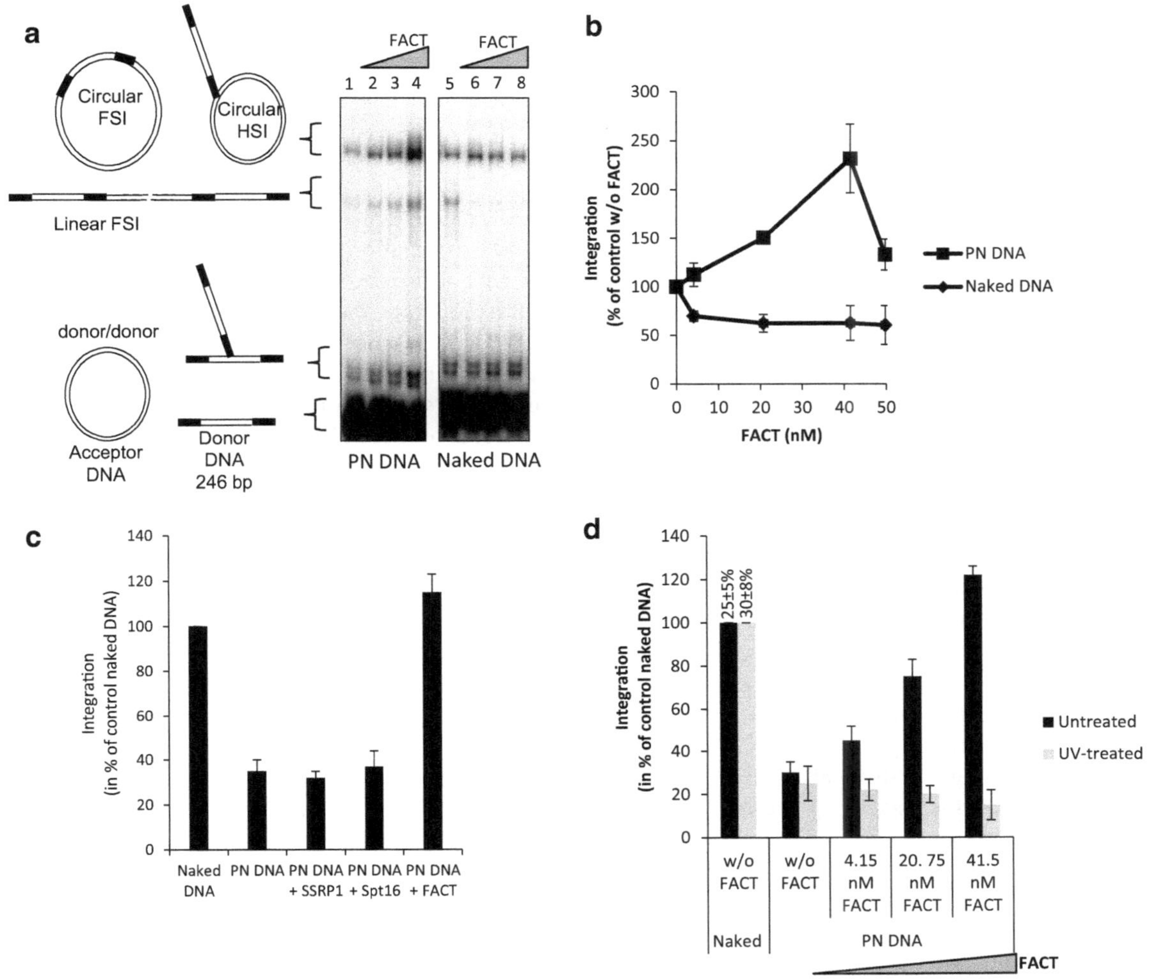

Fig. 2 Effect of FACT complex on in vitro HIV-1 integration. A concerted integration assay was performed using 200 nM of HIV-1 IN, 10 ng of donor DNA, 50 ng of p5S naked or chromatinized plasmid DNA (PN DNA), and increasing concentrations of FACT complex. The reaction products were loaded onto 1% agarose gels, and a representative set of experiments performed with the wild-type integrase is reported in (**a**). The positions and structures of the donor substrate and the different half-site (HSI), full-site (FSI) and donor/donor integration (d/d) products are shown. Quantification of the total integration products (FSI, HSI and donor/donor) was performed via gel detection and is reported in (**b**) as the percentage relative to activity without FACT. Effect of FACT complex or SSRP1 or Spt16 proteins [optimal FACT concentration as determined in (**b**)] on integration into PN DNA was analyzed and integration efficiency was reported in (**c**). Effect of FACT nucleosome remodeling activity on integration catalyzed by HIV-1 IN on chromatinized substrates was analyzed by comparing chromatinized p5S vector (histone/DNA ratio = 1) treated with UV or untreated as reported in "Methods" section. The treated or untreated substrates were then used in concerted integration assays without FACT or with increasing concentrations of the complex. Quantification of the total integration products detected on gel (see representative experiment in Additional file 4: Figure S4) (FSI, HSI and donor/donor) was performed via gel detection and is reported in (**d**) as the percentage relative to activity detected on naked DNA (pre-normalized data obtained in the control experiments using naked DNA and without FACT are also reported as the percentage of integrated substrate). All values are shown as the mean ± standard deviation (*error bars*) of at least three independent sets of experiments

while nucleosomes assembled as dense and stable chromatin were found to be refractory to HIV-1 integration (c.f. lanes 1 and 5), the addition of FACT restored integration in the chromatinized template (lanes 2–4) reaching a higher integration efficiency than found in naked DNA (c.f. lanes 4 and 5). Notably, optimal integration stimulation was obtained with a 1–2 nucleosome/FACT ratio which correlates well with the optimal ratio previously reported for the action of FACT on chromatin dissociation during transcription stimulation [17]. Increasing the FACT concentration led to a decrease in integration efficiency confirming that an optimal amount of the complex is required for the integration promotion (see Fig. 2b). Importantly, in contrast to what is observed on the chromatinized templates, the addition of FACT on the concerted integration assay using naked DNA induced a minor inhibition that may be due to competition with DNA (indeed, pre-incubation between FACT and naked acceptor DNA increased this inhibition effect, data not shown). Furthermore, the restoration effect was not observed when using SSRP1 or SPT16 alone (Fig. 2c). This result indicates that the restoration of integration observed on chromatinized DNA was dependent on the presence of nucleosomes and required a fully assembled active FACT complex.

To better understand the molecular mechanism of the integration restoration induced by FACT we investigated the structure of the chromatin after template treatment. To this purpose a typical Formaldehyde-Assisted Isolation of Regulatory Elements (FAIRE) approach was used as previously reported [28] and described in Additional file 4: Figure S4. Data show that FACT increases DNA accessibility in treated chromatinized templates suggesting that this chromatin remodeling activity could be responsible for the integration restoration. To confirm this hypothesis, we prevented the FACT-dependent histone/DNA dissociation by inducing protein-DNA crosslinks with UV in the acceptor DNA before integration assays. FAIRE analysis of templates pre-treated by UV and then submitted to FACT activity confirmed that UV crosslinking strongly inhibits FACT-dependent nucleosome remodeling (Additional file 4: Figure S4). Integration assays performed using these UV-treated templates showed that the crosslinking also abolished the capability of FACT to restore integration in PN (Fig. 2d and typical experiment in Additional file 4: Figure S4). Taken together, these data showed a strong correlation between the FACT-mediated restoration of HIV-1 integration and its nucleosome dissociation activity.

LEDGF/p75 potentiates the FACT-mediated restoration of HIV-1 integration

We next addressed whether the presence of LEDGF/p75 could modulate the effect of FACT on integration. As reported in Fig. 3a, the purified IN•LEDGF/p75 complex was also inhibited by the chromatinization of the acceptor template. Addition of FACT also restored integration catalyzed by the IN•LEDGF/p75 complex on PN templates and the observed restoration was more efficient than observed with IN alone. This result suggests that LEDGF/p75 potentiates the effect of FACT on integration. No stimulation of the FACT-mediated remodeling activity was observed in the presence of LEDGF/p75 suggesting that the potentiation effect was not due to the stimulation of FACT activity on chromatin (Additional file 4: Figure S4). To further determine whether the binding of LEDGF/p75 to IN was required for this potentiation we tested the isolated LEDGF/p75 integrase binding domain (IBD) which does not bind to FACT ([21], c.f. Fig. 1). As shown in Fig. 3b, the IBD did not induce the

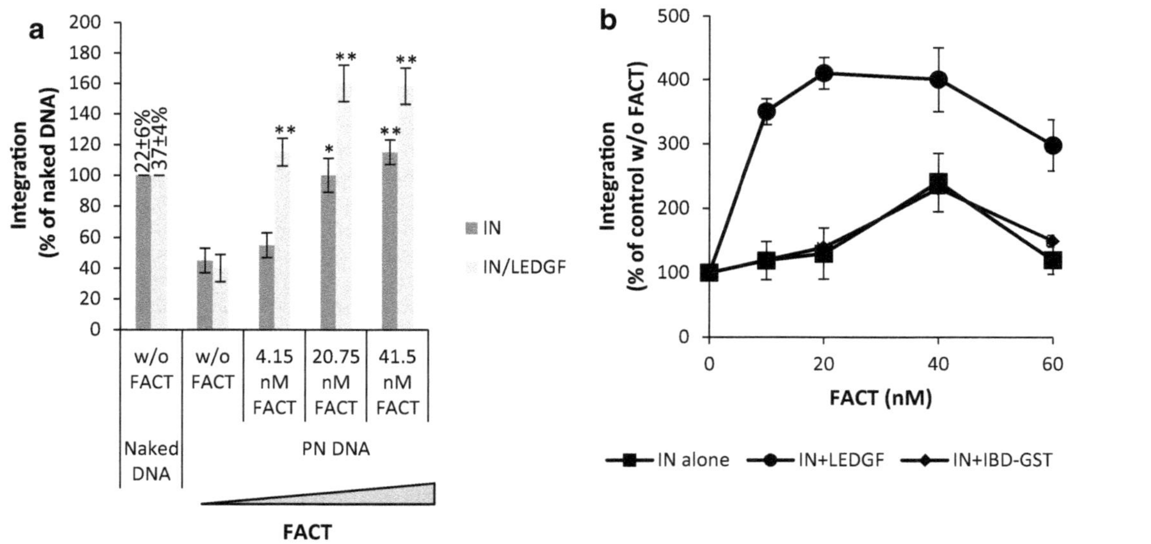

Fig. 3 Effect of LEDGF/p75 on FACT-mediated stimulation of in vitro HIV-1 integration. Concerted integration assays were performed as done in Fig. 2 but using IN•LEDGF complex, naked or chromatinized acceptor DNA and increasing concentrations of FACT (**a**). Results are reported as the percentage relative to activity detected on naked DNA without FACT and pre-normalized data obtained in the control experiments using naked DNA and without FACT are reported as the percentage of integrated substrate. Results obtained with full length LEDGF/p75 or the LEDGF 326–471 IBD-GST domains are reported in (**b**). All values are shown as the mean ± standard deviation (*error bars*) of at least three independent sets of experiments. The p-values were calculated by Student's t-test and are shown as $^*p < 0.05$ and $^{**}p < 0.005$ to represent the probability of obtaining significant differences compared with control experiments performed with PN DNA without FACT

FACT potentiation observed with the LEDGF/p75 full-length protein. Taken together these data strongly support the importance of both LEDGF/p75-FACT and LEDGF/p75-IN interactions in the FACT-mediated restoration of HIV-1 integration.

Interestingly, the integration efficiency in PN at the optimal FACT concentration was higher than that detected on naked DNA. These data suggest that the chromatin structures generated by the action of FACT on polynucleosomes are highly preferential for HIV-1 integration and are even better substrates than naked DNA in vitro. Consequently, we have investigated the impact of these FACT-induced chromatin structures on integration.

Reconstituted chromatin containing partially dissociated nucleosomes is a favored substrate for HIV-1 integration in vitro

The remodeling activity of the FACT complex has been extensively studied and some studies have shown the FACT-dependent eviction of H2A/H2B dimers from the native nucleosome [18]. We, thus, wondered whether the FACT-dependent activation of integration in reconstituted chromatin was due to a local increase in targeted sites mediated by the eviction of H2A/H2B dimers by the remodeling complex. Chromatin assembled with H3/H4 tetramers, instead of H2A/H2B/H3/H4 octamers, forms tetrasomal templates (PN Tetra) that share structures similar to the FACT remodeled products [29]. Thus, these PN Tetra templates allow to mimic the chromatin structures enriched in PolII transcribed regions, and to test their capacity as integration substrates. Consequently, we tested PN Tetra in HIV-1 integration and compared the integration efficiency with naked DNA or chromatin assembled with native histone octamers (PN Octa). The pBSK-601-Zeo vector (p601) containing 601 Widom repetitions for highly stable assembly of nucleosomes was used to generate these templates. As reported in Fig. 4a, polynucleosomes assembled with H3/H4 tetramers were always found to be better substrates for integration than chromatin fully assembled with histone octamers. Interestingly, these tetrasomal nucleosomes were also found to be better substrates than the corresponding naked DNA regardless of the ratios used (see quantification in Fig. 4b).

To confirm the preference of HIV-1 IN for partially dissociated nucleosomes, we performed an in vitro selectivity assay as set up previously [15] using a mixture of naked and chromatinized Octa or Tetra templates. Integration was clearly preferred in the naked DNA when mixed with PN Octa templates (Fig. 4c). In contrast, when a mixture of naked DNA and PN Tetra substrates was used (Fig. 4d), integration was preferred in the PN Tetra templates, thereby confirming the preference of HIV-1 for these structures over naked DNA or native nucleosome templates.

Since PFV intasome was shown to require the binding to H2A/H2B histones for efficient docking to the nucleosome [7], we next compared HIV-1 and PFV IN on the evicted and remodeled templates. As reported in Fig. 5a and b, while PN Tetra are good substrates for HIV-1 integration, they are rather ineffective with PFV. In addition, while FACT stimulates HIV-1 integration into PN Octa templates, its remodeling activity clearly inhibits PFV integration in the same templates (Fig. 5c). This inhibition confirmed the poor integration efficiency found for PFV IN on partially dissociated chromatin and the requirement of this retroviral system for native nucleosomes, as suggested by the structure of the PFV intasome/nucleosome complex [7].

Taken together, these results strongly suggest that in vitro efficient HIV-1 integration into chromatin requires the presence of partially dissociated nucleosomes as generated by FACT remodeling. We next investigated the effect of FACT in the context of infected cells.

Cells with enhanced FACT-mediated chromatin remodeling activity are more permissive to HIV-1 infection and integration

The FACT complex has a dual activity on nucleosomes during polII transcription since it induces a partial dissociation of the histones, thereby allowing the polymerase to travel across nucleosomes, and participates in their re-association after elongation [17, 18, 30, 31]. We thus investigated how these activities exerted in preferred integration regions could influence viral replication.

FACT-dependent chromatin remodeling can be enhanced by both genetic and pharmacologic approaches. Curaxins drugs, as CBLC137, have been previously shown to causes partial unwrapping of DNA from the nucleosome core leading to the dissociation of the H2A/H2B dimer and the exposure of the H3/H4 tetramer docking surface for FACT binding [32]. Binding of FACT to this pre-dissociated nucleosomal structure induces its trapping onto the nucleosome, inhibits its chaperone activity and enhances the nucleosome dissociation leading to a global increase in chromatin accessibility (29, and Fig. 6a). Consequently, curaxins treatment was used to study the effect of FACT-mediated chromatin remodeling on the early steps of viral replication. CBLC100 and CBLC137 compounds previously described to induce the FACT nucleosomal trapping [32, 33] were used. Both molecules were first assayed on in vitro integration catalyzed by recombinant HIV-1 IN in order to exclude direct effects on the protein. As shown in Additional file 5:

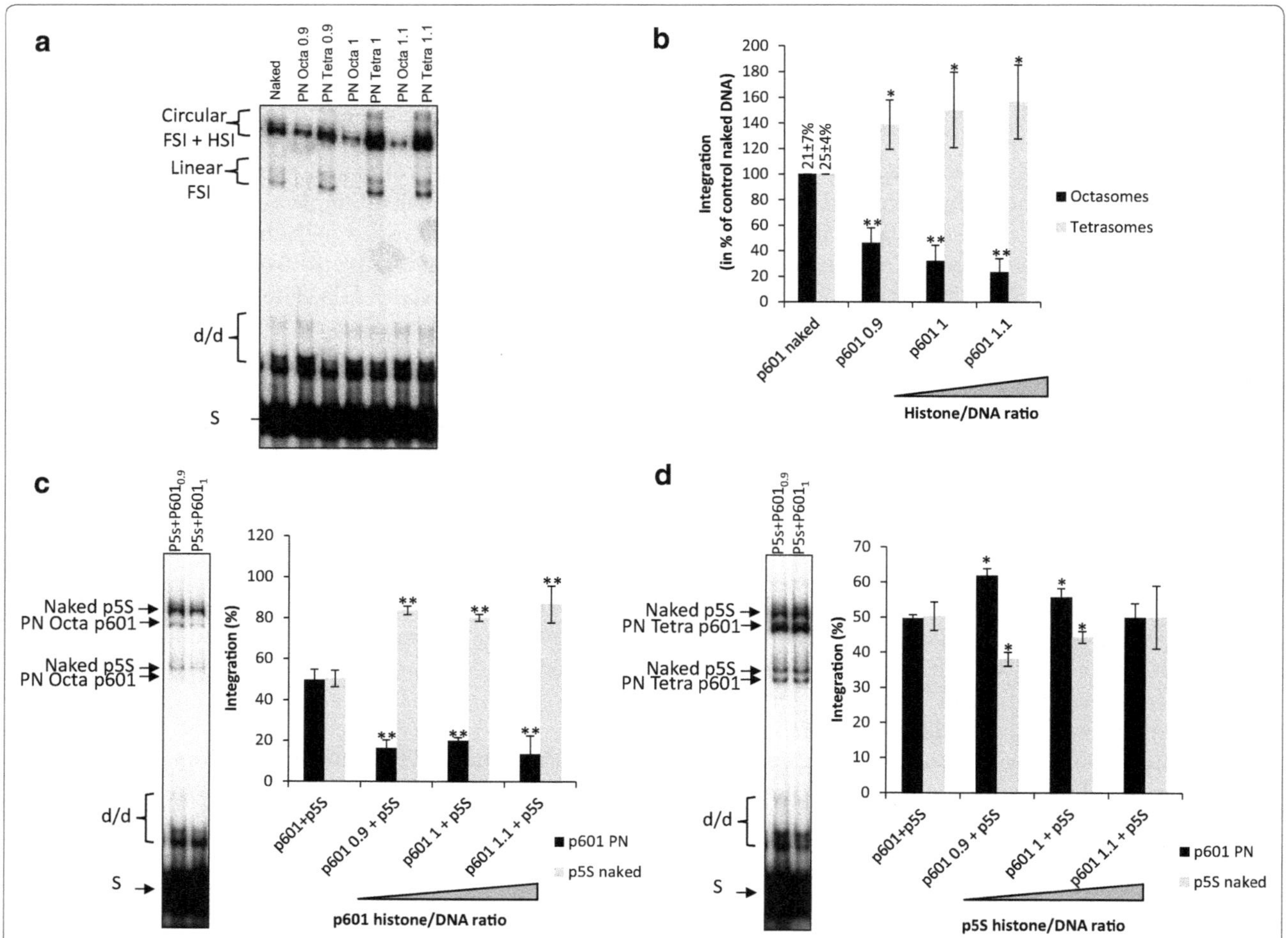

Fig. 4 Integration catalyzed by HIV-1 integrase onto dissociated nucleosomes in vitro. A concerted integration assay was performed using 200 nM of HIV-1 IN, 10 ng of donor DNA, 50 ng of p601 naked or p601 plasmid chromatinized either using native histone octamers (PN Oct) or H3/H4 tetramers (PN Tet) with increasing histone/DNA ratio as reported. Typical experiment is shown in (**a**) and quantification of the integration products is reported in (**b**). Results are reported as the percentage relative to activity detected on naked DNA and pre-normalized data obtained in the control experiments using naked DNA and without FACT are reported as the percentage of integrated substrate. Selectivity assays were performed under similar conditions except that a mixture of naked p5S and either p601 assembled with native octamers was used (**c**) or H3/H4 tetramers (**d**). Quantification of the integration products detected in each vector is reported as percentage of the total integration. All values are shown as the mean ± standard deviation (*error bars*) of at least three independent sets of experiments. The p-values were calculated by Student's t-test and are shown as *$p < 0.05$ and **$p < 0.005$ to represent the probability of obtaining significant differences compared with untreated conditions

Figure S5, while CBLC137 showed no or little effect on IN, CBLC100 inhibited directly the retroviral enzyme in the 0.1–1 μM range. Measurement of the cytotoxicity of the drugs in all our cell models showed a modest effect at concentrations below 200 nM, the CBLC137 compounds being the least effective (Additional file 5: Figure S5). For these reasons we selected the CBLC137 molecule for further analysis of its effect on the early steps of viral replication. We first analyzed the effect of curaxin treatment on the structure of cellular chromatin. To this end, the accessibility of genomic DNA was determined after curaxin treatment using the FAIRE approach as previously reported [28]. As shown in Fig. 6b, treatment of HEK-293T cells with CBLC137 led to a global increase in free chromatin DNA, as reported before [33], confirming the inhibition of the FACT histone chaperone activity probably due to its trapping onto dissociated nucleosomes as previously observed [32, 33]. Similar results were obtained in different cell lines as K562 and HeLa P4. We next analyzed the impact of curaxin treatment on the retroviral cycle. We first evaluated the infectivity of LAI wild type virus in HeLaP4 cells treated with CBLC137. As shown in Fig. 6c a stimulation of LTR-dependent β-galactosidase expression was observed in HeLa P4 cells treated with CBLC137 and infected with the virus. Quantification of viral DNA

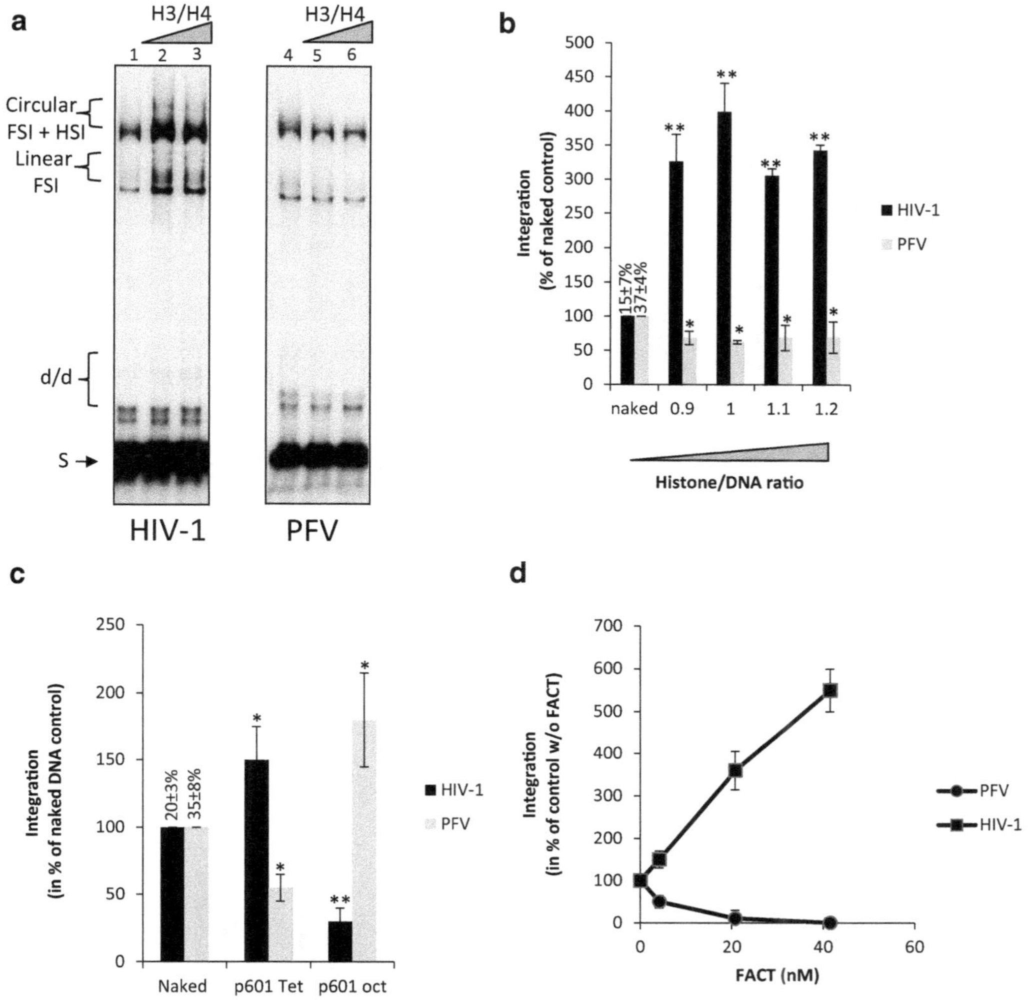

Fig. 5 Effect of nucleosome dissociation on integration catalyzed by HIV-1 and PFV integrases in vitro. Concerted integration assays were performed using either HIV-1 or PFV INs under similar conditions and using p601 vectors naked (lanes 1 and 4) or chromatinized with native histone octamers or H3/H4 tetramers with increasing histone/DNA ratios (lanes 2 and 5, 1.1 ratio; lanes 3 and 6, 1.2 ratio). Typical experiment is shown in (**a**) and quantification of the integration products is reported in (**b**). Comparison of the integration catalyzed by HIV-1 or PFV IN on naked and p601 vectors assembled with histone octamers or tetramers is shown in (**c**). Results are reported as the percentage relative to activity detected on naked DNA without FACT and pre-normalized data obtained in the control experiments using naked DNA are reported as the percentage of integrated substrate. Comparison of the effect of FACT on HIV-1 and PFV integration catalyzed on p601 chromatinized with histone octamers is shown in (**d**). All values are shown as the mean ± standard deviation (*error bars*) of at least three independent sets of experiments. The p-values were calculated by Student's t-test and are shown as *$p < 0.05$ and **$p < 0.005$ to represent the probability of obtaining significant differences compared with untreated conditions

population confirmed that integration was stimulated in this system (Fig. 6d and Additional file 6: Figure S6). In order to avoid biases linked to the previously reported regulation of the LTR-driven transcription by FACT [20, 21] and analyze specifically the effect of the complex on the integration step, we next used a single round 293T cells model transduced by a pRRLsin-PGK-eGFP-WPRE VSV-G pseudotyped lentiviral vector. In this system the eGFP expression is independent from the viral LTR and depends on PGK promoter that is not sensitive to FACT. Quantification of the number of eGFP-positive cells by flow cytometry showed a significant and reproducible 1.4 to 1.8-fold increase for cells treated by the drug (Fig. 6e). Quantification of the viral DNA populations 0–48 h post-transduction indicated that the amount of DNA integrated was increased while no significant change was detected in either the total DNA or the two LTR circles (Fig. 6f, and Additional file 6: Figure S6).

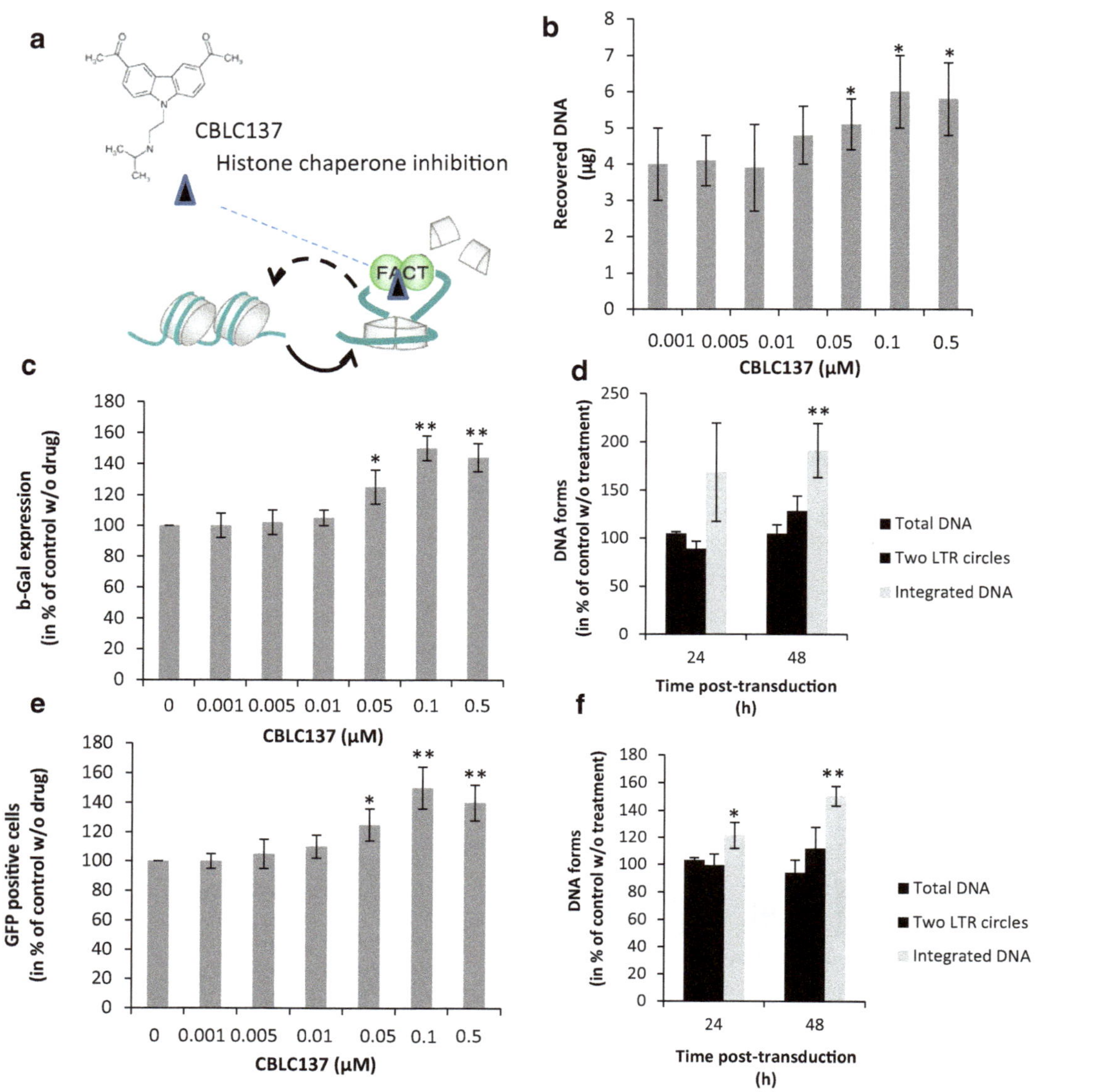

Fig. 6 Effect of FACT-mediated chromatin remodeling chemical promotion on early steps of HIV-1 LAI virus and derived lentiviral vectors. HeLa P4 and HEK293T cells were treated with CBLC137 curaxin 6 h before cell infection with LAI wild type virus or transduction with lentiviral vectors. The structure of the CBLC137 drug is shown in (**a**) as well as its effect on FACT. CDLC137 binds to nucleosomal DNA and forms structures with high FACT affinity leading to the trapping of the complex into dissociated nucleosomes, preventing its histone chaperone activity and enhancing the histone dissociation [32, 33]. This induces a global increase in chromatin accessibility that was evaluated by FAIRE (data obtained in HEK293T cells are reported in **b**). The effect on LAI virus replication was evaluated by quantifying the LTR-dependent β-galactodisase expression (**c**) and viral DNA populations at 0–48 h post-transduction (see 24 and 48 h time points in **d** and full time course analysis in Additional file 6: Figure S6). Early steps of lentiviral vector replication were evaluated by quantifying GFP-positive HEK293T cells (**e**) and viral DNA populations at 0–48 h post-transduction (see 24 and 48 h time points in **f** and full time course analysis in Additional file 6: Figure S6). Quantification of the viral DNA populations were done after optimal 0.1 µM curaxins treatment. All values are shown as the mean ± standard deviation (*error bars*) of at least three independent sets of experiments. The p-values were calculated by Student's t-test and are shown as *$p < 0.05$ and **$p < 0.005$ to represent the probability of obtaining significant differences compared with untreated conditions

Knockdown of the SSRP1 encoding gene by siRNA has also been shown to increase the amount of FACT-remodeled chromatin sites in highly transcribed genes [26]. Consequently, we also performed additional siRNA knockdown of the SSRP1 encoding gene to analyze the importance of active FACT amount in viral infectivity. SSRP1 encoding gene knock down was conducted essentially as previously described [31]. Two successive transfections of siRNA anti-SSRP1 performed as described in "Methods" section allowed us to achieve about 80%

of apparent SSRP1 extinction (Fig. 7a). SSRP1 knockdown also induced a global increase in free chromatin DNA (see FAIR analysis in Fig. 7b) accompanied with an increase of viral infectivity (Fig. 7c) and integration efficiency (Fig. 7d and Additional file 7: Figure S7). Interestingly, siRNA suboptimal concentrations (>20 nM) decreased the integration efficiency suggesting, as observed in vitro, that optimal FACT amounts could be required for the promotion of integration step.

The constant correlation found between the FACT-mediated increase in global accessibility to DNA detected in cells either treated by curaxin, or depleted in SSRP1, prior to transduction, and the efficiency of integration strongly suggests that this step is modulated by FACT chromatin remodeling activity.

Stimulation of FACT-mediated chromatin remodeling promotes HIV-1 integration by a LEDGF dependent mechanism

Since LEDGF/p75 has been shown to bind FACT and promotes its effect on in vitro integration we have investigated the importance of this factor in the FACT-mediated regulation of viral integration in cells. For this purpose we analyzed the viral infectivity in TZM cells and TZM cells knockout (KO) for LEDGF/p75 [34] and affected for FACT activity. Accumulation of LEDGF

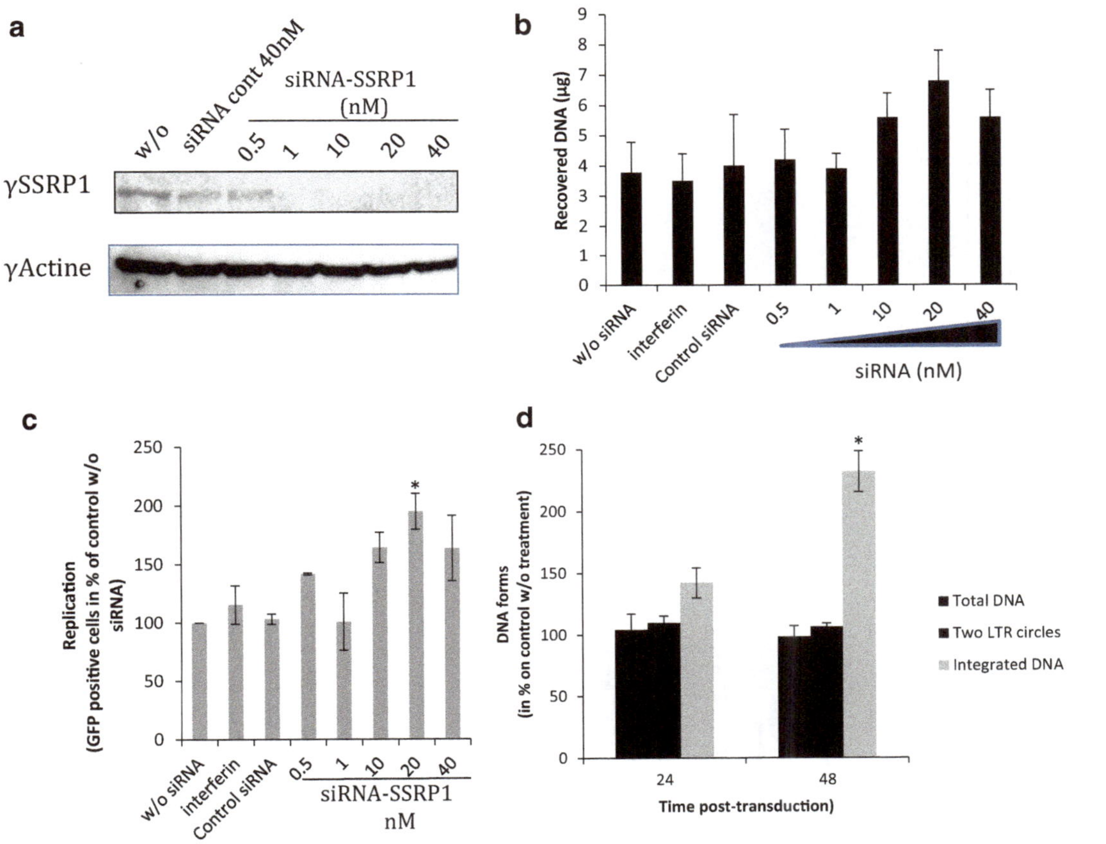

Fig. 7 Effect of FACT knock down on early steps of HIV-1 derived lentiviral vectors in HEK293T cells. SSRP1 encoding gene knock down was conducted essentially as previously described [31] by performing two successive siRNA transfections (see "Methods" section). The amount of SSRP1 protein monitored by western blotting is reported in (**a**). This method allowed us to achieve more than 80% of apparent SSRP1 extinction. After knock down, FAIRE analyses were performed as indicated in "Methods" section and the results obtained in cells depleted for SSRP1 are reported in (**b**). The early steps of replication of lentiviral vectors carrying the GFP encoding gene were determined by measuring GFP-positive cells with flux cytometry (**c**) and quantification of the viral DNA populations at 0–48 h post-transduction using quantitative PCR (data obtained after a 20 nM siRNA treatment are reported in **d** for the 24 and 48 h time points, see full time course analysis in Additional file 7: Figure S7). Data obtained with the interferin transfector agent alone and control siRNA are also reported in the figure. All values are shown as the mean ± standard deviation (*error bars*) of four independent sets of experiments. The p-values were calculated by Student's t-test and are shown as $*p < 0.05$ and $**p < 0.005$ to represent the probability of obtaining significant differences compared with untreated conditions

knockout and FACT knockdown by siRNA did not allow us to find conditions allowing the analysis of the viral infectivity in cells by this method without biases due to different cellular growth. However, the curaxin-based pharmacological approach allowed us to obtain conditions compatible with this analysis. Parental TZM cells and TZM cells knockout for LEDGF/p75 were thus treated by CBLC137 prior to transduction with a pRRL-sin-PGK-eGFP-WPRE VSV-G pseudotyped lentiviral vector. Viral infectivity was then measured. As reported in Fig. 8a, curaxin treatment also induced a stimulation of viral infectivity in TZM cells. In contrast no stimulation was detected in TZM cells LEDGF/p75 KO while CBLC137 treatment induced the expected increase in chromatin accessibility in both cell lines (Fig. 8b). The LEDGF/p75-dependent effect of curaxins on viral infectivity suggests that this factor is required for the integration modulation by FACT in cells. To confirm this we tested the effect of curaxin on the early steps of replication of PFV vector which is not expected to be dependent on LEDGF/p75 and whose integration was found to be inversely regulated by FACT in vitro (see Fig. 5). As reported in Fig. 8c, curaxin treatment did not stimulate PFV infectivity in HEK 293T cells in contrast to HIV-1 vector and even a slight inhibition was detected at high CBLC137 concentrations in agreement with the in vitro data.

Discussion

Here we report the role of the FACT histone chaperone complex in modulating HIV-1 integration by affecting the structure of chromatin. In cell, FACT controls nucleosome deposition on DNA and has been implicated in many processes involving chromatin, such as transcription, DNA replication, recombination and repair [18, 30, 35]. Interestingly, recent reports show that the FACT complex can

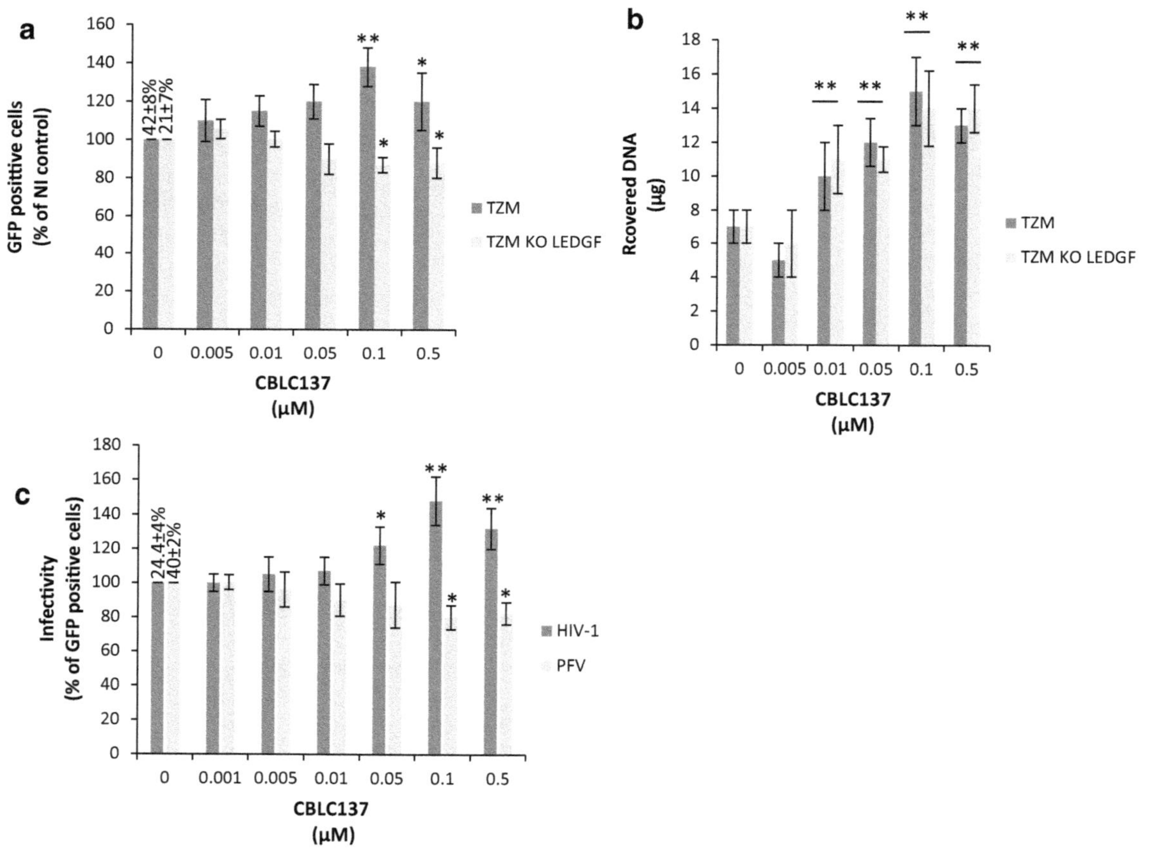

Fig. 8 Effect of FACT-mediated chromatin remodeling chemical promotion on early steps of HIV-1 in LEDGF/p75 knock-out context and on PFV infectivity. Similar experiments than described in Fig. 7 were performed in TZM and TZM knock out for LEDGF/p75 (**a**, **b**) or in HEK293T cells transduced either with HIV-1 or PFV vectors (**c**). The effect on the early steps of lentiviral vector replication was evaluated by quantifying GFP-positive cells. All values are shown as the mean ± standard deviation (*error bars*) of at least three independent sets of experiments. The p-values were calculated by Student's t-test and are shown as *$p < 0.05$ and **$p < 0.005$ to represent the probability of obtaining significant differences compared with untreated conditions

bind the LEDGF/p75 HIV-1 integrase cofactor via SSRP1 [21] and could participate in the modulation of viral gene expression [20]. In these studies no effect of FACT on the integration step could be detected. However, the analyses presented by these authors were focused on the viral LTR-dependent gene transcription. In contrast, possible regulation of the retroviral integration mechanisms by FACT has been recently proposed as SSRP1 depletion was shown to inhibit Avian Leukosis Virus (ALV) integration [36]. In this latest work no inhibition of HIV-1 integration by SSRP1 depletion was observed suggesting that FACT-dependent modulation of integration could depend on the retrovirus family. This different regulation of HIV-1 integration by FACT was demonstrated in our work showing a clear stimulation of this replication step when the FACT remodeling activity was stimulated. Interestingly, a stimulation of HIV-1 integration after SSRP1 knockdown was also observed by Winans et al. [36] but was found non-statistically relevant. However, these data were obtained in chicken DT40, which are not naturally permissive to HIV-1 infection, and collected at 24 h post transduction while our data indicate that the increase in integrated DNA amount becomes more relevant a 48 h than 24 h (see Additional file 7: Figure S7).

The possible difference in the retroviral integration regulation mechanism by FACT was confirmed by our data showing that HIV-1 integration regulation by this complex depends on the presence of nucleosomes and chromatin remodeling activity while the activation of integration observed for ALV does not depend on the presence of nucleosomes but on the direct binding of SSRP1 on ALV integrase. These observations contrast with the role of nucleosomes on FACT-dependent activation of HIV-1 integration and the absence of direct interaction between HIV-1 IN and FACT, as presented by our studies. These differences suggest that the mechanism of FACT modulation of integration may depend on the retrovirus and differs between HIV-1 and ALV. In the case of HIV-1, the FACT mediated activation of integration is also modulated by LEDGF/p75 IN partner. We show here that the IN•viral complex can pull the FACT complex down from a cellular protein extract in a LEDGF/p75-dependent manner and the IN•LEDGF complex is able to bind the recombinant FACT in contrast to IN alone. This indicates that IN/LEDGF/FACT interactions are not mutually exclusive and can thus occur simultaneously in the physiological integration complex. These data also suggest that the FACT complex can be loaded at the integration sites where integration and transcription machinery can meet. The stimulation of FACT effect on integration by LEDGF/p75 confirmed the modulation of FACT-dependent regulation of the viral DNA insertion by a mechanism that may involve the LEDGF/SSRP1 interaction. This is additionally supported by the lack of effect of the isolated IBD domain of LEDGF lacking the SSRP1 interaction domain.

We previously reported that while HIV-1 integration into isolated nucleosomes is efficient, their compaction in dense chromatin restricts the reaction and their remodeling can overcome this restriction [8, 14]. Since FACT is associated with the PolII-transcribed regions of chromatin and is closely linked to LEDGF/p75, we made the hypothesis that its remodeling activity could regulate the access of the HIV-1 Intasome to nucleosomes. The observed stimulation of HIV-1 integration in cells where FACT-mediated chromatin remodeling was promoted by curaxins treatment confirmed our hypothesis. As previously shown [32, 33], FACT trapping on chromatin induced by curaxins both reduces the re-association of nucleosomes after PolII machinery elongation and enhances the nucleosomes dissociation leading to an increase in the global amount of open chromatin, especially in the transcribed regions targeted by HIV-1 intasomes. Given the requirement of open chromatin structure for efficient HIV-1 integration, we propose that the activated integration observed in cells inhibited or knockdown for FACT, results from an increased access of nucleosomes for the HIV-1 intasome as supported by our in vitro analysis. Indeed, in vitro integration assays performed in PN templates showed that FACT remodeling activity allowed efficient HIV-1 integration into dense chromatin that was initially refractory to viral DNA insertion. Importantly, this FACT-mediated activation of integration was found to be nucleosome-dependent since no stimulation was detected on naked DNA, in contrast to what was recently observed for ALV [36]. Our data indicate that FACT generates chromatin structures that are highly preferential for in vitro integration. This restoration process was stimulated by the presence of LEDGF/p75, suggesting that its association with FACT may promote the restoration of integration. Based on the direct interaction between FACT and LEDGF/p75, we can speculate that this association may increase the local FACT concentration around integration sites leading to a coupling between nucleosome remodeling and viral DNA insertion.

Additionally, the use of tetrasomes that lack H2A/H2B dimers and mimic the FACT-generated nucleosome structures along transcribed genes allowed us to demonstrate that chromatin containing these partially dissociated forms is a preferential substrate for in vitro HIV-1 integration, and is even better than naked DNA. Interestingly, PFV integrates less efficiently in the dissociated nucleosomes than in the fully structured ones. These data correlate well with the requirement of direct interactions between the PFV intasome and H2A/H2B dimers for optimal integration [7] as well as with the preference of this virus to integrate more often in regions of high nucleosome density [8]. This validates the specificity of our model and highlights the different behavior of HIV-1 and PFV intasomes, as previously reported [8]. These models very likely require distinct constraints for integrating into chromatin and could thus have a distinct active structure at the surface of the targeted nucleosome. This is supported by the various intasome structures reported recently in the literature showing distinct 3D features [4, 35–39]. These distinct preferences for chromatin structures observed for the different retrovirus could also explain their different sensitivity to FACT chromatin remodeling as detected for HIV-1, PFV and ALV. Our biochemical data also suggest that an optimal FACT amount is required for reaching an equilibrium between the nucleosome dissociation and their re-association suitable for integration. This would explain why the FACT depletion in cell allows to reach a local concentration of the complex suitable for the nucleosome dissociation and the integration facilitation. This is fully supported by our FAIR results indicating that under the FACT knockdown or inhibition conditions the chromatin opening is favored.

Extensive data from the literature, in addition to the results obtained here for HIV-1 integration, suggest that HIV-1 intasomes are targeted to the PolII-transcribed region of chromatin, mostly thanks to LEDGF/p75 and H3K36me3 recognition, where FACT and LEDGF/p75 are both enriched due to the interaction between LEDGF/p75 and SSRP1. This is strongly supported by the enrichment of FACT in the transcription region of chromatin [26] and the enhancement of its recruitment by H3K36 trimethylation [40]. Collectively, these events should lead to an increase in FACT concentration in the vicinity of HIV-1 integration sites as induced in our in vitro integration assays. This was supported by the loss of integration stimulation by curaxin treatment in TZM cells KO for LEDGF/p75. Chromatin remodeling mediated by FACT and occurring in these HIV-1-targeted regions may generate partially dissociated nucleosomes, leading to chromatin structures that are preferential substrates for both RNA transcription and HIV-1 integration. This coupling between FACT remodeling and integration would thus allow efficient integration onto nucleosomes as detected in infected cells. Inhibition of FACT chaperone activity by curaxins or the extinction of SSRP1 expression would therefore be expected to induce an increase in open chromatin in the PolII regions and to stimulate integration, as observed in our experiments. After these effects on integration FACT may then exert its regulation function on viral gene transcription as previously observed [20, 21]. Based on the difference in their preference for distinct chromatin structure we propose that FACT deregulation could induce different and inverted effect depending on the retrovirus as observed by us with HIV-1 and PFV (see Fig. 8c) and by other authors using ALV model [36]. These putative functions of FACT on retroviral integration is recapitulated in Fig. 9.

The molecular mechanism allowing integration onto partially dissociated nucleosomes remains to be fully unraveled. Indeed, the action of FACT on chromatin could increase accessibility to both DNA and to histones. These two parameters could influence HIV-1 integration on nucleosomes, which may require additional protein/protein interactions as supported by the physical contacts between the PFV intasome and protein histones (H2A/H2B) reported before [7]. We can speculate that such interaction between HIV-1 IN and other histones could also occur and would require the local partial dissociation of the nucleosomes induced by FACT to be efficient. Interestingly, in compact chromatin several histone tails such as histone H4 are engaged in interaction with neighboring nucleosomes and are therefore less or maybe not accessible for association with incoming intasomes [41–43]. Local FACT-meditated remodeling of the chromatin could thus allow or promote both protein/protein and protein/DNA interactions within the targeted nucleosomes. This process would be one mechanism allowing the efficient HIV-1 integration into PolII regions in addition to LEDFG/p75 and CPSF6 targeting, histone modifications, and intrinsic chromatin dynamics found in these loci.

Conclusions

In addition to reporting a new cellular cofactor of HIV-1 integration, this study also demonstrates a potential link between the retroviral integration machinery and the PolII complex. This suggests that HIV-1 integration/transcription are closely coupled, as indicated by the IN/LEDGF/FACT interaction. This would open the way for a new understanding of these viral steps and could lead to new antiviral strategies targeting both integration and transcription.

Methods

Proteins, DNA substrates and chemicals

HIV-1 IN, PFV IN, LEDGF, IN•LEDGF/p75 complex and GST-fused proteins were purified following previously published protocols [8, 44–46]. Polynucleosome assembly was performed as previously reported using either recombinant H3, H4, H2A and H2B octamer or recombinant H3, H4 tetramers (New England Biolabs) by gradient salt dialysis on p5S vector described before [8, 14] or pBSK-Zeo-601 plasmid containing a succession of Widom-601 sequences. Nucleosome assembly

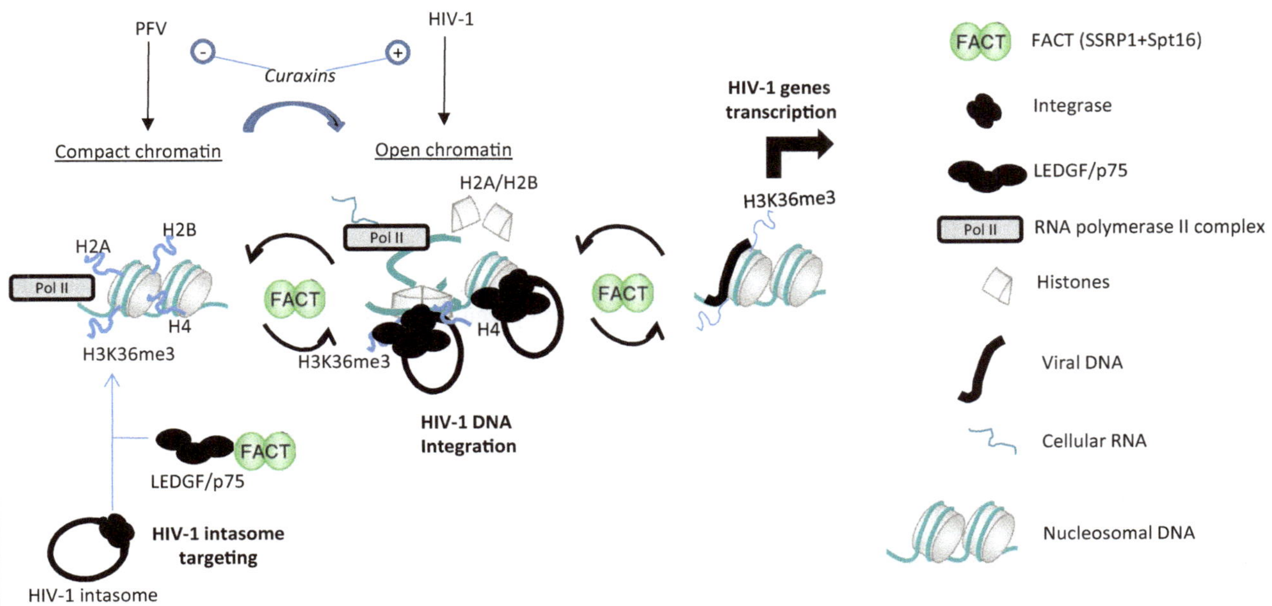

Fig. 9 Model for FACT action on HIV-1 integration steps. HIV-1 intasomes are targeted to the PolII-transcribed region of the chromatin thanks to the association of IN with LEDGF/p75 and binding to H3K36me3. In these regions the FACT complex and LEDGF/p75 are both enriched owing to the LEDGF/p75 and SSRP1 interaction. Chromatin remodeling mediated by FACT occurring in the vicinity of the targeted region may generate partially dissociated nucleosomes leading to chromatin structures that are preferential substrates for both RNA transcription and HIV-1 integration. FACT action on chromatin at the vicinity of the integration sites may lead to increased accessibility to nucleosomal DNA as well as histone tails. Integration could then occur efficiently on partially dissociated nucleosome or accessible native nucleosome. Further FACT-mediated regulation of the viral genes transcription can then occur. The expected effect of curaxin treatment on HIV-1 and PFV integration is reported on the basis of the drugs effect on FACT chromatin remodeling activity and on the retroviruses preferences for open or compact chromatin structure

was checked by DNase I protection, typical restriction enzyme assay (REA) and mono- and di-nucleosome gel shift in 0.8% native agarose gel, as done before [8]. FACT complex was purified as reported before [17, 18] and its activity was checked by chromatin remodeling assay as reported in Additional file 4: Figure S4. Polyclonal anti-HIV-1 IN antibodies were purchased from Bioproducts MD (Middletown, MD, USA). Polyclonal anti-SSRP1 were purchased from Abcam (ab21584) and anti-Spt16 were purchased from Santa-Cruz (sc-28734). Monoclonal anti-LEDGF/p75 were purchased from Bethyl (848A). CBLC100 and CBLC137 FACT curaxins inhibitor were a kind gift from Dr Gurova K.V. [33].

Selection of IN•viral DNA cofactors and in vitro interactions

Biotinylated IN•viral DNA-enriched fractions were generated by incubating recombinant pure IN with short DNA fragment corresponding to the 21 bp final nucleotides of the U5 viral ends biotin labeled in 5′ under optimized conditions allowing the formation of highly active IN•DNA complex, as done previously [22]. The generated complexes were checked by concerted integration (see Additional file 1: Figure S1) and then incubated with cellular protein extracts from HeLa P4 cells obtained by cell sonication sorted after counter-selection on beads containing only DNA. The cellular interactants were selected after 1 h of incubation at 37 °C with magnetized streptavidin beads coupled to the IN•DNA complexes in an interaction buffer (50 mM HEPES, pH 7.5; 1 μg/ml BSA; 1 mM DTT; 0.1% Tween 20; 10% glycerol; and 100 mM NaCl). After magnetization and washing with the buffer, the interacting proteins were eluted by adding Laemmli protein loading buffer and heating at 95 °C. The eluted proteins were loaded on 12% SDS-PAGE then stained with Silver Nitrate (ProteoSilver Silver stain kit from Promega). The bands corresponding to the selected proteins were digested as described by Allmann et al. [47].

Mass spectrometry analysis

Online nanoLC-MS/MS analyses were performed using an Ultimate 3000 system (Dionex, Amsterdam, The Netherlands) coupled to a nanospray LTQ Orbitrap XL mass spectrometer (Thermo Fisher Scientific, Bremen, Germany). Ten microliters of each peptide extract were loaded on a 300 μm ID × 5 mm PepMap C_{18} precolumn

(LC Packings, Dionex, USA) at a flow rate of 20 μl/min. After 5 min desalting, peptides were online separated on a 75 μm ID × 15 cm C_{18}PepMap™ column (LC packings, Dionex, USA) with a 2–40% linear gradient of solvent B (0.1% formic acid in 80% ACN) in 108 min. The separation flow rate was set at 200 nl/min. The mass spectrometer operated in positive ion mode at a 1.8 kV needle voltage and a 42 V capillary voltage. Data were acquired in a data-dependent mode alternating an FTMS scan survey over the range m/z 300–1700 with the resolution set to a value of 60,000 at m/z 400 and six ion trap MS/MS scans with Collision Induced Dissociation (CID) as activation mode. MS/MS spectra were acquired using a 3 m/z unit ion isolation window and normalized collision energy of 35. Mono-charged ions and unassigned charge-state ions were rejected from fragmentation. Dynamic exclusion duration was set to 30 s.

Database search and results processing

Mascot and Sequest algorithms through Proteome Discoverer 1.3 Software (Thermo Fisher Scientific Inc.) were used for protein identification in batch mode by searching against the UniProt *Homo sapiens* database (65,304 entries, Reference Proteome Set, Release 2012_03). Two missed enzyme cleavages were allowed. Mass tolerances in MS and MS/MS were set to 10 ppm and 0.6 Da. Oxidation of methionine was searched as dynamic modifications. Carbamidomethylation on cysteine was searched as fixed modification. Peptide validation was performed using Percolator algorithm [48] and only "high confidence" peptides were retained corresponding to a 1% False Positive Rate at peptide level. Only proteins with two minimum and distinct peptides were considered in the results.

In vitro co-precipitation

IN, LEDGF/p75 or IN•LEDGF/p75 (100 nM) were incubated with 80 nM of FACT complex in 10 μl interaction buffer (50 mM HEPES, pH 7.5; 1 μg/ml BSA;1 mM DTT; 0.1% Tween 20; 10% glycerol; and 50–240 mM NaCl) for 20 min on ice and then for 30 min at room temperature. A 15 μl aliquot of either Dynabeads M-280 sheep anti-Rabbit IgG (Invitrogen, ref. 11203D) previously coupled to polyclonal anti-IN antibodies or Dynabeads M-280 sheep anti-Mouse IgG (Invitrogen, ref. 11201D) previously coupled to monoclonal anti-LEDGF/p75, and washed was then added to a total volume of 300 μl interaction buffer and incubated at room temperature for 1 h under rotation. The beads were washed three times with 300 μl interaction buffer and the precipitated products were re-suspended in 8 μl of H_2O then 2 μl of 5× Laemmli buffer were added, after which they were separated on a 12% gel via SDS-PAGE. Interacting proteins were detected either by Western blot analysis using anti-HIV-1 IN, anti-SSRP1/SPT16 and/or LEDGF antibodies either by direct gel staining using colloidal blue. GST pull down were performed using 4 μg of proteins incubated in 10 μl interaction buffer (50 mM HEPES, pH 7.5; 1 μg/ml BSA; 1 mM DTT; 0.1% Tween 20; 10% glycerol; and 50–240 mM NaCl) for 20 min on ice and then for 30 min at room temperature. 15 μl of glutathione Sepharose 4B beads (GE Healthcare) were washed and diluted in 275 μl of interaction buffer and then incubated for 1 h at room temperature under rotation. The beads were washed three times with 800 μl interaction buffer and the precipitated products were re-suspended in 8 μl of H_2O then 2 μl of 5× Laemmli buffer were added, after which they were separated on a 12% gel via SDS-PAGE. Interacting proteins were detected either by Western blot using the corresponding antibodies or by direct gel staining using colloidal blue.

Immunoprecipitation in cells

FLAG-LEDGF/p75, HIV-1 IN-myc and SSRP1-myc expression plasmids [21, 49, 50] were transfected in HEK293T-derived, LEDGF/p75-deficient cells si1340/1428 cells [51] using the calcium-phosphate co-transfection method as described in [21]. Seventy-two hours after transfection cells (~3 × 10^6) were lysed in 300 μl of CSK I buffer (10 mM Pipes pH 6.8, 100 mM NaCl, 1 mM EDTA, 300 mM sucrose, 1 mM $MgCl_2$, 1 mM DTT, 0.5% Triton X-100) containing protease inhibitors (final concentration: leupeptine 2 μg/ml, aprotinin 5 μg/μl, PMSF 1 mM, pepstatin A 1 μg/ml). Cellular lysates were centrifuged at 1000*g* for 6 min at 4 °C and the pellet containing Triton X-100-insoluble proteins and chromatin-bound proteins was re-suspended in 20 μl of CSK II buffer (10 mM Pipes pH 6.8, 100 mM NaCl, 300 mM sucrose, 6 mM $MgCl_2$, 1 mM DTT) supplemented with protease inhibitors, 16 units of turbo DNase (Ambion™), 3.4 μl of $(NH_4)_2SO_4$, and 3.1 μl of 10× turbo DNase reaction buffer. DNase treatment was conducted at 37 °C for 30 min. After incubation, 300 μl of CSK I buffer was added to the DNase treated sample to dilute the $(NH_4)_2SO_4$ and centrifuged at 22,000*g* for 3 min. Then the supernatant (S2 fraction) was pre-clear twice with goat anti-mouse IgG-coated magnetic beads (magnetized beads, Thermo Scientific, Cat. No. 21354). Then pre-cleared lysates were incubated for 2 h at 4 °C with magnetized beads preloaded with anti-FLAG mAb (Sigma, F3165). Bead-bound proteins were eluted by boiling in Laemmli sample buffer after extensive washing. Then, immunoprecipitated proteins were analyzed by immunoblotting as described in [21]. FLAG-tagged LEDGF/p75 was detected with anti-FLAG mAb (1/500, M2, Sigma) and Myc-tagged SSRP1 and HIV-1 IN were detected with anti-Myc mAb (1/500, clone 9E10, Covance, MMS-150P).

In vitro integration assays

Typical concerted integration assays were performed as previously reported [8] using 200 nM of IN, 10 ng of donor DNA, and 50 ng of plasmid DNA (reaction solution: 20 mM HEPES pH 7, 15% DMSO, 8% PEG, 10 mM $MgCl_2$, 20 µM $ZnCl_2$, 100 mM NaCl, 10 mM DTT final concentrations). After the reaction, the resulting integration products were treated with proteinase K 1 mg/ml for 1 h at 55 °C and with phenol/chloroform/isoamyl alcohol (24/25/1, v/v/v). Aqueous phase was then loaded onto a 1% agarose gel. The gel was then dried and submitted to autoradiography. The bands corresponding to free substrate (S), donor/donor (d/d), linear FSI (FSI) and circular HSI + FSI (HSI + FSI) products were quantified by ImageJ software. The circular FSI products were quantified by cloning them into bacteria and determining the numbers of ampicillin-, kanamycin- and tetracycline-resistant clones as percentages of the integration reaction control, which was performed using the wild-type enzyme. UV crosslink was performed by placing 50 ng of chromatinized vectors in 15% DMSO, 8% PEG, 10 mM $MgCl_2$, 20 µM $ZnCl_2$, 100 mM NaCl, 10 mMDTT final concentration in a 96-well plate. The plate was irradiated under a UV lamp (Bioblock Scientifif) at 254 nM for 10 min at 4 °C before use in concerted integration.

Cellular procedures

Human Embryonic Kidney 293 (HEK-293T), HeLa and HeLa P4 are typical laboratory cell lines. TZMbl LEDGF knock-out cells [34] were a kind gift of Dr. José Esté from the AIDS Research Institute-Irsicaixa. Lentiviral transductions were performed using pRRLsin-PGK-eGFP-WPRE VSV-G pseudotyped lentiviral vectors as previously described [24]. PFV transductions were performed using single-cycle viruses produced by co-transfection of HEK293T cells (Cell Services, London Research Institute) with pMD9 (GFP reporter PFV vector) and codon-optimized foamy virus GAG, POL and ENV packaging constructs and as previously described [7]. An optimized multiplicity of infection (MOI) of 1 was used, which resulted in 25–35% of the cells containing one copy of proviral DNA as determined before. Fluorescence was quantified 10 days post-transduction by counting 10,000 cells on a FACSCalibur flow cytometer (Becton–Dickinson, San Jose, CA, USA). HIV-1 DNA species were quantified at 8, 24 and 48 h post-transduction as previously described [52]. The total and integrated HIV-1 DNA levels were determined as copy numbers per 10^6cells. Integrated cDNA and 2-LTR circles were expressed as a percentage of the total viral DNA.

HeLa P4 cells expressing CD4 and CXCR4 receptors, and carrying the stably integrated lacZ gene under the control of the HIV-1 LTR were infected by HIV-1 Lai (1.10^8 particles/ml, M.O.I = 0.4) as previously reported [24]. Under this system the β-galactosidase activity, whose expression is linked to the expression of the Tat protein, is proportional to HIV-1 integration. In the infection experiments HeLa P4 cells were plated in 48-multiwell plates at 50,000 cells/well using 400 µl of DMEM (Invitrogen, Carlsbad, CA) containing 10% (v/v) fetal calf serum (FCS, Invitrogen) and, 50 µg/ml of gentamycin (Invitrogen). After overnight incubation at 37 °C, medium was replaced with 400 µl of fresh DMEM containing either HIV-1 Lai (1.10^8 particles/ml, M.O.I = 0.4) produced as described in [53]. After 24 h at 37 °C each well was refilled with 400 µl of a reaction buffer containing 50 mM Tris–HCl pH 8, 100 mM β-mercaptoethanol, 0.05% Triton X-100 and 5 mM of 4-methylumbelliferyl-β-D galactoside (4-MUG) (Sigma, St. Louis, MO). The level of the reaction was measured in a fluorescence microplate reader (Cytofluor II; Applied Biosystems, Foster City, CA) at 360/460 nm Ex/Em after 24 h incubation.

siRNA transient transfection

293T and HeLa cells (40% confluence in 48 well plates) were transfected with 0.5, 1, 10, 20, and 40 nM of the scramble and SSRP1siRNA (Santa-Cruz sc-37877) using the INTERFERin Polyplus transfection agent. After 24 h (60% confluence), cells were retransfected with 0.5, 1, 10, 20, and 40 nM of the scramble and SSRP1siRNA using the INTERFERin Polyplus transfection agent. Cells were cultured in DMEM with 20% FCS for serum stimulation. The cells were harvested after 96 h for Western blotting. For Western blotting, the cells were lysed with RIPA buffer containing phenylmethylsulphonylfluoride (PMSF) protease inhibitor (0.1 mM) and were subjected to Western blot.

Formaldehyde-assisted isolation of regulatory elements (FAIRE)

In cellulo: Analysis was adapted from previously reported conditions [54]. Four independent cultures (biological replicates) of cells treated or untreated by curaxin or subjected to SSRP1 siRNA were grown in 245 × 245-mm plates to 90% confluence. Formaldehyde was added directly to the plates at room temperature (22–25 °C) to a final concentration of 1% and incubated for 1, 2, 4, or 7 min, respectively. Glycine was added to a final concentration of 125 mM for 5 min at room temperature to quench the formaldehyde. Cells were rinsed with phosphate-buffered saline containing PMSF, and the plate was scraped and rinsed two more times. The cells were spun at 2000 rpm for 4 min and snap-frozen. Cells were resuspended in 1 ml of lysis buffer (2% Triton X-100, 1% SDS, 100 mM NaCl, 10 mM Tris–Cl at pH 8.0, 1 mM EDTA) per 0.4 g of cells and lysed using glass bead disruption for five 1-min sessions with 2-min incubations on ice between sessions. Samples were then sonicated for five sessions of 60 pulses (1 s on/1 s off) using a Branson Sonifier at 15% amplitude. Cellular debris was cleared by spinning at 15,000 rcf for 5 min at 4 °C. DNA was isolated by adding an equal volume of phenol–chloroform (phenol, chloroform, and isoamyl alcohol 25:24:1 saturated with 10 mM Tris at pH 8.0, 1 mM EDTA), vortexing, and spinning at 15,000 rpm for 5 min at 4 °C. The aqueous phase was isolated and stored in a separate tube. An additional 500 µl of TE was added to the organic phase, vortexed, and spun again at 15,000 rpm for 5 min at 4 °C. The aqueous phase was isolated and combined with the first aqueous fraction, and a final phenol–chloroform extraction was performed on the pooled aqueous fractions to ensure that all protein was removed. The DNA was precipitated by addition of sodium acetate to 0.3 M, glycogen to 20 µg/ml, and two-fold the volume of 95% ethanol, and incubated at −20 °C overnight. The precipitate was spun at 15,000 rpm for 10 min at 4 °C, then the pellet was washed with 70% ethanol and dried in a Speed-Vac. The pellet was resuspended in dH_2O and treated with Rnase A (100 µg/ml) and incubated at 37 °C for 2 h. DNA concentration was evaluated in the pellet and the different supernatant by nanodrop. *In vitro.* 50–100 ng of PN DNA treated or not with FACT were incubated 30 min at 37 °C with 2% formaldehyde. DNA was then sonicated 3 × 30 s and extracted with 50 µl of phenol–chloroform (24/25 v/v) solution followed by a 10-min centrifugation at 13,000*g*. The free DNA concentration in the supernatant was quantified by nanodrop and loaded on 1% agarose gel.

Additional files

Additional file 1: Figure S1. Strategy for selection of cellular interactants of IN•viral DNA complex. Cellular extracts from HeLa P4 cells were first incubated with streptavidin beads coupled to the viral DNA fragments in order to avoid selecting proteins binding solely to the DNA. The elution was then incubated with streptavidin beads coupled to fractions enriched in active IN•viral DNA complexes (**A**). Formation of active IN•DNA complexes was checked in vitro concerted integration (**B**), the data obtained with increasing concentration of INs, 100, 200, 400 nM (lanes 1–3) are reported. The elution of the interacting proteins was loaded on 12% SDS-PAGE gel stained with silver nitrate and the bands were excised and electroeluted. A typical result from a selection performed with the IN•viral DNA and the control DNA alone is reported in (**C**).

Additional file 2: Table S2. List of additionally selected proteins. Cellular extracts from HeLa P4 cells were incubated with streptavidin beads coupled to fraction enriched in active IN•viral DNA complexes as shown in Additional file 1: Figure S1. The elution of the interacting proteins was loaded on 12% SDS-PAGE gel stained with silver nitrate and the bands were excised from gel and submitted to electroelution. The selected proteins were identified by MS–MS. A list of proteins found only associated to IN•viral DNA complexes but not to control DNA alone and linked to chromatin or transcription is provided here in addition to the factors reported in Table 1.

Additional file 3: Figure S3. *In vitro* interaction between HIV-1 IN, FACT and LEDGF/p75 variants. Immunoprecipitation or GST pull down were performed using recombinant cofactors, polyclonal anti-HIV-1 IN or anti-LEDGF antibodies. The interactions were monitored by direct gel staining using colloidal blue and quantified by Image J software.

Additional file 4: Figure S4. FAIRE analysis and effect of UV-crosslinking on chromatin FACT and integration activities. FAIRE analyses were performed as indicated in materials and methods section and schematized in (**A**, adapted from [54]). Analysis of nucleosome remodeling activity of FACT on chromatinized substrates in vitro was performed by quantifying the free DNA recovered after FAIRE assay performed on naked p5S or chromatinized p5S loaded on 1% agarose gel after treatment or not with FACT complex and UV-treated or untreated (**B**). The recovered DNA was then quantified using ImageJ software and data are shown as the mean ± standard deviation (error bars) of at least three independent sets of experiments (**C**). A typical concerted integration performed with DNA substrate pre-treated with UV before FACT addition is reported in (**D**). Analysis of nucleosome remodeling activity of FACT on chromatinized substrates in vitro in the presence of absence of LEDGF/p75 is reported in (**E**).

Additional file 5: Figure S5. Effect of FACT curaxin inhibitors on in vitro HIV-1 integration, cell viability and chromatin structure. The structure of CBL100 and CBL137 and their effect in typical concerted integration catalyzed by HIV-1 IN are shown in (**A**). The viability of the cells treated with curaxins was measured using a typical MTT assay and data are shown in (**B**). Effect of curaxin treatment on chromatin structure was analyzed by FAIR, as previously shown (**C**).

Additional file 6: Figure S6. Effect of FACT-mediated chromatin remodeling chemical promotion on early steps of HIV-1 LAI virus and derived lentiviral vectors. HeLaP4 and HEK293T cells were treated with CBLC137 curaxin (0.1 µM) 6 h before cell infection with LAI wild type virus or transduction with lentiviral vectors. The effect on LAI virus (**A**) and viral vector (**B**) reverse transcription and integration was evaluated by quantitative PCR performed on the different viral DNA populations at 0–48 h post-transduction. All values are shown as the mean ± standard deviation (error bars) of at least three independent sets of experiments done in duplicates.

Additional file 7: Figure S7. Effect of FACT knock down on early steps of HIV-1 derived lentiviral vectors in HEK293T cells. The early steps of replication of lentiviral vectors in cells knockdown for SSRP1 (see Fig. 7) were evaluated by quantification of the viral DNA populations at 0–48 h post-transduction using quantitative PCR (data obtained after a 20 nM siRNA treatment are reported here). All values are shown as the mean ± standard deviation (error bars) of four independent sets of experiments done in duplicates.

Authors' contributions

DL, EM, VP performed the in vitro pull down experiments. JM performed the siRNA experiments. PL purified the FACT complex and PFV IN protein. CC purified the HIV-1 IN protein. ML (Lavigne) purified the GST fused proteins. DL and VP performed the curaxins experiments. MSB and VP generated the chromatinized templates. VP performed the in vitro functional assays. APL and ML (Llano) performed the cellular pull down experiments and analyzed the data. OD performed the viral DNA quantifications and analyzed the data. MR purified the LEDGF/p75 and IN•LEDGF/p75 complex and analyzed the data. VP, MLA and ML (Lavigne) analyzed the data. VP wrote the manuscript. ML (Lavigne), MR, OD, PL edited the manuscript. All authors read and approved the final manuscript.

Author details

[1] Fundamental Microbiology and Pathogenicity Laboratory, UMR 5234 CNRS, University of Bordeaux, SFR TransBioMed, 146 rue Léo Saignat, 33076 Bordeaux Cedex, France. [2] Centre Génomique fonctionnelle Bordeaux, Plateforme Proteome, Université de Bordeaux, Bordeaux, France. [3] Department of Biological Sciences, University of Texas at El Paso, El Paso, TX, USA. [4] Département de Biologie Structurale Intégrative, UDS, U596 INSERM, UMR7104 CNRS, IGBMC

(Institut de Génétique et de Biologie Moléculaire et Cellulaire), Illkirch-Graffenstaden, France. [5] LBPA, UMR8113, CNRS, ENS-Cachan, Cachan, France. [6] Department of Virology, UMR 3569, CNRS, Institut Pasteur, Paris, France. [7] Institut Cochin-INSERM U1016-CNRS UMR8104, Université Paris Descartes, Paris, France. [8] International Associated Laboratory (LIA) of Microbiology and Immunology, CNRS/University de Bordeaux/Heinrich Pette Institute-Leibniz Institute for Experimental Virology, Bordeaux, France. [9] Viral DNA Integration and Chromatin Dynamics Network (DyNAVir), Paris, France.

Acknowledgements

The authors are deeply grateful to Dr. Simon Litvak and Dr. Eric Deprez for fruitful discussions and corrections of the manuscript. We acknowledge José Esté from the AIDS Research Institute-Irsicaixa for the kind gift of TZMbl KO LEDGF cells and Dr. Gurova K.V. for the gift of curaxins. The manuscript was edited by Prof Ray Cooke.

Competing interests

The authors declare that they have no competing interests.

Funding

This work was supported by the French National Research Agency (ANR, RETROSelect, jcjc2011 program); the French National Research Agency against AIDS (ANRS, AO2016); SIDACTION (AO2016, VIH20160721002); the French Infrastructure for Integrated Structural Biology (FRISBI) ANR-10-INSB-05-01; Instruct, a part of the European Strategy Forum on Research Infrastructures (ESFRI); the Centre National de la Recherche Scientifique (CNRS); the University of Bordeaux.

References

1. Lesbats P, Engelman AN, Cherepanov P. Retroviral DNA integration. Chem Rev. 2016;116:12730–57.
2. Demeulemeester J, Rijck JD, Gijsbers R, Debyser Z. Retroviral integration: site matters: mechanisms and consequences of retroviral integration site selection. BioEssays. 2015;37:1202.
3. Kvaratskhelia M, Sharma A, Larue RC, Serrao E, Engelman A. Molecular mechanisms of retroviral integration site selection. Nucleic Acids Res. 2014;42:5917–28.
4. Pruss D, Bushman FD, Wolffe AP. Human immunodeficiency virus integrase directs integration to sites of severe DNA distortion within the nucleosome core. Proc Natl Acad Sci USA. 1994;91:5913–7.
5. Pruss D, Reeves R, Bushman FD, Wolffe AP. The influence of DNA and nucleosome structure on integration events directed by HIV integrase. J Biol Chem. 1994;269:25031–41.
6. Pryciak MP, Varmus EH. Nucleosomes, DNA-binding proteins, and DNA sequence modulate retroviral integration target site selection. Cell. 1992;69:769–80.
7. Maskell DP, Renault L, Serrao E, Lesbats P, Matadeen R, Hare S, et al. Structural basis for retroviral integration into nucleosomes. Nature. 2015;523:366–9.
8. Benleulmi MS, Matysiak J, Henriquez DR, Vaillant C, Lesbats P, Calmels C, et al. Intasome architecture and chromatin density modulate retroviral integration into nucleosome. Retrovirology. 2015;12:13.
9. Naughtin M, Haftek-Terreau Z, Xavier J, Meyer S, Silvain M, Jaszczyszyn Y, et al. DNA physical properties and nucleosome positions are major determinants of HIV-1 integrase selectivity. PLoS ONE. 2015;10:e0129427.
10. Serrao E, Krishnan L, Shun MC, Li X, Cherepanov P, Engelman A, et al. Integrase residues that determine nucleotide preferences at sites of HIV-1 integration: implications for the mechanism of target DNA binding. Nucleic Acids Res. 2014;42:5164–76.
11. Pasi M, Mornico D, Volant S, Juchet A, Batisse J, Bouchier C, et al. DNA minicircles clarify the specific role of DNA structure on retroviral integration. Nucleic Acids Res. 2016;44:7830.
12. Craigie R, Bushman FD. Host factors in retroviral integration and the selection of integration target sites. Microbiol Spectr. 2014;2(6). doi:10.1128/microbiolspec.MDNA3-0026-2014.
13. Sowd GA, Serrao E, Wang H, Wang W, Fadel HJ, Poeschla EM, et al. A critical role for alternative polyadenylation factor CPSF6 in targeting HIV-1 integration to transcriptionally active chromatin. Proc Natl Acad Sci USA. 2016;113:E1054–63.
14. Lesbats P, Botbol Y, Chevereau G, Vaillant C, Calmels C, Arneodo A, et al. Functional coupling between HIV-1 integrase and the SWI/SNF chromatin remodeling complex for efficient in vitro integration into stable nucleosomes. PLoS Pathog. 2011;7:e1001280.
15. Lesbats P, Lavigne M, Parissi V. HIV-1 integration into chromatin: new insights and future perspective. Future Virol. 2011;6(9). doi:10.2217/fvl.11.84.
16. Wang GP, Ciuffi A, Leipzig J, Berry CC, Bushman FD. HIV integration site selection: analysis by massively parallel pyrosequencing reveals association with epigenetic modifications. Genome Res. 2007;17:1186–94.
17. Orphanides G, Wu WH, Lane WS, Hampsey M, Reinberg D. The chromatin-specific transcription elongation factor FACT comprises human SPT16 and SSRP1 proteins. Nature. 1999;400:284–8.
18. Belotserkovskaya R, Oh S, Bondarenko VA, Orphanides G, Studitsky VM, Reinberg D. FACT facilitates transcription-dependent nucleosome alteration. Science. 2003;301:1090–3.
19. Hsieh F-K, Kulaeva OI, Patel SS, Dyer PN, Luger K, Reinberg D, et al. Histone chaperone FACT action during transcription through chromatin by RNA polymerase II. Proc Natl Acad Sci USA. 2013;110:7654.
20. Huang H, Santoso N, Power D, Simpson S, Dieringer M, Miao H, et al. FACT proteins, SUPT16H and SSRP1, are transcriptional suppressors of HIV-1 and HTLV-1 that facilitate viral latency. J Biol Chem. 2015;290:27297–310.
21. Lopez AP, Kugelman JR, Garcia-Rivera J, Urias E, Salinas SA, Fernandez-Zapico ME, et al. The structure specific recognition protein 1 associates with lens epithelium-derived growth factor proteins and modulates HIV-1 replication. J Mol Biol. 2016;428:2814–31.
22. Lesbats P, Metifiot M, Calmels C, Baranova S, Nevinsky G, Andreola ML, et al. In vitro initial attachment of HIV-1 integrase to viral ends: control of the DNA specific interaction by the oligomerization state. Nucleic Acids Res. 2008;36:7043–58.
23. Suzuki Y, Chew ML. Role of host-encoded proteins in restriction of retroviral integration. Front Microbiol. 2012;3:227.
24. Cosnefroy O, Tocco A, Lesbats P, Thierry S, Calmels C, Wiktorowicz T, et al. Stimulation of the human RAD51 nucleofilament restricts HIV-1 integration in vitro and in infected cells. J Virol. 2012;86:513–26.
25. Mousnier A, Kubat N, Massias-Simon A, Segeral E, Rain JC, Benarous R, et al. von Hippel Lindau binding protein 1-mediated degradation of integrase affects HIV-1 gene expression at a postintegration step. Proc Natl Acad Sci USA. 2007;104:13615–20.
26. Belotserkovskaya R, Reinberg D. Facts about FACT and transcript elongation through chromatin. Curr Opin Genet Dev. 2004;14:139–46.
27. Llano M, Delgado S, Vanegas M, Poeschla EM. Lens epithelium-derived growth factor/p75 prevents proteasomal degradation of HIV-1 integrase. J Biol Chem. 2004;279:55570–7.
28. Solomon MJ, Varshavsky A. Formaldehyde-mediated DNA-protein crosslinking: a probe for in vivo chromatin structures. Proc Natl Acad Sci USA. 1985;82:6470.
29. Hamiche A, Carot V, Alilat M, De Lucia F, O'Donohue MF, Revet B, et al. Interaction of the histone (H3-H4)2 tetramer of the nucleosome with positively supercoiled DNA minicircles: potential flipping of the protein from a left- to a right-handed superhelical form. Proc Natl Acad Sci USA. 1996;93:7588–93.
30. Birch JL, Tan BC, Panov KI, Panova TB, Andersen JS, Owen-Hughes TA, et al. FACT facilitates chromatin transcription by RNA polymerases I and III. EMBO J. 2009;28:854–65.
31. Li Y, Zeng SX, Landais I, Lu H. Human SSRP1 has Spt16-dependent and -independent roles in gene transcription. J Biol Chem. 2007;282:6936–45.
32. Safina A, Cheney P, Pal M, Brodsky L, Ivanov A, Kirsanov K, et al. FACT is a sensor of DNA torsional stress in eukaryotic cells. Nucleic Acids Res. 2017;45:1925–45.

33. Gasparian AV, Burkhart CA, Purmal AA, Brodsky L, Pal M, Saranadasa M, et al. Curaxins: anticancer compounds that simultaneously suppress NF-kappaB and activate p53 by targeting FACT. Sci Transl Med. 2011;3:95ra74.
34. Badia R, Pauls E, Riveira-Munoz E, Clotet B, Esté JA, Ballana E. Zinc finger endonuclease targeting PSIP1 inhibits HIV-1 integration. Antimicrob Agents Chemother. 2014;58:4318–27.
35. Tan BC-M, Liu H, Lin C-L, Lee S-C. Functional cooperation between FACT and MCM is coordinated with cell cycle and differential complex formation. J Biomed Sci. 2010;17:11.
36. Winans S, Larue RC, Abraham CM, Shkriabai N, Skopp A, Winkler D, et al. The FACT complex promotes avian leukosis virus DNA integration. J Virol. 2017;91(7):e00082-17.
37. Passos DO, Li M, Yang R, Rebensburg SV, Ghirlando R, Jeon Y, et al. Cryo-EM structures and atomic model of the HIV-1 strand transfer complex intasome. Science. 2017;355:89–92.
38. Yin Z, Shi K, Banerjee S, Pandey KK, Bera S, Grandgenett DP, et al. Crystal structure of the Rous sarcoma virus intasome. Nature. 2016;530:362–6.
39. Ballandras-Colas A, Brown M, Cook NJ, Dewdney TG, Demeler B, Cherepanov P, et al. Cryo-EM reveals a novel octameric integrase structure for betaretroviral intasome function. Nature. 2016;530:358–61.
40. Carvalho S, Raposo AC, Martins FB, Grosso AR, Sridhara SC, Rino J, et al. Histone methyltransferase SETD2 coordinates FACT recruitment with nucleosome dynamics during transcription. Nucleic Acids Res. 2013;41:2881.
41. Song F, Chen P, Sun D, Wang M, Dong L, Liang D, et al. Cryo-EM study of the chromatin fiber reveals a double helix twisted by tetranucleosomal units. Science. 2014;344:376–80.
42. Dorigo B, Schalch T, Bystricky K, Richmond TJ. Chromatin fiber folding: requirement for the histone H4N-terminal tail. J Mol Biol. 2003;327:85–96.
43. Luger K, Mader AW, Richmond RK, Sargent DF, Richmond TJ. Crystal structure of the nucleosome core particle at 2.8 A resolution. Nature. 1997;389:251–60.
44. Michel F, Crucifix C, Granger F, Eiler S, Mouscadet JF, Korolev S, et al. Structural basis for HIV-1 DNA integration in the human genome, role of the LEDGF/P75 cofactor. EMBO J. 2009;28:980–91.
45. Botbol Y, Raghavendra NK, Rahman S, Engelman A, Lavigne M. Chromatinized templates reveal the requirement for the LEDGF/p75 PWWP domain during HIV-1 integration in vitro. Nucleic Acids Res. 2008;36:1237–46.
46. Shun MC, Botbol Y, Li X, Di Nunzio F, Daigle JE, Yan N, et al. Identification and characterization of PWWP domain residues critical for LEDGF/p75 chromatin binding and human immunodeficiency virus type 1 infectivity. J Virol. 2008;82:11555–67.
47. Allmann S, Mazet M, Ziebart N, Bouyssou G, Fouillen L, Dupuy J-W, et al. Triacylglycerol storage in lipid droplets in procyclic trypanosoma brucei. PLoS ONE. 2014;9:e114623.
48. Käll L, Canterbury JD, Weston J, Noble WS, MacCoss MJ. Semi-supervised learning for peptide identification from shotgun proteomics datasets. Nat Methods. 2007;4:923–5.
49. Bueno MTD, Garcia-Rivera JA, Kugelman JR, Morales E, Rosas-Acosta G, Llano M. SUMOylation of the lens epithelium-derived growth factor/p75 attenuates its transcriptional activity on the heat shock protein 27 promoter. J Mol Biol. 2010;399:221–39.
50. Garcia-Rivera JA, Bueno MTD, Morales E, Kugelman JR, Rodriguez DF, Llano M. Implication of serine residues 271, 273, and 275 in the human immunodeficiency virus type 1 cofactor activity of lens epithelium-derived growth factor/p75. J Virol. 2010;84:740–52.
51. Llano M, Vanegas M, Fregoso O, Saenz D, Chung S, Peretz M, et al. LEDGF/p75 determines cellular trafficking of diverse lentiviral but not murine oncoretroviral integrase proteins and is a component of functional lentiviral preintegration complexes. J Virol. 2004;78:9524–37.
52. Munir S, Thierry S, Subra F, Deprez E, Delelis O. Quantitative analysis of the time-course of viral DNA forms during the HIV-1 life cycle. Retrovirology. 2013;10:87.
53. Metifiot M, Faure A, Guyonnet-Duperat V, Bellecave P, Litvak S, Ventura M, et al. Cellular uptake of ODNs in HIV-1 human-infected cells: a role for viral particles in DNA delivery? Oligonucleotides. 2007;17:151–65.
54. Giresi PG, Kim J, McDaniell RM, Iyer VR, Lieb JD. FAIRE (Formaldehyde-Assisted Isolation of Regulatory Elements) isolates active regulatory elements from human chromatin. Genome Res. 2007;17:877–85.

2

Efficacies of Cabotegravir and Bictegravir against drug-resistant HIV-1 integrase mutants

Steven J. Smith[1], Xue Zhi Zhao[2], Terrence R. Burke Jr.[2] and Stephen H. Hughes[1*]

Abstract

Background: Integrase strand transfer inhibitors (INSTIs) are the class of antiretroviral (ARV) drugs most recently approved by the FDA for the treatment of HIV-1 infections. INSTIs block the strand transfer reaction catalyzed by HIV-1 integrase (IN) and have been shown to potently inhibit infection by wild-type HIV-1. Of the three current FDA-approved INSTIs, Dolutegravir (DTG), has been the most effective, in part because treatment does not readily select for resistant mutants. However, recent studies showed that when INSTI-experienced patients are put on a DTG-salvage therapy, they have reduced response rates. Two new INSTIs, Cabotegravir (CAB) and Bictegravir (BIC), are currently in late-stage clinical trials.

Results: Both CAB and BIC had much broader antiviral profiles than RAL and EVG against the INSTI-resistant single, double, and triple HIV-1 mutants used in this study. BIC was more effective than DTG against several INSTI-resistant mutants. Overall, in terms of their ability to inhibit a broad range of INSTI-resistant IN mutants, BIC was superior to DTG, and DTG was superior to CAB. Modeling the binding of CAB, BIC, and DTG within the active site of IN suggested that the "left side" of the INSTI pharmacophore (the side away from the viral DNA) was important in determining the ability of the compound to inhibit the IN mutants we tested.

Conclusions: Of the two INSTIs in late stage clinical trials, BIC appears to be better able to inhibit the replication of a broad range of IN mutants. BIC retained potency against several of the INSTI-resistant mutants that caused a decrease in susceptibility to DTG.

Keywords: HIV-1, Integrase, Infectivity, Potency, Susceptibility, Modeling, Resistance

Background

INSTIs are the class of antiretroviral (ARV) drugs most recently approved by the FDA to treat HIV-1 infections. INSTIs target the second reaction performed by HIV-1 Integrase (IN), strand transfer (ST), in which IN catalyzes the integration of the viral DNA into the cellular genome [1, 2]. INSTIs have a centralized pharmacophore, which contains a chelating functionality that interacts with the two catalytic Mg^{2+} ions at the IN active site [3, 4]. This central pharmacophore is joined to a halogenated benzyl moiety that interacts with the penultimate base at the 3′ end of the viral DNA [5]. Thus, INSTIs interact with both the enzyme and its nucleic acid substrate. The combination of these interactions allows the INSTIs to target and potently inhibit HIV-1 IN. Raltegravir (RAL) and Elvitegravir (EVG) are the first and second FDA-approved INSTIs, respectively. They potently inhibit WT HIV-1; however, resistant mutants can develop relatively quickly (Fig. 1). A partial list of the well-defined primary resistance mutations includes: Y143R, N155H, G140S/Q148H, T66I, and E92Q. Other mutations that confer resistance to RAL and EVG have been identified. In many cases, mutations selected by either RAL or EVG reduce the susceptibility of IN to the other INSTI, showing that RAL and EVG have overlapping resistance profiles [6–8].

In 2013, Dolutegravir (DTG) was approved by the FDA and it quickly became a preferred drug for combination antiretroviral therapy (cART) [9–12]. DTG differs from the first generation INSTIs in that its chelating motif is located on a tri-cyclic scaffold [13, 14]. In addition, the structural component that connects the central chelating

*Correspondence: hughesst@mail.nih.gov
[1] HIV Dynamics and Replication Program, National Cancer Institute-Frederick, National Institutes of Health, Frederick, MD, USA
Full list of author information is available at the end of the article

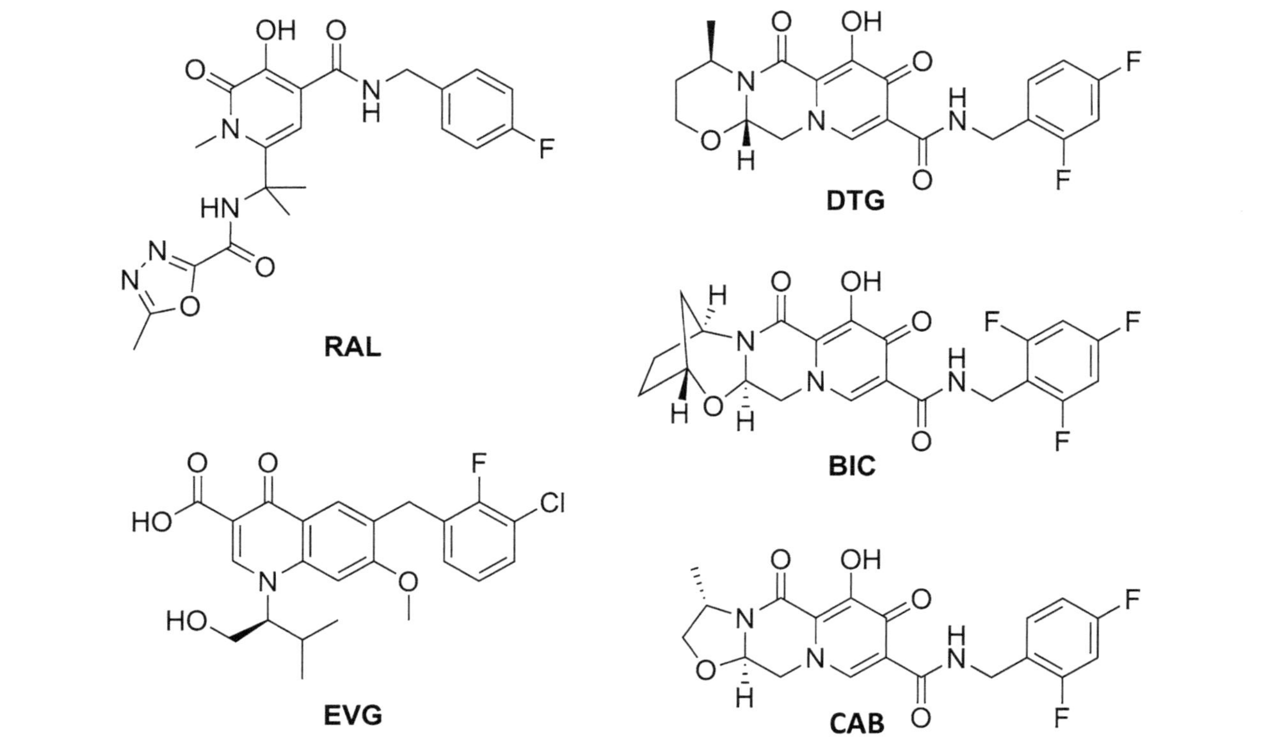

Fig. 1 Chemical structures of INSTIs. The chemical structures of RAL, EVG, DTG, BIC, and CAB are shown

moiety to the halogenated benzyl group is longer than it is in either RAL or EVG (Fig. 1) [15]. Not only do these structural differences allow DTG to be highly effective against WT HIV-1, but DTG is much more potent against IN mutants that confer resistance to the first generation INSTIs. Moreover, it has been difficult to select for DTG resistant mutants in cell culture and the treatment of HIV-1 patients using DTG has been, generally speaking, quite successful [16–21].

The usefulness of most ARV drugs is limited by the emergence of resistant mutants, and DTG will not be an exception. Recent in vitro selection studies with DTG have uncovered resistance mutations [22–24]. In clinical trials with INSTI-experienced subjects [25, 26] whose viruses had INSTI resistance mutations at the primary position Q148 plus at least one additional mutation at any of the secondary positions, L74, E138, G140, or G163, patients were put on a salvage regimen that included DTG. This change in therapy failed to lower HIV-1 below 50 copies/mL. Analysis of the virus present in the patients after the trial showed that additional mutations were selected in IN. These results showed that mutations that confer resistance to DTG can be selected in viruses that carry preexisting resistance mutations.

Recently, two new INSTIs, Cabotegravir (CAB) and Bictegravir (BIC), have been developed and these are currently in late phase clinical trials [13, 27, 28]. BIC and CAB, which are structurally similar to DTG, (both have tri-cyclic central pharmacophores), could offer therapeutic alternatives to HIV-1 patients (Fig. 1). Here, we describe evaluation of the antiviral potency of CAB and BIC against broad panels of well-characterized INSTI-resistant single and double mutants, and against the INSTI-resistant triple mutants identified in the VIKING clinical trials. Our objective was to determine how well these new INSTIs performed compared to DTG, the current standard of care.

Results

Initial screening of CAB and BIC against primary INSTI resistant mutants

The abilities of CAB and BIC to inhibit the replication of WT HIV-1 and INSTI-resistant mutants were determined in single-round viral replication assays. We initially screened CAB and BIC against a panel of primary INSTI-resistant mutants, which included: Y143R, N155H, G140S/Q148H, T66I, E92Q, H51Y, G118R, R263K, H51Y/R263K, and E138K/E263K. Y143R, N155H, and G140S/Q148H (Fig. 2; see also Additional file 1: Table S1A) were chosen because they have been selected in patients by treatment with RAL [29–31]; the T66I and E92Q mutants were selected by treatment with EVG

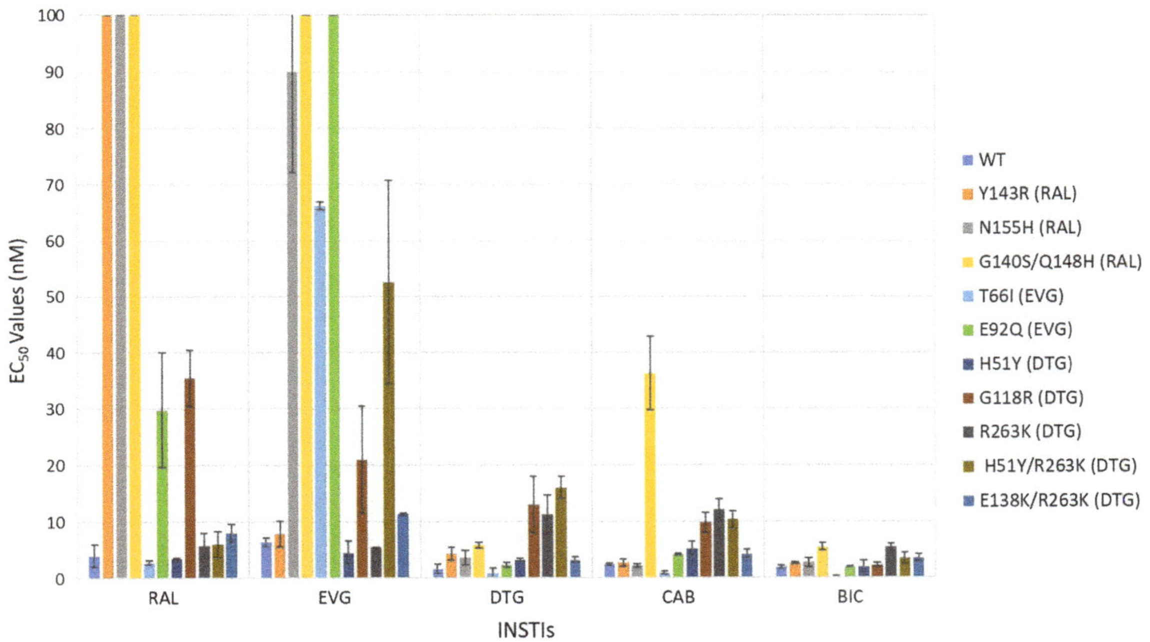

Fig. 2 Antiviral activities of BIC and CAB against primary INSTI-resistant mutants. The EC_{50} values were determined, in single round infection assays, using vectors that carry the INSTI-resistant IN mutants. Error bars represent the standard deviations in the data from independent experiments (n = 4). The EC_{50} values shown in the figure have a maximum of 100 nM. The EC_{50} values of RAL against Y143R, N155H, and G140S/Q148H, EVG versus G140S/Q148H, and E92Q primary INSTI-resistant mutants were all > 100 nM

[32–34]. The IN mutations H51Y, G118R, R263K, H51Y/R263K, E138K/R263K mutants have been selected with DTG in cell culture [22–24]. The R263K mutation has been selected in several treatment-experienced, INSTI-naïve patients undergoing DTG therapy [16]. Both CAB and BIC potently inhibited the infection of WT HIV-1 with EC_{50} values equivalent to the FDA-approved INSTIs (<3 nM). Moreover, CAB and BIC were minimally toxic in cell culture assays with CC_{50} values > 250 μM (data not shown), which is similar to the FDA-approved INSTIs. This demonstrates that these INSTIs have very favorable therapeutic indexes in cultured cells. Additionally, both CAB and BIC potently inhibited the RAL-resistant mutants Y143R and N155H; the EVG-resistant IN mutants T66I and E92Q, and the DTG-resistant IN mutant H51Y and E138K/R263K with EC_{50} values < 5 nM. However, only BIC potently inhibited the well-known RAL-resistant IN double mutant G140S/Q148H and the DTG-resistant IN mutants G118R, R263K, and H51Y/R263K with EC_{50} values ≤ 5 nM. The RAL-resistant IN mutant G140S/Q148H caused a substantial loss of CAB potency (36.3 ± 6.5 nM), while there was a smaller but still modest loss of potency against the DTG-resistant IN mutants G118R (12.1 ± 1.9 nM), R263K (13.4 ± 1.3 nM), and H51Y/R263K (10.4 ± 1.5 nM). These antiviral data were compared to previous screens, in which the antiviral potencies of RAL, EVG, and DTG were measured against the same INSTI-resistant primary mutants [35, 36]. When the antiviral profiles of the second generation INSTIs, DTG, CAB and BIC were compared to the FDA-approved INSTIs for WT HIV-1 and the RAL- and EVG-resistant mutants, all of the second generation INSTIs had antiviral profiles that were obviously superior to RAL and EVG. The differences were sufficiently clear cut that the comparisons between the first and second generation INSTIs were not subjected to statistical analysis. The more important question was whether either CAB or BIC was better than DTG, in terms of their ability to inhibit the IN mutants. To make the comparison objective, the statistical significance of the EC_{50} data for CAB, BIC, and DTG were analyzed using the Student's t test. The EC_{50} values for WT HIV for DTG, CAB and BIC were similar, which allowed us to compare the EC_{50} values for the mutants directly. In the initial screen, which included ten INSTI-resistant primary mutants, BIC was significantly better than CAB for seven of these ten primary mutants (four p values < 0.01 and three p values < 0.001; see Fig. 3 and Additional file 1: Table S1B). In addition, BIC was better than DTG against three of the primary mutants. In contrast, CAB was significantly better than DTG for one primary mutant and DTG was better than CAB for three of the primary mutants.

Antiviral activities of CAB and BIC against other common INSTI-resistant single mutants

We determined the antiviral profiles of CAB and BIC, as well as the FDA-approved INSTIs, against a second panel of additional INSTI-resistant single mutants to compare the strengths and weaknesses of the two new INSTIs and the FDA-approved INSTIs [37–39]. This panel of INSTI-resistant single mutants included: M50I, L74M, T97A, S119R, E138K, G140S, Q146L, Q146P, Q148H, Q148K, Q148R, and S153Y (Fig. 4; Additional file 1: Table S2A). BIC potently inhibited this entire panel of INSTI-resistant mutants with EC_{50} values below 5 nM, which was comparable to DTG. CAB also inhibited the majority of mutants in this panel. However, it lost some potency against the INSTI-resistant single mutants E138K (12.9 ± 1.0 nM), Q146P (10.3 ± 2.1 nM), and Q148H (6.8 ± 1.5 nM). Most of the INSTI-resistant single mutants in this panel caused significant drops in susceptibility to the first generation INSTIs, RAL and EVG, with the Q148H/K/R mutants having the greatest effect on the EC_{50} values. Based on the data obtained with the mutants in this panel, DTG was better than CAB and BIC (Fig. 3; Additional file 1: Table S2B). DTG was significantly better than CAB against six of the mutants and better than BIC against four mutants (two p values < 0.001). Conversely, BIC was better than CAB against five of these mutants.

Antiviral activities of CAB and BIC against a panel of INSTI-resistant double mutants having a primary mutation at position Q148

We next tested CAB, BIC, and the FDA-approved INSTIs against a panel of INSTI-resistant double mutants that included either one of the primary mutations at position Q148 (H/K/R), or Y143R or N155H. These primary mutations were combined with a secondary mutation at positions E138 (A/K) or G140 (A/C/S) (Fig. 5; Additional file 1: Table S3A). BIC potently inhibited (EC_{50} < 5 nM) the INSTI-resistant double mutants G140A/Q148H, Y143R/Q148H, Q148H/N155H, G140S/Q148K, E138A/Q148R, E138K/Q148R, and Q148R/N155H. Conversely, the INSTI-resistant double mutants G140A/Q148R (10 ± 2.5 nM), G140C/Q148R (6.4 ± 1.4 nM), G140S/Q148R (6.1 ± 1.3 nM) caused small losses in susceptibility to BIC, whereas the double mutants E138K/Q148K (59.3 ± 4.9 nM) and G140A/Q148K (137.1 ± 5.0 nM) resulted in substantial reductions in susceptibility to BIC. However, there was a large reduction in CAB potency against most of the double mutants in the panel. The double mutants Y143R/Q148H (6.0 ± 0.4 nM), E138A/Q148R (25.6 ± 0.8 nM), E138K/Q148R (24.1 ± 0.1 nM), and G140A/Q148R (13.7 ± 2.7 nM) caused a minimal loss in susceptibility to CAB, whereas E138K/Q148K (772.1 ± 72.2 nM), G140A/Q148K (393.1 ± 51.1 nM), G140S/Q148K (87.3 ± 7.6 nM), G140C/Q148R (66.6 ± 8.1 nM), and Q148R/N155H (50.5 ± 6.5 nM) caused large reductions in susceptibility to CAB. DTG was very effective across this panel of INSTI-resistant double mutants. However, it sustained a moderate loss in potency against the INSTI-resistant double mutant E138K/Q148K (25.0 ± 2.1 nM) and significant loss in potency against the INSTI-resistant double mutant G140A/Q148K (450.7 ± 58.8 nM). The first generation INSTIs RAL and EVG exhibited considerable loss of potency against all of the mutants in this panel of INSTI-resistant double mutants. BIC was significantly better than DTG against five of these double mutants (four p values < 0.01; Fig. 3; Additional file 1: Table S3B); however, DTG was better than BIC for four of the mutants (two p values < 0.001). CAB was not significantly better than either DTG or BIC against any double mutants.

Antiviral activities of CAB and BIC against a panel of INSTI-resistant double mutants that included the primary mutations T66I and N155H and additional mutations at other positions

We determined the antiviral profiles of CAB, BIC, and the FDA-approved INSTIs against the EVG-resistant double mutant T66I/E157Q and a panel of INSTI-resistant double mutants with a primary mutation at N155H and one of the following secondary mutations: E92Q, G140S, Y143H/R, or G163R (Fig. 6; Additional file 1: Table S4A). BIC, CAB, and DTG retained potency against the INSTI-resistant double mutant T66I/E157Q and the other INSTI-resistant double mutants (EC_{50} < 5 nM). The first generation INSTIs, RAL and EVG, failed to potently inhibit any of these double mutants. Based on this panel of mutants, the antiviral profiles of the three second generation INSTIs were similar to each other. CAB was significantly better than BIC against two of these double mutants (one p value < 0.001; see Figs. 3 and 7 and Additional file 1: Table S4B). DTG had better activity against one of the double mutants than CAB and BIC (p value < 0.01).

(See figure on next page.)

Fig. 3 Statistical significance of the antiviral data among DTG, CAB, and BIC. The Student's t test was used to calculate the statistical significance of the differences in the antiviral activities of the INSTIs. Because of multiple comparisons, p values < 0.025 were considered statistically significant when comparing the efficacies among DTG, CAB, and BIC

Figure and Supplementary Table	*p*-Value < 0.025	*p*-Value < 0.01	*p*-Value < 0.001	Overall Comparison among INSTIs for Table	
Figure 2 Antiviral Data and Supplementary Table 1	BIC - DTG (R263K)	DTG – CAB (G140S/Q148H) BIC – CAB (G140S/Q148H) BIC – CAB (T66I) DTG – CAB (E92Q) DTG – CAB (H51Y) BIC - DTG (H51Y) BIC - CAB (H51Y) BIC - CAB (G118R) CAB - DTG (H51Y/R263K)	BIC - CAB (E92Q) BIC - CAB (R263K) BIC - DTG (H51Y/R263K) BIC - CAB (H51Y/R263K)	BIC > CAB BIC > DTG DTG > CAB DTG > BIC CAB > BIC CAB > DTG	7 3 3 0 0 1
Figure 3 Antiviral Data and Supplementary Table 2	DTG - CAB (Q148R)	CAB - DTG (L74M) DTG – CAB (T97A) DTG – BIC (Q146L) DTG – CAB (Q146P) BIC - CAB (Q146P) DTG – CAB (Q148H) DTG – BIC (Q148H) BIC - CAB (Q148H) DTG - CAB (Q148K) BIC - CAB (Q148K) BIC - CAB (Q148R)	CAB – BIC (M50I) DTG – CAB (E138K) DTG – BIC (E138K) BIC - CAB (E138K) DTG – BIC (Q146P)	BIC > CAB BIC > DTG DTG > CAB DTG > BIC CAB > BIC CAB > DTG	5 0 6 4 1 1
Figure 4 Antiviral Data and Supplementary Table 3	BIC - DTG (Y143R/Q148H) DTG – BIC (G140A/Q148R)	BIC - DTG (G140A/Q148H) BIC - CAB (G140A/Q148H) DTG - CAB (Q148H/N155H) BIC - CAB (Q148H/N155H) BIC - DTG (G140A/Q148K) BIC - CAB (G140A/Q148K) DTG – CAB (G140A/Q148R) DTG – BIC (G140C/Q148R) BIC - DTG (G140S/Q148R) BIC - DTG (Q148R/N155H)	DTG – CAB (Y143R/Q148H) BIC - CAB (Y143R/Q148H) DTG – CAB (E138K/Q148K) DTG – BIC (E138K/Q148K) BIC - CAB (E138K/Q148K) DTG – CAB (G140S/Q148K) DTG – BIC (G140S/Q148K) BIC - CAB (G140S/Q148K) DTG - CAB (E138A/Q148R) BIC - CAB (E138A/Q148R) DTG - CAB (E138K/Q148R) BIC - CAB (E138K/Q148R) DTG - CAB (G140C/Q148R) BIC - CAB (G140C/Q148R) DTG - CAB (G140S/Q148R) BIC - CAB (G140S/Q148R) DTG - CAB (Q148R/N155H) BIC - CAB (Q148R/N155H)	BIC > CAB BIC > DTG DTG > CAB DTG > BIC CAB > BIC CAB > DTG	11 5 10 4 0 0
Figure 5 Antiviral Data and Supplementary Table 4		DTG – BIC (T66I/E157Q) DTG - CAB (G140S/N155H) CAB – BIC (Y143H/N155H)	CAB – BIC (T66I/E157Q) BIC - CAB (G140S/N155H)	BIC > CAB BIC > DTG DTG > CAB DTG > BIC CAB > BIC CAB > DTG	1 0 1 1 2 0
Figure 6 Antiviral Data and Supplementary Table 5	BIC - DTG (E138A/S147G/Q148R)	DTG - CAB (L74M/G140A/Q148R) BIC - CAB (L74M/G140A/Q148R) DTG - CAB (L74M/G140C/Q148R) BIC - DTG (L74M/G140C/Q148R) BIC - CAB (L74M/G140C/Q148R) DTG - BIC (E138K/G140C/Q148R)	DTG - CAB (T97A/Y143R/Q148H) BIC - CAB (T97A/Y143R/Q148H) DTG - CAB (E138K/G140A/Q148K) BIC - CAB (E138K/G140A/Q148K) DTG - CAB (E138K/G140C/Q148R) BIC - CAB (E138K/G140C/Q148R) BIC - CAB (E138A/S147G/Q148R)	BIC > CAB BIC > DTG DTG > CAB DTG > BIC CAB > BIC CAB > DTG	6 2 5 1 0 0
Figure 7 Antiviral Data and Supplementary Table 6	BIC - CAB (T66I/T97A/E157Q)	DTG - CAB (T66I/T97A/E157Q) DTG - CAB (G140S/Y143R/N155H) BIC - CAB (G140S/Y143R/N155H) BIC - DTG (E92Q/N155H/G163R)	DTG - CAB (T97A/Y143R/N155H) BIC - CAB (T97A/Y143R/N155H)	BIC > CAB BIC > DTG DTG > CAB DTG > BIC CAB > BIC CAB > DTG	3 1 3 0 0 0
Figure 8 Antiviral Data and Supplementary Table 7		CAB - DTG (T97A/G140S/Q148H) BIC - CAB (T97A/G140S/Q148H) DTG - CAB (E138K/G140S/Q148H) DTG - CAB (G140S/Y143R/Q148H) BIC - CAB (G140S/Y143R/Q148H) DTG - CAB (G140S/Q148H/N155H) BIC - CAB (G140S/Q148H/N155H) DTG - CAB (G140S/Q148H/G163K)	BIC - DTG (T97A/G140S/Q148H) DTG - CAB (E138A/G140S/Q148H) BIC - CAB (E138A/G140S/Q148H) BIC - DTG (E138K/G140S/Q148H) BIC- CAB (E138K/G140S/Q148H) BIC - DTG (G140S/Q148H/G163K) BIC -CAB (G140S/Q148H/G163K)	BIC > CAB BIC > DTG DTG > CAB DTG > BIC CAB > BIC CAB > DTG	6 3 5 0 0 1

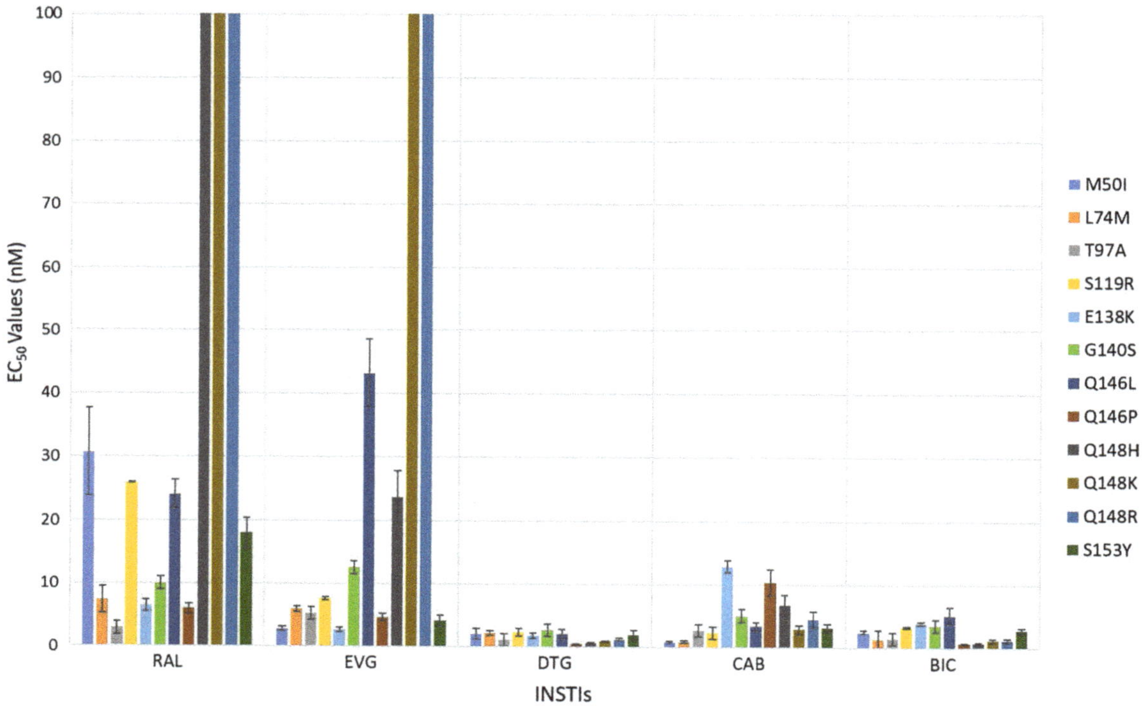

Fig. 4 Antiviral activities of BIC and CAB against common INSTI-resistant single mutants. The EC_{50} values were determined using vectors that carry the INSTI-resistant IN double mutants in single round infection assays. Error bars represent the standard deviations in the data from independent experiments (n = 4). The EC_{50} values shown in the figure have a maximum of 100 nM. The EC_{50} values of RAL against Q148H, Q148K, and Q148R and EVG versus Q148K and Q148R INSTI-resistant mutants were all > 100 nM

Antiviral activities of CAB and BIC against a panel INSTI-resistant triple mutants that included a primary mutation (Q148H/K/R) and two additional mutations

We determined the antiviral activities of CAB, BIC, and the FDA-approved INSTIs against a panel of INSTI-resistant triple mutants that included a primary mutation at Q148 (H/K/R) with two additional mutations at primary or secondary positions. The panel of INSTI-resistant triple mutants included: T97A/Y143R/Q148H, T97A/Q148H/N155H, E138K/G140A/Q148K, L74M/G140A/Q148R, L74M/G140C/Q148R, E138K/G140C/Q148R, and E138A/S147G/Q148R (Fig. 8; Additional file 1: Table S5A). Overall, DTG and BIC showed similar antiviral profiles against these triple mutants, and in some cases, CAB also retained potency. DTG, BIC, and CAB potently inhibited (EC_{50}s ≤ 5 nM) the T97A/Y143R/Q148H and E138A/S147G/Q148R INSTI-resistant triple mutants. The E138K/G140C/Q148R INSTI-resistant mutant caused only a small loss of potency to DTG (5.3 ± 1.0 nM) and BIC (8.2 ± 1.1 nM). This mutant showed a significant reduction in susceptibility to CAB (134.2 ± 0.3 nM). The L74M/G140C/Q148R triple mutant was moderately susceptible to BIC (6.1 ± 0.9 nM) and DTG (10.2 ± 1.3 nM). However, this mutant caused a massive loss in susceptibility to CAB (220.3 ± 41.2 nM). The L74M/G140A/Q148R triple mutant with a different mutation at position G140, caused a modest loss of susceptibility to both DTG (12.0 ± 2.1 nM) and BIC (11.7 ± 1.3 nM); however, this also caused a substantial loss in susceptibility to CAB (53.2 ± 14.8 nM). Finally, DTG, BIC, and CAB failed to retain potency against the E138K/G140A/Q148K INSTI-resistant triple mutant (EC_{50}s > 200 nM). The first generation INSTIs, RAL and EVG failed to retain potency against the entire panel of INSTI-resistant triple mutants, except for T97A/Y143R/Q148H, against which EVG showed modest inhibition, with an EC_{50} value of 41.6 ± 3.0 nM. BIC had significantly higher potencies against two of the INSTI-triple mutants than DTG (one p value < 0.01, see Fig. 3 and Additional file 1: Table S5B), compared to one for DTG versus BIC against the triple mutants in this panel. Both BIC and DTG were more effective than CAB. BIC had better efficacies than CAB against six of the triple mutants (four p values < 0.001); DTG was better than CAB for five of the mutants in this panel (three p values < 0.001).

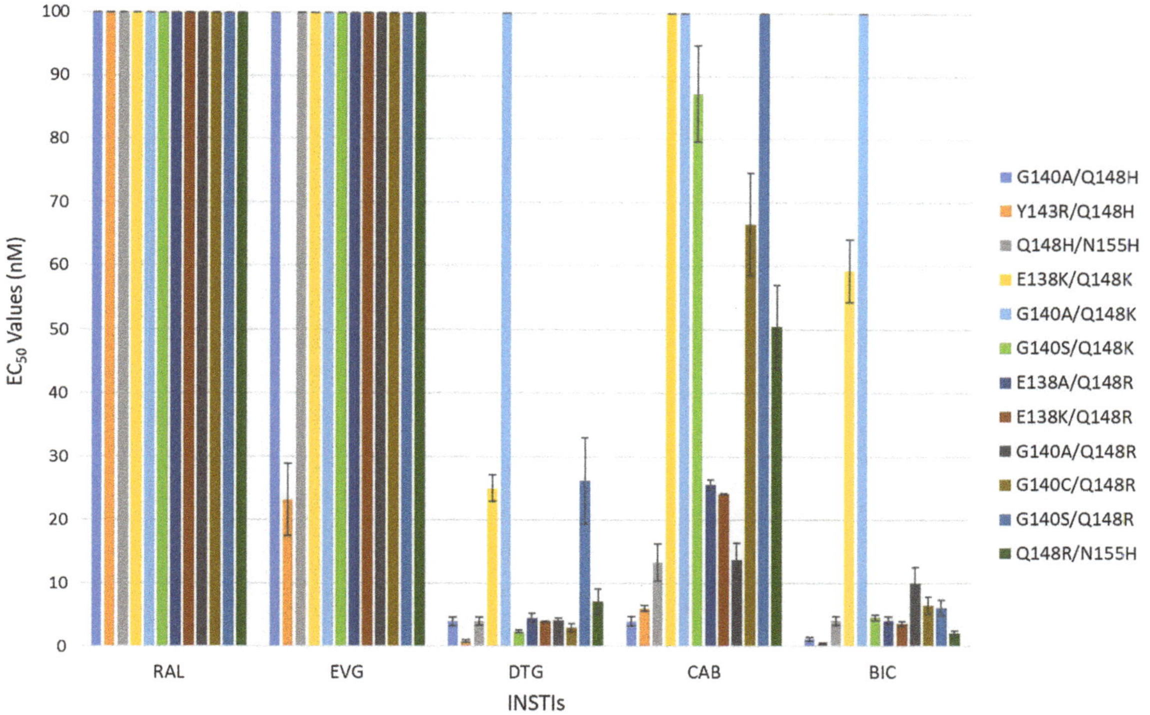

Fig. 5 Antiviral activities of BIC and CAB against a panel of INSTI-resistant double mutants that have a primary mutation at position Q148. The EC_{50} values were determined using vectors that carry the INSTI-resistant double mutants in single round infection assays. Error bars represent the standard deviations in the data from independent experiments (n = 4). The EC_{50} values shown in the figure have a maximum of 100 nM. The EC_{50} values of RAL and EVG against this entire panel (except for EVG versus Y143R/Q148H) were all > 100 nM. The EC_{50} values of DTG against G140A/Q148K, CAB versus E138K/Q148K, G140A/Q148K, G140S/Q148R, and BIC against G140A/Q148K were all > 100 nM

Antiviral activities of CAB and BIC versus a panel of INSTI-resistant triple mutants that consists of a primary mutation at T66I and N155H with additional secondary mutations

We examined CAB, BIC, and the FDA-approved INSTIs, against a panel of INSTI-triple mutants that included T66I/T97A/E157Q, T97A/Y143R/N155H, G140S/Y143R/N155H, and E92Q/N155H/G163R (Fig. 9; Additional file 1: Table S6A). The triple mutant T66I/T97A/E157Q is an EVG-resistant mutant and, as expected, this mutant showed a substantial decrease in potency to EVG (69.4 ± 11.8 nM) and a lesser loss of potency to RAL (33.5 ± 8.7 nM). In contrast, DTG (0.5 ± 0.1 nM), BIC (0.4 ± 0.2 nM), and CAB (0.8 ± 0.1 nM) retained full potency against this triple mutant. Additionally, DTG, BIC, and CAB retained high antiviral potencies against the E92Q/N155H/G163R INSTI-resistant triple mutant (EC_{50} < 5 nM). The G140S/Y143R/N155H triple mutant was susceptible to both DTG (2.6 ± 0.3 nM) and BIC (2.1 ± 0.1 nM), but it caused a moderate loss in potency to CAB (20.0 ± 3.5 nM). Both DTG and BIC retained significant potency against the T97A/Y143R/N155H triple mutant, 8.5 ± 1.5 nM and 8.2 ± 1.7 nM, respectively, whereas CAB lost substantial potency (142.2 ± 8.3 nM). RAL and EVG both failed to potently inhibit the T97A/Y143R/N155H, G140S/Y143R/N155H, and E92Q/N155H/G163R INSTI-resistant triple mutants (EC_{50}s > 90 nM). Both DTG and BIC were more effective than CAB; each one had a significantly higher potency than CAB against 3 of the triple mutants in this panel (at least one p value < 0.001; see Fig. 3 and Additional file 1: Table S6B). BIC was significantly better than DTG against one of the triple mutants (p value < 0.01).

Antiviral activities of CAB and BIC against a panel of INSTI-resistant triple mutants that include the well-characterized RAL-resistant double mutant G140S/Q148H and an additional secondary mutation

We tested the antiviral potencies of CAB, BIC, and the FDA-approved INSTIs against a panel of INSTI-resistant triple mutants that included the RAL-resistant G140S/Q148H double mutations with an additional mutation: T97A, E138A/K, Y143R, N155H, or G163K (Fig. 10; Additional file 1: Table S7A). As expected, both of the first generation INSTIs, RAL and EVG, were ineffective against this panel of six INSTI-resistant triple mutants

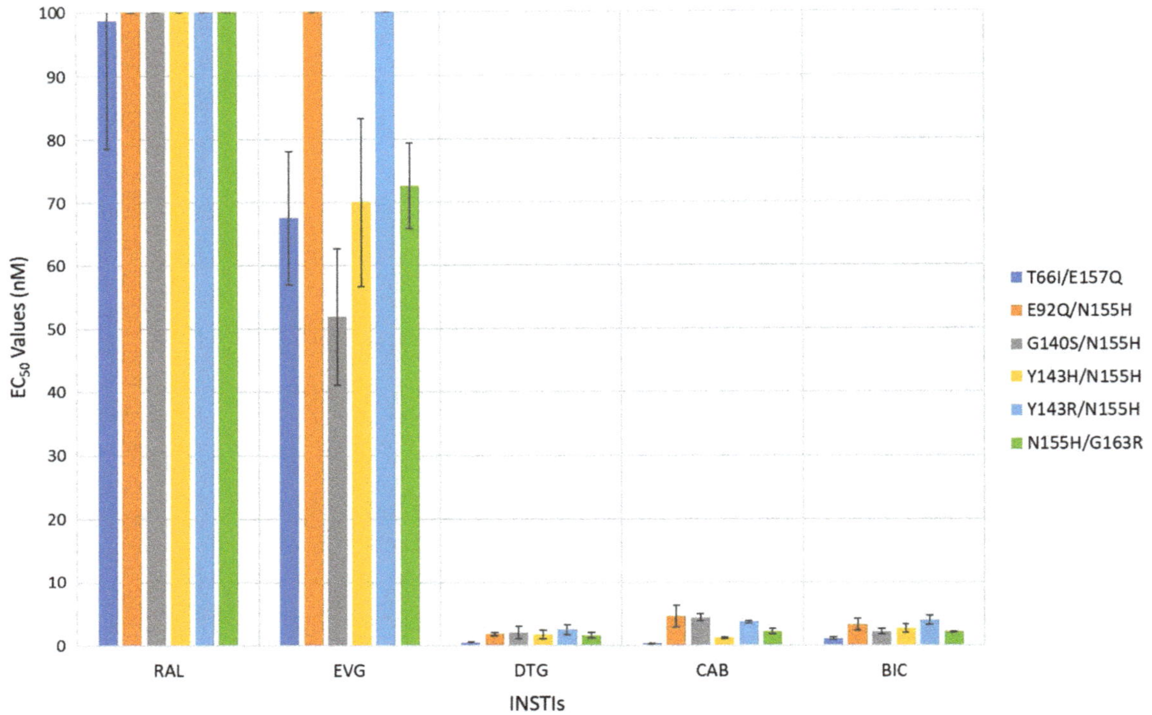

Fig. 6 Antiviral activities of BIC and CAB against a panel of INSTI-resistant double mutants that included the primary mutations T66I and N155H with additional mutations at other positions. The EC_{50} values were determined using vectors that carry the INSTI-resistant double mutants in single round infection assays. Error bars represent the standard deviations in the data from independent experiments (n = 4). The EC_{50} values shown in the figure have a maximum of 100 nM. The EC_{50} values of RAL against this entire panel of INSTI-resistant double mutants were all > 100 nM. The EC_{50} values of EVG against E92Q/N155H and Y143R/N155H were all > 100 nM

(EC_{50}s > 5000 nM). In addition, DTG, which potently inhibited the G140S/Q148H INSTI-resistant double mutant (EC_{50} < 5 nM) showed a loss of potency against the INSTI-resistant triple mutants. The E138A/G140S/Q148H, G140S/Y143R/Q148H, and G140S/Q148H/G163K triple mutants caused modest drops in potency, from 13.8 ± 4.8 nM, to 7.7 ± 2.0 nM, and 24.3 ± 1.1 nM, respectively. However, the INSTI-resistant triple mutants T97A/G140S/Q148H, E138K/G140S/Q148H, and G140S/Q148H/N155H caused substantial reductions in potency (EC_5s ≥ 55 nM). CAB was not broadly active against these INSTI-resistant triple mutants; most of the mutants caused significant drops in susceptibility to CAB. Conversely, for this panel of mutants, BIC was more effective than DTG in retaining potency. BIC showed at most a modest loss in potency against E138A/G140S/Q148H, E138K/G140S/Q148H, G140S/Y143R/Q148H, and G140S/Q148H/G163K (EC_{50}s < 10 nM. However, the INSTI-resistant triple mutant T97A/G140S/Q148H caused a larger reduction in susceptibility to BIC (29.5 ± 4.4 nM). Thus, BIC was superior to the other INSTIs in terms of its ability to retain antiviral activity against this set of triple mutants. BIC was the superior INSTI against this panel of triple mutants, as it had significantly better potencies against five INSTI-resistant triple mutants than CAB (2 p values < 0.01 and 3 p values < 0.001, see Fig. 3 and Additional file 1: S7B) and three triple mutants than DTG (p values < 0.001). DTG was a better INSTI than CAB as it had higher efficacies against five triple mutants than CAB.

Homology modeling of the binding of BIC and CAB into the HIV-1 IN active site using PFV intasome structural data

Using the previously reported crystal structure of the PFV intasome with DTG bound at the catalytic site (PDB ID: 3S3M) [15] and the structure of the HIV-1 IN strand transfer complex (STC) solved by electron microscopy as a template (PDB ID: 5U1C) [40], homology models were prepared of BIC and CAB bound to the HIV-1 intasome (Fig. 11, panels B and C). Modeling allowed us (1) to understand better how BIC and CAB bind in the HIV-1 IN active site and (2) to identify structural features that may help (or hinder) these INSTIs in overcoming INSTI-resistant mutants. The chelating motifs of BIC and CAB aligned similarly to

Overall Comparison	BIC > CAB	39
	BIC > DTG	14
	DTG > CAB	33
	DTG > BIC	10
	CAB > BIC	3
	CAB > DTG	3
***p*-Value < 0.025**	BIC > CAB	1
	BIC > DTG	3
	DTG > CAB	1
	DTG > BIC	1
	CAB > BIC	0
	CAB > DTG	0
***p*-Value < 0.01**	BIC > CAB	17
	BIC > DTG	7
	DTG > CAB	18
	DTG > BIC	5
	CAB > BIC	1
	CAB > DTG	3
***p*-Value < 0.001**	BIC > CAB	21
	BIC > DTG	4
	DTG > CAB	14
	DTG > BIC	4
	CAB > BIC	2
	CAB > DTG	0

Fig. 7 Overall Comparison of the statistical significance of the antiviral data among DTG, CAB, and BIC. The Student's *t* test was used to calculate the statistical significance of the differences in the antiviral activities of the INSTIs. The *p* values < 0.025, < 0.01, and < 0.001 between DTG, CAB, and BIC were used to decide which INSTIs were more broadly efficacious against the mutants

DTG (Fig. 11, panel D), as did the halobenzyl moieties, which have π–π hydrophobic stacking interactions with the penultimate cytosine on the 3′ ends of the viral DNA. However, it is the "left" side of the tricyclic ring system of BIC, which is the portion of the INSTI that is distal to the end of the viral DNA, and has an oxazepine ring which features a methylene bridge and lacks a methyl group, that appears to distinguish BIC from DTG. This cyclic modification can, in the model, adopt and maintain a different configuration from the components on the left side of DTG and CAB. Both the methyl-modified oxazine ring of DTG and the methyl-modified oxazole ring of CAB appear to be more constrained than the oxazepine ring of BIC. The greater flexibility of the oxazepane ring allows it to bend backwards or forwards, depending on the exact geometry of the active site, which can be modified by nearby mutations. Thus, the apparent greater conformational flexibility of the oxazepine ring could allow BIC to bind tightly to the various INSTI-resistant mutants, such as G118R and S119R, which affect the periphery of the IN active site, and limit the modifications that can be added to the "left" side of the INSTI scaffold distal to the end of the viral DNA (unpublished observations). Conversely, the oxazole ring of CAB is pointed out and away from the position occupied by the corresponding oxazine ring of DTG (Fig. 11, panel B), and its methyl group does not appear to be in a position to make an important contribution to binding, which is in good agreement with the data of Yoshinaga et al. [41], which appeared when this manuscript was in review. This could account for the fact that, although CAB and DTG adopt similar spatial orientations when bound to the IN active site, DTG is much more broadly effective against INSTI-resistant mutants (see "Discussion").

Discussion

The relatively recent development of INSTIs as potent and effective HIV-1 inhibitors permits improved treatment strategies for HIV-1 infected patients. In general, INSTIs are minimally toxic and work well in combination with other ARV drug classes [20, 42–44]. In addition, DTG appears not to readily select for resistance mutations. DTG is now widely used in therapies for the treatment of both naïve and experienced patients [11, 12, 16, 18, 19]. However, patients in advanced clinical trials that were previously on a RAL-based therapy, who switched to a DTG-based salvage therapy, have shown signs of virological failure. In some cases, additional resistance mutations were selected [25, 26]. Therefore, there is a need for new INSTIs that can overcome emerging INSTI-resistant mutants.

BIC and CAB are now in late stage clinical trials [45–48]. Based on our antiviral analysis of the ability of these new INSTIs to inhibit previously identified INSTI-resistant single, double, and triple mutants in a single round replication assay, it appears that both BIC and CAB are both more broadly effective than either of the first generation INSTIs, RAL and EVG. However, in terms of their ability to inhibit the fifty-seven INSTI-resistant mutants we tested, BIC was significantly better than DTG against fourteen out of the mutants (Fig. 7; seven featured p values < 0.01, whereas four had p values < 0.001). BIC was also better than CAB against thirty-nine of the mutants (twenty-one had p values < 0.001). Conversely, DTG was better than BIC against ten of the mutants tested (four with p values < 0.001) and better than CAB

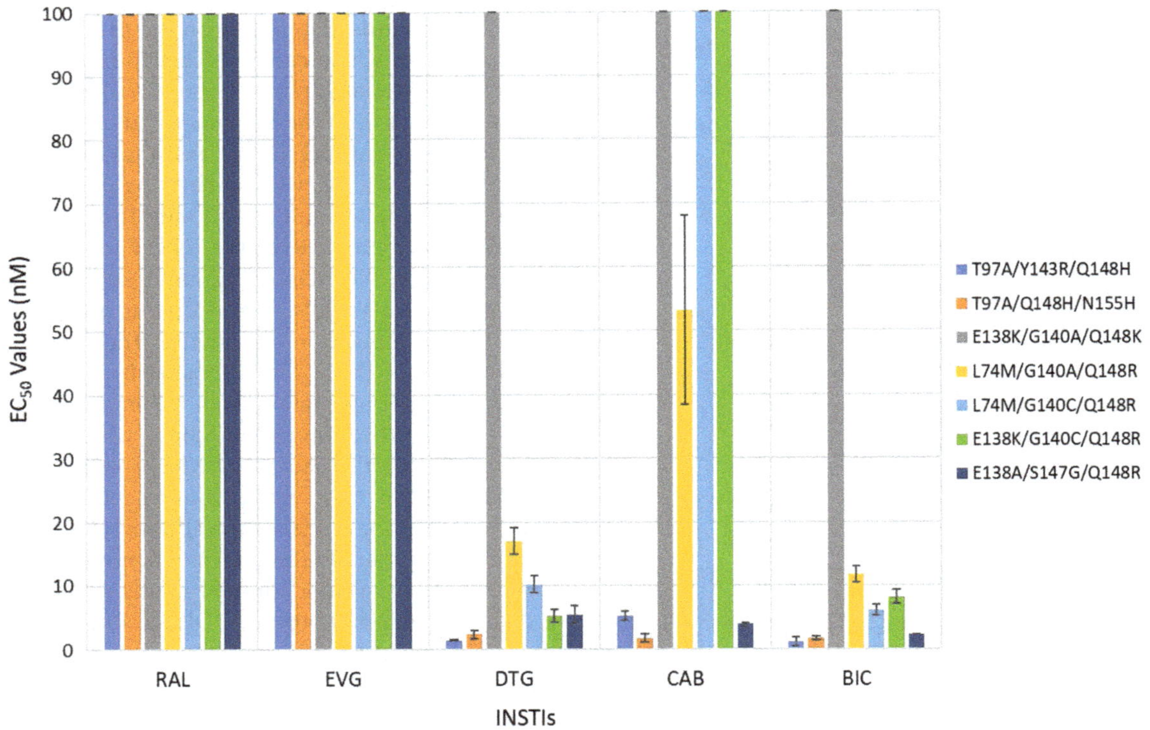

Fig. 8 Antiviral activities of BIC and CAB against a panel INSTI-resistant triple mutants that included a primary mutation (Q148H/K/R) and two additional mutations. The EC_{50} values were determined using vectors that carry the INSTI-resistant triple mutants in single round infection assays. Error bars represent the standard deviations in the data from independent experiments (n = 4). The EC_{50} values shown in the figure have a maximum of 100 nM. The EC_{50} values of RAL and EVG versus this entire panel of INSTI-resistant triple mutants were all > 100 nM. The EC_{50} values of DTG against E138K/G140A/Q148K, CAB versus E138K/G140A/Q148K, L74M/G140C/Q148R, and E138K/G140C/Q148R, and BIC against E138K/G140A/Q148K were all > 100 nM

for thirty-three of the mutants (fourteen with p values < 0.001). CAB was better than BIC and DTG for three mutants each (all three p values < 0.01 for DTG and two p values < 0.001 for BIC). Overall, our conclusions concerning the relative efficacies of the new INSTIs against mutants are in good agreement with the data of Yoshinaga et al. [41] and Neogi et al. [49], which appeared when this manuscript was in review.

Given the complexities of pharmacology, a significant difference in the behavior of a drug against a particular mutant (or mutants) may or may not translate directly into a desirable clinical outcome. However, given the problems that arise with drug resistance, it is likely that, among related compounds, those that are more broadly effective against resistant viruses will have an advantage in the clinic. In addition, in comparing the potencies of the compounds, the single round assay allows us to directly compare the efficacies of the new INSTIs against INSTI-resistant mutants in a reproducible and accurate manner. The single round assay avoids the issue of the effects of the mutations on the replication capacity, which, in turn, affects the number of viral life cycles in assays done with replication competent viruses, and by extension, can affect the EC_{50}s.

Having a better understanding of how INSTIs bind to HIV-1 IN is an important part of developing more effective new drugs. However, superpositioning the available PFV and HIV IN structures has revealed differences in the PFV and HIV-1 IN active sites [15, 40]. Notably, the β4α2 loops are in different positions relative to the IN active site and there are differences in the structures and locations of the C-terminal domains (CTDs) near the IN active site. Thus, the contacts and interactions between INSTIs and the PFV intasome might not correspond exactly to the related contacts in the HIV-1 intasome. Until the structure of the HIV-1 intasome with these INSTIs bound is solved, HIV-1 IN models based on the structures of the PFV template with bound INSTIs and the available HIV-1 strand transfer (STC) structures can be used to predict how new INSTIs will bind to the HIV-1 intasome. DTG, BIC, and CAB are similar chemically and structurally. Not surprisingly, based on the model we built using the available structural information, all three compounds adopt similar configurations within

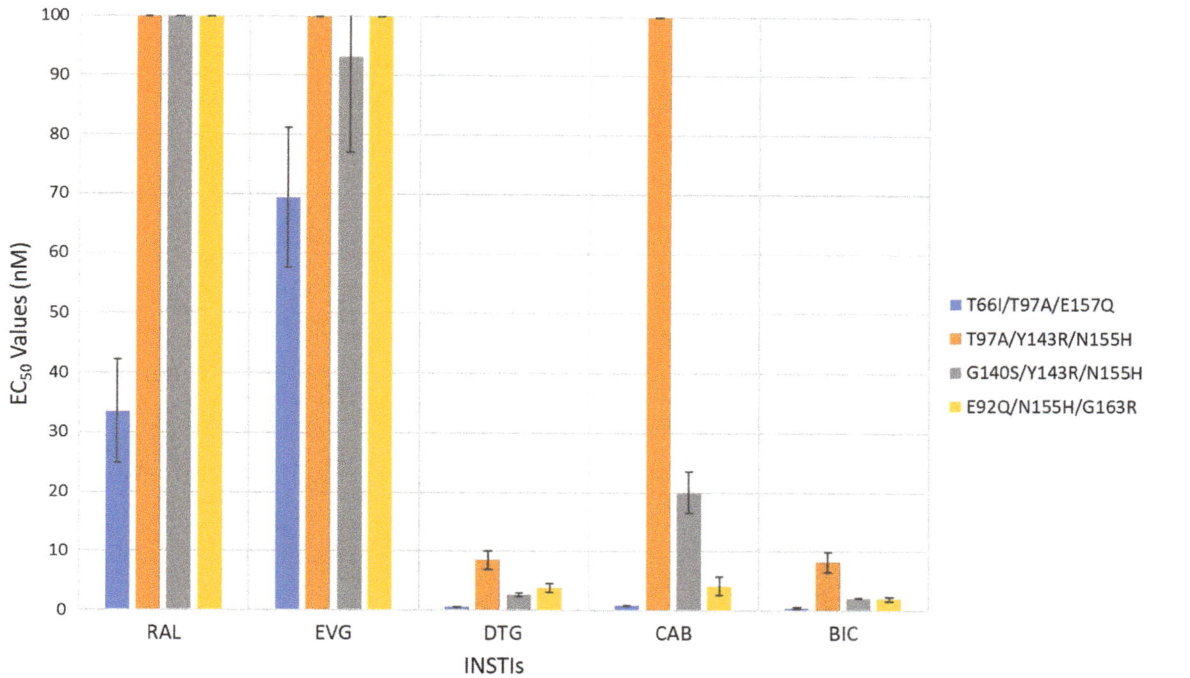

Fig. 9 Antiviral activities of BIC and CAB versus a panel of INSTI-resistant triple mutants that consists of a primary mutation at T66I and N155H with additional secondary mutations. The EC_{50} values were determined using vectors that carry the INSTI-resistant triple mutants in single round infection assays. Error bars represent the standard deviations of the data from independent experiments (n = 4). The EC_{50} value shown in the figure have a maximum of 100 nM. The EC_{50} values of RAL against T97A/Y143R/N155H, G140S/Y143R/N155H, and E92Q/N155H/G163R were all > 100 nM. The EC_{50} values of EVG versus T97A/Y143R/N155H and E92Q/N155H/G163R and CAB against T97A/Y143R/N155H were all > 100 nM

the active site of HIV-1 IN. It appears that the structural differences on the "left side" of these INSTIs, the part of the pharmacophore away from the 3′ end of the viral DNA, are largely responsible for their different resistance profiles.

Although there are similarities, as noted above, BIC is better than DTG, and DTG is better than CAB, in terms of their respective abilities to broadly inhibit the known IN mutants. We think it is likely that BIC is more broadly effective in its ability to inhibit a range of INSTI-resistant mutants, because it is better able to adjust its conformation, in response to the changes in the shape of the active site caused by the various resistance mutations. Thus, as has been proposed for the binding of non-nucleoside reverse transcriptase inhibitors (NNRTIs) to HIV-1 RT [50–53], the most broadly potent compounds are those that are able to adjust their binding mode and/or their configuration in response to changes in and around the IN active site. As briefly discussed earlier, BIC has an oxazepine ring with a methylene bridge appended to its chelating scaffold, which differs from the oxazine ring of DTG. It appears, based on the models, that the oxazepine ring of BIC is more flexible, which would allow it to be more conformationally adaptable. The introduction of resistance mutations in residues in and around the IN active site may cause alterations in the active site geometry. These changes could potentially affect the binding of relatively rigid compounds, giving rise to resistance. The greater flexibility of the extended ring system of BIC may help it adapt to changes in the geometry in the IN active site, allowing BIC to overcome many of the known IN resistance mutations. However, the details of the binding of BIC, particularly to the INSTI-resistant forms of HIV-1 IN, will require additional high resolution structural data. Conversely, the methyl-modified oxazole ring of CAB does not appear to be in a favorable position to interact with WT IN. In addition, it does not appear to be conformationally adaptable. As a consequence, CAB may have difficulty overcoming the changes in the geometry of the active site of HIV IN caused by resistance mutants.

Generally speaking, the second generation INSTIs (DTG, BIC, and CAB) are much more proficient at inhibiting these INSTI-resistant mutants than RAL and EVG. Based on the data from our panel of mutants, DTG and BIC are more broadly effective against the mutants than CAB (Fig. 3). The potency of the second generation INSTIs can be affected by triple mutants which arise when mutations at G140 (A/C/S) and Q148 (H/R) are

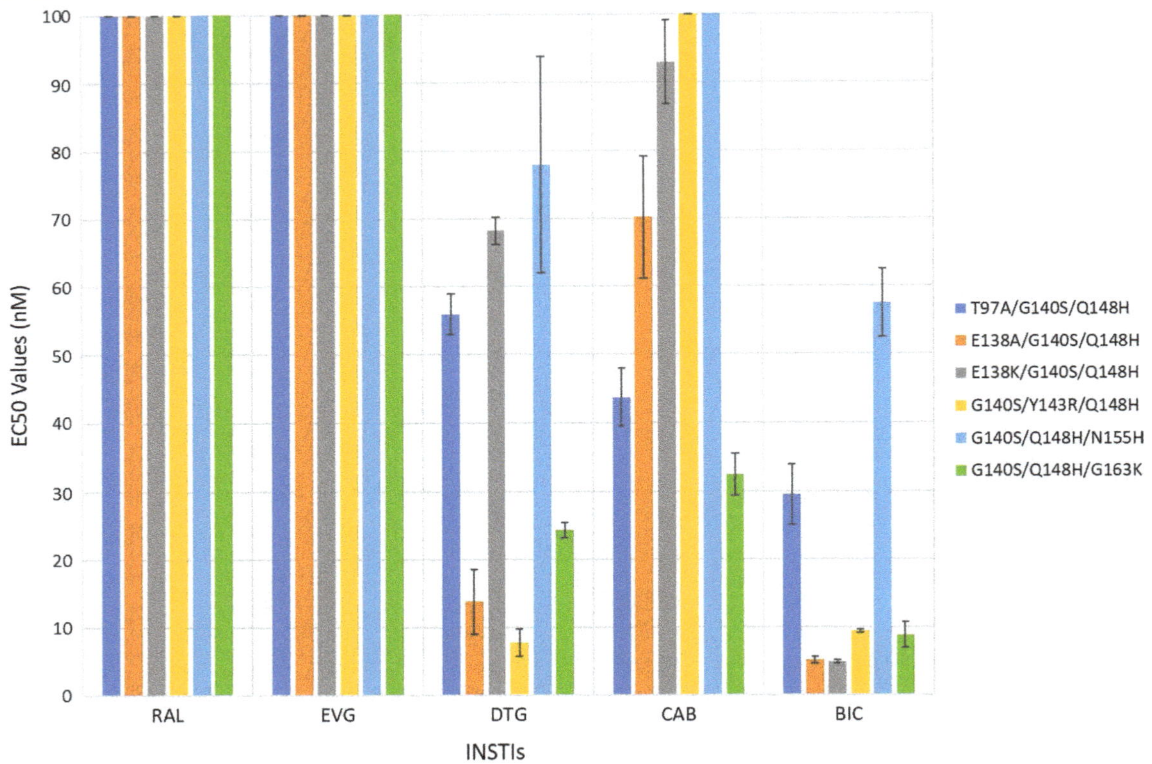

Fig. 10 Antiviral activities of CAB and BIC against a panel of INSTI-resistant triple mutants that include the well-characterized RAL-resistant double mutant G140S/Q148H plus an additional secondary mutation. The EC_{50} values were determined using vectors that carry the INSTI-resistant triple mutants in single round infection assays. Error bars represent the standard deviations in the data from independent experiments (n = 4). The EC_{50} values shown in the figure have a maximum of 100 nM. The EC_{50} values of RAL and EVG against this entire panel of INSTI-resistant triple mutants were all > 100 nM. The EC_{50} values of CAB against G140S/Y143R/Q148H and G140S/Q148H/N155H were all > 100 nM

combined with the polymorphic mutation at L74M or T97A. Although it is not entirely clear how frequently L74M and T97A occur in either B or non-B HIV-1 subtypes in INSTI-naïve patients, it is possible that, when these polymorphisms are present, that they could affect the development of resistance.

Conclusions

Based on these results, BIC appears to be a very promising INSTI. CAB has obvious disadvantages in terms of its breadth of antiviral potency relative to both BIC and DTG. However, CAB may have other advantages; it can be formulated as a long-acting compound that can be injected into patients once every 2–3 months [46, 47]. Nonetheless, based on experience with previous ARV drugs, in the long-term, resistant viruses will emerge. Thus, it is likely that at least some of the compounds that are broadly effective against the known mutants will be successful. This idea is underscored by the fact that, currently, in Washington DC, where the levels of HIV infection is still high, approximately 20% of new cases involve a HIV strain that has at least one drug resistance mutation [54].

Methods

INSTI synthesis

Acquisition of RAL, EVG, and DTG was previously described [15, 55, 56]. BIC was obtained from PharmaBlock (Cat. No. PBLJ8958) and CAB was obtained from AbovChem LLC (Cat. No. HY-15592).

Cell-based assays

Single-round viral infectivity assays, using HIV-1 vectors that express either WT or mutant forms of IN, were used to determine antiviral potencies (EC_{50} values) of the compounds as previously described [57]. The Student's *t* test was used to calculate the *p* values used to determine statistical significance.

Vector constructs

The vector pNLNgoMIVR-ΔENV.LUC has been described previously [36]. To produce the new IN mutants used in this study, the IN open reading frame

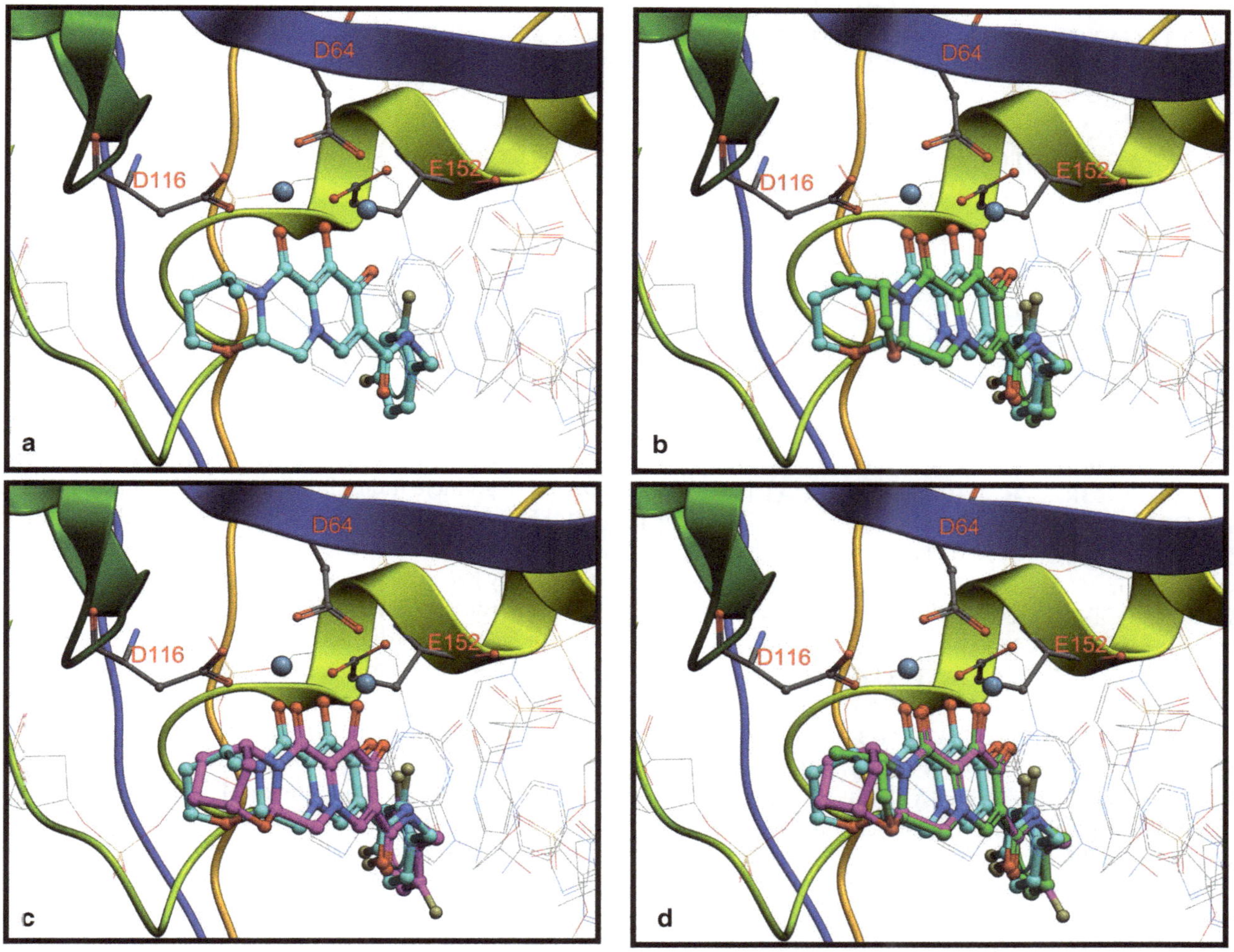

Fig. 11 Modeling BIC and CAB into the PFV Intasome. The four panels show models of DTG, CAB, or BIC bound in the active site of HIV-1 IN. The upper left panel **a** shows a model of DTG (cyan) bound to HIV-1 IN. The upper right panel **b** shows a model of CAB (green) bound to the active site of the HIV-1 IN using DTG (cyan) as the template. The lower left panel **c** shows a model of BIC (magenta) bound to the active site of HIV-1 IN using DTG (cyan) as the template. The lower right panel **d** shows an overlay of the binding of DTG (cyan), BIC (magenta), and CAB (green) to HIV-1 IN, specifically showing how the "left-side" of these three INSTIs, the part of pharmacophores distal from the end of the viral DNA, interact with HIV-1 IN. All four panels show the Mg^{2+} cofactors rendered in space-filling format (slate gray) interacting with the chelating motifs of each of the INSTIs. HIV-1 IN is depicted in multi-colored ribbons with active site residues D64, D116, and E152 labeled in red and rendered in dark gray ball and stick format

was removed from pNLNgoMIVR-ΔENV.LUC by digestion with KpnI and SalI and resulting fragment was inserted between the KpnI and SalI sites of pBluescript KS+. Using that construct as the wild-type template, we prepared the following HIV-1 IN mutants using the QuikChange II XL site directed mutagenesis kit (Agilent Technologies, Santa Clara, CA) protocol: M50I, L74M, T97A, S119R, E138K, G140S, Q146L, Q146P, Q148H, Q148K, Q148R, S153Y, T66I/E157Q, E92Q/N155H, T124A/153Y, E138A/Q148R, E138K/Q148K, E138K/Q148R, E138K/263K, G140A/Q148H, G140A/Q148K, G140A/Q148R, G140C/Q148R, G140S/Q148K, G140S/Q148R, G140S/N155H, Y143H/N155H, Y143R/Q148H, Y143R/N155H, Q148H/N155H, Q148R/N155H, N155H/G163R, T66I/T97A/E157Q, L74M/G140A/Q148R, L74M/G140C/Q148R, E92Q/N155H/G163R, T97A/G140S/Q148H, T97A/Y143R/Q148H, T97A/Y143R/N155H, T97A/Q148H/N155H, E138A/G140S/Q148H, E138A/S147G/Q148R, E138K/G140A/Q148K, E138K/G140C/Q148R, E138K/G140S/Q148H, G140S/Y143R/Q148H, G140S/Y143R/N155H, G140S/Q148H/N155H, and G140S/Q148H/G163K. The following sense oligonucleotides were used with matching cognate antisense oligonucleotides (not shown) (Integrated DNA Technologies, Coralville, IA) in the mutagenesis: M50I, 5′-CAG CTAAAAGGGGAAGCCATTCATGGACAAGTAGAC

TGT-3′; T66I, 5′- ATATGGCAGCTAGATTGTATT CATTTAGAAGGAAAAGTT-3′; L74M, 5′- TTAGAA GGAAAAGTTATCATGGTAGCAGTTCATGTAGCC-3′; E92Q, 5′- GCAGAAGTAATTCCAGCACAAACA GGGCAAGAAACAGCA-3′; T97A, 5′- GCAGAGACA GGGCAAGAAGCTGCATACTTCCTCTTAAAA-3′; S119R, 5′- GTACATACAGACAATGGCCGTAATTTC ACCAGTACTACA-3′; E138A, 5′- TGGGCGGGGATC AAGCAGGCTTTTGGCATTCCCTACAAT-3′; E138K, 5′- TGGGCGGGGATCAAGCAGAAATTTGGCATT CCCTACAAT-3′; G140A, 5′- GGGATCAAGCAGGAA TTTGCTATTCCCTACAATCCCCAA-3′; G140C, 5′- GGGATCAAGCAGGAATTTTGTATTCCCTACAAT CCCCAA-3′; G140S, 5′-GGGATCAAGCAGGAATTT TCCATTCCCTACAATCCCCAA-3′; Y143H, 5′- CAG GAATTTGGCATTCCCCATAATCCCCAAAGTCAA GGA-3′; Y143R, 5′-CAGGAATTTGGCATTCCCAGA AATCCCCAAAGTCAAGGA-3′; Q146L, 5′- GGCATT CCCTACAATCCCTTAAGTCAAGGAGTAATAGAA-3′; Q148H, 5′-TACAATCCCCAAAGTCACGGAGTA ATAGAATCT-3′; Q148K, 5′- CCCTACAATCCCCAA AGTAAAGGAGTAATAGAATCTATG-3′; Q148R, 5′- CCCTACAATCCCCAAAGTCGTGGAGTAATAGAA TCTATG-3′; S153Y, 5′- AGTCAAGGAGTAATAGAA TATATGAATAAAGAATTAAAG-3′; N155H, 5′-GGA GTAATAGAATCTATGCATAAAGAATTAAAGAAA ATT-3′; 5′-E157Q, 5′-ATAGAATCTATGAATAAACAA TTAAAGAAAATTATAGGA-3′; G163K, 5′- GAATTA AAGAAAATTATAAAACAGGTAAGAGATCAGGCT -3′, G163R, 5′ -GAATTAAAGAAAATTATACGTCAG GTAAGAGATCAGGCT -3′, E138K for G140S/Q148H, 5′- TGGTGGGCGGGGATCAAGCAGAAATTTTCC ATTCCCTACAATCCC-3′; S147G for E138A/Q148R, 5′- ATTCCCTACAATCCCCAAGGTCGTGGAGTAATA GAATCT-3′; E138K for G140C/Q148R, 5′- TGGGCG GGGATCAAGCAGAAATTTTGTATTCCCTACAAT-3′; E138K for G140A/Q148K, 5′- TGGGCGGGGATC AAGCAGAAATTTGCTATTCCCTACAAT-3′; E138A for G140S/Q148H, 5′- TGGGCGGGGATCAAGCAG GCATTTTCCATTCCCTACAAT-3′; Y143R for Y143R/Q148H, 5′-CAGGAATTTGGCATTCCCAGAAAT CCCCAAAGTCACGGA-3′; Y143R for G140S/Q148H, 5′-CAGGAATTTTCCATTCCCAGAAATCCCCAA AGTCACGGA-3′; Y143R for G140S/N155H, 5′-CAG GAATTTTCCATTCCCAGAAATCCCCAAAGTCAA GGA-3′.

The following IN mutants from Fig. 2 (Additional file 1: Tables S1A and S1B), H51Y, T66I, E92Q, G118R, Y143R, N155H, R263K, H51Y/R263K, and G140S/Q148H have been described [35]. The remaining E138K/R263K double mutant was made using the previously constructed E138K mutant and the appropriate listed R263K oligonucleotides, which were used to add the second mutation.

The IN mutants from Fig. 4 (Additional file 1: Tables S2A and S2B), which includes M50I, L74M, T97A, S119R, E138K, G140S, Q146L, Q146P, Q148H, Q148K, Q148R, and S153Y, were constructed as described above using the appropriate listed oligonucleotides.

The IN mutants from Fig. 5 (Additional file 1: Tables S3A and S3B), were made as followed. The E138A/Q148R and E138K/Q148R double mutants were made using the previously generated Q148R mutant and the E138A and E138K oligonucleotides, respectively, to add the second mutation. The E138K/Q148K double mutant was constructed using the previously made E138K mutant and the appropriate Q148K oligonucleotides, which were used to add the second mutation. The G140A/Q148H and G140A/Q148K double mutants were made with the previously constructed G140A mutant and the appropriate oligonucleotides for the second mutation either Q148H or Q148K, respectively. The double mutants G140A/Q148R and G140C/Q148R were made with the previously generated Q148R mutant and the oligonucleotides for the second mutation, either G140A or G140C, respectively. The double mutants G140S/Q148K and G140S/Q148R were made using the previously generated G140S mutant and appropriate oligonucleotides for the second mutation, either Q148K or Q148R, respectively. The double mutants Q148H/N155H and Q148R/N155H were made using the previously generated N155H mutant and appropriate oligonucleotides for the second mutation, either Q148H or Q148R, respectively. The double mutant Y143R/Q148H was made using the previously generated Q148H mutant and appropriate oligonucleotides to introduce the second mutationY143R.

The IN mutants from Fig. 6 (Additional file 1: Tables S4A and S4B), were made as followed. The T66I/E157Q double mutant was made using the previously generated T66I mutant and the appropriate E157Q oligonucleotides to add the second mutation. The double mutants E92Q/N155H, G140S/N155H, Y143H/N155H, Y143R/N155H, and N155H/G163R were made using the previously generated N155H mutant and appropriate oligonucleotides for the second mutation, either E92Q, G140S, Y143H, Y143R, or G163R, respectively.

The IN mutants from Fig. 8 (Additional file 1: Tables S5A and S5B), were constructed as followed. The L74M/G140A/Q148R triple mutant was made using the previously generated G140A/Q148R double mutant and the oligonucleotides for the third mutation, L74M. The triple mutant L74M/G140C/Q148R was made with the previously generated G140C/Q148R double mutant and the oligonucleotides for the third mutation, L74M. The triple mutant T97A/Y143R/Q148H was constructed using the previously generated Y143R/Q148H double mutant and the appropriate oligonucleotides for the third mutation,

T97A. The triple mutant E138K/G140C/Q148R was made using the previously generated G140C/Q148R double mutant and the appropriate oligonucleotides to create the third mutation, E138K. The triple mutant T97A/Q148H/N155H was made using the previously constructed Q148H/N155H double mutant and the appropriate oligonucleotides for the third mutation, T97A. The triple mutant E138A/S147G/Q148R was made with the previously generated E138A/Q148R double mutant and oligonucleotides to make the third mutation, S147G. The triple mutant E138K/G140A/Q148K was made using the previously constructed double mutant G140A/Q148K double mutant and the appropriate oligonucleotides to make the third mutation, E138K.

The IN mutants from Fig. 9 (Additional file 1: Tables S6A and S6B) were made as followed. The T66I/T97A/E157Q triple mutant was made using the previously generated T66I/E157Q double mutant and the oligonucleotides for the third mutation, T97A. The E92Q/N155H/G163R triple mutant was made using the previously generated E92Q/N155H double mutant and the oligonucleotides for the third mutation, G163R. The triple mutant G140S/Y143R/N155H was made using the previously constructed G140S/N155H double mutant and the correct oligonucleotides to create the third mutation, Y143R. The triple mutant T97A/Y143R/N155H was made with the previously generated Y143R/N155H double mutant and the appropriate oligonucleotides for the third mutation, T97A.

The IN mutants from Fig. 10 (Additional file 1: Tables S7A and S7B), were constructed as followed. The triple mutants T97A/G140S/Q148H, G140S/Q148H/N155H, and G140S/Q148H/G163K were each made with the previously generated G140S/Q148H double mutant and the appropriate oligonucleotides for the third mutation, either T97A, N155H, or G163K, respectively. The triple mutant was E138A/G140S/Q148H was made using the previously constructed G140S/Q148H double mutant and oligonucleotides to make the third mutation E138A. The triple mutant E138K/G140S/Q148H was made using the previously generated G140S/Q148H double mutant and the correct oligonucleotides to make the third mutation, E138K. The triple mutant G140S/Y143R/Q148H was made using the previously constructed G140S/Q148H double mutant and the appropriate oligonucleotides to make the third mutation Y143R.

The DNA sequence of each construct was verified independently by DNA sequence determination. The mutated IN coding sequences from pBluescript KS+ were then subcloned into pNLNgoMIVR-ΔEnv.LUC (between the KpnI and SalI sites) to produce mutant HIV-1 constructs; the sequence of the final construct was checked by DNA sequencing.

Computer modeling

All modeling was conducted using MOE 2016.0802 (Chemical computing group, Montreal, Quebec, Canada). The sequences and structures of DTG bound in the PFV intasome (PDB ID: 3S3M) and HIV-1 IN (PDB ID: 5U1C) served as the structural templates to construct a HIV-1 IN model with DTG bound in the active site. First, the N-terminal portions of the NTD, CCD, and CTD domains of the PFV and HIV-1 IN were used to align the domains properly. Next the sequences and structures of HIV-1 and PFV INs were aligned so that the HIV IN sequence was matched to superpose the HIV-1 and PFV IN. The coordinates of the HIV-1 IN structure (PDB ID: 5U1C) from the aforementioned alignment were used as the IN template to construct the HIV-1 IN model. This structure was modified to fit the structural coordinates of DTG, Mg^{2+} cofactors, and the viral DNA from the PFV intasome (PDB ID: 3S3M). The model of the HIV-1 intasome with DTG bound was energy minimized using a PFROSST forcefield with relative field solvation as recommended by the manufacturer. The new HIV-1 IN model was then aligned (structure only) with the HIV-1 IN structure (PDB ID: 5U1C) from the aforementioned alignment with PFV IN (PDB ID: 3S3M) and aligned to a RMSD value of 0.82 Å. Additionally, the new HIV-1 IN model was aligned with the previously solved HIV-1 IN structure (PDB ID: 5U1C) and aligned to a RMSD value of 1.12 Å. The surface (Van der Waals) of DTG was determined to locate possible steric clashes with the active site residues in the model. To identify potential contacts with CAB and BIC, both INSTIs were docked using DTG as the template. CAB or BIC were placed using the triangle matcher method and scored with London dG with approximately 30 poses and then the putative ligand poses were further refined using the rigid receptor method in MOE and scored with the GBVI/WSA dG function. If the expected ligand poses were not created, a pharmacophore editor tool in the docking function was used to add certain features that made the appropriate docking of CAB or BIC to DTG easier to view and the resulting structures were refined in the manner described above. The poses with the best docking scores were selected based on how well the bound compounds overlay with the DTG scaffold, bound to Mg^{2+}, and how well their halogenated benzyl moiety interacted hydrophobically through π–π stacking with the penultimate cytosine on the 3′ end of the bound viral DNA. Docking poses images were refined using MolSoft ICM Pro software version 3.8-5 (MolSoft LLC, San Diego, CA).

Abbreviations

INSTI's: integrase strand transfer inhibitors; ARV: antiretroviral; STC: strand transfer complex; HIV: human immunodeficiency virus; PFV: prototype foamy virus; NNRTI's: non-nucleoside reverse transcriptase inhibitors; FDA: Food and Drug Administration; EC_{50}: half maximal inhibitory concentration.

Author details

[1] HIV Dynamics and Replication Program, National Cancer Institute-Frederick, National Institutes of Health, Frederick, MD, USA. [2] Chemical Biology Laboratory, National Cancer Institute-Frederick, National Institutes of Health, Frederick, MD, USA.

Acknowledgements

The authors would like to thank Teresa Burdette for help in preparing the manuscript, Alan Kane for help with the figures, and Brian Luke for help with the statistical analysis. This research was supported by the Intramural Research Programs of the National Cancer Institute and the Intramural AIDS Targeted Antiviral Program.

Authors' contributions

SS performed the experiments. SS and SH designed the experiments. SS, XZ, TB, and SH analyzed the data. XZ and TB contributed with compounds and imaging software. SS and SH drafted the manuscript. All authors read and approved the final manuscript.

Competing interests

The authors declare that they have no competing interests.

Funding

This research was supported by the Intramural Research Programs of the National Cancer Institute.

References

1. Bushman FD, Craigie R. Activities of human immunodeficiency virus (HIV) integration protein in vitro: specific cleavage and integration of HIV DNA. Proc Natl Acad Sci USA. 1991;88:1339–43.
2. Engelman A, Mizuuchi K, Craigie R. HIV-1 DNA integration: mechanism of viral DNA cleavage and DNA strand transfer. Cell. 1991;67:1211–21.
3. Hazuda DJ, Felock P, Witmer M, Wolfe A, Stillmock K, Grobler JA, Espeseth A, Gabryelski L, Schleif W, Blau C, Miller MD. Inhibitors of strand transfer that prevent integration and inhibit HIV-1 replication in cells. Science. 2000;287:646–50.
4. Wainberg MA, Mesplede T, Raffi F. What if HIV were unable to develop resistance against a new therapeutic agent? BMC Med. 2013;11:249.
5. Hare S, Gupta SS, Valkov E, Engelman A, Cherepanov P. Retroviral intasome assembly and inhibition of DNA strand transfer. Nature. 2010;464:232–6.
6. Metifiot M, Johnson B, Smith S, Zhao XZ, Marchand C, Burke T, Hughes S, Pommier Y. MK-0536 inhibits HIV-1 integrases resistant to raltegravir. Antimicrob Agents Chemother. 2011;55:5127–33.
7. Zhao XZ, Smith SJ, Metifiot M, Johnson BC, Marchand C, Pommier Y, Hughes SH, Burke TR Jr. Bicyclic 1-hydroxy-2-oxo-1,2-dihydropyridine-3-carboxamide-containing HIV-1 integrase inhibitors having high antiviral potency against cells harboring raltegravir-resistant integrase mutants. J Med Chem. 2014;57:1573–82.
8. Zhao XZ, Smith SJ, Metifiot M, Marchand C, Boyer PL, Pommier Y, Hughes SH, Burke TR Jr. 4-amino-1-hydroxy-2-oxo-1,8-naphthyridine-containing compounds having high potency against raltegravir-resistant integrase mutants of HIV-1. J Med Chem. 2014;57:5190–202.
9. Ballantyne AD, Perry CM. Dolutegravir: first global approval. Drugs. 2013;73:1627–37.
10. Shah BM, Schafer JJ, Desimone JA Jr. Dolutegravir: a new integrase strand transfer inhibitor for the treatment of HIV. Pharmacotherapy. 2014;34:506–20.
11. Walmsley SL, Antela A, Clumeck N, Duiculescu D, Eberhard A, Gutierrez F, Hocqueloux L, Maggiolo F, Sandkovsky U, Granier C, et al. Dolutegravir plus abacavir-lamivudine for the treatment of HIV-1 infection. N Engl J Med. 2013;369:1807–18.
12. Clotet B, Feinberg J, van Lunzen J, Khuong-Josses MA, Antinori A, Dumitru I, Pokrovskiy V, Fehr J, Ortiz R, Saag M, et al. Once-daily dolutegravir versus darunavir plus ritonavir in antiretroviral-naive adults with HIV-1 infection (FLAMINGO): 48 week results from the randomised open-label phase 3b study. Lancet. 2014;383:2222–31.
13. Johns BA, Kawasuji T, Weatherhead JG, Taishi T, Temelkoff DP, Yoshida H, Akiyama T, Taoda Y, Murai H, Kiyama R, et al. Carbamoyl pyridone HIV-1 integrase inhibitors 3. A diastereomeric approach to chiral nonracemic tricyclic ring systems and the discovery of dolutegravir (S/GSK1349572) and (S/GSK1265744). J Med Chem. 2013;56:5901–16.
14. Kobayashi M, Yoshinaga T, Seki T, Wakasa-Morimoto C, Brown KW, Ferris R, Foster SA, Hazen RJ, Miki S, Suyama-Kagitani A, et al. In Vitro antiretroviral properties of S/GSK1349572, a next-generation HIV integrase inhibitor. Antimicrob Agents Chemother. 2011;55:813–21.
15. Hare S, Smith SJ, Metifiot M, Jaxa-Chamiec A, Pommier Y, Hughes SH, Cherepanov P. Structural and functional analyses of the second-generation integrase strand transfer inhibitor dolutegravir (S/GSK1349572). Mol Pharmacol. 2011;80:565–72.
16. Cahn P, Pozniak AL, Mingrone H, Shuldyakov A, Brites C, Andrade-Villanueva JF, Richmond G, Buendia CB, Fourie J, Ramgopal M, et al. Dolutegravir versus raltegravir in antiretroviral-experienced, integrase-inhibitor-naive adults with HIV: week 48 results from the randomised, double-blind, non-inferiority SAILING study. Lancet. 2013;382:700–8.
17. Min S, Sloan L, DeJesus E, Hawkins T, McCurdy L, Song I, Stroder R, Chen S, Underwood M, Fujiwara T, et al. Antiviral activity, safety, and pharmacokinetics/pharmacodynamics of dolutegravir as 10-day monotherapy in HIV-1-infected adults. AIDS. 2011;25:1737–45.
18. Raffi F, Jaeger H, Quiros-Roldan E, Albrecht H, Belonosova E, Gatell JM, Baril JG, Domingo P, Brennan C, Almond S, et al. Once-daily dolutegravir versus twice-daily raltegravir in antiretroviral-naive adults with HIV-1 infection (SPRING-2 study): 96 week results from a randomised, double-blind, non-inferiority trial. Lancet Infect Dis. 2013;13:927–35.
19. Raffi F, Rachlis A, Stellbrink HJ, Hardy WD, Torti C, Orkin C, Bloch M, Podzamczer D, Pokrovsky V, Pulido F, et al. Once-daily dolutegravir versus raltegravir in antiretroviral-naive adults with HIV-1 infection: 48 week results from the randomised, double-blind, non-inferiority SPRING-2 study. Lancet. 2013;381:735–43.
20. van Lunzen J, Maggiolo F, Arribas JR, Rakhmanova A, Yeni P, Young B, Rockstroh JK, Almond S, Song I, Brothers C, Min S. Once daily dolutegravir (S/GSK1349572) in combination therapy in antiretroviral-naive adults with HIV: planned interim 48 week results from SPRING-1, a dose-ranging, randomised, phase 2b trial. Lancet Infect Dis. 2012;12:111–8.
21. Wainberg MA, Han YS. Will drug resistance against dolutegravir in initial therapy ever occur? Front Pharmacol. 2015;6:90.
22. Quashie PK, Mesplede T, Han YS, Oliveira M, Singhroy DN, Fujiwara T, Underwood MR, Wainberg MA. Characterization of the R263K mutation in HIV-1 integrase that confers low-level resistance to the second-generation integrase strand transfer inhibitor dolutegravir. J Virol. 2012;86:2696–705.
23. Quashie PK, Mesplede T, Han YS, Veres T, Osman N, Hassounah S, Sloan RD, Xu HT, Wainberg MA. Biochemical analysis of the role of G118R-linked dolutegravir drug resistance substitutions in HIV-1 integrase. Antimicrob Agents Chemother. 2013;57:6223–35.
24. Mesplede T, Osman N, Wares M, Quashie PK, Hassounah S, Anstett K, Han Y, Singhroy DN, Wainberg MA. Addition of E138K to R263K in HIV integrase increases resistance to dolutegravir, but fails to restore activity of the HIV integrase enzyme and viral replication capacity. J Antimicrob Chemother. 2014;69:2733–40.
25. Castagna A, Maggiolo F, Penco G, Wright D, Mills A, Grossberg R, Molina JM, Chas J, Durant J, Moreno S, et al. Dolutegravir in antiretroviral-experienced patients with raltegravir- and/or elvitegravir-resistant HIV-1: 24-week results of the phase III VIKING-3 study. J Infect Dis. 2014;210:354–62.

26. Eron JJ, Clotet B, Durant J, Katlama C, Kumar P, Lazzarin A, Poizot-Martin I, Richmond G, Soriano V, Ait-Khaled M, et al. Safety and efficacy of dolutegravir in treatment-experienced subjects with raltegravir-resistant HIV type 1 infection: 24-week results of the VIKING Study. J Infect Dis. 2013;207:740–8.
27. Yoshinaga T, Kobayashi M, Seki T, Miki S, Wakasa-Morimoto C, Suyama-Kagitani A, Kawauchi-Miki S, Taishi T, Kawasuji T, Johns BA, et al. Antiviral characteristics of GSK1265744, an HIV integrase inhibitor dosed orally or by long-acting injection. Antimicrob Agents Chemother. 2015;59:397–406.
28. Tsiang M, Jones GS, Goldsmith J, Mulato A, Hansen D, Kan E, Tsai L, Bam RA, Stepan G, Stray KM, et al. Antiviral activity of bictegravir (GS-9883), a novel potent HIV-1 integrase strand transfer inhibitor with an improved resistance profile. Antimicrob Agents Chemother. 2016;60:7086–97.
29. Cooper DA, Steigbigel RT, Gatell JM, Rockstroh JK, Katlama C, Yeni P, Lazzarin A, Clotet B, Kumar PN, Eron JE, et al. Subgroup and resistance analyses of raltegravir for resistant HIV-1 infection. N Engl J Med. 2008;359:355–65.
30. Malet I, Delelis O, Valantin MA, Montes B, Soulie C, Wirden M, Tchertanov L, Peytavin G, Reynes J, Mouscadet JF, et al. Mutations associated with failure of raltegravir treatment affect integrase sensitivity to the inhibitor in vitro. Antimicrob Agents Chemother. 2008;52:1351–8.
31. Fransen S, Gupta S, Danovich R, Hazuda D, Miller M, Witmer M, Petropoulos CJ, Huang W. Loss of raltegravir susceptibility by human immunodeficiency virus type 1 is conferred via multiple nonoverlapping genetic pathways. J Virol. 2009;83:11440–6.
32. Goethals O, Clayton R, Van Ginderen M, Vereycken I, Wagemans E, Geluykens P, Dockx K, Strijbos R, Smits V, Vos A, et al. Resistance mutations in human immunodeficiency virus type 1 integrase selected with elvitegravir confer reduced susceptibility to a wide range of integrase inhibitors. J Virol. 2008;82:10366–74.
33. Shimura K, Kodama E, Sakagami Y, Matsuzaki Y, Watanabe W, Yamataka K, Watanabe Y, Ohata Y, Doi S, Sato M, et al. Broad antiretroviral activity and resistance profile of the novel human immunodeficiency virus integrase inhibitor elvitegravir (JTK-303/GS-9137). J Virol. 2008;82:764–74.
34. Margot NA, Hluhanich RM, Jones GS, Andreatta KN, Tsiang M, McColl DJ, White KL, Miller MD. In vitro resistance selections using elvitegravir, raltegravir, and two metabolites of elvitegravir M1 and M4. Antiviral Res. 2012;93:288–96.
35. Zhao XZ, Smith SJ, Maskell DP, Metifiot M, Pye VE, Fesen K, Marchand C, Pommier Y, Cherepanov P, Hughes SH, Burke TR Jr. HIV-1 integrase strand transfer inhibitors with reduced susceptibility to drug resistant mutant integrases. ACS Chem Biol. 2016;11:1074–81.
36. Zhao XZ, Smith SJ, Maskell DP, Metifiot M, Pye VE, Fesen K, Marchand C, Pommier Y, Cherepanov P, Hughes SH, Burke TR Jr. Structure-guided optimization of HIV integrase strand transfer inhibitors. J Med Chem. 2017;60:7315–32.
37. Markowitz M, Morales-Ramirez JO, Nguyen BY, Kovacs CM, Steigbigel RT, Cooper DA, Liporace R, Schwartz R, Isaacs R, Gilde LR, et al. Antiretroviral activity, pharmacokinetics, and tolerability of MK-0518, a novel inhibitor of HIV-1 integrase, dosed as monotherapy for 10 days in treatment-naive HIV-1-infected individuals. J Acquir Immune Defic Syndr. 2006;43:509–15.
38. Metifiot M, Marchand C, Maddali K, Pommier Y. Resistance to integrase inhibitors. Viruses. 2010;2:1347–66.
39. Johnson VA, Calvez V, Gunthard HF, Paredes R, Pillay D, Shafer RW, Wensing AM, Richman DD. Update of the drug resistance mutations in HIV-1: March 2013. Top Antivir Med. 2013;21:6–14.
40 Passos DO, Li M, Yang R, Rebensburg SV, Ghirlando R, Jeon Y, Shkriabai N, Kvaratskhelia M, Craigie R, Lyumkis D. Cryo-EM structures and atomic model of the HIV-1 strand transfer complex intasome. Science. 2017;355:89–92.
41. Yoshinaga T, Seki T, Miki S, Miyamoto T, Suyama-Kagitani A, Kawauchi-Miki S, Kobayashi M, Sato A, Stewart E, Underwood M, Fujiwara T. Novel secondary mutations C56S and G149A confer resistance to HIV-1 integrase strand transfer inhibitors. Antiviral Res. 2018;152:1–9.
42. Steigbigel RT, Cooper DA, Teppler H, Eron JJ, Gatell JM, Kumar PN, Rockstroh JK, Schechter M, Katlama C, Markowitz M, et al. Long-term efficacy and safety of Raltegravir combined with optimized background therapy in treatment-experienced patients with drug-resistant HIV infection: week 96 results of the BENCHMRK 1 and 2 Phase III trials. Clin Infect Dis. 2010;50:605–12.
43. Eron JJ Jr, Rockstroh JK, Reynes J, Andrade-Villanueva J, Ramalho-Madruga JV, Bekker LG, Young B, Katlama C, Gatell-Artigas JM, Arribas JR, et al. Raltegravir once daily or twice daily in previously untreated patients with HIV-1: a randomised, active-controlled, phase 3 non-inferiority trial. Lancet Infect Dis. 2011;11:907–15.
44. Sax PE, DeJesus E, Mills A, Zolopa A, Cohen C, Wohl D, Gallant JE, Liu HC, Zhong L, Yale K, et al. Co-formulated elvitegravir, cobicistat, emtricitabine, and tenofovir versus co-formulated efavirenz, emtricitabine, and tenofovir for initial treatment of HIV-1 infection: a randomised, double-blind, phase 3 trial, analysis of results after 48 weeks. Lancet. 2012;379:2439–48.
45. Gallant JE, Thompson M, DeJesus E, Voskuhl GW, Wei X, Zhang H, White K, Cheng A, Quirk E, Martin H. Antiviral activity, safety, and pharmacokinetics of bictegravir as 10-Day monotherapy in HIV-1-infected adults. J Acquir Immune Defic Syndr. 2017;75:61–6.
46. Margolis DA, Gonzalez-Garcia J, Stellbrink HJ, Eron JJ, Yazdanpanah Y, Podzamczer D, Lutz T, Angel JB, Richmond GJ, Clotet B, et al. Long-acting intramuscular cabotegravir and rilpivirine in adults with HIV-1 infection (LATTE-2): 96-week results of a randomised, open-label, phase 2b, non-inferiority trial. Lancet. 2017;390:1499–510.
47. Markowitz M, Frank I, Grant RM, Mayer KH, Elion R, Goldstein D, Fisher C, Sobieszczyk ME, Gallant JE, Van Tieu H, et al. Safety and tolerability of long-acting cabotegravir injections in HIV-uninfected men (ECLAIR): a multicentre, double-blind, randomised, placebo-controlled, phase 2a trial. Lancet HIV. 2017;4:e331–40.
48. Sax PE, DeJesus E, Crofoot G, Ward D, Benson P, Dretler R, Mills A, Brinson C, Peloquin J, Wei X, et al. Bictegravir versus dolutegravir, each with emtricitabine and tenofovir alafenamide, for initial treatment of HIV-1 infection: a randomised, double-blind, phase 2 trial. Lancet HIV. 2017;4:e154–60.
49. Neogi U, Singh K, Aralaguppe SG, Rogers LC, Njenda DT, Sarafianos SG, Hejdeman B, Sonnerborg A. Ex-vivo antiretroviral potency of newer integrase strand transfer inhibitors cabotegravir and bictegravir in HIV type 1 non-B subtypes. AIDS. 2018;32:469–76.
50. Das K, Clark AD Jr, Lewi PJ, Heeres J, De Jonge MR, Koymans LM, Vinkers HM, Daeyaert F, Ludovici DW, Kukla MJ, et al. Roles of conformational and positional adaptability in structure-based design of TMC125-R165335 (etravirine) and related non-nucleoside reverse transcriptase inhibitors that are highly potent and effective against wild-type and drug-resistant HIV-1 variants. J Med Chem. 2004;47:2550–60.
51. Das K, Lewi PJ, Hughes SH, Arnold E. Crystallography and the design of anti-AIDS drugs: conformational flexibility and positional adaptability are important in the design of non-nucleoside HIV-1 reverse transcriptase inhibitors. Prog Biophys Mol Biol. 2005;88:209–31.
52. Smith SJ, Pauly GT, Akram A, Melody K, Ambrose Z, Schneider JP, Hughes SH. Rilpivirine and doravirine have complementary efficacies against NNRTI-resistant HIV-1 mutants. J Acquir Immune Defic Syndr. 2016;72:485–91.
53. Smith SJ, Pauly GT, Akram A, Melody K, Rai G, Maloney DJ, Ambrose Z, Thomas CJ, Schneider JT, Hughes SH. Rilpivirine analogs potently inhibit drug-resistant HIV-1 mutants. Retrovirology. 2016;13:11.
54. Kassaye SG, Grossman Z, Balamane M, Johnston-White B, Liu C, Kumar P, Young M, Sneller MC, Sereti I, Dewar R, et al. Transmitted HIV drug resistance is high and longstanding in metropolitan Washington. DC. Clin Infect Dis. 2016;63:836–43.
55. Varadarajan J, McWilliams MJ, Hughes SH. Treatment with suboptimal doses of raltegravir leads to aberrant HIV-1 integrations. Proc Natl Acad Sci USA. 2013;110:14747–52.
56. Varadarajan J, McWilliams MJ, Mott BT, Thomas CJ, Smith SJ, Hughes SH. Drug resistant integrase mutants cause aberrant HIV integrations. Retrovirology. 2016;13:71.
57. Smith SJ, Hughes SH. Rapid screening of HIV reverse transcriptase and integrase inhibitors. J Vis Exp. 2014;86:51400.

3

Structural alteration of DNA induced by viral protein R of HIV-1 triggers the DNA damage response

Kenta Iijima[1], Junya Kobayashi[2] and Yukihito Ishizaka[1*]

Abstract

Background: Viral protein R (Vpr) is an accessory protein of HIV-1, which is potentially involved in the infection of macrophages and the induction of the ataxia-telangiectasia and Rad3-related protein (ATR)-mediated DNA damage response (DDR). It was recently proposed that the SLX4 complex of structure-specific endonuclease is involved in Vpr-induced DDR, which implies that aberrant DNA structures are responsible for this phenomenon. However, the mechanism by which Vpr alters the DNA structures remains unclear.

Results: We found that Vpr unwinds double-stranded DNA (dsDNA) and invokes the loading of RPA70, which is a single-stranded DNA-binding subunit of RPA that activates the ATR-dependent DDR. We demonstrated that Vpr influenced RPA70 to accumulate in the corresponding region utilizing the LacO/LacR system, in which Vpr can be tethered to the LacO locus. Interestingly, RPA70 recruitment required chromatin remodelling via Vpr-mediated ubiquitination of histone H2B. On the contrary, Q65R mutant of Vpr, which lacks ubiquitination activity, was deficient in both chromatin remodelling and RPA70 loading on to the chromatin. Moreover, Vpr-induced unwinding of dsDNA coincidently resulted in the accumulation of negatively supercoiled DNA and covalent complexes of topoisomerase 1 and DNA, which caused DNA double-strand breaks (DSBs) and DSB-directed integration of proviral DNA. Lastly, we noted the dependence of Vpr-promoted HIV-1 infection in resting macrophages on topoisomerase 1.

Conclusions: The findings of this study indicate that Vpr-induced structural alteration of DNA is a primary event that triggers both DDR and DSB, which ultimately contributes to HIV-1 infection.

Background

Combined antiretroviral therapy (cART) for human immunodeficiency virus-1 (HIV-1)-positive patients suppresses viral replication to a non-detectable level, thereby preventing immunodeficiency caused by T cell depletion. Unfortunately, the interruption of the cART regimen allows the expansion of viral replication from long-lived reservoir cells [1]. Because macrophages comprise the major cell population involved in the formation of viral reservoirs [2, 3], understanding the mode of viral infection in resting macrophages is crucial.

Viral protein R (*Vpr*) is an accessory gene of HIV-1, which encodes a virion-associated nuclear protein that is made up of 96 amino acids (aa) [4] and can facilitate viral infection in resting macrophages [5–8]. The induction of cell-cycle abnormality at the G_2/M phase is a well-investigated function of Vpr among its pleiotropic activities [9]. Notably, Vpr induces a DNA damage response (DDR) involving ATR/ATRIP-Chk1 activation, the phosphorylation of histone H2AX (γH2AX) and the formation of BRCA1, 53BP1, RAD51 and FANCD2 foci [10–14]. Structural analyses revealed that Vpr contains three alpha helices with self-dimerisation properties along with a flexible carboxy (C)-terminal region with a basic stretch that is involved in DNA binding [15, 16], both of which were attributable to the cell-cycle abnormalities at the G_2/M phase [17]. Originally, Vpr-induced G_2/M arrest was proposed to contribute to viral infection by delaying

*Correspondence: zakay@ri.ncgm.go.jp
[1] Department of Intractable Diseases, National Center for Global Health and Medicine, 1-21-1 Toyama, Shinjuku-ku, Tokyo 162-8655, Japan
Full list of author information is available at the end of the article

cell death, thereby providing longer periods for viral replication [18]; however, its biological significance in resting macrophages remains largely unclear.

Regarding the upstream events that potentially trigger Vpr-induced DDR, several lines of evidence suggest that Vpr expression provokes replication stress, thereby inducing chromatin loading of the replication protein A 70-kDa subunit (RPA70), which is a DNA-binding subunit of the single-strand DNA (ssDNA) binding heterotrimeric protein [12, 19]). Considering that DNA binding of RPA70 triggers the ATR/ATRIP-Chk1 pathway responsible for G_2/M cell-cycle arrest [20], it is imperative to determine the mechanism of RPA70 loading onto the chromatin. Furthermore, cellular ubiquitination by Vpr is also required for DDR activation: the Q65R mutant of Vpr, which cannot bind DDB1/VprBP, an adaptor protein for Cul4 E3-ligase, is defective for the G_2/M checkpoint activation and cellular ubiquitination [21–24]. Although these findings imply that Vpr modulates RPA70 loading onto the chromatin in close functional association with the DDB1/VprBP-Cul4-dependent ubiquitination pathway, no cellular targets of Vpr-dependent ubiquitination have yet been identified.

For determining whether Vpr influences the structural alteration of DNA, Kichler et al. [25] performed an electron microscopic study and proposed that the C-terminal moiety of Vpr can aggregate plasmid DNA. Moreover, Lyonnais et al. [26] found that Vpr mediates the bridging and stretching of DNA helices. These observations strongly suggest that Vpr is capable of altering DNA structures. Generally, aberrant DNA structures interfere with DNA replication and transcription and may cause DNA damage. Such molecular consequences have been well-characterised in the context of DNA topological stress induced by camptothecin (CPT) and etoposide [27], which inhibit topoisomerase 1 (Topo1) and Topo2, respectively. Topo1 is involved in the relaxation of excess supercoils on dsDNA, whereas Topo2 is involved in the disentangling of dsDNA strands through cycles of cleavage and re-ligation. Chemical inhibition of re-ligation after relief of DNA distortions results in the formation of covalent complexes between Topo1 or Topo2 and cleaved DNA ends, which potentially induce DNA double-strand breaks (DSBs) due to subsequent collision of replication or transcription [27, 28].

Because Vpr itself does not possess DNA cleavage activity, Vpr must induce the DDR with the aid of host factor(s) [16]. Several cellular proteins, including UNG2, HTLF and SLX4 have been suggested as candidates for participation in Vpr-induced DDR [14, 29, 30]. UNG2 and HTLF are DNA repair proteins, and suppression of their functions leads to aggravation of DNA damage [29, 30]. On the other hand, Laguette et al. [14] suggested that Vpr causes the premature activation of the SLX4-Mus81/Eme1 complex, which is a structure-specific nuclease complex, and promotes the cleavage of its target DNA structures, including DNA replication and recombination intermediates [14]. These observations suggest that cellular proteins involved in replication stress or its repair contribute to Vpr-induced DDR induction.

The findings of this study revealed that Vpr-induced structural alteration of DNA lead to RPA70 loading and negative supercoil formations. We demonstrated lines of evidence supporting that such Vpr-induced structural alteration of DNA is an upstream event that leads to DDR induction as well as DSB formation. Similar activity was observed at the cellular level; when Vpr was artificially recruited onto a specific chromosomal locus, it induced RPA70 loading, accumulation of negative supercoils, formation of a Topo1-DNA covalent complex (Topo1-cc) and DSBs in the corresponding region. Notably, Vpr also induced the ubiquitination of histone H2B, thereby increasing histone mobility and promoting RPA70 loading onto the chromatin. Together with the data regarding the Vpr-induced promotion of viral DNA integration in resting macrophages in a Topo1-dependent manner, we proposed that the structural alteration of DNA by Vpr acts as a trigger event for DDR and DSBs induction, which ultimately contributes to HIV-1 infection.

Results

Vpr unwinds supercoiled DNA and alters its topological configuration

To elucidate the activity of Vpr on dsDNA, we first performed atomic force microscopy (AFM) to analyse the structural alteration of dsDNA. In these experiments, supercoiled dsDNA was incubated in a buffer solution with recombinant Vpr protein (rVpr), attached onto a freshly cleaved mica surface through a gentle Ni^{2+}-mediated interaction, and then subjected to analysis [31]. AFM under liquid conditions can avoid potential artefacts resulting from the fixation or staining of DNA. Typical morphological changes and height profiles are shown in Fig. 1a. Markedly, treatment with chloroquine (Chlq), which is a DNA intercalator, resulted in the conversion of dsDNA into a highly expanded and de-condensed form (Fig. 1a, compare upper and middle panel and Additional file 1: Figure S1a). Interestingly, rVpr induced similar morphological changes in DNA (bottom panel). Each height profile was measured, and the root-mean-square (RMS) of the roughness (Rq) was calculated on the basis of the integrated data; Rq indicates the status and bulkiness of condensed dsDNA. As shown in Fig. 1b, the Rq value was significantly reduced by treatment with Chlq and rVpr (see also Additional file 1: Figure S1b). In contrast, *ΔC12*, which is a Vpr mutant that lacked the

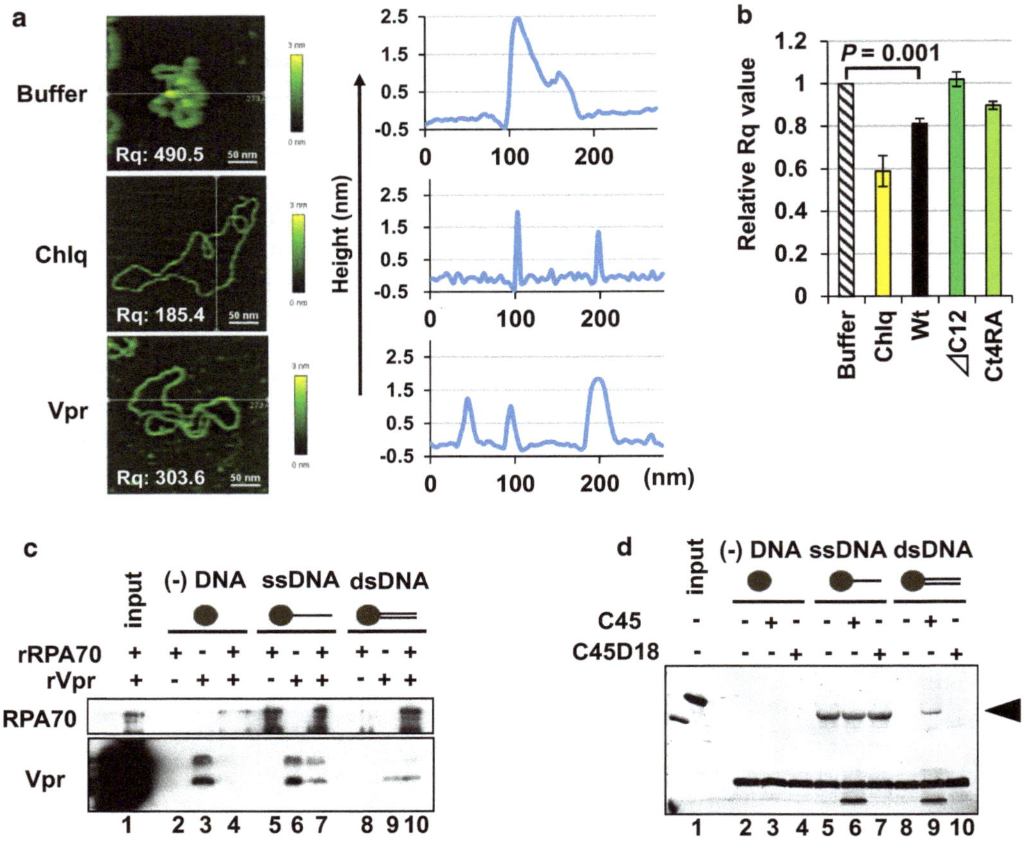

Fig. 1 Vpr-induced structural alteration of DNA. **a** Representative AFM images of dsDNA and height profiles. Right panels depict height profiles detected in each cross section shown by white lines in the left panels. **b** rVpr decreased the Rq value. The experiments were repeated at least three times. Error bars indicate ± SEM. **c** RPA70 bound dsDNA after treatment of rVpr. RPA70 was pulled down using beads conjugated with DNA and detected by Western blotting (WB). **d** T4gp32 bound dsDNA after treatment with C45. T4gp32 was pulled down with DNA-bound beads in the presence of C45 or C45D18 peptide. Arrowhead, T4gp32

C-terminal 12 aa [16], did not affect the Rq value (Fig. 1b and Additional file 1: Figure S1b). Similarly, the Ct4RA mutant, in which all four arginines (R) of the RxRRxR motif of the C-terminal region were replaced with alanines (A), showed lower activity than Vpr-Wt (Fig. 1b and Additional file 1: Figure S1b, c). These data suggest that Vpr influences the structural alteration of dsDNA by interacting via positively charged residues in its C-terminal stretch. Notably, other Vpr mutants, including Q65R, R77Q and R80A, showed similar activity with Vpr-Wt (Additional file 1: Figure S1c).

Our initial experiment using AFM suggested that Vpr generates an ssDNA stretch in the dsDNA molecule. To confirm this, we tested whether RPA70 formed an association with dsDNA when treated with rVpr. As shown in Fig. 1c, Western blotting (WB) after a pull-down procedure detected RPA70 when dsDNA and RPA70 were incubated with rVpr (compare lanes 8–10). To further confirm the dependence of RPA70 loading on the C-terminal stretch of Vpr, we performed experiments using C45, which is a peptide made up of 45 aa of the C-terminal Vpr [32]. In this experiment, we used T4gp32 (T4 gene 32 protein), a well-established ssDNA-binding protein of the T4 phage [33, 34]. Consistent with the first experiment, we observed that T4gp32 bound dsDNA in the presence of C45 (Fig. 1d, lane 9). In contrast, no association was detected between dsDNA and T4gp32 when C45D18, which is a truncated form of C45 lacking the C-terminal 18 aa, was incubated (lane 10). Because Vpr did not interact with RPA70 or T4gp32 (Additional file 2: Figure S2a for RPA70 and Figure S2b for T4gp32), the data suggest that Vpr-induced unwinding of dsDNA occurred through its positively charged C-terminal region.

When circular-dsDNA is relaxed in one part of a dsDNA, topological changes are aroused in the other (Fig. 2a, step-1). Such dsDNA with differential topologies can be detected using a DNA supercoiling assay [35], in which dsDNA is treated with *Escherichia coli* Topo1, followed by deproteinisation with proteinase K (ProK). On treatment with *E. coli* Topo1, negative supercoiling was removed through a nicking and re-ligation cycle (Fig. 2a, step-2), whereas treatment with ProK removed Vpr and Topo1, and generated relaxed forms of dsDNA (RFs) (Fig. 2a, step-3). Since agarose electrophoresis can separate each topoisomer, rVpr induced the formation of a slower-migrating species of DNA, representing RFs (Fig. 2b, lanes 6–8). Similar experiments were performed using various mutants of Vpr (Fig. 2c), indicating that Vpr-Wt increased the RFs of dsDNA (lane 5), whereas ⊿C12 or Ct4RA did not (lanes 17 and 20). In contrast, Vpr mutants (Q65R, R77Q and R80A) showed activities comparable with that of Vpr-Wt (Fig. 2c, lanes 8, 11 and 14, respectively).

Expression of Vpr induces the Topo1-cc formation

Topo1-cc formation induces SSBs (DNA single-strand breaks) and DDR [27], while collision of a replication fork or transcription with the complex triggers DSB [27, 28]. To maintain genomic stability, Topo1-cc is removed by tyrosyl-DNA phosphodiesterase 1 (TDP1) and degraded by the SUMO/ubiquitin-mediated proteasomal degradation pathway [27, 36, 37]. For characterising the functional link between Vpr and Topo1 at the cellular level, we examined the Topo1 protein using Mit-23 cells, in which the tetracycline promoter tightly regulates Vpr expression [17]. When doxycycline (Dox) is added to these cells, Vpr is expressed at a level comparable with that in HIV-1 infected cells [38]. As observed in Fig. 3a, Vpr expression led to reduction in the level of Topo1 (relative level decreased to 0.71); an immunohistochemical analysis revealed that the nuclear

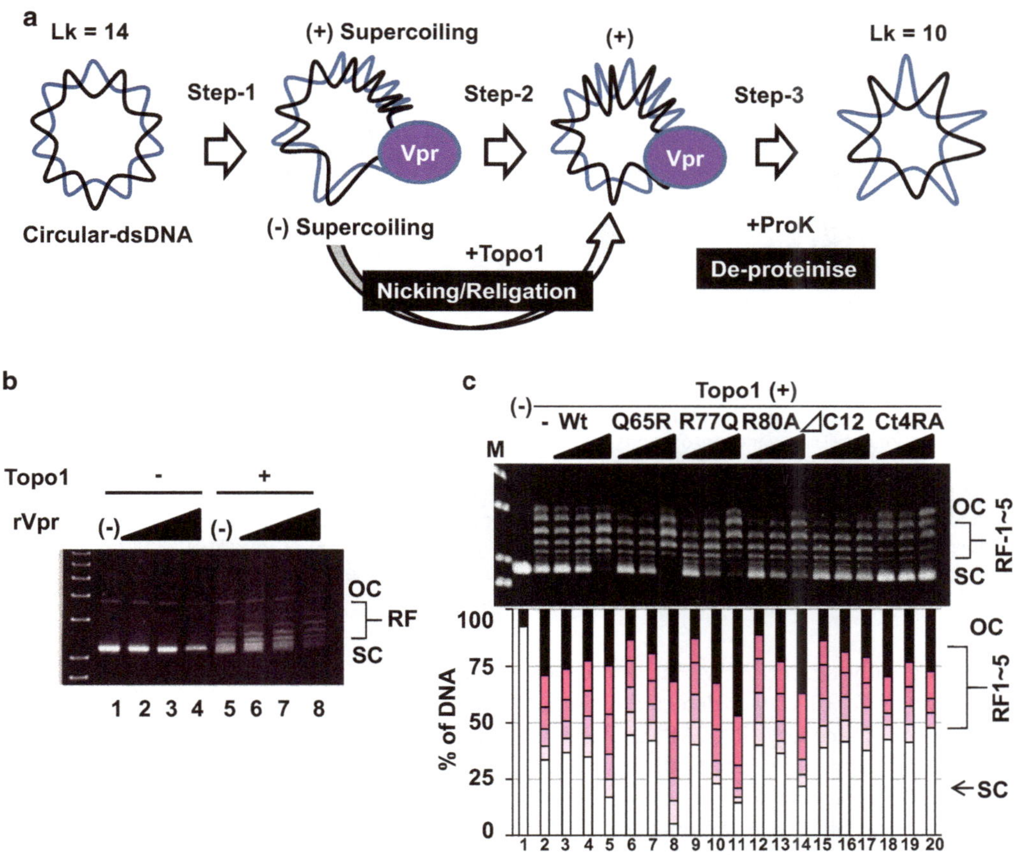

Fig. 2 Vpr induced topological changes on DNA. **a** Schematic representation of experimental procedure for DNA supercoiling assay. Possible induction of partial unwinding of dsDNA by Vpr induces negative and positive supercoiling (Step-1). In the presence of *E. coli* Topo1, negative supercoiling (lower side) is relieved by nicking/relegation activity (Step-2). After the de-proteinisation, net amounts of linking number (Lk) are decreased (Step-3). **b** DNA supercoiling assay with rVpr-Wt. The effects of various amounts of rVpr were tested using *E. coli* Topo1. *OC* open circular, *RF* relaxed form, *SC* supercoiled. **c** DNA supercoiling assay with mutants of Vpr. Intensities of each topoisomer were quantified, and their relative amounts are shown in the bottom graph

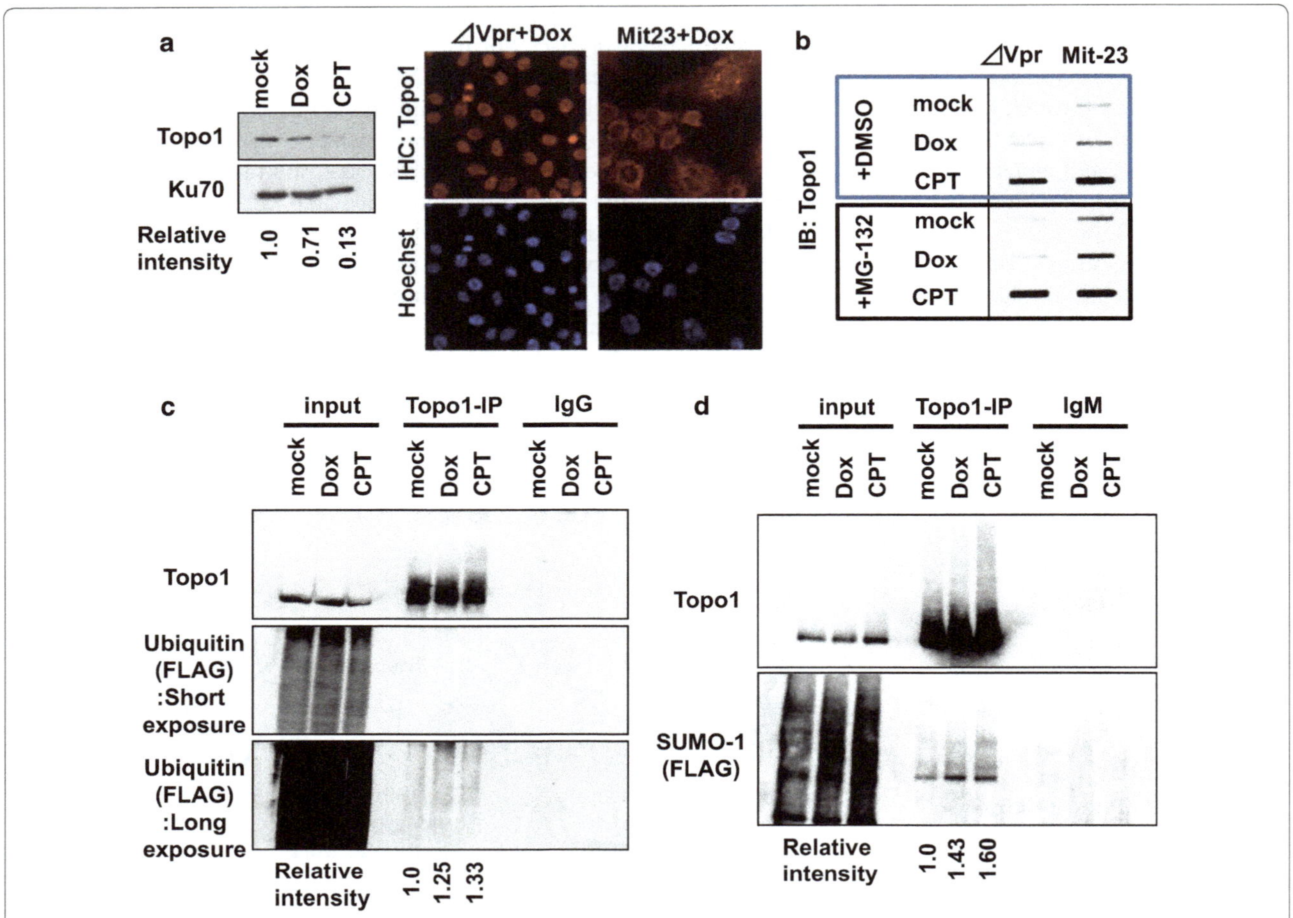

Fig. 3 Vpr expression induces Topo1 stress. **a** Reduced expression of Topo1 under Vpr expression. Mit-23 cells were treated with Dox (3 μg/ml, 2 days) or CPT (20 μM, 1 h). By WB analysis, Topo1 and Ku70 as an internal control were detected. Relative intensities of Topo1 were calculated by normalizing the amounts of Topo1 by Ku70 (Left panel). Immunohistochemical analysis detected Topo1 and DNA (Hoechst-33258) (Right panel). ΔVpr cells were used as a control cell line without exogenous Vpr. **b** Formation of Topo1-cc under Vpr expression. RADAR analysis was performed on Mit-23 and ΔVpr cells, which were treated with Dox or CPT in the presence or absence of MG-132 (50 μM, 2 h). **c** Ubiquitination of Topo1 under Vpr expression. Mit-23 cells transfected with 3×FLAG/6×His-Ubiquitin were treated with Dox or CPT in the presence of MG-132, and immunoprecipitated with α-Topo1 antibody. Relative intensities of ubiquitinated Topo1 in the immunoprecipitates are shown. **d** Vpr induces SUMOylation of Topo1. Immunoprecipitation using α-Topo1 was performed in Mit-23 cells transfected with 3×FLAG/6×His-SUMO-1. CPT and Vpr expression induced modification of Topo1 with SUMO-1. Relative intensities of SUMOylated Topo1 in the immunoprecipitates are shown

signal of Topo1 was greatly reduced (Fig. 3a). Moreover, an increase in the level of Topo1-cc was detected by Rapid approach to DNA adducts recovery (RADAR) analysis, in which DNA–protein adducts were specifically recovered in the presence of chaotropic ion and detergents under denaturing conditions (Fig. 3b and Additional file 3: Figure S3a, b) [39]. Vpr-induced accumulation of Topo1-cc was enhanced by treatment with MG-132, which is a proteasomal inhibitor (Fig. 3b). Moreover, the Vpr expression increased the susceptibility of Topo1 to ubiquitination (Fig. 3c) and SUMOylation (Fig. 3d).

To demonstrate the involvement of Topo1 in Vpr-induced DDR, we performed RNA interference (RNAi) experiment (knockdown efficiency is shown in Additional file 4: Table S1). Down-regulation of *Topo1* by siRNA reduced the number of γH2AX-positive cells (Figs. 4a, b, $P = 0.007$), whereas the over-expression of TDP1 decreased the number of Vpr-induced γH2AX-positive cells (Additional file 5: Figure S4a, b). Notably, down-regulation of *DDB1* and *VprBP* dramatically reduced cell-cycle arrest (Fig. 4c, d), whereas that of *Topo1* partially attenuated the Vpr-induced G_2/M arrest. Because accumulation of Topo1-cc can lead to DSB formation [27, 28], we examined whether DSBs were induced in cells under Vpr expression. As shown in Fig. 4e, f, a neutral comet assay, which is a highly sensitive method to detect DSBs, revealed that Vpr induced DSB. In this case, the down-regulation of *Topo1* significantly attenuated Vpr-induced DSB (Fig. 4f).

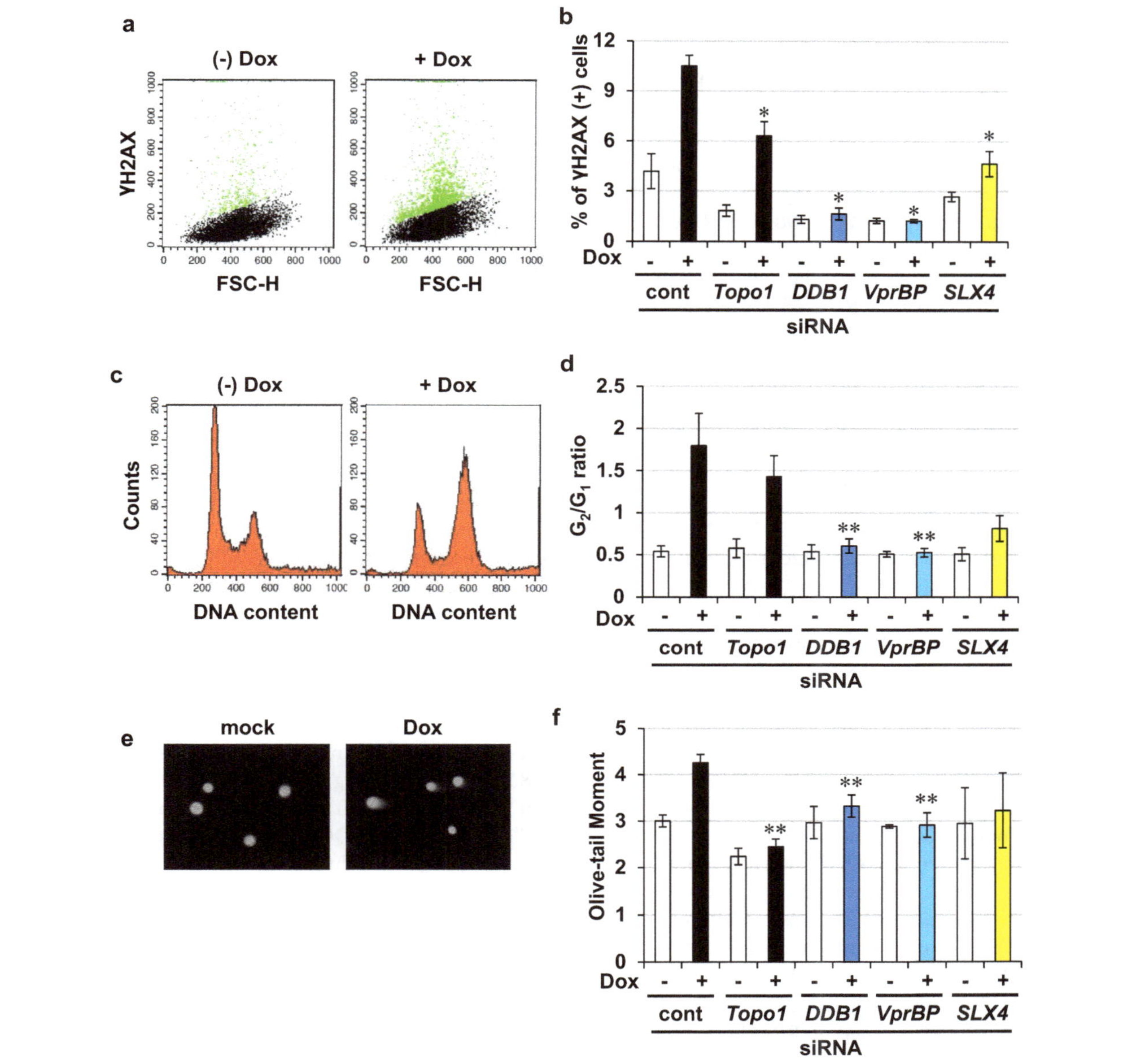

Fig. 4 Vpr expression induces Topo1-mediated DDR and DSB. **a, b** Involvement of Topo1 in Vpr-induced DDR. Flow cytometry analysis was performed in Mit-23 cells transfected with indicated siRNAs to measure the Vpr-induced phosphorylation of H2AX. Representative scatter plots of γH2AX are shown in (**a**). Green colored plots were gated as γH2AX positive cells. Mean percentages of γH2AX positive cells are shown in (**b**). Error bars indicate ± SEM. Data obtained from more than three independent experiments. $^{*}P < 0.01$. **c, d** Effects of *Topo1* siRNA on Vpr-induced cell-cycle abnormality. DNA contents were measured in Mit-23 cells transfected with indicated siRNAs by flow cytometry. Representative histograms of DNA contents (**c**) and average of ratios of G_2 to G_1 cells (**d**) are shown. $^{**}P < 0.05$. **e, f** Involvement of Topo1 on Vpr-induced DSB. Neutral comet assay was performed in Mit-23 cells transfected with indicated siRNAs to quantitate the amounts of DSB. Representative images of each comet (**e**) and average of Olive-tail moment (**f**) are shown

Forced accumulation of Vpr on a targeted DNA locus induces structural alteration of DNA and DDR

To investigate Vpr-induced structural alteration of DNA in the cell, we used U2OS/2-6-3, a human osteosarcoma cell line containing 200 copies of a construct composed of 256 *lac operator* (LacO) repeats at a single locus on 1p36 [40]. This cell line can be used to analyse the molecular events in a specific region of chromosomal DNA. When a fused protein of Vpr and Lac repressor (LacR) was expressed, Vpr was recruited to the LacO repeats (Fig. 5a)

[40–42], which was confirmed by detecting mCherry-focus accumulated on a single distinct region in the nucleus (Fig. 5b, left panel). In these cells, mCherry-positive focus was precisely co-localised with all DDR-related molecules when a fused molecule of Cherry-LacR-Vpr was expressed: the following DDR-related molecules were examined: γH2AX (Fig. 5b, middle panels and Additional file 6: Figure S5a), phosphorylated ataxia telangiectasia mutated (ppATM-Ser1981), ppRPA32-Ser33, ppRAD17-Ser645 and FK2-stained mono- and polyubiquitin conjugates (Additional file 6: Figure S5b–e). On the other hand, Cherry-LacR fused Ovalbumin (OVA), which is a control molecule, generated no DDR signals overlapping with the mCherry-positive focus (Fig. 5b, upper panels and Additional file 6: Figure S5).

We next attempted to detect the structural alteration of DNA in the LacO repeats of U2OS/2-6-3 cells. For this purpose, we first used a psoralen, which is a DNA-intercalating cross-linker used for detecting regions containing negatively supercoiled DNA [43, 44]. Cells were incubated with a biotinylated psoralen and irradiated with UV to cross-link with DNA. Then, the psoralen-conjugated DNA fragments were recovered by a pull-down procedure with streptavidin beads. A qPCR analysis of the recovered DNA (Fig. 5a, primers shown as arrows below the boxes) revealed that the expression of LacR-fused Vpr-Wt increased the amount of psoralen-bound DNA corresponding with the LacO repeat region (Fig. 5c, $P = 0.0058$). Surprisingly, the Q65R mutant also increased psoralen binding to the LacO repeats (Fig. 5c, $P = 0.034$), whereas Ct4RA did not.

Further investigation of RPA70 loading on the LacO repeat region using a chromatin immunoprecipitation (ChIP) assay revealed that the expression of Cherry-LacR fused Vpr-Wt significantly increased RPA70 loading (Fig. 5d). In contrast, both Q65R and Ct4RA mutants increased RPA70 loading, but less potently than Vpr-Wt, which implies the dependence of RPA70 loading on the dual functional properties of Vpr; that is, DNA binding through the C-terminal region of Vpr and ubiquitination defective in the Q65R mutant. To test this notion, we monitored RPA70 loading with the Q65R/Ct4RA double mutant and observed that this mutant completely lost the ability to load RPA70 (Fig. 5d, $P = 0.22$). To evaluate the level of TopoI-cc, we recovered DNA without cross-linking under denaturing conditions, as in the RADAR analysis, and subjected the samples to ChIP assay done with the α-TopoI antibody. Vpr-Wt significantly induced TopoI-cc accumulation on the LacO repeat region, whereas the Q65R and Ct4RA mutants did not (Fig. 5e, $P = 0.049$ for Vpr-Wt). As both Q65R and Ct4RA were severely defective in TopoI-cc accumulation, the coordinated functions of Vpr were required for provoking such downstream events.

We next quantified DSBs at the region where Vpr accumulated using a ligation-mediated (LM)-PCR (see Methods and Additional file 7: Figure S6) [41]. Surprisingly, Vpr-Wt induced DSBs at a level comparable with that by Fok1 endonuclease, which was used as a positive control (Fig. 5f). In contrast, the DSB level was not elevated in the Q65R and C-terminal mutants of Vpr. Requirement of TopoI for Vpr-induced DSBs was confirmed also by a Cre/loxP-mediated Cherry-LacR-Vpr expression system, in which expression of Cherry-LacR-Vpr in U2OS/2-6-3 cells is induced by Cre expression (Additional file 8: Figure S7a, b). Obtained data again showed that down-regulation of *TopoI* suppressed TopoI-cc and DSB formation (Additional file 8: Figure S7c, d).

Vpr-mediated chromatin remodelling is required for RPA70 loading

For elucidating the functional link between Vpr-dependent ubiquitination and RPA70 loading, we measured the mobility changes of histone H2B using a fluorescence recovery after photo-bleaching (FRAP) assay [45]. Intriguingly, Vpr expression enhanced the recovery of H2B-GFP after photo-bleaching (Fig. 6a, $P = 0.018$). Moreover, this mobility change was reduced when H2B was mutated to a non-ubiquitinated form (K120R) (Fig. 6b, right column). Consistently, we observed higher levels of ubiquitination of H2B in cells expressing Vpr-Wt (Fig. 6c, lane 4), but not in cells expressing Q65R, R77Q, R80A and Ct4RA (Fig. 6c, lanes 6, 8, 10 and 12). Notably, the enhanced mobility of H2B-GFP was not detected in cells expressing the Q65R mutant (Fig. 6d, right column), implying that the reduction of RPA70 loading was caused by defective chromatin remodelling by Q65R mutant. Consistently, treatment with trichostatin A (TSA), a HDAC inhibitor that opens chromatin [46], successfully recovered RPA70 loading in the Q65R-expressing cells (Fig. 6e).

Forced accumulation of Vpr induces proviral DNA integration at the targeted chromatin region

To examine the association between Vpr-induced DSB and viral integration, we analysed HIV-1 integration in the Vpr-accumulated LacO region using qPCR (Fig. 7a and see "Methods" and Additional file 9: Figure S8). The frequency of proviral DNA integration in LacO repeats increased when LacR-fused Fok1 or Vpr-Wt was co-expressed at the time of HIV-1 infection (Additional file 10: Figure S9, Additional file 11: Figure S10). For demonstrating the direct effects of virion-associated Vpr, we prepared lentiviral particles composed of defective integrase (IN-D64A) and Cherry-LacR-fused Vpr (CLV) or Cherry-Vpr (CV) (Fig. 7b), infected them into U2OS/2-6-3 cells and performed qPCR. When CLV

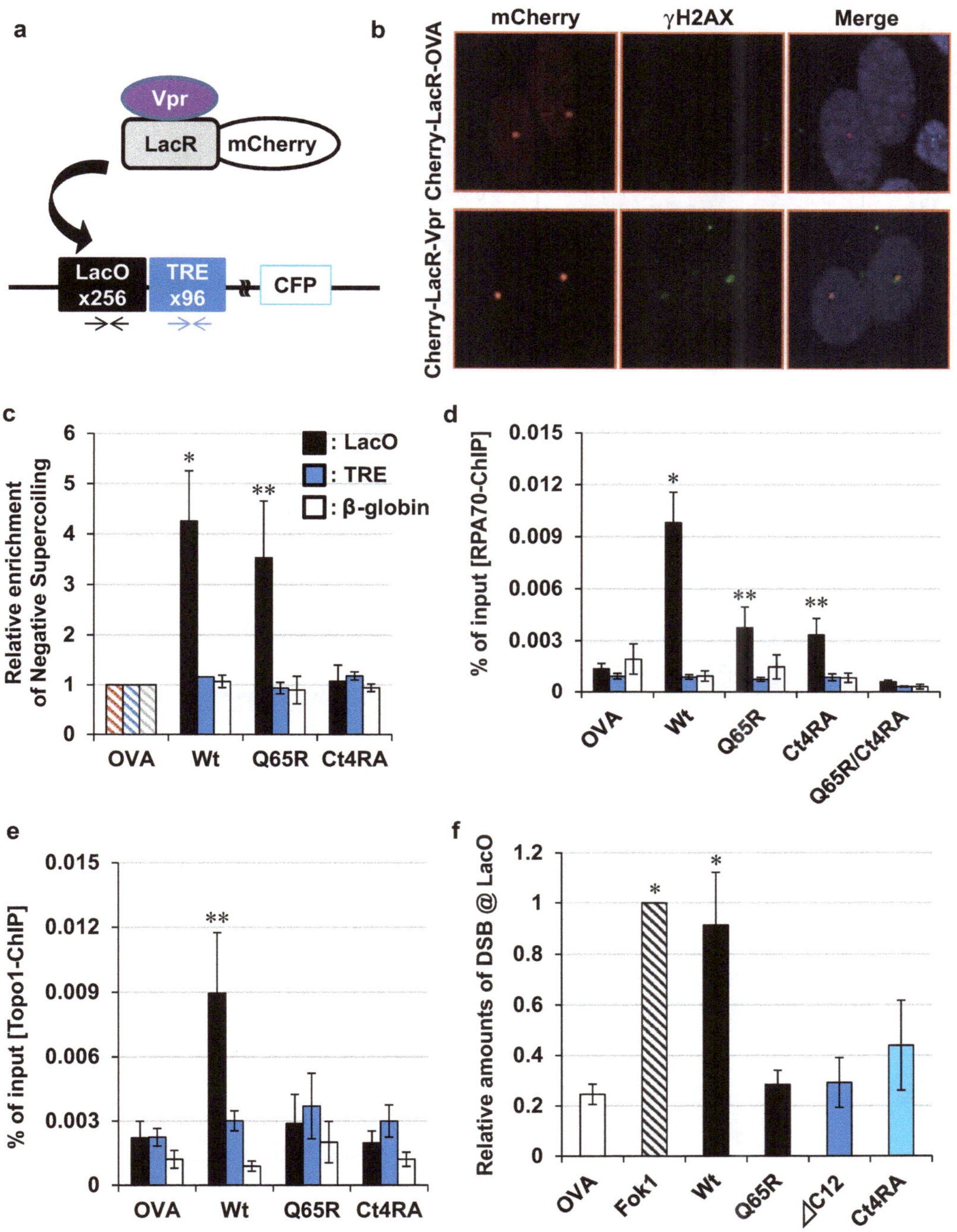

Fig. 5 Forced accumulation of Vpr induces DDR in the targeted region. **a** Schematic of experiments using the LacO/LacR system. In U2OS/2-6-3 cells, 200 copies of p3216PECMS2β, which contains 256 copies of the LacO sequence and 96 copies of the tetracycline response element (TRE) upstream of the CFP coding region are integrated into a single site. In this cell lines, the LacR-Vpr fusion is forcibly recruited to the LacO repeat region (curved arrow). Arrows indicate PCR primers used for amplification of the LacO (red) and TRE (blue) regions (nucleotide sequence is shown in Additional file 20: Table S3). **b** Co-localised signals of Cherry-LacR-Vpr and DDR marker. Phosphorylated H2AX (γH2AX) was detected as a representative DDR marker after 2 days of transfection with indicated construct. **c** Formation of negatively supercoiled DNA in the Vpr-accumulated region. Two days after transfection, a psoralen-bound DNA was recovered and subjected to qPCR analysis with the primers shown in Fig. 5a. The relative copy numbers of LacO (red column), TRE (blue column), and β-globin (as a negative control; blank column) are shown. Error bars indicate ± SEM. **P* < 0.01; ***P* < 0.05. **d** Enhanced loading of RPA70 by Vpr. Loading of RPA70 was quantitated by ChIP assay. **e** Formation of Topo1-cc by Vpr. ChIP assay was performed without cross-linking. MG-132 was treated for 2 h at 50 μM. **f** DSB induction by Vpr. LM-PCR was performed to measure the amounts of DSB-ends in the LacO repeat region. Each column indicates the amount of DSBs relative to LacR-fused Fok1 (positive control)

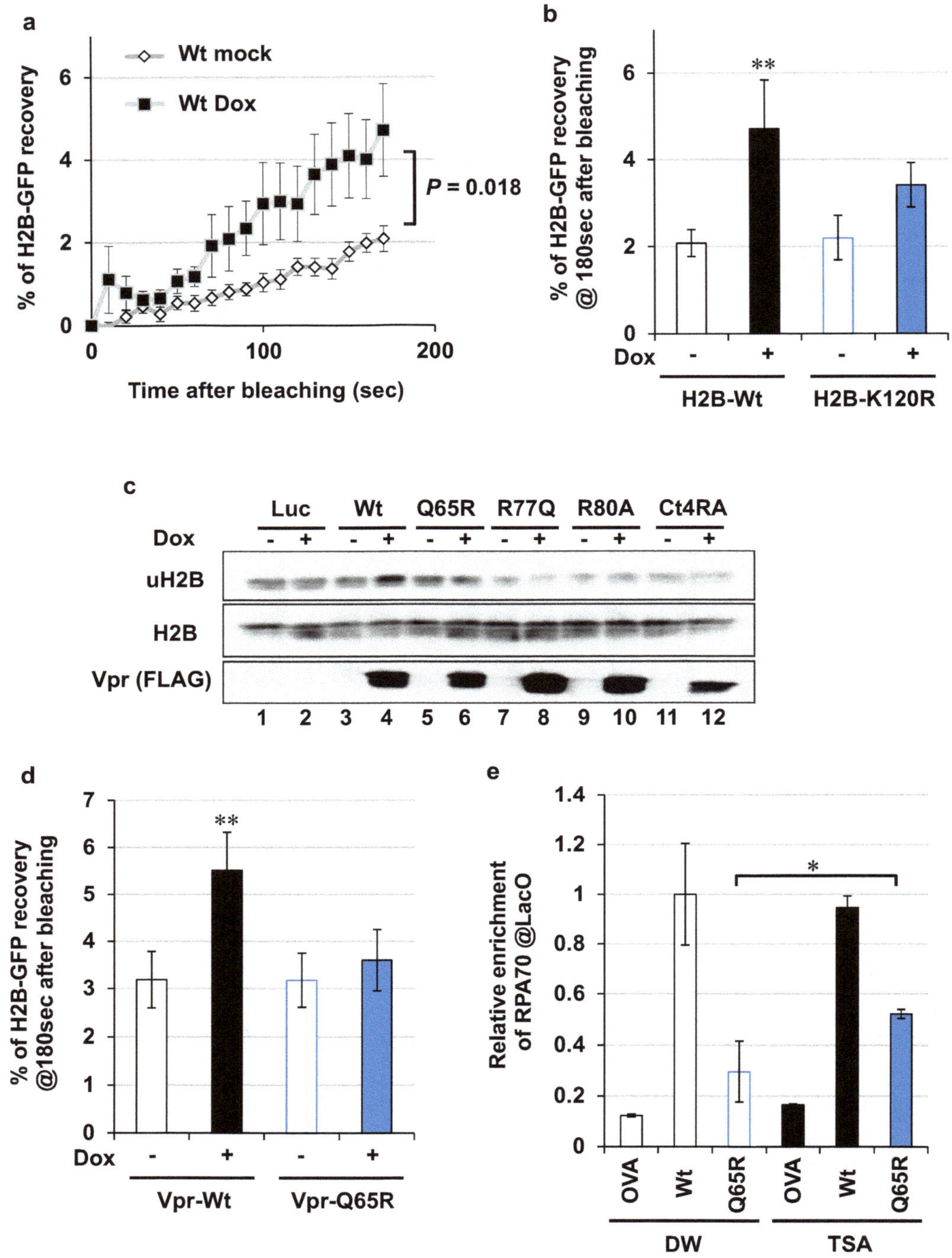

Fig. 6 Vpr induces chromatin remodelling through histone H2B ubiquitination. **a** Increased mobility of H2B under Vpr expression. The FRAP assay was performed using Mit-23 cells with H2B-GFP. The collected values for the recovery rate of each GFP signal were subjected to statistical analysis. Error bars indicate ± SEM. **b** Vpr-induced mobilization of H2B depends on ubiquitination. Mean percentages of H2B-GFP recovery after 180 s of photo-bleaching were compared. $^{**}P < 0.05$. **c** H2B was ubiquitinated by Vpr-Wt expression. For this experiment, we newly established cell lines (HT1080vRxt-Vpr), in which expression of Vpr-Wt (lanes 3 and 4), Vpr-Q65R (lanes 5 and 6), Vpr-R77Q (lanes 7 and 8), Vpr-R80A (lanes 9 and 10) and Vpr-Ct4RA (lanes 11 and 12) can be controlled by the tetracycline promoter, and ubiquitination of H2B on K120 was examined after Dox treatment (5 µg/ml, 2 days). As a negative control, a clone expressing Luciferase (Luc) was also included (lanes 1 and 2). Total H2B and Dox-induced expression of Vpr were detected. **d** Mobility of H2B was not increased by the Q65R mutant of Vpr. Graph shows mean percentages of H2B-GFP recovery after 180 s of photo-bleaching. **e** Defect of the Q65R mutant for RPA70 loading was rescued by TSA. After treatment with TSA (50 nM, 16 h), RPA70 loading onto the LacO repeats was quantitated by ChIP assay. $^{*}P < 0.01$

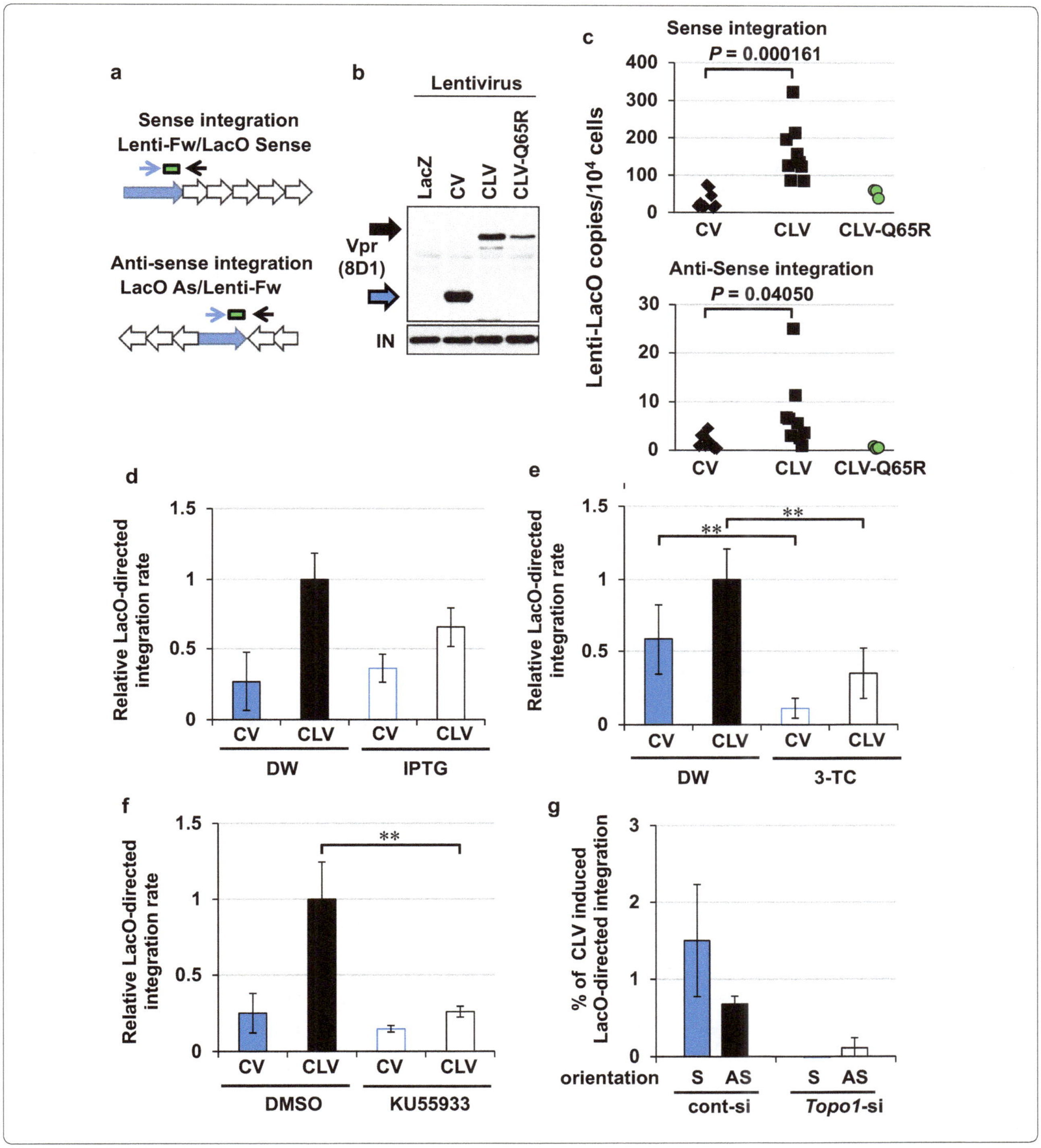

virus was infected, the frequency of proviral DNA integration into the LacO repeats significantly increased without marked effects on the overall viral infectivity (Fig. 7c and Additional file 12: Figure S11). In striking contrast, the infection of CLV-Q65R virus, a lentivirus with Cherry-LacR fused to Vpr mutant of Q65R, did not induce the LacO-directed integration, indicating that Vpr-induced ubiquitination and DSB is required for these integrations (Fig. 7b, c). The frequency of Vpr-induced LacO-directed integration was reduced by isopropyl β-D-1-thiogalactopyranoside (IPTG), which blocks LacO/LacR binding (Fig. 7d), as well as

(See figure on previous page.)
Fig. 7 Proviral DNA integration in Vpr-accumulated sites. **a** Schematic of LacO-directed integration of proviral DNA. The qPCR analysis of the copy number of LacO-integrated proviral DNA was performed using PCR primers targeting the 3′-LTR (blue arrow) and LacO repeat (red arrow). The green box indicates the position of the TaqMan probe for the 3′-LTR (nucleotide sequence is shown in Additional file 20: Table S3). **b** Production of lentiviral particles with CV, CLV, and CLV-Q65R. Integrase (IN) was detected as an internal control. *LacZ*-coding lentivirus (LacZ) was included as a negative control. **c** Site-specific integration of proviral DNA. CV, CLV, and CLV-Q65R incorporated lentivirus was infected into U2OS/2-6-3 cells and subjected to qPCR analysis (Fig. 7a) at 2dpi (n = 9, 9, 3 for CV, CLV, and CLV-Q65R, respectively). Similar infectivity of each virus was confirmed by colony-formation assay (Additional file 12; Figure S11). **d** Effects of LacO/LacR inhibition on site-specific integration of proviral DNA. Cells were pretreated with IPTG (15 mM, 1 h) prior to the lentivirus infection. Error bars indicate ± SEM. Data were obtained from more than three independent experiments. **e** A reverse transcriptase inhibitor blocked the infection. Cells were pretreated with 3-TC (50 μM, 1 h) prior to lentivirus infection. $^{**}P < 0.05$. **f** Site-specific integration of proviral DNA depended on ATM activity. ATM inhibitor, KU55933 (10 μM) was added to culture medium 1 h before infection. **g** Topo1 is important for site-specific integration of proviral DNA. A U2OS/2-6-3 subclone transduced with pCAL-loxP-CLV (263/loxP-CLV) was infected with Cre- or LacZ-expressing adenovirus for 2 days under down-regulation of *Topo1*. The cells were then infected with NL4-3/D64A/R− virus and subjected to qPCR analysis at 2dpi. The overall integration rate was quantitated by *Alu-gag* two-step nested qPCR (Additional file 11: Figure S10) to estimate the percentage of LacO-directed integration. *S* sense integration, *AS* anti-sense integration

by the reverse-transcriptase inhibitor 2′,3′-dideoxy-3′-thiacytidine (3-TC) (Fig. 7e). In addition, these site-directed integrations were abrogated by the treatment with an ATM inhibitor (KU55933) (Fig. 7f), which is consistent with the results of a previous work showing that ATM activity is required for DSB-directed integration [8]. Moreover, RNAi experiment indicated that proviral DNA is integrated in the vicinity of Vpr-accumulated sites in a Topo1-dependent manner (Fig. 7g).

We also explored the lentiviral integration sites using linear amplification mediated (LAM)-PCR for estimating the frequency of selective integration into the LacO repeats (Additional file 13: Figure S12). Two out of 96 analysed samples (2%) contained LacO-dependent integrations in the genome.

Topo1 requirement for Vpr-mediated enhancement of viral infection in non-dividing cells

Lastly, we examined the role of Vpr-induced DSB and Topo1 in the viral infection of resting macrophages. First, we prepared Vpr-proficient (R+) and deficient (R−) viruses, infected them into MonoMac-6 (MM-6) cells and applied the cells to neutral comet assay. In this experiment, we used IN-D64A mutant virus to exclude the effects of IN-dependent DSB [47, 48]. MM-6 is a monocyte leukaemia-derived macrophage-like cell line that can differentiate into resting macrophages following phorbol 12-myristate 13-acetate (PMA) treatment. A neutral comet assay revealed that infection of the R+ virus provoked DSBs in differentiated MM-6 cells (Fig. 8a). Furthermore, RNAi experiments revealed that DSB induction by the R+ virus was Topo1-dependent (Fig. 8b and Additional file 4: Table S1 for knockdown efficiency). Moreover, the addition of rVpr in the culture medium of MM-6 cells also induced DSBs that depended on *Topo1*, *DDB1* and *SLX4* (Fig. 8c). Notably, down-regulation of *Topo1* significantly suppressed the viral infectivity of the R+ virus, although the originally low infectivity of R− was unaffected (Fig. 8d, R+, $P < 0.0001$; R−, $P = 0.78$ and Additional file 14: Figure S13 for other sets of Topo1 siRNA). Vpr-dependent upregulation of viral infectivity was confirmed using luciferase assay (Additional file 15: Figure S14a) and EGFP expression (Additional file 15: Figure S14b). Similarly, down-regulation of *DDB1* and *SLX4* also led to decrease in the viral infectivity of the R+ virus. Finally, we performed similar experiments using primary monocyte-derived macrophages (MDMs) prepared from two healthy donors and confirmed that Vpr-dependent increase in viral infection depended on Topo1 (Fig. 8e and Additional file 4: Table S1 for knockdown efficiency).

Discussion

Studies aimed at elucidating the mode of Vpr-induced DDR identified multiple Vpr-interacting cellular factors; however, the initial triggering event remains elusive. Here we presented evidence that structural alteration of DNA is the most upstream event of Vpr-induced DDR. Interestingly, our data suggest that structural alteration of DNA also induce DSB and contribute to HIV-1 infection in resting macrophages.

First, AFM observations revealed that supercoiled DNA shifted to a relaxed form when rVpr was added to DNA in an aqueous solution. This structural alteration was confirmed with the detection of negatively supercoiled DNA using the supercoiling assay with *E. coli* Topo1. Previous reports demonstrating that negative supercoiling of dsDNA occurs when RNA polymerase unwinds dsDNA or the nucleosome assembles on dsDNA [49, 50] together suggest that similar change in the topological configuration of dsDNA was induced by Vpr. This idea was also supported by Vpr-dependent loading of the ssDNA-binding proteins onto dsDNA.

Vpr-induced structural alteration of DNA was also observed in vivo in an experimental series using the LacO/LacR system in U2OS/2-6-3 cells. When a chimeric fusion protein of Vpr and LacR accumulated on the LacO region, we observed the formation of activated

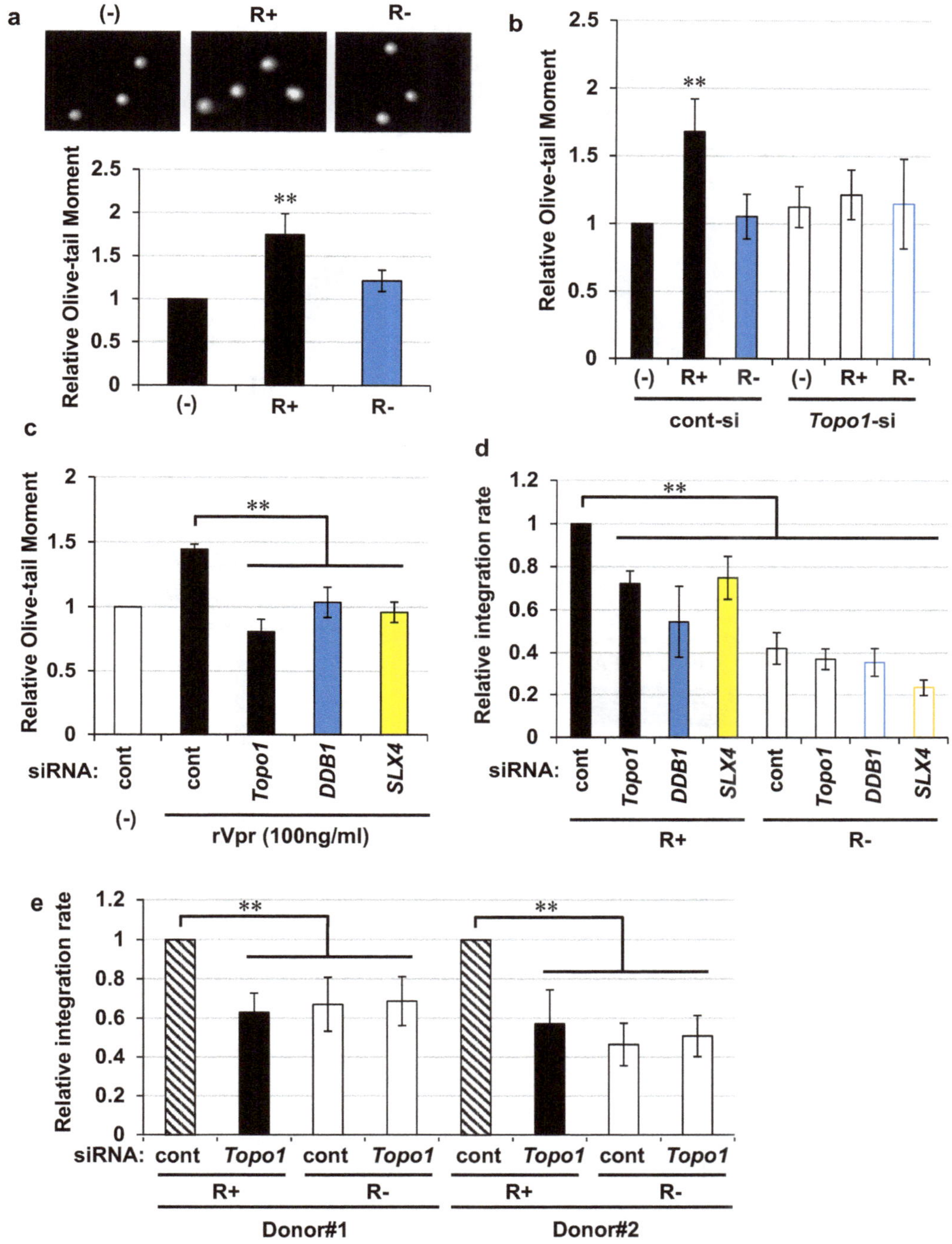

Fig. 8 Vpr-induced DSB stimulates viral infection in resting macrophages. **a** Vpr induced DSBs in resting macrophages. Differentiated MM-6 cells were infected with NL4-3/D64A virus for 1 day, and then subjected to neutral comet assay. Representative images of each comet are shown (upper panel). The relative Olive-tail moment was evaluated (lower panel). Error bars indicate ± SEM calculated based on data obtained from more than three independent experiments. Data from non-infected cells (−), Vpr-proficient virus (R+), and Vpr-deficient virus (R−) are shown. $^{**}P < 0.05$. **b** Involvement of Topo1 in Vpr-induced DSB. Neutral comet assay was performed as in **a** after 3 days of *Topo1* targeting siRNA transfection. **c** DSB induced by rVpr was blocked by down-regulation of *Topo1*. Three days after transfection of indicated siRNA, cells were treated with rVpr (100 ng/ml, 16 h). **d** Integration of proviral DNA was blocked by down-regulation of *Topo1*. MM-6 cells were infected with Vpr-proficient (R+) or deficient (R−) NL4-3 virus after transfection with siRNA, and the integration rate was quantitated by *Alu-gag* two-step nested qPCR at 2dpi; relative integration rates are shown. **e** Vpr-dependent increase of viral infectivity required Topo1 in MDMs. MDMs from two healthy donors were infected with Vpr-proficient (R+) or deficient (R−) NL4-3 virus 2 days after transfection with *Topo1* targeting siRNA, and the integration rates were quantitated as in **d** at 2dpi

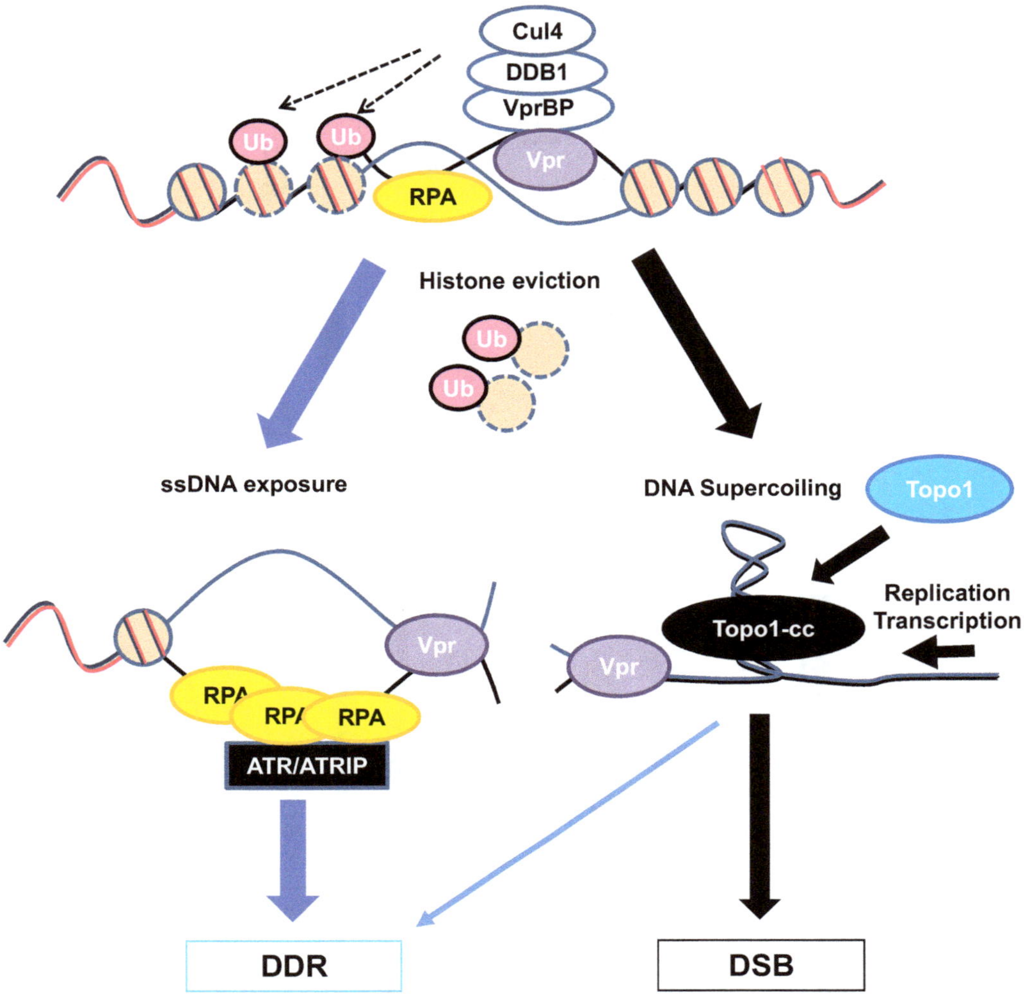

Fig. 9 Hypothetical model of DDR and DSB induction by Vpr. Vpr unwinds dsDNA and allows limited loading of RPA70. Simultaneously, Vpr induces ubiquitination of histone H2B, and histone eviction occurs in the vicinity. Chromatin remodeling by histone eviction promotes efficient loading of RPA70, leading to G_2/M checkpoint activation by ATR (left side). Vpr-induced unwinding of dsDNA in turn causes accumulation of supercoiling of DNA and formation of Topo1-cc (right side). In conjunction with DNA replication or transcription, Topo1-cc induces DSB formation, and proviral DNA is integrated at the DSB sites

forms of multiple DDR-related molecules, chromatin loading of RPA70, accumulation of negative supercoiling, generation of Topo1-cc and DSBs in the same region foci (Fig. 5 and Additional file 6: Figure S5). RPA70 loading onto chromatin by Vpr-induced unwinding of dsDNA effectively explains the mechanism of Vpr-induced DDR.

Notably, our data suggested that Vpr-induced RPA70 loading was modulated at least two steps (Fig. 9). In the first step, the DNA-binding activity of Vpr changes the superhelicity of the DNA and partially unwinds dsDNA. This function was demonstrated using in vitro experiments (Fig. 1c, d), which also revealed the importance of the positively charged amino acids in the C-terminal stretch. In the second step, Vpr-mediated ubiquitination is required for chromatin remodelling, which enables the recruitment of cellular factors, including RPA70, to chromatin. This step was demonstrated via experiments using well-characterised Q65R mutant, which is defective in the ubiquitination process [21, 22]. Intriguingly, when the Q65R mutant was forced to accumulate on the LacO repeats, it could not stimulate RPA70 loading (Fig. 5d), although the RPA70 loading defect of the Q65R mutant was complemented by TSA (Figs. 5e, 6e). To elucidate this phenomenon, we investigated whether Vpr modulated chromatin remodelling and induced ubiquitination of histone H2B [51, 52]. Interestingly, the FRAP assay revealed that the recovery of H2B after photobleaching was more rapid in cells expressing Vpr-Wt than in those expressing the Q65R mutant. Similarly, a non-ubiquitinated H2B mutant exhibited reduced mobility in the Vpr-expressing cells. Moreover, the ubiquitination of H2B was promoted by Vpr-Wt, but not by the Q65R mutant. Considering that Cul4 regulates H2B ubiquitination for facilitating the DDR [53] and that the Q65R mutant is ubiquitination-defective due to its inability to bind DDB1/VprBP [21, 22], it is plausible that H2B is a target of Vpr-mediated ubiquitination, which is critical for Vpr-induced RPA70 loading. In addition to histone H2B, the association of Vpr with several chromatin modification factors, including p300, SNF2 h, NuRD and HDAC1 [38, 54–56], may also contribute to the efficient reorganization of chromatin. Our data suggested that the concerted actions of structural alteration of DNA and chromatin remodelling are required for efficient RPA70 loading by Vpr.

Notably, we found that Vpr-induced structural alteration of DNA could induce both DDR activation and DSB formation. In the latter process, Topol played a key role in creating DNA breaks (Figs. 4f, 8b, c; Additional file 8: Figure S7d), albeit its subtle contribution to Vpr-induced DDR activation (Fig. 4d). We also detected Vpr-induced DSB in resting macrophages in a Topol-dependent manner (Fig. 8b, c), which suggested that Topol-cc-mediated DSB could also arise in non-proliferating cells through interference with transcription [28]. Here, although we focussed on Topol, we could not exclude the possibility of involvement of the other DNA structure modification factors such as Topo2, structure-specific nucleases including SLX4-Mus81/Eme1, and nucleotide excision repair (NER) factors in the processing of Vpr-induced aberrant DNA structures. Consistent with this possibility, we observed that the down-regulation of *SLX4* or *XPG* (a nuclease acting in NER) by siRNA suppressed Vpr-induced DSB (Fig. 4f and Additional file 16: Figure S15c, d). In addition, these factors also function in Topol-mediated DSB induction. SLX4 causes DSBs through Topol-bound DNA strand incision during the replication process [57, 58] and XPG induces DSBs through the processing of R-loops (DNA/RNA hybrids), which are formed when Topol inhibition blocks the transcription [59, 60]. In resting macrophages, it is plausible that R-loop formation is a major trigger event of Vpr-induced DSBs. Consistently, we observed an increase in R-loops at the Vpr-accumulated region (Additional file 17: Figure S16) and a decrease in Vpr-induced γH2AX-positive cells by the over-expression of RNaseH1, an RNase that degrades R-loops (Additional file 5: Figure S4).

We confirmed the requirement of SLX4 for Vpr-induced DDR, which was in concordance with the report of Laguette et al. [14], but in contrast with another study that reported SLX4 to be dispensable [61]. These discrepant observations are attributable to the different levels of residual activity of SLX4 under down-regulation by siRNA and gene disruption. In addition, differentially reorganised DNA repair pathway in SLX4 deficient cells leads to different cellular responses to Vpr.

Several lines of evidence indicate that DSBs increase the efficiency of HIV-1 infection, especially in resting cells [8, 18, 62, 63]. The finding that Vpr-induced DSBs are important for viral infection in resting macrophages effectively explains the previous observations that Vpr is required for the infection of resting cells [5–8]. Furthermore, the observation that DSB sites are directly targeted by proviral DNA integration supports the relevance of Vpr-induced DSB in viral infection [8].

The mechanism by which Vpr induces structural alteration of DNA remains unclear. In this study, we analysed DNA supercoiling using recombinant Vpr and Vpr-derived peptides, which suggested the necessity of C-terminal stretch of Vpr for DNA unwinding. Notably, we also detected that a higher concentration of Vpr-derived peptides (0.15–1.5 Vpr-peptides/bp) resulted in strongly underwound DNA (Additional file 18: Figure S17) in the DNA supercoiling assay. In a previous electron microscopic study, higher concentrations of Vpr (at least 4.5 Vpr-peptides/bp) induced the aggregation of DNA [26]. Because plectonemic or toroidal DNA structures are the compacted forms of negatively supercoiled DNA, it is possible that our current findings reflect the same phenomena as that reported in a previous study [26]. Siddiqui et al. [64] reported that Vpr-induced DDR required helical domain II (37–50 aa), but not the C-terminal region. These contradictory observations in the function of the C-terminal region of Vpr could be derived from the methods adopted for evaluating the Vpr-induced DNA damages. Siddiqui et al. [64] assessed the chromosomal abnormalities as Vpr-induced DNA damages, whereas we measured DSB itself by neutral comet assay and LM-PCR (Figs. 4f, 5f).

HIV treatment with cART can effectively suppress viral replication. However, complete eradication of HIV has not been successfully achieved, largely due to long-lived reservoir cells [1]. Recent observations indicate that viral sanctuary sites are distributed in various organs, including in the gut-associated lymphoid tissue and the central nervous system, in which persistent infections are observed [2, 3]. Vpr has reportedly been detected in the cerebrospinal fluid of HIV-positive patients [65, 66]. Notably, Topol-mediated DNA damage is responsible for several neurodegenerative disorders [67], suggesting that Vpr may be involved in the development of HIV associated neurocognitive disorder

[68] by exacerbating Topo1 insults. Our results provide a rationale for developing anti-Vpr compounds, which can significantly contribute to the improvement of the therapeutic regimen for HIV-positive patients under the current cART.

Conclusions

We discovered that Vpr induces DNA structural alteration as an initial trigger of DDR and DSB. Notably, Vpr modulates chromatin remodelling by ubiquitinating histone H2B, and it facilitates RPA70 loading on unwound supercoiled DNA. We believe that our findings provide an answer to the long-standing question on how Vpr facilitates the infection of HIV-1 into macrophages.

Methods

Key resources (cell lines, chemicals, reagents, antibodies, recombinant DNA)

List of resources is included in Additional file 19: Table S2.

Cell lines and cell culture

Cell lines used were MonoMac-6 (MM-6) (DSMZ), HEK293T (RIKEN Cell Bank), HT1080 (the Healthy Science Research Resources Bank), *Δ*Vpr, Mit-23 (derived from HT1080) [38], U2OS/2-6-3 (provided by Dr. David Spector, Cold Spring Harbor Laboratory, Cold Spring Harbor) [40], U2OS/2-6-3+pCAL-loxP-CLV (263/loxP-CLV), HT1080vRxt (retroviral RetroX-tet-ONE)-Luc or -Vpr, Mit-23+H2B-GFP and HT1080vRxt-Vpr+H2B-GFP. We confirmed no contamination of mycoplasma by periodically checking with Hoechst-33258 (Sigma-Aldrich) staining. MM-6 cells were cultured in RPMI 1640 with 10% fetal bovine serum (FBS) (Gibco) supplemented with 1% NEAA (Gibco) and 1% OPI media (Sigma). For differentiating into macrophage-like state, MM-6 cells were treated for 3 days with 50 nM of PMA. HEK293T and HT1080 cells were cultured in DMEM with 10% FBS, and *Δ*Vpr and Mit-23 cells were cultured in DMEM with 10% tetracycline free FBS (Hyclone) supplemented with HygromycinB (12.5 μg/ml) and G418 (100 μg/ml). U2OS/2-6-3 cells were cultured in DMEM with 10% FBS supplemented with HygromycinB (12.5 μg/ml). 263/loxP-CLV cells, U2OS/2-6-3 derived newly established cell line stably transduced with pCAL-loxP-CLV, were cultured in DMEM with 10% FBS supplemented with HygromycinB (12.5 μg/ml) and G418 (100 μg/ml). HT1080vRxt-Luc or -Vpr cells, HT1080 derived newly established cell lines stably transduced with retroviral RetroX-tet-ONE-Luc or -Vpr, were cultured in DMEM with 10% tetracycline free FBS supplemented with puromycin (1 μg/ml). All cell lines stably transduced with pEF-Bos-H2B-GFP (+H2B-GFP) were maintained in BlasticidineS (2 μg/ml) containing medium. All cells were grown at 37 °C in humidified atmosphere containing 5% CO_2.

Preparation of monocyte derived macrophages (MDMs)

Experimental procedures were approved by the internal review board. Peripheral blood mononuclear cells (PBMCs) of healthy volunteers who gave informed consent were isolated by Ficoll density gradient separation using Lymphoprep (Axis-Shield). Monocytes were isolated using monocyte isolation kit II (Miltenyi) by depletion of non-monocytes. For preparation of MDMs, isolated monocytes were cultured for 7 days in RPMI1640 supplemented with 10% FBS in the presence of 25–50 ng/ml of recombinant human macrophage-colony stimulating factor (rhM-CSF) (R&D).

Bacterial strains

DH5α chemical competent *E. coli* (TOYOBO) was used for standard transformation. For the transformation of retrovirus, lentivirus, and HIV-1 vector DNA, Stbl3 chemically competent *E. coli* (Life Technologies) was used. In the case of protein expression, BL21DE3 (Stratagene) was used for transformation.

rVpr purification from Wheat germ extract

By Wheat germ extract (WGE) cell free transcription/translation system (CellFree Sciences), FLAG-Strep-tag2 (F/S) fused to the N-terminus of Vpr (pNL4-3, GenBank: AF324493.2) was expressed by pEU-F/S-Vpr as template. After the protein synthesis, WGE was diluted in 3× volume of StA binding Buffer (50 mM Tris–Cl [pH 8.0], 500 mM NaCl, 1 mM EDTA, 1% NP40, 0.5% Brij-35) supplemented with RNaseA (100 μg/ml). After the incubation with Strep-Tactin agarose beads overnight at 4 °C on rotating platform, beads were washed with StA binding buffer, and eluted by 1×SA elution buffer (Calbiochem) supplemented with 1% NP40. The eluate was incubated with FLAG-M2 agarose beads overnight at 4 °C on rotating platform. The beads were extensively washed with phosphate buffer saline (PBS) (−), and eluted by 0.1 M HEPES [pH 2.5] and immediately neutralized by 1 M HEPES [pH 8.0], and adjusted to 50 mM of NaCl. The concentration of Vpr in the eluted fraction was determined by α-Vpr ELISA (MBL) [13] or Coomassie Brilliant Blue (CBB) gel staining compared to known amounts of Bovine serum albumin (BSA). The rVpr produced by WGE was used in experiments except Figs. 1c, 2c. Comparable activity of F/S-Vpr with rVpr from *E. coli* was confirmed by neutral comet assay.

Purification of rVpr from *E. coli*

rVpr was purified as previously reported [13] with some modification. Briefly, BL21DE3 codon plus was transformed with pGEX-Vpr, in which Vpr was fused to C-terminus of GST followed by PreScission protease cleavage site, and cultured in 2×YTG supplemented with 100 μg/ml Ampicillin (2×YTG-Amp) over night at 30 °C. On the next day, the culture was inoculated to 1:50 ratio into fresh 2×YTG-Amp, and incubated at 37 °C until OD600 reached to 0.8. And then, 1 mM of IPTG was added to the bacterial culture medium, and further incubated for 3 h at 25 °C. Collected pellet from 500 ml culture was suspended in 20 ml of Binding buffer (20 mM sodium-phosphate buffer [pH 7.6], 150 mM NaCl, 0.5% TritonX-100, 10% glycerol) supplemented with 1 mM PMSF. After sonication and centrifugation, lysate was filtered through 5 μm filter and incubated with 500 μl of glutathione Sepharose (GSH) beads overnight at 4 °C on rotating platform. On the next day, beads were sequentially washed with Binding buffer supplemented with 500 mM NaCl or 1% TritonX-100, and Cleavage buffer (50 mM Tris–HCl [pH 7.0], 150 mM NaCl, 1 mM EDTA, 10% glycerol) twice, and then suspended in Cleavage buffer supplemented with 1 mM dithiothreitol (DTT) and 20 μl of PreScission protease. After overnight incubation on rotating platform at 4 °C, GSH beads were collected (Vpr remains bound to GSH beads), and washed with Binding buffer. Then, Vpr protein was eluted with Binding buffer supplemented with 0.1% TritonX-100. The eluate was incubated with new GSH beads to remove the un-cleaved GST-Vpr and PreScission protease for 1 h, and then supernatant was incubated with α-Vpr-antibody (8D1) conjugated CNBr-agarose beads overnight at 4 °C on rotating platform. The beads were extensively washed with PBS (−), and Vpr was eluted by 0.1 M HEPES [pH 2.5] and immediately neutralized by 1 M HEPES [pH 8.0]. The concentration of Vpr in eluted fraction was determined by α-Vpr ELISA [13]. The rVpr expressed in bacteria was used in experiments for Figs. 1c and 2c.

rRPA70 purification

BL21DE3 was transformed with pET15-6×His-RPA70 (in which hRPA70 was tagged with hexa-histidine followed by PreScission protease cleavage site on N-terminus) and cultured in 2×YTG-Amp overnight at 30 °C. On the next day, the culture was inoculated to 1:50 ratio into fresh 2×YTG-Amp, and incubated at 37 °C until OD600 reached to 0.8. And then, 1 mM of IPTG was added to bacterial culture medium, further incubated overnight at 16 °C. Collected pellet from 500 ml culture was suspended in 20 ml of Buffer-A (50 mM Tris–HCl [pH 7.5], 500 mM NaCl, 0.5% TritonX-100, 10 mM mercaptoethanol, 10% glycerol) supplemented with 1 mM PMSF. After sonication and centrifugation, lysate was filtered through 5 μm filter and incubated with 10 ml of Affi-gel Blue resin (BioRad) overnight at 4 °C on rotating platform. Beads were washed with 20 ml of Buffer-A twice, and eluted with 10 ml of lysis buffer supplemented with 2.5 M NaCl. Eluate was diluted by 4-fold volume of Buffer-B (50 mM Tris–HCl [pH 8.0], 0.5% TritonX-100, 0.625% Empigen, 10% glycerol, 10 mM mercaptoethanol, 12.5 mM imidazole), followed by overnight incubation with Ni–NTA beads (Invitrogen). Beads were washed with Buffer-C (50 mM Tris–HCl [pH 8.0], 1 M NaCl, 0.5% Empigen, 0.5% TritonX-100, 10 mM mercaptoethanol, 10% glycerol, 10 mM imidazol) and Buffer-D (50 mM Tris–HCl [pH 8.0], 150 mM NaCl, 1% TritonX-100, 0.5% Empigen, 10% glycerol). Subsequently, beads were further washed with Buffer-E (50 mM Tris–HCl [pH 7.0], 150 mM NaCl, 0.1% Empigen, 10% glycerol) twice, and suspend in Buffer-E supplemented with 1 mM DTT, 1 mM EDTA, and 20 μl of PreScission protease. After overnight incubation on rotating platform at 4 °C, GSH beads were added to the eluate to remove the PreScission protease, and mixed for 1 h, and then supernatant was used as purified rRPA70 fraction. The concentration of RPA70 was estimated on CBB gel staining compared to known amounts of BSA.

Sample preparation for AFM analysis

Fifty ng of pUC18 was incubated with 5 mM of Chlq or 0.5 μM of rVpr (1000-fold excess amount of Vpr to pUC18; 0.37 Vpr molecule/bp) in 50 μl of AFM reaction buffer (10 mM Tris–HCl [pH 8.0], 50 mM NaCl, 1 mM $MgCl_2$) for 30 min at 37 °C. Prior to the binding of DNA, freshly cleaved mica were treated with AFM binding buffer (AFM reaction buffer supplemented with 5 mM $NiCl_2$) for at least 10 min, and quickly blew out by N_2 gas just before sample loading. And then, appropriate amounts of DNA sample (typically 10 μl) was dropped on the Ni^{2+}-treated mica surface and placed for 10 min. Following quick washes of mica surface with AFM binding buffer twice, 100 μl of AFM binding buffer was loaded on mica for in liquid AFM observation.

AFM measurement in aqueous solution

All measurements were carried out with a JPK NanoWizard ULTRA Speed AFM (JPK Instruments) on inverted microscope (IX71, Olympus) equipped with acoustic hood and active vibration isolation (Micro40, Accurion). Ultrashort cantilever (USC-F0.3-k0.3, NanoWorld) was used with 110–120 kHz drive frequency for high-speed, high-resolution imaging in liquid environment. The images were scanned in an intermittent contact mode (AC mode) for 2.0 μm × 2.0 μm area (512 × 512 pixels) at scan rate of 2.0 Hz, Z-range of 1.3 μm. Data processing was performed by JPK SPM Data Processing software (JPK Instruments). For processing images, the same parameters were used for all samples analysed in the same day. To obtain a root mean square of roughness (Rq) value, DNA molecule was manually surrounded by the minimal rectangle. For each treatment, at least 30 randomly selected DNA molecules were subjected to Rq value measurements, and independent experiments were performed more than three times.

RPA70 Pull down assay by DNA-bound beads

At first, 10 pmol of biotinylated 80-mer ss/dsDNA was incubated with 10 μl of Dynabeads M280 Streptavidine (SA-beads) in SA binding buffer (10 mM Tris–HCl [pH 8.0], 1 M NaCl, 1 mM EDTA) at room temperature on rotating platform. After 15 min, beads were washed with SA binding buffer three times, and kept in IP buffer (50 mM Tris–HCl [pH 8.0], 150 mM NaCl, 1 mM EDTA, 1% NP40). Subsequently, 10 pmol of rRPA70 and rVpr was added to suspension of DNA-bound beads, and incubated for 1 h at 4 °C on the rotating platform. Beads were washed with IP buffer three times, and proteins recovered by pull-down procedures (pulled-down proteins) were analysed by WB.

T4gp32 Pull down assay by DNA-bound beads

The beads bound with ss/dsDNA was prepared as above, and beads were suspended in 1×CutSmart buffer (NEB) supplemented with 1% NP40. Subsequently, 50 pmol of T4gp32 (NEB) and 100 pmol of C45 or C45D18 peptide was added to suspension of DNA-bound beads, and incubated for 1 h at room temperature on rotating platform. Beads were washed by 1×CutSmart buffer supplemented with 1% NP40 three times. Pulled-down proteins were separated on SDS-PAGE, followed by staining with Oriole fluorescent gel staining (BioRad), and visualized by LAS400 with UV transilluminator.

Supercoiling assay

Supercoiled plasmid DNA, pBluescriptII (500 ng, Fig. 2b) or pUC18 (100 ng, Fig. 2c), was incubated in 20 μl of 1×CutSmart buffer with 0.025 U of *E. coli* Topoisomerase1 (NEB) and increasing amounts of rVpr (0.71, 2.36, 7.1 pmol and 0.63, 2, 6.32 pmol for Fig. 2b, c, respectively) for 30 min at 37 °C. After that, samples were heat-inactivated for 20 min at 80 °C, and incubated for 20 min at 55 °C with 0.5% SDS and Proteinase K (1 mg/ml) for deproteinisation. Following phenol–chloroform extraction, purified DNA was separated on 0.8% agarose gel, and stained with ethidium bromide or 1×SyBr Gold (Life technologies) (Fig. 2b, c, respectively). Images were captured by GelDoc Ez (BioRad), and the intensity of each topoisomers was analysed by Image Lab software (BioRad).

Immunohistochemistry (IHC)

Immunohistochemistry was performed as a standard protocol. Briefly, cells were fixed with 4% paraformaldehyde, and permeabilized with 0.5% TritonX-100. For the staining of Topo1, we performed pre-extraction with following buffer (20 mM HEPES [pH 7.5], 50 mM NaCl, 300 mM sucrose, 0.1% TritonX-100, 3 mM $MgCl_2$) before fixing. After blocking with 5% skim milk/TBS with 0.1% Tween20, indicated antibodies were used. After that, optimal α-IgG antibody conjugated with Alexa fluorescent dye was used as a secondary antibody. Nuclei were stained by 1 μM of Hoechst-33258 and observed with fluorescent microscope (BX51, Olympus; BZ-X710, Keyence).

DNA–protein covalent complex recovery (RADAR)

DNA–protein covalent complex was recovered according to a protocol previously reported [40]. Briefly, Mit-23 cells were treated with Dox (3 μg/ml) for 1 day or CPT (20 μM) for 1 h. In the case of MG-132 treatment, MG-132 (50 μM) was added to the culture medium 2 h prior to recovery. After treatment, cells were directly lysed in Buffer-M (6 M guanidine thiocyanate, 10 mM Tris–HCl [pH 6.5], 20 mM EDTA, 4% TritonX-100, 1% *N*-lauroylsarcosine, 1% DTT), and ethanol precipitation was performed by addition of half volume of 99% ethanol and centrifugation. Following washing with 70% ethanol twice, pellets were dissolved and sonicated in 8 mM NaOH solution. Amounts of DNA were quantitated by Picogreen (Invitrogen) with InfiniteM1000 PRO plate reader. For slot-blotting, equal amounts of DNA were absorbed to nitrocellurose membrane (BioRad) by BioDot-SF microfiltration apparatus (BioRad) with TBS buffer (10 mM Tris [pH 7.5], 150 mM NaCl). Topo1-cc or dsDNA were detected by respective antibodies.

Immunoprecipitation (IP)

Immunoprecipitation for detecting post-translational modifications of Topo1, Mit-23 cells were transfected with 3×FLAG/6×His-tagged ubiquitin or SUMO-1 expression vector. On the next day, cells were suspended in medium with or without of Dox (3 μg/ml) and incubated for 1 day. Then, MG-132 (50 μM) was added to all samples for 2 h. For CPT-treatment, cells were treated for 1 h before recovery. For detecting the ubiquitination, cells were lysed in 0.5×RIPA buffer (50 mM Tris–HCl [pH 8.0], 150 mM NaCl, 1 mM EDTA, 1% NP40, 0.25% sodium deoxychorate (DOC), 0.05% SDS) supplemented with 1× protease inhibitor cocktail (Roche) and 10 mM *N*-ethylmaleimide (NEM), and subjected to sonication. After that, samples were incubated with 50 U of Benzonase and 2.5 mM $MgCl_2$ for 1 h at 16 °C. Following centrifugation, equal amounts of supernatant were incubated for 16 h with α-Topo1 polyclonal antibody or Rabbit-IgG at 4 °C. Immunocomplex was recovered by DynaBeads ProteinA and washed with 0.5×RIPA buffer, and subjected to WB analysis. For detecting SUMO-1 modification, cells were boiled in Hot-Lysis buffer (10 mM Tris–HCl [pH 8.0], 150 mM NaCl, 2% SDS) supplemented with 1× protease inhibitor cocktail and 10 mM NEM for 10 min, and diluted in 2-fold volume of Hot-Lysis Dilution buffer (10 mM Tris–HCl [pH 8.0], 150 mM NaCl, 1 mM EDTA, 1% TrironX-100) supplemented with 10 mM NEM. After the sonication, samples were incubated for 1 h with 50 U of Benzonase and 2.5 mM $MgCl_2$ at 16 °C. Following centrifugation, samples were further diluted in 19-fold volume of Hot-Lysis Dilution buffer (at this time concentration of SDS was 0.05%), and equal amounts of supernatant were incubated for 16 h with α-Topo1 monoclonal antibody or Mouse-IgM at 4 °C. Immunocomplex was recovered by ProteinL magnetic beads (Thermo Scientific) and washed with Hot Lysis Wash buffer (10 mM Tris–HCl [pH 8.0], 250 mM NaCl, 1 mM EDTA, 1% NP40), and subjected to WB analysis.

Flow cytometry (FCM) analysis

Two days before Dox treatment, Mit-23 cells were transfected with indicated siRNA (50 nM) or plasmid DNA by Lipofectamine RNAiMax or Lipofectamine 2000 transfection reagent, respectively. After culturing cells in the presence of Dox (3 μg/ml) for 1 day, we harvested cells and suspended in freshly prepared 70% ethanol. Recovered cells were kept for 2 h at −20 °C. After that, cells were suspended in FCM buffer [PBS (+) supplemented with 4% FBS and 0.1% TritonX-100] and left at 4 °C for 30 min. For γH2AX-staining, cells were incubated with 100 μl of FCM buffer supplemented with 0.5 μg of α-γH2AX-antibody FITC conjugated (Millipore). After incubation for 2 h at 4 °C, samples were washed three time with FCM buffer, and then nuclear staining was achieved by propidium iodide (1 μg/ml)/PBS (−) supplemented with RNaseA (200 μg/ml). After treatment for 1 h at room temperature in the dark place, FCM analysis was performed by FACSCalibur on at least 10,000 cells.

Neutral comet assay

Neutral comet assay was performed following the manufacturer's instruction (Trevigen) with some modification. Briefly, Mit-23 cells were harvested by 2 mM EDTA/PBS (−). Differentiated MM-6 cells were detached by Accutase (Innovative Cell Technologies) and collected. After resuspension in PBS (−), cells were embedded in Low-Melting Agarose (Trevigen), spread and solidified over the Comet Slides (Trevigen) on ice. The slides were immersed in Lysis Buffer (Trevigen) for at least 1 h, and then incubated in neutral electrophoresis buffer (0.1 M Tris-Ac [pH 9.0], 0.3 M $NaOAc{\cdot}3H_2O$) for 30 min at 4 °C. After electrophoresis for 1 h at 0.75 V/cm at 4 °C, samples were fixed in precipitation buffer (1 M NH_4Ac, 85% ethanol) and 70% ethanol. After staining nuclei by SyBr gold, images were captured by (BZ-X710, Keyence) with 20× objectives utilizing function of Z-stacks and image stitching for obtaining with high resolution image of large field. Integrated images were analysed by Comet Assay IV software (Perceptive Instruments) using "Olive-tail moment" as a parameter of extent of DSB. Olive-tail moment generally indicate the extent of DNA damages, which calculated by product of the tail length and the fraction of total DNA in the tail. In Comet Assay IV software (Perceptive Instruments), "Olive-tail moment" is defined as: the product of the proportion of tail intensity and the displacement of tail center of mass relative to the center of the head. In all experiments, at least 80 nuclei were subjected to analysis.

Pull-down assay of negative supercoiled DNA

U2OS/2-6-3 cells were transfected with indicated plasmid DNA by Viafect transfection reagent (Promega). After 2 days of transfection, 5×10^6 cells were suspended in 250 μl of RSB (10 mM Tris [pH 7.5], 10 mM NaCl, 3 mM $MgCl_2$) and combined with 3.5 ml of RSB with 0.1% NP40. After gentle mixing, cells were centrifuged at 500×*g* for 10 min at 4 °C. Nuclear pellet was suspended in PBS (−) supplemented with 5 μM of Ez-Link Psoralen-PEG_3-Biotin (bPso) (Thermo scientific) for 30 min at 4 °C. After centrifugation at 500×*g* at 4 °C for 10 min, pellet was suspended in 100 μl of PBS (−) and removed to 96-well plate. For cross-linking the bPso to DNA, 365 nm wavelength of UV (UVL-21, UVP) was irradiated for 30 min on ice, and then collected nuclei were lysed in sonication buffer (50 mM Tris [pH 7.5], 140 mM NaCl, 1 mM EDTA, 1 mM EGTA, 1% TritonX-100, 0.1% DOC, 0.1% SDS) supplemented with 1× protease inhibitor cocktail. To obtain the DNA fragment with average size of 250 bp, sonication was carried out following conditions: 20 cycles of 30 s-ON/30 s-OFF, using Bioruptor UCD-250 (Cosmo Bio). After that, samples were centrifuged at 16,000×*g* for 30 min at 4 °C, and obtained supernatants was gently mixed with SA-beads blocked with 0.5% BSA and 100 μg/ml salmon sperm DNA at least 12 h. SA-beads were sequentially washed with RIPA buffer (50 mM Tris–HCl [pH 8.0], 150 mM NaCl, 1 mM EDTA, 1% NP40, 0.5% DOC, 0.1% SDS) for 10 min, High salt wash buffer (50 mM Tris–HCl [pH 8.0], 500 mM NaCl, 1 mM EDTA, 1% NP40, 0.5% DOC, 0.1% SDS) for 15 min, LiCl wash buffer (50 mM Tris–HCl [pH 8.0], 250 mM LiCl, 1 mM EDTA, 1% NP40, 0.5% DOC) for 15 min, and 1×TE buffer (10 mM Tris–HCl [pH 8.0], 1 mM EDTA) for 10 min twice. All processes were performed at 4 °C on rotating platform. After washing, SA-beads was treated with RNaseA (1 mg/ml) in 1×TE buffer for 30 min at 37 °C, and ProteinaseK (100 μg/ml) with 0.5% SDS for 1 h at 55 °C.

And then supernatant was purified by phenol–chloroform extraction (fraction-1). Furthermore, residual DNA on SA-beads was further extracted by 95% formamide with 10 mM EDTA for 10 min at 90 °C (fraction-2). Combined fractions were purified by ethanol precipitation, and purified DNA was dissolved in 1×TE buffer and subjected to qPCR analysis with SyBr Premix ExTaq Tli RNaseH plus (TaKaRa) by StepOne Real-time PCR system. Oligonucleotides used in this procedure are listed in (Additional file 20: Table S3).

Chromatin immunoprecipitation (ChIP)

U2OS/2-6-3 cells were transfected with indicated plasmid DNA by Viafect transfection reagent. After the 2 days of transfection, cells were cross-linked with 1% paraformaldehyde for 10 min, followed by quenching with 0.125 M glycine for 5 min at room temperature. After washing with PBS (−) twice, cells were suspended in Buffer-1 (10 mM HEPES [pH 7.5], 10 mM EDTA, 0.5 mM EGTA, 0.75% TritonX-100) supplemented with 1×protease inhibitor cocktail and incubated for 10 min on ice. After the centrifugation at 1700×*g* for 10 min at 4 °C, collected cells were resuspended in Buffer-2 (10 mM HEPES [pH 7.5], 200 mM NaCl, 1 mM EDTA, 0.5 mM EGTA) supplemented with 1× protease inhibitor cocktail and incubated for 5 min on ice. Subsequently, cells were recovered by centrifugation (1700×*g* for 10 min at 4 °C), and lysed in SDS lysis buffer (50 mM Tris–HCl [pH 8.0], 10 mM EDTA, 1% SDS) supplemented with 1× protease inhibitor cocktail. After fragmentation to average length of 250 bp, samples were centrifuged at 16,000×*g* for 30 min at 4 °C, and obtained supernatants were diluted in 9-fold volume of ChIP Dilution buffer (50 mM Tris–HCl [pH 8.0], 167 mM NaCl, 1.1% TritonX-100, 0.11% DOC). After pre-clearing with ProteinG Sepharose beads and IgG Sepharose beads, 2 µg of α-RPA70 monoclonal-antibody or mouse-IgG were added to each aliquot of chromatin lysate, and incubated on rotating platform for over 12 h at 4 °C. Immunocomplex was recovered by Dynabeads ProteinG blocked with 0.5% BSA and 100 µg/ml salmon sperm DNA. After 1 h incubation, beads were extensively washed as described in "Pull-down assay of negative supercoiled DNA", and incubated for 6 h at 65 °C in 1×TE buffer supplemented with 250 mM NaCl and 0.5% SDS, and treated with RNaseA (1 mg/ml) for 30 min at 37 °C, and ProteinaseK (100 µg/ml) for 1 h at 55 °C. At this time, input DNA was treated as same way. For eluting DNA, beads were incubated with 0.1 M $NaHCO_3$ and 1% SDS for 30 min at 65 °C, and recovered DNA was purified by phenol–chloroform extraction and ethanol precipitation. Purified DNA was dissolved in 1×TE buffer and subjected to qPCR analysis. Oligonucleotides used in this procedure are listed in (Additional file 20: Table S3).

Native-ChIP

For detecting Topo1-cc (covalent complex of Topo1 and DNA), ChIP assay without cross-linking was performed. U2OS/2-6-3 cells were transfected with indicated plasmid DNA by Viafect transfection reagent. After 2 days of transfection, cells were treated with MG-132 (50 µM) for 2 h, and directly lysed in Buffer-M. DNA–protein covalent complex was recovered by ethanol precipitation as described. Pellets were dissolved in SDS lysis buffer supplemented with 1× protease inhibitor cocktail and sonicated. Following dilution, immunoprecipitation was performed with α-Topo1 polyclonal-antibody or Rabbit IgG. Immunocomplex was recovered by DynaBeads ProteinA blocked with 0.5% BSA and 100 µg/ml salmon sperm DNA, and washed as with the ChIP protocol. Collected beads were incubated in 1×TE with RNaseA (1 mg/ml) for 30 min at 37 °C, ProteinaseK (100 µg/ml) with 0.5% SDS for 1 h at 55 °C, and 0.1 M $NaHCO_3$ with 1% SDS for 30 min at 65 °C. Eluted DNA was purified, and analysed by qPCR as in standard ChIP assay. Oligonucleotides used in this procedure are listed in (Additional file 20: Table S3).

DNA/RNA hybrids immunoprecipitation (DRIP)

Genomic DNA was purified by DNeasy Blood & Tissue Kit (Qiagen) or Buffer-M according to RADAR protocol [39]. Following sonication, samples were treated with or without of RNaseH (20 U, NEB) in 1×RNaseH Reaction buffer overnight at 37 °C. After purification, DNA pellets were dissolved in DRIP buffer (50 mM Tris–HCl [pH 8.0], 150 mM NaCl, 1 mM EDTA, 0.05% TritonX-100), and incubated with S9.6 antibody (KeraFast) or mouse IgG for 12 h at 4 °C on rotating platform. Following processes were performed as with Native-ChIP, and analysed by qPCR as in standard ChIP assay. The promotor and terminator regions of *Actin* gene were analysed by the qPCR analysis, as negative and positive control of DRIP assay, respectively [69]. Oligonucleotides used in this procedure are listed in (Additional file 20: Table S3).

Ligation-mediated (LM)-PCR

LM-PCR experiments were performed according to a protocol previously reported [41] with some modifications. In brief, after 2 days of transfection, cells were directly lysed in HMW buffer (10 mM Tris–HCl [pH 7.5], 100 mM NaCl, 1 mM EDTA, 0.5% SDS) supplemented with ProteinaseK (200 µg/ml), and incubated overnight at 55 °C without agitation. After phenol–chloroform extraction, same volume of isopropanol was added to sample, and DNA was gently picked up and rinsed with 70% ethanol. Recovered DNA was suspended in 1×TE supplemented with RNaseA (100 μg/ml). Equal amounts of DNA was treated with Quick-Blunting Kit (NEB) and ligated with blunt-end linkers (JW-Linker: annealed JW102 with JW103). Following purification, equal amounts of DNA were subjected to qPCR analysis with LacO-Rev and JW102 primers using SyBr Premix ExTaq Tli RNaseH plus. For estimating the amounts of LacO/JW-linker junction, corresponding DNA fragment was cloned into pMD20, and used as standard sample. The qPCR against to *β-globin* was carried out in parallel to normalize the amounts of input DNA. Oligonucleotides used in this procedure are listed in (Additional file 20: Table S3).

Fluorescence recovery after photobleaching (FRAP) assay

For FRAP assay, newly established cell lines: Mit-23 with H2B(Wt)-GFP or H2B(K120R)-GFP, HT1080vRxt-Vpr-Wt or -Q65R with H2B(Wt)-GFP, in which pEF-Bos-H2B-GFP was stably transduced to Mit-23 or HT1080vRxt-Vpr, were treated with Dox (3 μg/ml) for 1.5 days. Images were obtained by a Leica TCS SP5 confocal microscope with a Leica HCX APO 100×/1.40–0.60 oil immersion lens, and obtained data were analysed by Leica LAS AF Lite software (Leica). For GFP excitation, we used the 488 nm line of an Argon laser and fluorescence emission was collected between 500 and 530 nm. All experiments were done on warmed stage at 37 °C. In experiments, pre-bleach image was acquired by three consecutive images. Then a single square on the nucleus was bleached with five times of laser pulses of 1.318 s at 100% power. Images were then collected at 10 s intervals for 180 s. For calculating the rate of FRAP, the background signals (ROI^B) were subtracted from the region of interest for bleached area (ROI^+) and un-bleached area (ROI^-). The corrected signal of ROI^+ was normalized by ROI^- at each time point, and relative signal intensity compared to pre-bleaching (ROI^0) was used to evaluate the change of GFP intensity, according to the equation;

$$\text{Relative signal intensity} = \{(ROI^+ - ROI^B) \div (ROI^- - ROI^B)\} \div (ROI^0 - ROI^B)$$

The recovery of GFP signal intensity was expressed by subtracting the relative signal intensity of immediately after bleaching from that of each time point. In all experiments, at least 10 cells were subjected to analysis.

Production of pseudotyped HIV-1

For production of Vesicular Stomatitis Virus Glycoprotein (VSV-G) pseudotyped HIV-1, HEK293T cells were co-transfected with pNL4-3/E- and pHIT-VSV-G using FuGENE6 transfection reagent (Promega). On the next day, culture medium was changed to DMEM supplemented with 0.1% FBS. On the day 2 after transfection, culture supernatant containing the virus were recovered and filtered through a 0.45 μm-filter, and viral titer was determined by p24 ELISA kit (Zeptometrics). NL4-3-Luc/E- and NL4-3-EGFP/E-, which contains *Luciferase* and *EGFP* in-frame of *Nef*, were prepared as same manner. Viral samples were treated with DNase I (Takara) before infection.

Lentivirus production

Lentivirus was produced using Virapower Lentiviral packaging mix (Invitrogen) according to a manufacturer's instruction with some modification. Briefly, HEK293T cells were co-transfected with pLenti6, pLP1-D64A, pLP2, and pLP-VSV-G using Lipofectamine 2000 transfection reagent. Culture medium was replaced to fresh medium at 16 h after transfection. On the day 2 after transfection, culture supernatant containing the virus was filtered through a 0.45 μm-filter and ultra-centrifuged at 40,000 rpm for 1 h at 4 °C (Optima TLX ultracentrifuge, Beckman Coulter). After removing supernatant, viral pellet was dissolved in Opti-MEM (Life technologies), and viral titer was determined by p24 ELISA kit. Viral samples were treated with DNase I before infection.

HIV-1 infection to MM-6 cells

For pseudotyped HIV-1 infection, 2.5 × 10^5 MM-6 cells were differentiated by 50 nM of PMA for 3 days in 12-well plate. In RNAi experiments, cells were transfected with 300 nM of indicated siRNA by Nucleofector Kit-V and suspended in PMA containing medium. After 3 days, NL4-3 viral solution (p24, 50 ng) was added, and incubated for 2 h. Following washing twice with pre-warmed medium, cells were cultured in PMA containing medium for additional 1 or 2 days, for neutral comet assay or *Alu-gag* two-step nested qPCR analysis, respectively. For luciferase assay, cell cultures (50,000 cells/well, in 96well plate) at 3 days post-infection (dpi) were directly lysed in equal volume of One-Glo Luciferase assay system (Promega) solution, and relative light units (RLU) were measured by microplate luminometer (Veritas). For FCM analysis of EGFP, cells were fixed with 1% paraformaldehyde at least 30 min at room temperature at 3dpi, and analysed by FACSCalibur on at least 10,000 cells.

HIV-1 infection to MDMs

For pseudotyped HIV-1 infection, 2.5 × 10^5 MDMs were differentiated by 25–50 ng/ml of rhM-CSF for 5 days in 12-well plate. In RNAi experiments, cells were transfected with 50 nM of indicated siRNA by Lipofectamine RNAiMax transfection reagent for 4 h and replaced to rhM-CSF containing medium. After 2 days of transfection, NL4-3 viral solution (p24, 12.5 ng) was added, and incubated for 2 h. Following washing twice with pre-warmed medium, cells were cultured in rhM-CSF containing medium. *Alu-gag* two-step nested qPCR analysis was performed at 2dpi.

HIV-1 infection to U2OS/2-6-3 cells and analysis of LacO-directed integration

U2OS/2-6-3 cells of 5×10^5 were transfected with indicated plasmid DNA by Viafect transfection reagent. After 2 days of transfection, cells were incubated with NL4-3 viral solution (p24, 100 ng) for 2 days, and subjected to qPCR analysis. Genomic DNA was extracted using Quickgene Nucleic acid isolation system (KURABO). The qPCR was done with LacO-(sense or antisense), Lenti-5237F, and TaqMan probe (pLenti6-LTR probe) using SsoAdvance universal probe Supermix by StepOne Real-time system. To measure copy numbers of integrated viral DNA that possessed the LacO/proviral DNA junction, corresponding DNA fragment was cloned into pMD20 and used as standard samples. In parallel, qPCR for *β-globin* was performed for normalization of input DNA. At this time, *Alu-gag* two-step nested qPCR was performed and measuring the overall infectivity. Oligonucleotides used in this procedure are listed in (Additional file 20: Table S3).

Alu-gag two-step nested qPCR

Genomic DNA was extracted using Quickgene Nucleic acid isolation system, and 100 ng of DNA was subjected to 12 cycles of *Alu-gag* 1st PCR using AmpliTaq gold 360 with Alu-F/R and 1st-gag-R. And then, qPCR was performed with 2-LTR-S and 2nd tag-R, and TaqMan probe (Probe-2). For determining the frequency of the integration, gDNA containing 0.485 copies of HIV integration/cell [8] was subjected to amplification at the same time. The qPCR against to *β-globin* was carried out in parallel to normalizing the amounts of input DNA. Oligonucleotides used in this procedure are listed in (Additional file 20: Table S3).

Lentivirus infection to U2OS/2-6-3 cells and analysis of LacO-directed integration

Lentiviral solution (p24, 100 ng) was infected to 2.5×10^5 U2OS/2-6-3 cells for 2 days. To quantitate the LacO-directed integration, gDNA was subjected to qPCR analysis with LacO-(sense or antisense)/Lenti-5237F, and TaqMan probe (pLenti6-LTR probe). For measuring the overall infectivity, cells were cultured in the presence of BlasticidineS (10 μg/ml) for 2 weeks. For normalizing the plating efficiency, colonies were formed in normal culture medium. Colonies were fixed with 70% ethanol and stained with Giemsa's staining solution, and enumerated. Oligonucleotides used in this procedure are listed in (Additional file 20: Table S3).

Cre mediated expression of Cherry-LacR-Vpr

U2OS/2-6-3+pCAL-loxP-CLV (263/loxP-CLV), in which expression of LacR-fused Vpr was switched on when Cre-recombinase was expressed, were first established by G418 selection. Then, cells were infected with adenovirus for LacZ (control) or Cre (inducing CLV expression) expression at multiplicity of infection (MOI) of 100 for 2 days, and subjected to live cell imaging and ChIP assay for Vpr (Additional file 8: Figure S7a, b). For siRNA experiments, 2.5×10^5 cells were transfected with control or *Topo1* siRNA (50 nM) by Lipofectamine RNAiMax transfection reagent. After 1.5 days of siRNA transfection, adenovirus was infected for 2 days, and cells were subjected to following analysis; LacO-directed integration, native-ChIP assay for Topo1-cc and LM-PCR (Fig. 7g and Additional file 8: Figure S7c, d, respectively). Oligonucleotides used in this procedure are listed in (Additional file 20: Table S3).

Linear amplification mediated-PCR

LAM-PCR was performed, as previously described [8] with some modification. Briefly, 1st liner PCR was performed using biotinylated-Lenti-5203F primer by two round of amplification with 50 cycles using Taq polymerase (Qiagen). After purification by SA-beads, dsDNA was synthesized with Klenow fragment (NEB) with random hexamer. Following extensive washes, dsDNA on SA-beads were digested with *Msp*I or *Nla*III, and ligated with corresponding dsDNA linker cassettes (LC). After that, 1st PCR was performed with biotinylated-M667/LC1, followed by SA-beads purification. And then, 2nd exponential PCR was performed with EV984/LC2. Size selection of amplified DNA was performed on agarose gel, and cloned into pGEM-T Easy vector. Sequencing was carried out by ABI3130x (Applied Biosystems), and integration sites were determined. Oligonucleotides used in this procedure are listed in (Additional file 20: Table S3).

Quantitative reverse-transcription-PCR (qRT-PCR)

Total RNA was purified by RNeasy kit (Qiagen), and cDNA was generated by HighCapacity cDNA reverse transcription kit (Invitrogen). Quantitative-PCR was performed using SyBr premix ExTaq Tli RNaseH plus. The expression levels of each gene were normalized by the relative amounts of *β-Actin*. The primers used for qRT-PCR are listed in (Additional file 21: Table S4).

Additional files

Additional file 1: Figure S1. Vpr induces structural alteration of DNA. **a** Representative AFM images of dsDNA in 1.0 μm × 1.0 μm field. **b** Representative raw data of Rq values of dsDNA. The Rq values of each dsDNA are shown as a plot, and the medians are indicated by red bars. **c** Relative Rq values of Vpr mutants. C-terminal mutants of Vpr were defective in DNA structural alteration. Data were obtained from more than three independent experiments. Error bar indicates ± SEM. In Fig. 1b, data of Buffer, Chlq, Wt, ΔC12 and Ct4RA are depicted.

Additional file 2: Figure S2. The interaction between Vpr and RPA70 (a) or T4gp32 (b) was not detected. **a** FLAG-Strep-Vpr (0.1, 0.317. 1.0 pmol) was incubated with rRPA70 (1.0 pmol), and pulled-down with Streptavidine M280 beads. Proteins were analysed by WB with indicated antibodies. **b** FLAG-Strep-Vpr (1, 3.17, 10 pmol) was incubated with T4gp32 (100 pmol), and pulled-down with Streptavidine M280 beads. Proteins were visualized by Oriole fluorescent gel staining.

Additional file 3: Figure S3. Vpr provokes Topo1 stress. **a** RADAR analysis detecting covalently bound DNA and Topo1. HEK293T cells were transfected with indicated Topo1-HA construct. Y723F is a catalytically inactive mutant of Topo1. A covalent complex of Topo1 and DNA was formed in the cells that were first transduced with Topo1-Wt, and then treated with CPT (20 μM, 1 h) or paraformaldehyde (PFA; 1 mM, 2 h), whereas the complex was only detected in cells with Topo1-Y723F when treated with PFA, but not with CPT. The same membrane was reprobed with α-dsDNA antibody after stripping. **b** Different amounts of DNA were blotted in the RADAR analysis.

Additional file 4: Table S1. Knockdown efficiency of siRNA target genes. Relative expression levels compared to cont-si (100%: shaded) are shown. Data were obtained from at least three independent experiments.

Additional file 5: Figure S4. Vpr-induced DNA damages are suppressed by TDP1 or RNaseH1 expression. **a** Exogenous expression of RNaseH1 (RNH1) and TDP1. Mit-23 cells were transfected with indicated vector (pFLAG-CMV2 based vector), and whole cell extracts were subjected to WB analysis. Arrowhead (black), TDP1; arrow, RNaseH1; arrowhead (red), Vpr; asterisk, non-specific bands. **b** RNH1 and TDP1 suppresses Vpr induced DDR. Mit-23 cells transfected with RNH1 or TDP1 showed reduced level of Vpr-induced phosphorylation of H2AX. A representative result out of two independent experiments is depicted.

Additional file 6: Figure S5. Forced accumulation of Vpr induced DDR in the vicinity of chromatin. **a–e** Forced accumulation of Vpr induces phosphorylation of H2AX at Ser139 (a), ATM at Ser1981 (b), RPA32 at Ser33 (c), Rad17 at Ser645 (d), and accumulation of mono- and poly-ubiquitin conjugates (e), on surrounding region.

Additional file 7: Figure S6. LM-PCR for detecting DSB-ends in the LacO repeats. **a**. Schematic of LM-PCR. Genomic DNA samples prepared from cells, which were transduced with LacR-construct were treated with T4 polymerase and T4 PNK for blunting the DSB ends, and ligated with blunt-end JW-linkers. LacO/JW-linker junctions were amplified by specific primers, shown in (b). **b**. Diagram of LacO/JW-linker amplification product. Arrows above the box indicate the PCR primers for JW-linker and LacO. **c** Sequencing chromatogram of LacO/JW-linker junction. Black arrow and red arrow indicate JW-linker and LacO repeats, respectively. **d** Representative amplification plot of LacO/JW-qPCR. Cherry-LacR fused-OVA, Black; -Fok1, blue; -Vpr, red curve. Gray curve show the standard samples with indicated copy numbers.

Additional file 8: Figure S7. *Topo1* is involved in Vpr-induced DDR and DSB. **a** Live cell imaging of CLV inducible cell line. To confirm effect of Vpr on Topo1-cc formation, we transfected pCAL-loxP-CLV to U2OS/2-6-3 cells, and obtained a cell line (263/loxP-CLV), in which expression of a Cherry-LacR fused Vpr (CLV) can be switched on after Cre expression. In the experiment, 263/loxP-CLV cells were infected with adenoviruses expressing LacZ (Ad-LacZ) or Cre (Ad-Cre) for 2 days at MOI of 100. Single focus of mCherry was distinctly observed in Ad-Cre infected cells (lower panel), suggesting that CLV was expressed in Cre-dependent manner. Scale bar indicates 20 μm. **b** Specific accumulation of CLV on the LacO repeats. ChIP assay with α-Vpr antibody (8D1) was performed in 263/loxP-CLV cells infected with Ad-LacZ or -Cre at 2dpi. **c** Down regulation of *Topo1* reduces Vpr induced Topo1-cc on the LacO repeats. After downregulation of *Topo1*, 263/loxP-CLV cells, which were infected with Ad-Cre, was subjected to native-ChIP assay with α-Topo1 antibody. A representative result out of two independent experiments is depicted. **d** Topo1 is required for Vpr-induced DSB on the LacO repeats. After downregulation of *Topo1*, 263/loxP-CLV cells were infected with Ad-LacZ or -Cre, and subjected to LM-PCR analysis to measure the extent of DSB on the LacO repeats. Vpr-induced DSB was calculated by subtracting the amounts of DSB observed in Ad-LacZ infected cells from those observed in Ad-Cre infected cells. Data were obtained from three independent experiments.

Additional file 9: Figure S8. Representative data of Lenti-LacO qPCR for detecting the LacO-directed integration. **a, b** Amplification plots of sense integration (a) and anti-sense integration (b), respectively. Blue (light blue ~ dark blue) and red (pink ~ red) curves show CV- and CLV-virus infected samples with different MOIs shown by amounts of p24, respectively. Gray curves indicate standard samples with indicated copy numbers. **c, d** Sequencing chromatograms of sense (c) and anti-sense (d) integration product. Blue arrows and red arrows indicate 3′-LTR and LacO repeats, respectively.

Additional file 10: Figure S9. Proviral integration in the vicinity of Vpr-induced DSB sites. Targeting of HIV-1 proviral DNA to CLV-induced DSB sites. U2OS/2-6-3 cells were first transfected with indicated construct, and then infected with NL-4-3/D64A/R− virus. The percentages of LacO directed integration per overall integration (*Alu-gag* two-step qPCR) are shown. Data were obtained from three independent experiments. Error bar indicates ± SEM. In Cherry-Vpr, Cherry-LacR-Vpr, and Fok1-Cherry-LacR, the *P*-value was 0.37, 0.051, and 0.015 for sense integration, respectively. The *P*-value for antisense-integration was 0.44, 0.001, and 5.21×10^{-5}, respectively.

Additional file 11: Figure S10. Schematic of *Alu-gag* two-step nested qPCR. First PCR to amplify *Alu*-proviral DNA junction was performed using PCR primers targeting the *Alu* (black arrow) and *gag* (pink arrows). Red wavy line fused to *gag*-primer indicates the tag-sequence for 2nd qPCR primer binding. In second qPCR, viral DNA fragments were amplified by LTR primer (blue arrow) and tag-primer (red arrow). The green box indicates the position of the TaqMan probe for *gag*.

Additional file 12: Figure S11. Incorporation of Cherry-LacR-Vpr does not affect overall viral infectivity. U2OS/2-6-3 cells infected with indicated lentivirus, which had Blasticidine-resitance gene, were subjected to Blasticidine (Bsd) selection. The infected cells obtain Bsd resistance by the successful lentiviral integration. Data were obtained from three independent experiments. Error bar indicates ± SEM.

Additional file 13: Figure S12. A representative sequencing chromatogram of the LacO/proviral DNA junction, obtained by the LAM-PCR. Blue arrows and red arrows indicate 3′-LTR and LacO repeats, respectively.

Additional file 14: Figure S13. Topo1 is required for Vpr-dependent upregulation of viral infection. **a** MM-6 cells were infected with Vpr proficient (R+) or deficient (R−) NL4-3 viruses under down-regulation of *Topo1* by three species of siRNAs. The integration rate was quantitated by *Alu-gag* two-step nested qPCR at 2dpi; relative integration rates are shown. Data were obtained from more than three independent experiments. Error bar indicates ± SEM. $^{**}P < 0.05$ **b** Knockdown efficiency of each siRNA. Relative levels of *Topo1* expression are shown. Topo1 siRNA#3 was used in other experiments.

Additional file 15: Figure S14. Vpr upregulates viral infection in differentiated MM-6 cells. **a** MM-6 cells were infected with Vpr proficient (R+) or deficient (R−) NL4-3-Luc/E- viruses. Luciferase assay was performed at 3dpi. Data were obtained from more than three independent experiments. Error bar indicates ± SEM. **b** MM-6 cells were infected with Vpr proficient (R+) or deficient (R−) NL4-3-EGFP/E- viruses. Percentage of EGFP positive cells was determined by FCM at 3dpi. Representative FCM data are shown in bottom panels. Green colored plots were gated as EGFP positive cells.

Additional file 16: Figure S15. XPG is required for Vpr-induced DDR and DSB. **a, b** Effects of downregulation of XPG on Vpr-induced DDR. The expression of *XPG* in Mit-23 cells was first downregulated by siRNA, and then Vpr expression was initiated on day 2 after introduction of *XPG* targeting siRNA. On day 1 after Vpr expression was started, phosphorylation of H2AX (a) and G_2/M checkpoint activation (b) was analysed by flow cytometry. Downregulation of *XPG* significantly reduced Vpr-induced phosphorylation of H2AX ($P = 4.4 \times 10^{-5}$), and the G_2/M arrest ($P = 0.025$). Data were obtained from three independent experiments. Error bar indicates ± SEM. **c** XPG is required for Vpr-induced DSBs. Neutral comet assay was performed using Mit-23 cell, the *XPG* expression of which was down-regulated. Data were obtained from three independent experiments. Error bar indicates ± SEM. $P = 0.037$. **d** XPG is required for rVpr-induced DSB in resting macrophages. Differentiated MM-6 cells were treated with 100 ng/ml of rVpr under the down-regulation of *XPG*, and subjected to neutral comet assay. Data were obtained from three independent experiments. Error bar indicates ± SEM. $P = 0.001$. **e** XPG is required for Vpr-induced upregulation of viral infection in resting macrophages. Differentiated MM-6 cells were infected with Vpr proficient (R+) or deficient (R−) NL4-3 viruses after transfection of siRNA, and the integration rate was quantitated by *Alu-gag* two-step nested qPCR; relative integration rates are shown. A representative result out of two independent experiments is depicted.

Additional file 17: Figure S16. Forced accumulation of Vpr increases formation of R-loops. **a** U2OS/2-6-3 cells were transfected with indicated LacR fused constructs. Two days after transfection, DNA samples were prepared and subjected to DRIP assay. DRIP assay was done with α-DNA/RNA hybrids antibody (S9.6). RNaseH treatments were performed to confirm the specificity of DRIP assay. Analysis of promotor (green column) and terminator (yellow column) regions of *Actin* gene were included in the qPCR analysis, as negative and positive control of DRIP assay, respectively. A representative result out of two independent experiments is depicted. **b** 263/loxP-CLV cells were infected with Ad-LacZ or -Cre. DNA samples were prepared and subjected to DRIP assay at 2dpi. Data were obtained from more than three independent experiments. Error bar indicates ± SEM.

Additional file 18: Figure S17. C-terminus of Vpr is required for topoisomer induction activity. DNA supercoiling assay was performed with increasing amounts of peptides (1, 10, 100 pmol; 0.015, 0.15, 1.5 Vpr molecules/bp). In the presence of C45, faster migrating topoisomers appeared.

Additional file 19: Table S2. List of resources used in this study.

Additional file 20: Table S3. List of oligonucleotides used in this study.

Additional file 21: Table S4. List of primers for qRT-PCR.

Abbreviations

AFM: atomic force microscopy; ATR: ataxia-telangiectasia and Rad3-related; cART: combined antiretroviral therapy; ChIP: chromatin immunoprecipitation; CPT: camptothecin; DDR: DNA damage response; DSB: DNA double-strand break; FRAP: fluorescence recovery after photo-bleaching; HIV-1: human immunodeficiency virus-1; LM-PCR: ligation-mediated (LM)-PCR; MDMs: monocyte derived macrophages; MOI: multiplicity of infection; RADAR: rapid approach to DNA adducts recovery; RPA70: replication protein A 70-kDa subunit; SSB: DNA single-strand break; Topo1: topoisomerase 1; Topo1-cc: Topo1-covalent complex; Vpr: viral protein R.

Authors' contributions

KI conducted all experiments, and JK performed FRAP analysis. KI and YI designed the experiments and wrote the paper. All authors read and approved the final manuscript.

Author details

[1] Department of Intractable Diseases, National Center for Global Health and Medicine, 1-21-1 Toyama, Shinjuku-ku, Tokyo 162-8655, Japan. [2] Department of Genome Repair Dynamics, Radiation Biology Center, Kyoto University, Yoshidakonoe-cho, Sakyo-ku, Kyoto 606-8501, Japan.

Acknowledgements

We thank Dr. Mari Shimura for Mit-23 cells, Dr. David Spector for U2OS/2-6-3 cells, Dr. Evi Soutoglou for the Cherry-LacR construct, and Dr. Kenzo Tokunaga for the pNL4-3 vector. FRAP and AFM analyses were supported by the Radiation Biology Center Cooperative Research Program (Kyoto University), and One-stop Sharing Facility Center for Future Drug Discovery (The University of Tokyo), respectively.

Competing interests

The authors declare that they have no competing interests.

Funding

This work was supported in part by JSPS KAKENHI Grant Number JP26860313, Grant for National Center for Global Health and Medicine (21A-129), Grant-in-Aid for Research on HIV/AIDS Project from the Ministry of Health, Labor and Welfare of Japan, and the National Center for Global Health and Medicine (25A-108).

References

1. Trono D, Van Lint C, Rouzioux C, Verdin E, Barré-Sinoussi F, Chun TW, et al. HIV persistence and the prospect of long-term drug-free remissions for HIV-infected individuals. Science. 2010;329:174–80.
2. Iglesias-Ussel MD, Romerio F. HIV reservoirs: the new frontier. AIDS Rev. 2011;13:13–29.
3. Abbas W, Tariq M, Iqbal M, Kumar A, Herbein G. Eradication of HIV-1 from the macrophage reservoir: an uncertain goal? Viruses. 2015;7:1578–98.
4. Cohen EA, Dehni G, Sodroski JG, Haseltine WA. Human immunodeficiency virus vpr product is a virion-associated regulatory protein. J Virol. 1990;64:3097–9.
5. Balliet JW, Kolson DL, Eiger G, Kim FM, McGann KA, Srinivasan A, et al. Distinct effects in primary macrophages and lymphocytes of the human immunodeficiency virus type 1 accessory genes vpr, vpu, and nef: mutational analysis of a primary HIV-1 isolate. Virology. 1994;200:623–31.
6. Connor RI, Chen BK, Choe S, Landau NR. Vpr is required for efficient replication of human immunodeficiency virus type-1 in mononuclear phagocytes. Virology. 1995;206:935–44.
7. Eckstein DA, Sherman MP, Penn ML, Chin PS, De Noronha CM, Greene WC, et al. HIV-1 Vpr enhances viral burden by facilitating infection of tissue macrophages but not nondividing CD4+ T cells. J Exp Med. 2001;194:1407–19.
8. Koyama T, Sun B, Tokunaga K, Tatsumi M, Ishizaka Y. DNA damage enhances integration of HIV-1 into macrophages by overcoming integrase inhibition. Retrovirology. 2013;10:21.
9. Guenzel CA, Hérate C, Benichou S. HIV-1 Vpr-a still "enigmatic multitasker". Front Microbiol. 2014;5:127.
10. Roshal M, Kim B, Zhu Y, Nghiem P, Planelles V. Activation of the ATR-mediated DNA damage response by the HIV-1 viral protein R. J Biol Chem. 2003;278:25879–86.
11. Zimmerman ES, Chen J, Andersen JL, Ardon O, Dehart JL, Blackett J, et al. Human immunodeficiency virus type 1 Vpr-mediated G2 arrest requires Rad17 and Hus1 and induces nuclear BRCA1 and gamma-H2AX focus formation. Mol Cell Biol. 2004;24:9286–94.
12. Lai M, Zimmerman ES, Planelles V, Chen J. Activation of the ATR pathway by human immunodeficiency virus type 1 Vpr involves its direct binding to chromatin in vivo. J Virol. 2005;79:15443–551.
13. Nakai-Murakami C, Shimura M, Kinomoto M, Takizawa Y, Tokunaga K, Taguchi T, et al. HIV-1 Vpr induces ATM-dependent cellular signal with enhanced homologous recombination. Oncogene. 2007;26:477–86.
14. Laguette N, Brégnard C, Hue P, Basbous J, Yatim A, Larroque M, et al. Premature activation of the SLX4 complex by Vpr promotes G2/M arrest and escape from innate immune sensing. Cell. 2014;156:134–45.

15. Morellet N, Bouaziz S, Petitjean P, Roques BP. NMR structure of the HIV-1 regulatory protein VPR. J Mol Biol. 2003;327:215–27.
16. Tachiwana H, Shimura M, Nakai-Murakami C, Tokunaga K, Takizawa Y, Sata T, et al. HIV-1 Vpr induces DNA double-strand breaks. Cancer Res. 2006;66:627–31.
17. Shimura M, Tanaka Y, Nakamura S, Minemoto Y, Yamashita K, Hatake K, et al. Micronuclei formation and aneuploidy induced by Vpr, an accessory gene of human immunodeficiency virus type 1. FASEB J. 1999;6:621–37.
18. Groschel B, Bushman F. Cell cycle arrest in G2/M promotes early steps of infection by human immunodeficiency virus. J Virol. 2005;79:5695–704.
19. Zimmerman ES, Sherman MP, Blackett JL, Neidleman JA, Kreis C, Mundt P, et al. Human immunodeficiency virus type 1 Vpr induces DNA replication stress in vitro and in vivo. J Virol. 2006;80:10407–18.
20. Zou L, Elledge SJ. Sensing DNA damage through ATRIP recognition of RPA-ssDNA complexes. Science. 2003;300:1542–8.
21. DeHart JL, Zimmerman ES, Ardon O, Monteiro-Filho CM, Argañaraz ER, Planelles V. HIV-1 Vpr activates the G2 checkpoint through manipulation of the ubiquitin proteasome system. Virol J. 2007;4:57.
22. Belzile JP, Duisit G, Rougeau N, Mercier J, Finzi A, Cohen EA. HIV-1 Vpr-mediated G2 arrest involves the DDB1-CUL4AVPRBP E3 ubiquitin ligase. PLoS Pathog. 2007;3:e85.
23. Belzile JP, Richard J, Rougeau N, Xiao Y, Cohen EA. HIV-1 Vpr induces the K48-linked polyubiquitination and proteasomal degradation of target cellular proteins to activate ATR and promote G2 arrest. J Virol. 2010;84:3320–30.
24. Tan L, Ehrlich E, Yu XF. DDB1 and Cul4A are required for human immunodeficiency virus type 1 Vpr-induced G2 arrest. J Virol. 2007;81:10822–30.
25. Kichler A, Pages JC, Leborgne C, Druillennec S, Lenoir C, Coulaud D, et al. Efficient DNA transfection mediated by the C-terminal domain of human immunodeficiency virus type 1 viral protein R. J Virol. 2000;74:5424–31.
26. Lyonnais S, Gorelick RJ, Heniche-Boukhalfa F, Bouaziz S, Parissi V, Mouscadet JF, et al. A protein ballet around the viral genome orchestrated by HIV-1 reverse transcriptase leads to an architectural switch: from nucleocapsid-condensed RNA to Vpr-bridged DNA. Virus Res. 2013;171:287–303.
27. Pommier Y, Sun Y, Huang SN, Nitiss JL. Roles of eukaryotic topoisomerases in transcription, replication and genomic stability. Nat Rev Mol Cell Biol. 2016;17:703–21.
28. Sordet O, Redon CE, Guirouilh-Barbat J, Smith S, Solier S, Douarre C, et al. Ataxia telangiectasia mutated activation by transcription- and topoisomerase I-induced DNA double-strand breaks. EMBO Rep. 2009;10:887–93.
29. Eldin P, Chazal N, Fenard D, Bernard E, Guichou JF, Briant L. Vpr expression abolishes the capacity of HIV-1 infected cells to repair uracilated DNA. Nucleic Acids Res. 2014;42:1698–710.
30. Lahouassa H, Blondot ML, Chauveau L, Chougui G, Morel M, Leduc M, et al. HIV-1 Vpr degrades the HLTF DNA translocase in T cells and macrophages. Proc Natl Acad Sci USA. 2016;113:5311–6.
31. Murphy PJ, Shannon M, Goertz J. Visualization of recombinant DNA and protein complexes using atomic force microscopy. J Vis Exp. 2011;53:3061.
32. Mizoguchi I, Ooe Y, Hoshino S, Shimura M, Kasahara T, Kano S, et al. Improved gene expression in resting macrophages using an oligopeptide derived from Vpr of human immunodeficiency virus type-1. Biochem Biophys Res Commun. 2005;338:1499–506.
33. Shamoo Y, Friedman AM, Parsons MR, Konigsberg WH, Steitz TA. Crystal structure of a replication fork single-stranded DNA binding protein (T4 gp32) complexed to DNA. Nature. 1995;376:362–6.
34. Jose D, Weitzel SE, Baase WA, von Hippel PH. Mapping the interactions of the single-stranded DNA binding protein of bacteriophage T4 (gp32) with DNA lattices at single nucleotide resolution: gp32 monomer binding. Nucleic Acids Res. 2015;43:9276–90.
35. Nitiss JL, Soans E, Rogojina A, Seth A, Mishina M. Topoisomerase assays. Curr Protoc Pharmacol. 2012;3:3.3.1–3.27.
36. Desai SD, Zhang H, Rodriguez-Bauman A, Yang JM, Wu X, Gounder MK, et al. Transcription-dependent degradation of topoisomerase I-DNA covalent complexes. Mol Cell Biol. 2003;23:2341–50.
37. Steinacher R, Osman F, Lorenz A, Bryer C, Whitby MC. Slx8 removes Pli1-dependent protein-SUMO conjugates including SUMOylated topoisomerase I to promote genome stability. PLoS ONE. 2013;8:e71960.
38. Shimura M, Toyoda Y, Iijima K, Kinomoto M, Tokunaga K, Yoda K, et al. Epigenetic displacement of HP1 from heterochromatin by HIV-1 Vpr causes premature sister chromatid separation. J Cell Biol. 2011;194:721–35.
39. Kiianitsa K, Maizels N. A rapid and sensitive assay for DNA-protein covalent complexes in living cells. Nucleic Acids Res. 2013;41:e104.
40. Janicki SM, Tsukamoto T, Salghetti SE, Tansey WP, Sachidanandam R, Prasanth KV, et al. From silencing to gene expression: real-time analysis in single cells. Cell. 2004;116:683–98.
41. Soutoglou E, Dorn JF, Sengupta K, Jasin M, Nussenzweig A, Ried T, Danuser G, Misteli T. Positional stability of single double-strand breaks in mammalian cells. Nat Cell Biol. 2007;9:675–82.
42. Soutoglou E, Misteli T. Activation of the cellular DNA damage response in the absence of DNA lesions. Science. 2008;320:1507–10.
43. Naughton C, Avlonitis N, Corless S, Prendergast JG, Mati IK, Eijk PP, et al. Transcription forms and remodels supercoiling domains unfolding large-scale chromatin structures. Nat Struct Mol Biol. 2013;20:387–95.
44. Anders L, Guenther MG, Qi J, Fan ZP, Marineau JJ, Rahl PB, et al. Genome-wide localization of small molecules. Nat Biotechnol. 2014;32:92–6.
45. Kimura H, Cook PR. Kinetics of core histones in living human cells: little exchange of H3 and H4 and some rapid exchange of H2B. J Cell Biol. 2001;153:1341–53.
46. Tóth KF, Knoch TA, Wachsmuth M, Frank-Stöhr M, Stöhr M, Bacher CP, et al. Trichostatin A-induced histone acetylation causes decondensation of interphase chromatin. J Cell Sci. 2004;117:4277–87.
47. Daniel R, Ramcharan J, Rogakou E, Taganov KD, Greger JG, Bonner W, et al. Histone H2AX is phosphorylated at sites of retroviral DNA integration but is dispensable for postintegration repair. J Biol Chem. 2004;279:45810–4.
48. Cooper A, García M, Petrovas C, Yamamoto T, Koup RA, Nabel GJ. HIV-1 causes CD4 cell death through DNA-dependent protein kinase during viral integration. Nature. 2013;498:376–9.
49. Beard P, Hughes M, Nyfeler K, Hoey M. Unwinding of the DNA helix in simian virus 40 chromosome templates by RNA polymerase. Eur J Biochem. 1984;143:39–45.
50. Sekulic N, Bassett EA, Rogers DJ, Black BE. The structure of $(CENP-A-H4)_2$ reveals physical features that mark centromeres. Nature. 2010;467:347–51.
51. Pavri R, Zhu B, Li G, Trojer P, Mandal S, Shilatifard A, et al. Histone H2B monoubiquitination functions cooperatively with FACT to regulate elongation by RNA polymerase II. Cell. 2006;125:703–17.
52. Nakamura K, Kato A, Kobayashi J, Yanagihara H, Sakamoto S, Oliveira DV, et al. Regulation of homologous recombination by RNF20-dependent H2B ubiquitination. Mol Cell. 2011;41:515–28.
53. Zeng M, Ren L, Mizuno K, Nestoras K, Wang H, Tang Z, et al. CRL4(Wdr70) regulates H2B monoubiquitination and facilitates Exo1-dependent resection. Nat Commun. 2016;7:11364.
54. Taneichi D, Iijima K, Doi A, Koyama T, Minemoto Y, Tokunaga K, et al. Identification of SNF2 h, a chromatin-remodeling factor, as a novel binding protein of Vpr of human immunodeficiency virus type 1. J Neuroimmune Pharmacol. 2011;6:177–87.
55. Maudet C, Sourisce A, Dragin L, Lahouassa H, Rain JC, Bouaziz S, et al. HIV-1 Vpr induces the degradation of ZIP and sZIP, adaptors of the NuRD chromatin remodeling complex, by hijacking DCAF1/VprBP. PLoS ONE. 2013;8:e77320.
56. Romani B, Baygloo NS, Hamidi-Fard M, Aghasadeghi MR, Allahbakhshi E. HIV-1 Vpr protein induces proteasomal degradation of chromatin-associated class I HDACs to overcome latent infection of macrophages. J Biol Chem. 2016;291:2696–711.
57. Deng C, Brown JA, You D, Brown JM. Multiple endonucleases function to repair covalent topoisomerase I complexes in Saccharomyces cerevisiae. Genetics. 2005;170:591–600.
58. Kim Y, Spitz GS, Veturi U, Lach FP, Auerbach AD, Smogorzewska A. Regulation of multiple DNA repair pathways by the Fanconi anemia protein SLX4. Blood. 2013;121:54–63.
59. Sollier J, Stork CT, García-Rubio ML, Paulsen RD, Aguilera A, Cimprich KA. Transcription-coupled nucleotide excision repair factors promote R-loop-induced genome instability. Mol Cell. 2014;56:777–85.
60. Sollier J, Cimprich KA. Breaking bad: R-loops and genome integrity. Trends Cell Biol. 2015;25:514–22.
61. Fregoso OI, Emerman M. Activation of the DNA damage response is a conserved function of HIV-1 and HIV-2 Vpr that is independent of SLX4 recruitment. MBio. 2016;7:e01433-16.

62. Smith JA, Daniel R. Up-regulation of HIV-1 transduction in nondividing cells by double-strand DNA break-inducing agents. Biotechnol Lett. 2011;33:243–52.
63. Ebina H, Kanemura Y, Suzuki Y, Urata K, Misawa N, Koyanagi Y. Integrase-independent HIV-1 infection is augmented under conditions of DNA damage and produces a viral reservoir. Virology. 2012;427:44–50.
64. Siddiqui K, Del Valle L, Morellet N, Cui J, Ghafouri M, Mukerjee R, et al. Molecular mimicry in inducing DNA damage between HIV-1 Vpr and the anticancer agent, cisplatin. Oncogene. 2008;27:32–43.
65. Levy DN, Refaeli Y, MacGregor RR, Weiner DB. Serum Vpr regulates productive infection and latency of human immunodeficiency virus type 1. Proc Natl Acad Sci USA. 1994;91:10873–7.
66. Power C, Hui E, Vivithanaporn P, Acharjee S, Polyak M. Delineating HIV-associated neurocognitive disorders using transgenic models: the neuropathogenic actions of Vpr. J Neuroimmune Pharmacol. 2012;7:319–31.
67. Katyal S, Lee Y, Nitiss KC, Downing SM, Li Y, Shimada M, et al. Aberrant topoisomerase-1 DNA lesions are pathogenic in neurodegenerative genome instability syndromes. Nat Neurosci. 2014;17:813–21.
68. Saylor D, Dickens AM, Sacktor N, Haughey N, Slusher B, Pletnikov M, et al. HIV-associated neurocognitive disorder—pathogenesis and prospects for treatment. Nat Rev Neurol. 2016;12:234–48.
69. Skourti-Stathaki K, Kamieniarz-Gdula K, Proudfoot NJ. R-loops induce repressive chromatin marks over mammalian gene terminators. Nature. 2014;516:436–9.

4

An RNA-binding compound that stabilizes the HIV-1 gRNA packaging signal structure and specifically blocks HIV-1 RNA encapsidation

Carin K. Ingemarsdotter[1], Jingwei Zeng[1], Ziqi Long[1], Andrew M.L. Lever[1,2] and Julia C. Kenyon[1,3,4*]

Abstract

Background: NSC260594, a quinolinium derivative from the NCI diversity set II compound library, was previously identified in a target-based assay as an inhibitor of the interaction between the HIV-1 (ψ) stem-loop 3 (SL3) RNA and Gag. This compound was shown to exhibit potent antiviral activity. Here, the effects of this compound on individual stages of the viral lifecycle were examined by qRT-PCR, ELISA and Western blot, to see if its actions were specific to the viral packaging stage. The structural effects of NSC260594 binding to the HIV-1 gRNA were also examined by SHAPE and dimerization assays.

Results: Treatment of cells with NSC260594 did not reduce the number of integration events of incoming virus, and treatment of virus producing cells did not affect the level of intracellular Gag protein or viral particle release as determined by immunoblot. However, NSC260594 reduced the incorporation of gRNA into virions by up to 82%, without affecting levels of gRNA inside the cell. This reduction in packaging correlated closely with the reduction in infectivity of the released viral particles. To establish the structural effects of NSC260594 on the HIV-1 gRNA, we performed SHAPE analyses to pinpoint RNA structural changes. NSC260594 had a stabilizing effect on the wild type RNA that was not confined to SL3, but that was propagated across the structure. A packaging mutant lacking SL3 did not show this effect.

Conclusions: NSC260594 acts as a specific inhibitor of HIV-1 RNA packaging. No other viral functions are affected. Its action involves preventing the interaction of Gag with SL3 by stabilizing this small RNA stem-loop which then leads to stabilization of the global packaging signal region (psi or ψ). This confirms data, previously only shown in analyses of isolated SL3 oligonucleotides, that SL3 is structurally labile in the presence of Gag and that this is critical for the complete psi region to be able to adopt different conformations. Since replication is otherwise unaffected by NSC260594 the flexibility of SL3 appears to be a unique requirement for genome encapsidation and identifies this process as a highly specific drug target. This study is proof of principle that development of a new class of antiretroviral drugs that specifically target viral packaging by binding to the viral genomic RNA is achievable.

Keywords: HIV-1, Antiretroviral drugs, Packaging, RNA structure

*Correspondence: jck33@cam.ac.uk
[1] Department of Medicine, University of Cambridge, Addenbrookes Hospital, Box 157, Level 5, Hills Rd, Cambridge CB2 0QQ, UK
Full list of author information is available at the end of the article

Background

For HIV there is now a range of effective drugs targeting different stages of viral replication yet mutational escape from all of these has been achieved by the virus. Between the stages of integration of the provirus and virion maturation however there is a therapeutic gap wherein many of the processes are not virus specific. Exceptions include compounds targeting viral assembly binding to the HIV-1 Gag protein, or its NC domain, and interfering with genome encapsidation or virion maturation [1–3]. Others work by destabilizing and ejecting zinc from the two zinc knuckles in the NC domain. Almost all of them target the viral protein [4, 5].

The HIV-1 5′UTR is around 330 nucleotides (nt) long and contains a series of overlapping *cis*-acting signals that interact with viral and cellular factors to control the viral lifecycle. It is far more highly conserved than the coding regions that follow it, with little or no variation observed between hundreds of sequenced clinical isolates in some sections [6]. This degree of conservation against the background of a high viral mutation rate indicates the essential nature of these sequences, the structures they subtend and the inability of the virus to tolerate mutation within them. Compounds that target the Gag-RNA interaction by binding to highly conserved sequences of the RNA instead of the cognate protein may provide far less scope for mutational escape. One such region that shows extremely limited variation is a 14 nt stretch just upstream of the Gag translation initiation site. This RNA folds into a stable helix known as SL3, which has been shown to be of importance during the viral packaging process, by specifically binding the Gag protein in a high-affinity interaction [7, 8].

NSC260594 (hereafter referred to as NSC) is a quinolinium derivative that we identified in a high-throughput screen for compounds that interfere with the interaction between SL3 and Gag [9]. By NMR analysis this compound was shown to bind specifically to the terminal loop of SL3 where Gag is known to interact. NSC was also shown to inhibit production of infectious HIV in vitro at micromolar concentrations. The SL3 sequence has few predicted counterparts in the human genome, and the compound displays low cytotoxicity. This suggested that within infected cells its effect is related to it inhibiting genome encapsidation although this was not formally shown.

However, despite circumstantial evidence that NSC binds to SL3 and interferes with Gag binding in vitro, the HIV-1 5′UTR contains multiple *cis*- and *trans*-acting functions, including other protein binding sites and there is evidence that it undergoes a range of conformational changes in order to facilitate viral replication [10]. It is therefore possible that a compound that binds to SL3 exerts its antiviral effects at other and/or multiple stages of the HIV lifecycle. NSC has also been shown in a recent screen to be a protein binding inhibitor of the dengue virus 2 protease [11].

To clarify the true effect of NSC in a replicating viral system and to attempt to validate the potential for development of RNA targeting antiretrovirals, and antipackaging inhibitors, we examined the specific role of NSC in inhibiting multiple stages of the HIV-1 lifecycle. We show that NSC specifically targets the viral packaging process, resulting in significantly lower presence of genomic viral RNA inside the virions. It does not affect the steady state levels of viral unspliced genomic RNA (gRNA) or Gag protein inside cells. It does not interfere with viral reverse transcription or integration. When assessed by immunoblot we find that NSC does not significantly reduce viral particle production.

To further evaluate the mechanism by which NSC interferes with gRNA encapsidation the structural changes induced by NSC binding to the viral leader were studied by SHAPE chemistry. We show that NSC binding to SL3 has effects on several parts of the leader RNA, including SL3 itself. These changes were not observed when NSC was incubated with an SL3 deleted RNA. These findings suggest that NSC disrupts the interaction of SL3 with Gag but also, through stabilizing SL3, it leads to reduced conformational flexibility of the wider backbone structure of the RNA in the 5′leader. These observations reveal the critical nature of structural change in the RNA during the encapsidation process and provide proof-of-principle that specific RNA-targeting anti-packaging compounds are an achievable therapeutic aim.

Methods

Reagents

NSC260594 (Benzamide, 4-[(1-methyl-6-nitro-4(1H)-quinolinylidene)amino]-*N*-[4-[(1-methyl-4(1H)-pyridinylidene)amino]phenyl]-, molecular weight 505.0) was supplied by NCI-DTP (www.dtp.cancer.gov). It was dissolved in DMSO at 20 mM and stored at −20 °C. Plasmids pSVC21ΔBglII [12] pSVC21ΔBglIIΔp1 [13, 14] pVSVG [15] and pBluescript were grown in *E. coli* DH5α strain and purified using Maxiprep columns (Qiagen), and used to produce HIVΔEnv (WT) and HIVΔEnvΔpI (ΔpI) viruses. The ΔpI mutation is a 19-nt deletion of the SL3 region. Jurkat cells were maintained in RPMI 1640 medium (Gibco, Life Technologies) supplemented with 10% fetal bovine serum (Gibco, Life Technologies), penicillin [100 U/ml] and streptomycin [100 μg/ml] (Gibco, Life Technologies). 293T and TZMbl cells were maintained in DMEM containing 10% (v/v) FCS, 100 U/mL penicillin and 100 μg/mL streptomycin (w/v). TZMbl is a reporter cell line derived from HeLa cells expressing

CXCR4, engineered to express CD4 and CCR5 [16], the firefly luciferase gene and the *E. coli* β-galactosidase gene under the control of the HIV-1 LTR [16, 17]. It was obtained from the NIH AIDS Research and Reference Reagent Program. Full-length LAI virus was produced by transfection of pLAI, an infectious molecular clone of HIV-1 LAI, into 293T cells for 48 h and the supernatant harvested and cleared of cell debris by centrifugation at 3488×*g* for 5 min. An LAI plasmid-free virus stock was produced by infecting Jurkat cells with virus-containing supernatant followed by serial passaging in Jurkat cells. The LAI containing supernatant was cleared of cell debris by centrifugation at 2671×*g* for 5 min prior to the integration assay described below.

qRT-PCR to measure intracellular and extracellular HIV RNA levels

293T cells were plated 24 h before transfection in six-well plates at a density of 10^6 cells/well. Transfections were performed with TransIT-LT1 (Mirus) using 150 ng of either pSVC21ΔBglII or pSVC21ΔBglIIΔp1 and 50 ng of pVSVG and following manufacturer's instructions. pBluescript up to 1 μg was added to all wells as a bulking agent for transfection and to ensure even amounts of DNA were added to each well. 6 h post transfection cells were treated with NSC dissolved in DMSO to a final concentration of 50 μM or an equal volume of DMSO only. 24 h post treatment, supernatants and cellular fractions were collected. RNA was extracted from cells using the Qiagen RNeasy kit according to manufacturer's instructions, but without DNase treatment on the columns. DNA was subsequently removed by DNase treatment of a smaller volume of the purified nucleic acid with TURBO DNase (Thermo Fisher Scientific) in DNase buffer for 2 h at 37 °C and the RNA was recovered with phenol–chloroform extraction and ethanol precipitation. Supernatants were harvested for purification of RNA from virions. Initially supernatants were clarified at 5600×*g* for 10 min to pellet cell debris. Virions were then purified by centrifuging a 2:1 ratio of supernatant: 8.4% Optiprep in PBS at 21,500×*g* for 90 min. Virions were resuspended in 10% of the original supernatant volume of PBS followed by the addition of 10× that volume of Proteinase K buffer (50 mM Tris–Cl pH 7.5, 100 mM NaCl, 10 mM EDTA, 1% SDS, 100 μg/ml proteinase K, 100 μg/ml yeast tRNA) and incubation for 30 min at 37 °C. RNA was recovered with phenol–chloroform extraction and ethanol precipitation, resuspended in 1× DNase buffer and treated with 1/10th volume of TURBODNase (Thermo Fisher Scientific) for 60 min at 37 °C, phenol–chloroform extracted and ethanol precipitated once more and resuspended in water.

Cytoplasmic RNA extraction

293T cells were transfected and treated with NSC as described above. Cytoplasmic RNA was extracted with the RNeasy Mini kit (Qiagen) following the manufacturer's supplementary protocol with minor modifications. In brief, 24 h post-NSC treatment, 293T cells were washed in PBS followed by cell lysis with 175 μl chilled RLN buffer (50 mM Tris–Cl, pH8, 140 mM NaCl, 1.5 mM $MgCl_2$, 0.5% NP-40 (v/v)). Cells were removed by scraping and incubated on ice for 5 min, followed by centrifugation for 2 min at 300×*g*. The supernatant was removed and mixed with 600 μl RLT buffer (Qiagen), followed by incubation for 15 min at room temperature then vortexed. Samples were temporarily stored at −80 °C, and 430 μl 96–100% ethanol was added to each sample prior to loading of samples onto RNeasy spin columns according to the manufacturer's instructions without on-column DNase digestion. The RNA was eluted twice in 30 μl nuclease free H_20. Extracted RNA was subjected to DNase treatment with the Turbo DNA-*free*™ Kit (Thermo Fisher Scientific). In brief, 1 μg RNA was treated with 1 μl TURBO DNase in 5 μl 10× TURBO DNase Buffer in a total volume of 50 μl for 1 h at 37 °C. After 1 h, 5 μl DNase inactivation reagent was added to each reaction and incubated at room temperature for 5 min. The reactions were spun at 10,000×*g* for 1.5 min and the supernatant containing DNase-treated RNA was transferred to a fresh tube.

HIV *gag* qRT-PCR

DNase-treated RNA was reverse transcribed with the High Capacity cDNA Reverse Transcription Kit (Thermo Fisher Scientific) according to the manufacturer's instructions with RNase inhibitor added to the reaction, or with the addition of RNasin® Ribonuclease Inhibitor (Promega). The PCR cycling conditions were; 25 °C for 10 min, 37 °C for 120 min, 85 °C for 5 min followed by a holding step at 4 °C. The cDNA was further diluted to 1–4 ng and subjected to HIV *gag* qPCR with 25–50 nM HIV *gag* forward primer; HIV *gag* 6F: 5′-CATGTTTTCAGCATTATCAGAAGGA-3′, and 25–50 nM HIV *gag* reverse primer; HIV *gag* 84R: 5′-TGCTTGATGTCCCCCCACT-3′ and 100 nM HIV gag probe: 5′-[6FAM]CCACCCCACAAGATTTAAACACCATGCTAA[BHQ1]-3′ (Sigma) [18] in a reaction buffer containing 2× Taqman Fast Advanced Master Mix, (Thermo Fisher Scientific). The HIV *gag* qPCR conditions were as follows: Step 1. 50 °C for 2 min. Step 2, 95 °C for 20 s. Step 3. 95 °C, 3 s followed by 60 °C for 30 s, for 40 cycles.

To determine the relative amounts of HIV gRNA and β-actin in each intracellular sample, C_t values were interpreted against standard curves generated on the same qRT-PCR plate under the same conditions, using known quantities of β-actin standard DNA template reagents (401970 TaqMan, Thermo Fisher Scientific) or pSVC21ΔBglII. The β-actin primer sequences and qPCR cycling conditions are described below. β-Actin cDNA template concentrations were in the range of 500 pg–4 ng. Intracellular HIV-1 gRNA levels were then normalized against the β-actin RNA level in each sample versus the average β-actin level found in WT samples, using the following formula: average β-actin level of WT samples/β-actin level of sample. This value was then used to multiply the HIV gRNA content of each sample. The HIV gRNA value of each extracellular sample was then divided by the normalized HIV gRNA level in the corresponding intracellular sample. Outliers were defined as values > the value of the third quartile plus 1.5× the interquartile range, or < the value of the first quartile minus 1.5× the interquartile range [19] 8% of values, evenly distributed between WT, NSC and Δp1 samples, fell into this range and were excluded from further analyses. All data are presented as levels relative to the wild-type average, which was set to 1.

Western blots

293T cells were transfected and treated with NSC as above. Supernatants were removed and clarified to remove cellular debris at 16,000×*g* for 2 min. Viral particles were inactivated with 0.1% Empigen for 30 min at 56 °C and supernatants were stored at − 80 °C. Cells were washed with PBS and lysed with 500 µL 1× CCLR (cell culture lysis reagent, Promega) for 15 min with gentle rocking, transferred to microfuge tubes and clarified at 13,800×*g* for 1 min before storage at − 80 °C. Samples were denatured at 95 °C for 5 min in a 1:1 (v/v) ratio of Laemmli loading buffer (Biorad) containing 710 mM 2-mercaptoethanol and loaded onto SDS-PAGE gradient gels (Biorad). Gels were electrophoresed at 180 V for 1 h and transferred to nitrocellulose membranes for 20 min at 15 V using Biorad semi-dry transfer apparatus and Towbin transfer buffer (25 mM Tris, 192 mM glycine, 20% (v/v) methanol, pH 8.3). Membranes were blocked in PBST (PBS with 0.05% Tween-20) and 5% (w/v) nonfat milk powder for 30 min at room temperature and incubated with monoclonal anti-Gag p55/p24 (ARP 313, CFAR) in 2.5% (w/v) nonfat milk powder in PBST overnight at 4 °C. Membranes were washed for 4 × 10 min with PBST and incubated with secondary antibody, anti-mouse HRP antibody, (Cell Signalling Technology) for 30–60 min at room temperature, before washing with PBST and PBS and developing with ECL reagent (Thermo Fisher Scientific) according to the manufacturer's instructions. Bands were quantitated using Image J analysis. Intracellular Gag levels were normalized against GAPDH levels by stripping and reprobing the same blots with anti-GAPDH antibody (Abcam).

Gag p24 ELISA

96-well microplates (white, half-area, high binding, Greiner Bio-One) were coated overnight with 25 µl of 10 µg/ml anti-HIV-1 p24 Gag antibody (D7320, Aalto Bio Reagents) in 0.1 M $NaHCO_3$. The following day, the coating antibody was removed and plates were blocked for 1 h in 5% Bovine Serum Albumin (Sigma-Aldrich) in 1× TBS, and then washed four times in 1× TBS. Plates were loaded with 25 µl of inactivated sample and a standard curve of HIV-1 p24 antigen (AG6054, Aalto Bioreagents), ranging from 0.1 to 10 ng/ml, diluted in 0.05% Empigen-TBS and incubated for 1.5 h. Plates were washed in TBS as above and incubated with anti-HIV-1-p24 alkaline phosphatase conjugate (BC1071-AP) (Aalto Bioreagents) diluted 1:16,000 in 2% milk, 20% sheep serum (Sigma) 0.05% Tween-20 in 1× TBS for 1 h. Plates were washed 4× in PBS-0.1% Tween 20 and incubated with Lumiphos Plus (Lumigen) for 30 min in the dark. Luminescence levels were measured on a GloMax Multi + Multimode Reader instrument (Promega).

Infectivity assay

To measure infectivity, 6×10^4 TZMbl cells were plated per well in 24-well plates. The following day, 70 µl of virus-containing medium, or medium only, was mixed with 12.5 µl DEAE dextran (50 µg/µl) and added to each well containing TZMbl cells in a total volume of 500 µl DMEM containing 10% (v/v) FCS, 100 U/ml penicillin and 100 µg/ml streptomycin (w/v). 24-h post-transduction, the cells were lysed in 150 µl 1× CCLR buffer (cell culture lysis reagent, Promega) for 15 min at room-temperature. 5 µl of cell lysates was added to a white, half area, medium binding, 96-well microplate (Grainer Bio-One) and 25 µl of Luciferase Assay Reagent (Promega) was added through injectors with a GloMax Multi + Multimode Reader instrument (Promega) and luminescence levels were measured with the following settings; 0.5 s delay, speed 100 µl/s and a 2 s integration time. The average background reading from control cells transduced with medium only was subtracted from the reading from each experimental sample. These background values were less than 2% of the value of the experimental readings. Data were normalized to the average of the wild-type in each independent experiment.

Integration assay

Jurkat cells were prepared to a density of 5×10^5 cells/well in 24-well plates on day 1 in 500 µl medium. On day 2, Jurkat cells were pre-treated with NSC or raltegravir (NIH AIDS Reagent Program) for 6 h in a final concentration of 100 or 1 μM respectively in triplicate wells. After 6 h, 500 μl pLAI-containing supernatant at a p24 concentration of 365 ng/ml was added to each well, or culture medium was added to control wells. The initial drug concentration was thereby reduced by half during the infection step and the infection was continued for 24 h in a final concentration of 50 μM NSC or 500 nM raltegravir (NIH AIDS Reagent Program). 24 h post- infection, cells were harvested by centrifugation for 5 min at 664×*g* and the DNA was extracted with DNeasy Blood and Tissue kit (Qiagen), according to the manufacturer's instructions with the following minor modifications; The incubation step with Buffer AL, was performed for 30 min at 56 °C, and the DNA was eluted in nuclease-free water (Thermo Fisher Scientific).

Alu-Gag PCR

Integrated HIV DNA was amplified with the *Alu-gag* PCR based on a previously published method [20]. 1 μg of extracted DNA was used as template in each PCR containing 1× Colorless GoTaq Reaction buffer (Promega), 100 nM Alu Forward primer: 5′GCCTCCCAAAGTGCTGGGATTACA-3′ and 600 nM *gag* Reverse primer: 5′GTTCCTGCTATGTCACTTCC-3′ [20] PCR nucleotide mix (Promega) [200 μM], Go Taq G2 DNA polymerase (M7841 Promega) 2.5 U, in a final reaction volume of 50 μl. Samples were heated to 95 °C for 2 min. The DNA was amplified during the 40 PCR cycles as follows: 95 °C, 15 s; 50 °C, 15 s; 72 °C, 5 min.

The *Alu-gag* PCR products were purified with the QIAquick PCR purification kit (Qiagen) according to the manufacturer's instructions. The purified PCR products were eluted in 30 μl nuclease free water (Ambion).

Late RT PCR

Late RT PCR products were amplified by PCR using the pR 5′-AGACCAGATCTGAGCCTGGGAG-3′ and pMA′ 5′-CTGACGCTCTCGCACCC-3′ primers at a final concentration of 100 nM for each primer [21]. 400 ng DNA, extracted from LAI-infected or uninfected control cells, was used as template in each PCR reaction containing 1× Green GoTaq Reaction buffer (Promega), 200 μM PCR nucleotide mix (Promega), 100 nM pR primer, 100 nM pMA′ primer, and Go Taq G2 DNA polymerase (M7841 Promega) at a final concentration of 1.25 U, in a reaction volume of 50 μl. The PCR cycling conditions were as follows; 95 °C, 2 min; 40 cycles at 95 °C, 30 s; 64 °C, 30 s; 73 °C, 30 s; followed by 73 °C for 5 min.

HIV-1 *gag* PCR

To quantify the amount of integrated HIV-1 DNA, 2 μl of the purified *Alu-gag* PCR products were used as template in qPCR in a final volume of 10 μl and subjected to HIV *gag* qPCR as described above with 25 nM HIV *gag* forward primer and HIV 25 nM *gag* reverse primer and 100 nM HIV *gag* probe. 20–40 ng non-*Alu-gag* amplified DNA template was included as negative control for background levels.

Beta-actin qPCR

20 ng of input DNA for the *Alu-gag* PCR was used as template to amplify *β-actin* with the following primers: *β-actin* forward primer: 5′-GAGCGGTTCCGCTGCCCTGAGGCACTC-3′ and *β-actin* reverse primer: 5′-GGGCAGTGATCTCCTTCTGCATCCTG-3′ at 40 nM each (final concentration) [22] in a qPCR reaction containing 2× Fast SYBR Green Master Mix (Applied Biosystems). *β-Actin* was amplified with the following qPCR program on a 7500 Fast Real Time PCR System (Applied Biosystems): Step 1. 50 °C for 2 min. Step 2. 95 °C for 20 s. Step 3. 95 °C for 3 s followed by 60 °C for 30 s. Step three was repeated for 40 cycles. A dissociation stage was included at: 95 °C for 15 s, 60 °C for 20 s, 95 °C for 15 s, 60 °C for 15 s.

In vitro dimerization assays

Nucleotides 1–411 of the genome were prepared by in vitro transcription using DNA templates and purified as described previously [23]. For each sample, 10.5 pmol RNA was renatured by heating to 95 °C for 2 min, snap-cooling on ice for 2 min and diluting to 1 μM in 10 mM Tris–HCl, 140 mM KCl, 1 mM NaCl, 1 mM $MgCl_2$. NSC was added at 10× molar ratio either before or immediately after snap-cooling. RNA was incubated at 37 °C for 1–2 h and resolved by native agarose gel electrophoresis alongside an RNA ladder (Riboruler low range, Thermofisher Scientific) and ethidium bromide staining. Bands were visualized on a UV transilluminator and quantified using Image J software. Each independent experiment was performed using a different in vitro transcribed RNA preparation.

SHAPE analysis of RNA

RNA was transcribed and purified as described [23]. For each sample, 21 pmol of wild-type or Δp1 packaging signal RNA was renatured by heating to 95 °C for 2 min, snap-cooling on ice for 2 min and diluting to 1 μM in 10 mM Tris–HCl, 140 mM KCl, 1 mM NaCl, 1 mM $MgCl_2$. RNA was incubated at 37 °C for 2 h and each sample was divided into two equal volumes. One half was treated with a 10× molar ratio of NSC in DMSO for 10 min at room temperature (RT) and the other received an equal volume of DMSO only. Tubes were again divided into two equal volumes before addition of

10 mM (final concentration) NMIA in DMSO or DMSO only. The RNA samples were purified by ethanol precipitation after 45 min incubation at 37 °C and reverse transcribed as described previously. Samples were analyzed by capillary electrophoresis in an ABI 3730 XL sequencer and quantified with SHAPEfinder software [19]. Data were normalized as described previously [24]. Outliers were defined as above and were removed from the dataset. These constituted less than 8% of values. The average reactivities were then used as SHAPE constraints in RNAstructure [25] to predict structural models and their free energies.

Results

Early phases of the viral lifecycle from entry through integration are unaffected by NSC

NSC was previously observed to prevent Gag interacting with SL3 in vitro and to stabilize the RNA. Structural rearrangement of the RNA, as well as its interaction with viral proteins are integral events during reverse transcription [26–28] and potentially vulnerable to an intervention that prevents protein interaction and stabilizes SL3 structure. Conceivably a nucleotide binding agent might also interfere with the integration step. Perturbation of either of these processes would be detectable by infecting NSC treated/untreated cells with wild-type virions and quantitating integration frequency. Jurkat cells were thus pre-treated for 6 h with 100 µM NSC in 500 µl medium or untreated, and then mixed with an equal volume of medium containing HIV LAI. Viral integration events were determined by *Alu-gag* PCR and HIV-1 *gag* qPCR, based on a previously described method to assay for HIV-1 integration [20]. No statistically significant difference in integration events was seen in Jurkat cells in the presence of NSC compared to untreated pLAI-infected cells 24 h post-infection (Fig. 1). pLAI-infected cells treated with the integrase inhibitor raltegravir were included as a positive control for inhibition of integration, and this led to a greater than 95% reduction in integration events ($p < 0.05$ by t test). HIV-1 integration is thus unaffected by NSC. A similar trend was observed in two further independent experiments. As NSC is not affecting the levels of integrated provirus seen when cells are pre-treated with the compound by implication it is unlikely that the compound interferes with the reverse transcription process, since this should also lead to a decrease in the number of integration events. We confirmed this by briefly investigating reverse transcription by PCR on extracted DNA using the primers pR and pMA' as previously described [21] to detect late reverse transcription products and did not see a change in reverse transcription upon treatment with NSC (data not shown).

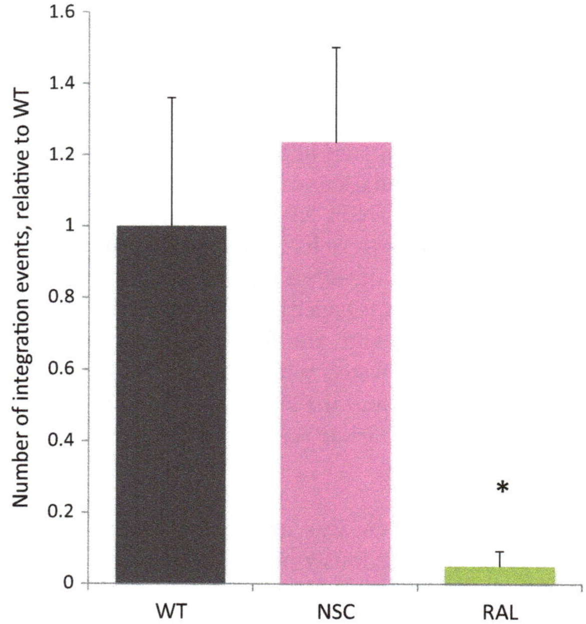

Fig. 1 Viral integration levels determined in the presence and absence of NSC and integration inhibitor. Jurkat cells were pre-treated with NSC [100 µM] or raltegravir [1 µM], and infected with LAI 6 h post-treatment in the presence or absence of 50 µM NSC or 500 nM raltegravir. 24 h post-infection, cells were harvested and subjected to DNA extraction and HIV-1 integration was assayed by *Alu-gag* PCR followed by quantification of integration events by HIV-1 *gag* qPCR. The number of integration events relative to LAI-infected cells (WT) in the presence of NSC or raltegravir is shown. For each sample, variations in the amount and quality of cellular DNA input was accounted for by dividing the number of HIV-1 integration events detected by the HIV-1 *gag* qPCR by the relative actin level detected in the input DNA for the *Alu-gag* PCR. All samples were then normalized relative to one another such that the average WT value was 1. The average of three samples is shown for each condition. Error bars represent SD. *Statistically significant from WT by Student's t test, $p < 0.05$)

Viral Gag production and budding are not significantly affected by NSC treatment

In the previous study ELISA assays on levels of supernatant p24 suggested the existence of lowered virion release from the cell in the presence of NSC. We reinvestigated this finding as the previous study did not take into account the intracellular p24 levels in each sample. Initial ELISA experiments, performed as previously, corroborated an apparent reduction in particle release as they showed around a 40% level of p24 in virions purified from equal volumes of supernatant in the presence of NSC versus wild-type. Strikingly this level was similar to that seen with the Δp1 virus; both results being statistically significant to $p < 0.05$ (Additional file 1). However, when intracellular p24 production was accounted

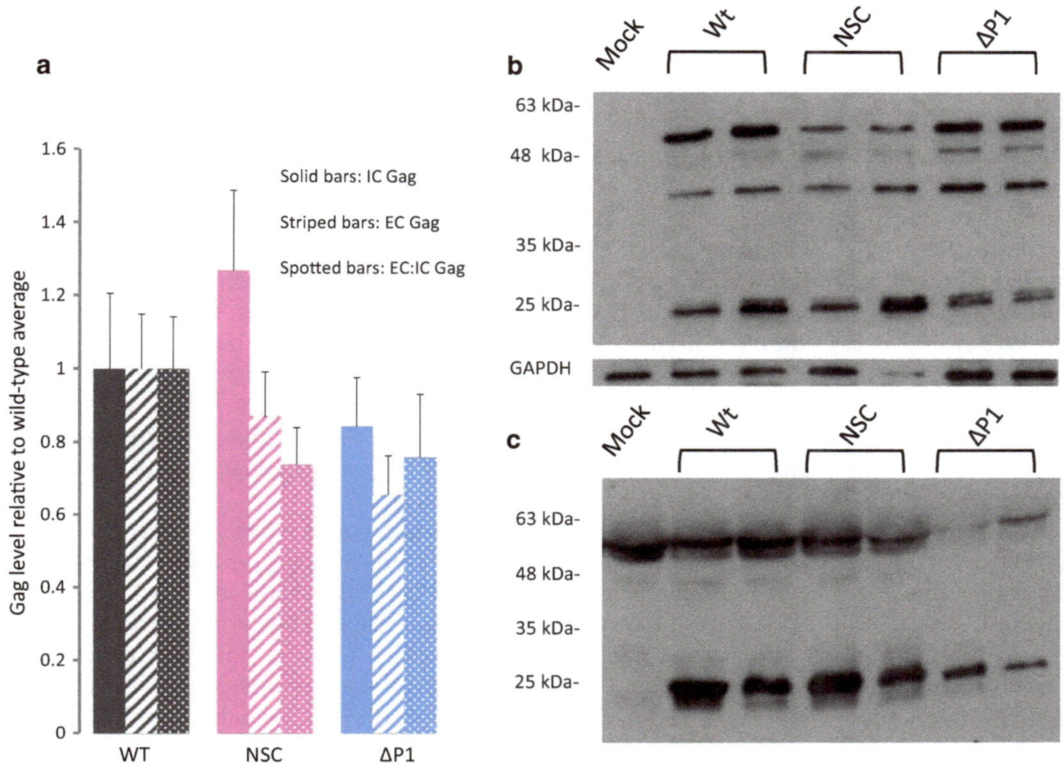

Fig. 2 Viral Gag production in the presence and absence of NSC. 293T cells were transfected with HIVΔEnv (WT and NSC) or HIVΔp1ΔEnv (Δp1) and VSV-G expression plasmids. 6 h post-transfection, cells were treated with 50 μM NSC (NSC) or the equivalent quantity of DMSO (WT and Δp1). 24 h post treatment, supernatants and cells were harvested. Equal volumes of samples were electrophoresed on SDS-PAGE gels and immunoblots performed using anti-p24/p55 antibody. Cellular blots were stripped and reprobed with anti-GAPDH antibody. **a** Results of cellular, supernatant and supernatant:cellular Gag expression levels, quantified using densitometry. Results show an average of four independent experiments, each containing 3–6 replicates (with each replicate examining an independently transfected well of a six-well plate) of each experimental condition. **b** Anti-Gag Western blot showing intracellular Gag expression levels, and anti-GAPDH control. Two experimental replicates, from independently transfected wells are shown (**c**) anti-Gag Western blot showing extracellular Gag expression. Two experimental replicates from independently transfected wells are shown

for (Additional file 1: IC and EC:IC bars) the data were no longer statistically significant. As discussed below detection of p24 levels depends on effective Gag processing and HIV RNA structure can affect the efficiency of viral protease cleavage of the Gag protein during particle maturation [10, 29]. Accordingly, we also assessed the production of Gag protein inside and its export outside cells by Western blot in addition to ELISA to ensure that we were detecting all Gag protein species regardless of their conformation. Using identical preparatory conditions to our qRT-PCR samples, the total amount of Gag observed in cellular fractions was measured by densitometry and normalized to GAPDH levels (Fig. 2a solid colour bars and Fig. 2b). The amount of Gag present in equal volumes of supernatant (Fig. 2a striped bars and Fig. 2c) was also measured, and the ratio of extracellular to intracellular Gag compared (Fig. 2a, spotted bars). Figure 2a shows an average of four separate experiments each containing 3–4 independent transfections of each of wild-type, NSC-treated and Δp1 extracts. No statistically significant difference from wild-type was detected when assessed in any of these independent experiments by *t* test and, crucially, no trend was apparent either. As levels of Gag protein were variable between samples of all three types, as can be seen in Fig. 2b, c, we compiled data from all four independent experiments to ascertain whether a difference would be observed with a larger sample size. Again, no significant difference was detected when data from all four experiments were compared, as presented in Fig. 2a, although on average the relative quantity of viral particles in the supernatant after NSC treatment was 0.87 of that of wild-type, and 0.65 for SL3 deletion virus. This suggests that there may be a small defect in particle release upon NSC treatment, and a larger defect upon SL3 deletion, but that neither of these is statistically significant using sample sizes of up to 13 replicates. Although solution based methods of quantitation are often thought to be more robust, the main difference in detection mechanism between ELISA and Western blot assays is that Western blots analyze all Gag products, regardless of their 3D conformation whereas the ELISA assay measures the major Capsid protein, p24, in a specific 3D structure. The ELISA assay may thus be influenced by structural perturbations in the Gag protein, in a way that the Western blot assay is not and, in these circumstances, Western blots provide more accurate data on the level of Gag protein released from cells.

In the presence of NSC, the level of viral RNA in the cytoplasm is unaffected but levels released into the supernatant are reduced

The interaction of Gag with SL3 has been proposed by many groups to be important in the packaging process [13, 30–32]. In the absence of NSC, Gag protein was shown to interact with an isolated SL3 RNA stem loop in vitro causing it to unwind and unmasking a fluorophore attached to the 5′ end from a quencher on the 3′ end of the RNA stem-loop. The presence of NSC prevented this unwinding [7, 9]. In a single round infectivity experiment it led to reduced levels of infectious virion production as determined by a Tat-sensitive reporter cell-based assay [9]. In order to ascertain whether, by preventing the native interaction of Gag with SL3, NSC is able to specifically lower gRNA levels inside viral particles, we measured gRNA levels in cellular lysates and purified virions by qRT-PCR in the presence or absence of NSC, with the Δp1 virus serving as a positive control for inhibition of packaging [13]. Samples transfected with pBluescript alone were included as a negative control in each experiment and these did not generate a detectable signal by qRT-PCR, as expected (data not shown). The level of gRNA found in cells was unaffected by the presence of NSC or by deletion of SL3, as shown in Fig. 3a, with no statistically significant differences from wild-type observed by *t* test. However, when the ratio of gRNA incorporated into virions was compared to intracellular levels a clear reduction was observed in the presence of NSC to a similar level to that seen in the SL3 deletion virus: Genomic RNA was reduced to approximately 35% of wild type levels in the presence of NSC compared to 34% for the SL3 deletion virus (Fig. 3b). Δp1 results were very similar to those observed previously by RPA assay [14]. It is possible that these results could reflect a difference in nuclear RNA metabolism or export, as a quinolinium derivative such as NSC could conceivably diffuse into the nucleus and affect these cellular functions. To ascertain whether the decrease in virion gRNA level upon NSC treatment we observed was specific to the packaging process we performed a nuclear:cytoplasmic fractionation and measured cytoplasmic gRNA levels against virion gRNA levels by qRT-PCR as before. The ratio of virion:cytoplasmic gRNA was 18.5% upon NSC treatment and similar to that of the Δp1 virus (22.3%) (Fig. 3c), suggesting that the compound acts after the cytoplasmic phase of the viral lifecycle.

The greater than five fold reduced level of gRNA in the presence of NSC compared to the wild type virus clearly cannot be accounted for by the possible minor reduction in export of Gag. The similarity to the phenotype of an established packaging defective mutant supports that NSC is primarily an inhibitor of RNA encapsidation. Taken together, these results suggest that NSC is specifically lowering the amount of viral RNA inside viral particles.

To further confirm this we measured the infectivity of virions produced in the presence of NSC. The reporter cell line TZMbl was inoculated with equivalent volumes of virus-containing medium and the infectivity measured compared to WT and Δp1 virus 24 h post-infection by luciferase assay. The reduction in infectivity in the presence of NSC was of a similar magnitude to the reduction in gRNA levels detected in virions in the supernatant (26% of wt, Fig. 3d). The effect of the drug was to decrease viral infectivity to a similar level to that seen in the Δp1 virus.

NSC does not prevent gRNA dimerization, but does have widespread stabilizing effects on the 5′UTR and packaging signal RNA structure

gRNA dimerization and RNA encapsidation are intimately linked and interdependent. Although NSC interacts with the loop region of SL3, and SL3 has not been implicated as an important component of the dimerization process, it is possible that NSC is causing inhibition of packaging as a secondary effect by interfering with the ability of the viral gRNA to dimerize. In order to ascertain whether NSC acts by preventing dimerization we performed in vitro dimerization assays in the presence and absence of NSC. Packaging signal RNA (nts 1–411 of the genome) was in vitro transcribed and refolded under dimerization inducing conditions as previously described [23, 33] and examined by native agarose electrophoresis and densitometry. NSC was added either before RNA snap-cooling and refolding (data shown in Fig. 4a, b) or after snap-cooling and before refolding (data not shown). During the in vitro dimerization process, the formation of dimer is a dynamic process, with the amount of dimer increasing over time. Thus we also examined dimerization at an earlier timepoint, (Fig. 4c) to verify that the levels of dimer that form are changing with time in a similar

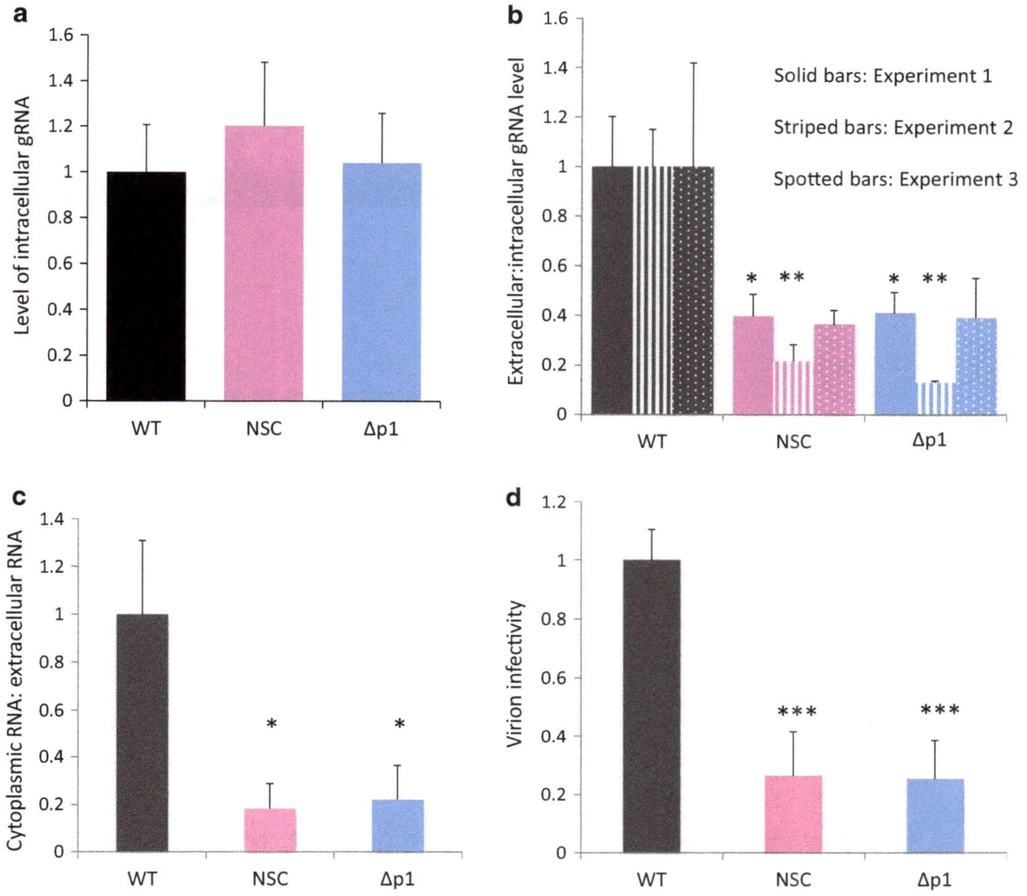

Fig. 3 Viral gRNA levels analyzed by qRT-PCR, in the presence and absence of NSC. 293T cells were treated as in Fig. 2. 24 h post treatment, supernatants and cells were harvested, virions purified from 500 μL supernatant and RNA extracted. **a** gRNA levels in cells, analyzed by qRT-PCR. Each sample was first normalized against β-actin levels and then normalized against the average wild-type gRNA levels. Data shown are the average of 14–18 samples taken from four independent experiments. Error bars represent SD. **b** Ratio of gRNA levels in purified virions from equivalent vo umes of supernatant to gRNA levels inside cells. Intracellular levels were first normalized against β-actin; these values were then used to divide the virion RNA levels for each sample. Data were then normalized against the wild-type average. Data are shown as three independent experiments, each containing 4–9 replicates. **c** Ratio of gRNA levels in purified virions from equivalent volumes of supernatant to cytoplasmic gRNA. Data are representative of two independent experiments. **d** Ability of virions produced in the presence or absence of NSC to transduce cells, measured by luciferase assay. Error bars represent the SD. *$p<0.05$; **$p<0.01$ by Student's *t* test. Data are representative of two independent experiments

manner, with or without NSC treatment. No significant differences were observed, suggesting that NSC is not blocking the ability of the viral RNA to form dimers efficiently.

NSC was originally identified as a potential antiviral by it inhibiting the ability of Gag to unwind a synthetic SL3 RNA molecule. However it was unclear whether the major mechanism of action of NSC was prevention of Gag binding to SL3, or from prevention of the unwinding of SL3 although there was evidence of the latter effect from a rise in melting temperature of SL3 when NSC was bound [9]. To further investigate the structural effects of NSC on RNA structure of the viral 5′UTR we performed SHAPE (selective 2′OH acylation analyzed by primer extension) analysis of the dimeric RNA in the presence and absence of the compound. SHAPE compounds react with the RNA backbone where it is flexible, which occurs mainly in single-stranded regions, and are used to inform structural modeling. In vitro transcribed RNAs were renatured under conditions that favor dimerization, as before. RNAs were treated with NSC in DMSO or DMSO only before probing with NMIA (N-methyl isatoic anhydride) and fragment analysis as performed in previous studies [23, 24]. 12 renatured samples were divided into two to be probed with/without NSC and divided into two again to be probed with/without NMIA,

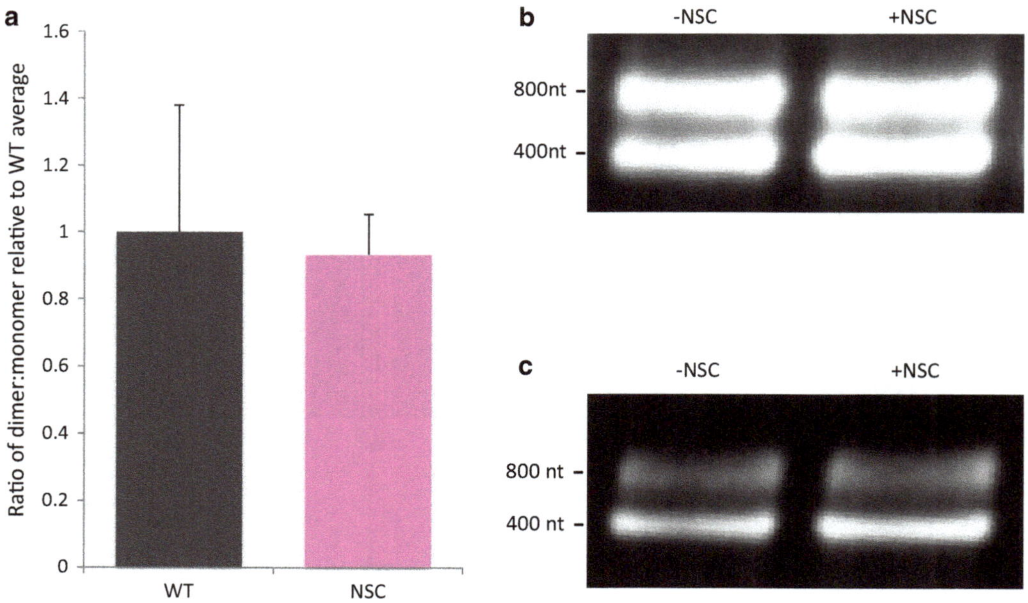

Fig. 4 Effects of NSC on dimerization of viral RNA. In vitro dimerization assays in the presence and absence of 10× molar ratio of NSC. RNA was heated to 95 °C and NSC in DMSO, or DMSO only, was added before snap-cooling on ice. RNA was then incubated for 1–2 h before native gel electrophoresis and densitometry analysis. **a** Dimerisation levels relative to the untreated RNA. Data are an average of five independent experiments, containing 1–3 samples of untreated/treated RNA each. Error bars show SD. **b** Representative samples from RNA renatured in the absence or presence of NSC, with (**b**) 2 h incubation at 37 °C or (**c**) 1 h incubation at 37 °C

using 6 different in vitro transcription preparations of RNA. Data were analyzed with SHAPEfinder software. Structures were modeled using RNAstructure [25] and are shown in Fig. 5a–d, with numerical data in the Additional file 2: Table S1. Backbone acylation sensitivity at each nucleotide is shown in Fig. 5a for wild-type RNA without NSC and for wild-type RNA with NSC in Fig. 5b. These data show that the action of NSC on the packaging signal RNA structure is to stabilize the sugar–phosphate backbone structure, reducing its conformational flexibility. This is best illustrated by the free energy change upon NSC treatment (− 155.3 to − 165.3 kcal/mole). Although most acylation differences upon NSC treatment are small, the vast majority of nucleotides are less reactive to NMIA after NSC treatment, and this extends across the RNA, as is shown in Fig. 5c, which illustrates the reactivity differences before and after NSC treatment. This stabilization occurs at SL3 in particular, as shown in Fig. 5d by the larger reduction in SHAPE reagent acylation of five of the SL3 nucleotides, but additional effects can be seen extending across the entire psi region. To ascertain whether these results were due to off-target, nonspecific effects of NSC binding to other regions of the RNA, we repeated the SHAPE experiments using the Δp1 RNA, with and without NSC treatment. This RNA has a 19 nt deletion at SL3, but is otherwise isogenic. In contrast to the wild type RNA, NSC did not stabilize this transcript (Additional file 3: Fig. S2; Additional file 2: Table S1). This implies that the stabilization of SL3 is responsible for a propagated increase in structural rigidity rather than NSC being a non-specific RNA stabilizing agent.

Discussion

We investigated the ability of NSC to block HIV-1 replication, focusing on individual life cycle stages. When virus is produced in the presence of NSC, this RNA-binding compound specifically reduces the amount of gRNA in virions in the supernatant fraction, compared to the amount of gRNA available for packaging in the cells. This is possibly associated with a non-significant minor decrease in the amount of Gag protein detected in the supernatant in the presence of NSC. The dominant mechanism of action of NSC is to specifically prevent the incorporation of viral gRNA into virus particles with a comparable defect to that produced in an established packaging defective viral mutant, the Δp1 virus. The infectivity of the virions produced upon NSC treatment correlated well with the reduction in gRNA packaging. No effects were seen on other stages of the viral lifecycle again confirming that NSC has a specific mechanism of action at the packaging stage of the viral lifecycle.

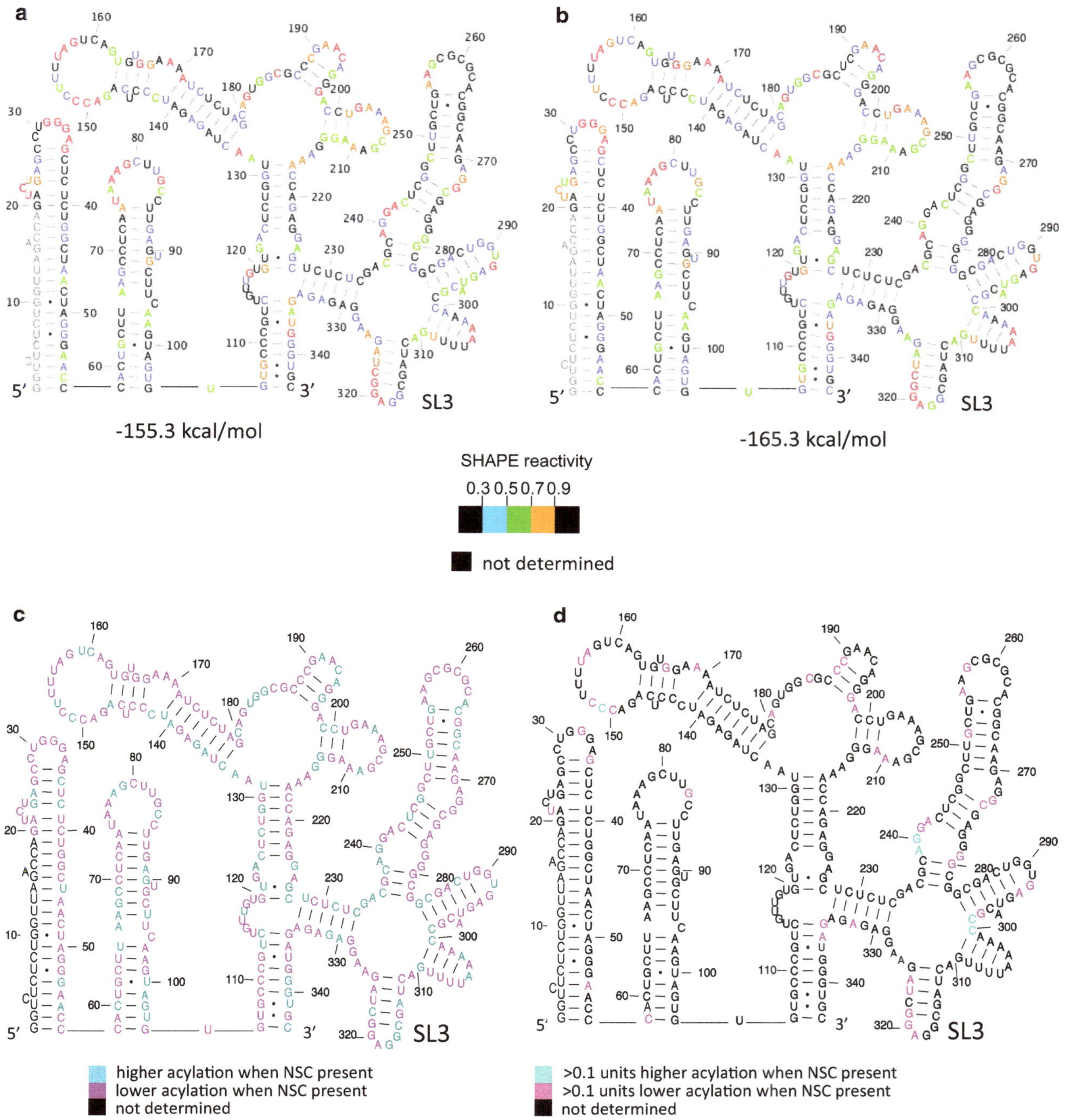

Fig. 5 Structural effects of NSC on the gRNA. The first 411 nts of the gRNA, containing the packaging signal region and beginning of the *gag* ORF were in vitro transcribed and refolded. RNA was treated with NSC in DMSO or DMSO only. Acylation sensitivity at each nucleotide was used to model the structure and predicted free energy. The most stable structure is shown. **a** Predicted RNA structure and stability of DMSO treated RNA. **b** Predicted RNA structure and stability of NSC treated RNA. **c, d** Acylation sensitivity difference upon NSC treatment. **c** Differences > or < 0 and **d** differences > or < 0.1 reactivity units

We cannot fully exclude that equal amounts of gRNA are initially incorporated into virions but that NSC renders the particles less stable, and they disintegrate faster. Although possible, we think this explanation is less plausible, as a previous study using a SL3 deletion virus showed no reduction of p24 protein in the

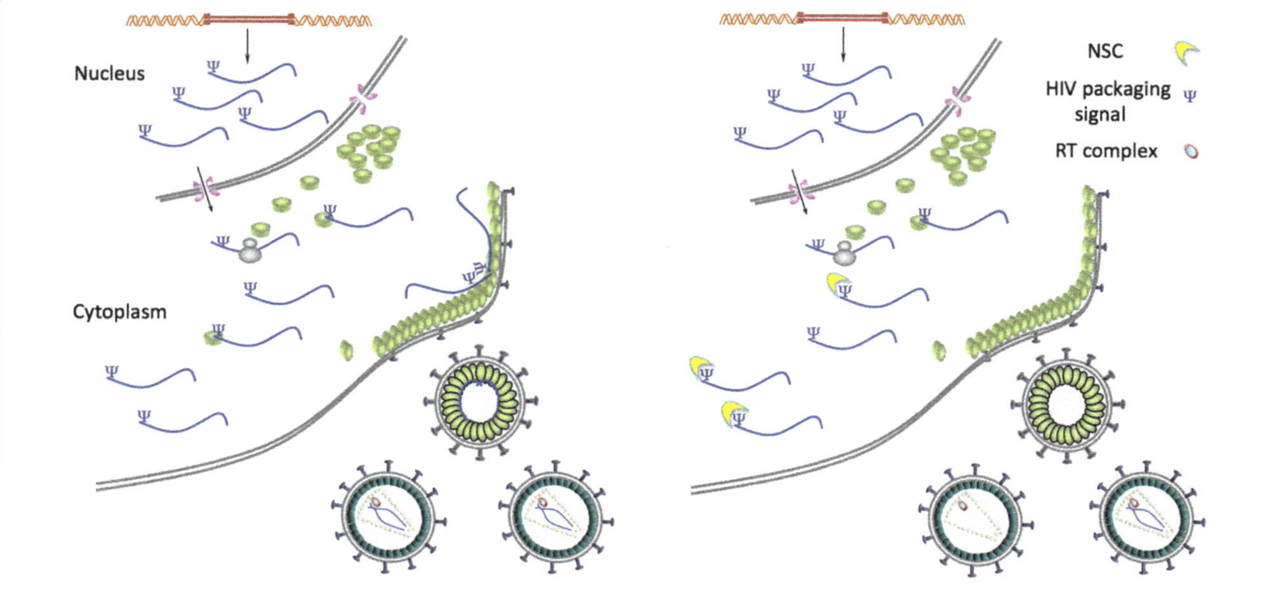

Fig. 6 Schematic diagram showing the RNA packaging impairment upon NSC treatment. NSC may also be present inside virions

virion- associated fraction, versus wild-type virus [34]. It is also possible that NSC does not prevent the initial stages of packaging, but instead destabilizes the RNA inside virions, leading to degradation of the RNA. In practical terms the effect is the same and gRNA levels in virions are lowered, which would be an equally desirable therapeutic outcome.

Our findings are summarized schematically in Fig. 6.

We observed that ELISA and Western blot can give differing results for Gag protein levels in the presence of RNA binding agents or RNA structural deletion mutants. These findings resonate with the known effects of RNA on structural stability of HIV virions [35] and may possibly relate to effects of RNA on Gag processing [36]. This observation has implications for investigation of viral processes that involve different structural changes in Gag and suggests that results should be validated by Western blots, or by alternative analyses relying on quantification of denatured proteins rather than relying only on methods detecting the native state. This is particularly pertinent as the viral maturation process involves multiple cleavages and structural rearrangements of the Gag protein, such as the formation of a new beta-hairpin structure in CA, necessary for proper core formation [37, 38]. Although large and small-scale structural rearrangements of the Gag polyprotein have been proposed to occur [38, 39] little is known about the structures of the cleavage intermediates. In addition, incorrect cleavage order has been shown to affect the viability of virions [40, 41].

HIV-1 RNA has been suggested to adopt a range of different conformers, with each one playing a different role in the viral lifecycle [10, 42, 43]. The SHAPE data presented here illustrate that NSC not only enhances the stability of SL3 but that this also stabilizes the entire packaging signal region, reducing its conformational flexibility. This result suggests that by stabilizing SL3, the 'BMH' form [44] of the RNA is stabilized. In this way, a structural stabilization at one site propagates across the structure, having wide-reaching effects. Acylation sensitivity differences within single-stranded regions are likely to be due to stabilization of the backbone into more or less reactive noncanonically paired conformations, such as by differences in sugar-pucker or noncanonical interactions [45]. Recent publications indicated that although TAR-binding compounds had small longer-range structural effects outside TAR [46, 47], they did not have such far-reaching stabilizing effects. The structure of SL3 is usually similar between structural models of the HIV-1 leader region produced by different groups [48–50] and it has not been proposed to be an integral part of conformational switches in the viral RNA. However, the stabilization of a BMH-like structure by NSC binding to SL3 suggests that unwinding or stabilization of the SL3 site may play a pivotal role in controlling structural changes within the HIV-1 leader RNA.

SL3 has recently been proposed to be of lower importance for Gag binding and viral packaging than nearby motifs [51], however, we show here that a compound that specifically binds to it has reasonably large effects on viral packaging efficiency, making a case for its importance in the packaging process.

Interestingly, our findings indicate that an HIV-1 RNA which is stabilized at SL3 does not show defects at other stages of the lifecycle, suggesting that structural flexibility imparted by this region is not necessary for other lifecycle stages.

Although our results show that in vitro, NSC does not influence the ability of the viral RNA to form dimers, we cannot exclude that it affects the structure or the dynamic stability of the dimers that do form in vivo. A recent model suggests that it is not simply SL1 that remodels into an intermolecular duplex, but also extensive regions around SL1, including the U5:AUG interaction, which was previously thought to be intramolecular [52]. SL3 is positioned in the middle of this proposed intermolecular interaction and its remodeling may be important for formation of the extended intermolecular complex.

Conclusions

By binding to the SL3 RNA in ψ, NSC specifically inhibits the incorporation of HIV gRNA into virions. It does this without affecting earlier processes of the viral lifecycle such as reverse transcription, integration, production and nuclear export of viral gRNA or the steady-state intracellular Gag protein levels. NSC260594 is thus a first-in-class RNA-binding anti-packaging compound. As a small molecule that affects RNA structure and function, it can be used to interrogate the role of RNA structure in the viral lifecycle without the need for mutation of the viral genome.

Additional files

Additional file 1: Figure S1. p24 concentration measured by ELISA. 293T cells were transfected and treated as for Fig. 2. Error bars represent standard deviation. Data shown is for three independent experiments $*p < 0.05$.

Additional file 2: Table S1. Table: Numerical representation of SHAPE reactivity differences at each nucleotide position. Colours shown are the same as those in Fig. 5 and Additional file 3: Fig. S2.

Additional file 3: Figure S2. Structural effects of NSC on the Δp1 RNA. Δp1 RNA was in vitro transcribed, refolded and treated with DMSO only or NSC in DMSO as for Fig. 5. Acylation sensitivity at each nucleotide was used to model the structure and predicted free energy. The most stable structure is shown. **a** Predicted RNA structure and stability of DMSO treated RNA. **b** Predicted RNA structure and stability of NSC treated RNA.

Authors' contributions

CKI, AMLL and JCK designed the study, CKI, JZ, ZL and JCK performed the experiments, CKI, AMLL and JCK analysed the data. CKI, AMLL and JCK wrote the manuscript. All authors read and approved the final manuscript.

Author details

[1] Department of Medicine, University of Cambridge, Addenbrookes Hospital, Box 157, Level 5, Hills Rd, Cambridge CB2 0QQ, UK. [2] Department of Medicine, National University of Singapore, Singapore, Singapore. [3] Department of Microbiology and Immunology, National University of Singapore, Singapore, Singapore. [4] Homerton College, University of Cambridge, Cambridge, UK.

Acknowledgements

The following reagent was obtained through the NIH AIDS Reagent Program, Division of AIDS, NIAID, NIH: Raltegravir (Cat # 11680) from Merck and Company, Inc.

Competing interests

The authors declare that they have no competing interests.

Funding

Funding was provided by the Clinical Academic Reserve, the Biomedical Research Centre (Grant No. RCCT.EFPO) and Medical Research Council (Grant No. RCAG/565)

References

1. Tang C, Loeliger E, Kinde I, Kyere S, Mayo K, Barklis E, et al. Antiviral inhibition of the HIV-1 capsid protein. J Mol Biol. 2003;327(5):1013–20.
2. Tang J, Jones SA, Jeffrey JL, Miranda SR, Galardi CM, Irlbeck DM, et al. Discovery of a novel and potent class of anti-HIV-1 maturation inhibitors with improved virology profile against gag polymorphisms. Bioorg Med Chem Lett. 2017;27(12):2689–94.
3. Adamson CS, Sakalian M, Salzwedel K, Freed EO. Polymorphisms in Gag spacer peptide 1 confer varying levels of resistance to the HIV-1 maturation inhibitor bevirimat. Retrovirology. 2010;7:36.
4. Spearman P. HIV-1 gag as an antiviral target: development of assembly and maturation inhibitors. Curr Top Med Chem. 2016;16(10):1154–66.
5. Waheed AA, Freed EO. HIV type 1 Gag as a target for antiviral therapy. AIDS Res Hum Retroviruses. 2012;28(1):54–75.
6. Los Alamos HIV sequence compendium 2016. https://www.hiv.lanl.gov/content/sequence/HIV/COMPENDIUM/2016/hiv1dna.pdf. Accessed 20 May 2017.
7. Bell NM, Kenyon JC, Balasubramanian S, Lever AM. Comparative structural effects of HIV-1 Gag and nucleocapsid proteins in binding to and unwinding of the viral RNA packaging signal. Biochemistry. 2012;51(15):3162–9.
8. Lever AM. HIV-1 RNA packaging. Adv Pharmacol. 2007;55:1–32.
9. Bell NM, L'Hernault A, Murat P, Richards JE, Lever AM, Balasubramanian S. Targeting RNA–protein interactions within the human immunodeficiency virus type 1 lifecycle. Biochemistry. 2013;52(51):9269–74.
10. Hellmund C, Lever AM. Coordination of genomic RNA packaging with viral assembly in HIV-1. Viruses. 2016;8(7):192.
11. Brecher M, Li Z, Liu B, Zhang J, Koetzner CA, Alifarag A, et al. A conformational switch high-throughput screening assay and allosteric inhibition of the flavivirus NS2B-NS3 protease. PLoS Pathogens. 2017;13(5):e1006411.
12. Richardson JH, Kaye JF, Child LA, Lever AM. Helper virus-free transfer of human immunodeficiency virus type 1 vectors. J Gen Virol. 1995;76(Pt 3):691–6.
13. Lever A, Gottlinger H, Haseltine W, Sodroski J. Identification of a sequence required for efficient packaging of human immunodeficiency virus type 1 RNA into virions. J Virol. 1989;63(9):4085–7.
14. Harrison GP, Miele G, Hunter E, Lever AM. Functional analysis of the core human immunodeficiency virus type 1 packaging signal in a permissive cell line. J Virol. 1998;72(7):5886–96.
15. Aiken C. Pseudotyping human immunodeficiency virus type 1 (HIV-1) by the glycoprotein of vesicular stomatitis virus targets HIV-1 entry to an endocytic pathway and suppresses both the requirement for Nef and the sensitivity to cyclosporin A. J Virol. 1997;71(8):5871–7.

16. Platt EJ, Wehrly K, Kuhmann SE, Chesebro B, Kabat D. Effects of CCR5 and CD4 cell surface concentrations on infections by macrophagetropic isolates of human immunodeficiency virus type 1. J Virol. 1998;72(4):2855–64.
17. Wei X, Decker JM, Liu H, Zhang Z, Arani RB, Kilby JM, et al. Emergence of resistant human immunodeficiency virus type 1 in patients receiving fusion inhibitor (T-20) monotherapy. Antimicrob Agents Chemother. 2002;46(6):1896–905.
18. Palmer S, Wiegand AP, Maldarelli F, Bazmi H, Mican JM, Polis M, et al. New real-time reverse transcriptase-initiated PCR assay with single-copy sensitivity for human immunodeficiency virus type 1 RNA in plasma. J Clin Microbiol. 2003;41(10):4531–6.
19. Vasa SM, Guex N, Wilkinson KA, Weeks KM, Giddings MC. ShapeFinder: a software system for high-throughput quantitative analysis of nucleic acid reactivity information resolved by capillary electrophoresis. RNA. 2008;14(10):1979–90.
20. Liszewski MK, Yu JJ, O'Doherty U. Detecting HIV-1 integration by repetitive-sampling Alu-gag PCR. Methods. 2009;47(4):254–60.
21. Shen N, Jette L, Liang C, Wainberg MA, Laughrea M. Impact of human immunodeficiency virus type 1 RNA dimerization on viral infectivity and of stem-loop B on RNA dimerization and reverse transcription and dissociation of dimerization from packaging. J Virol. 2000;74(12):5729–35.
22. Duffy SaC A. Analysis of HIV-1 RNA splicing. In: Stamm S, Smith CWJ, Lührmann R, editors. Alternative pre-mRNA splicing: theory and protocols. Weinheim: Wiley-VCH Verlag GmbH & Co. KGaA; 2012. https://doi.org/10.1002/9783527636778.ch41.
23. Kenyon JC, Prestwood LJ, Le Grice SF, Lever AM. In-gel probing of individual RNA conformers within a mixed population reveals a dimerization structural switch in the HIV-1 leader. Nucleic Acids Res. 2013;41(18):e174.
24. Kenyon JC, Tanner SJ, Legiewicz M, Phillip PS, Rizvi TA, Le Grice SF, et al. SHAPE analysis of the FIV Leader RNA reveals a structural switch potentially controlling viral packaging and genome dimerization. Nucleic Acids Res. 2011;39(15):6692–704.
25. Reuter JS, Mathews DH. RNAstructure: software for RNA secondary structure prediction and analysis. BMC Bioinform. 2010;11:129.
26. Isel C, Westhof E, Massire C, Le Grice SF, Ehresmann B, Ehresmann C, et al. Structural basis for the specificity of the initiation of HIV-1 reverse transcription. EMBO J. 1999;18(4):1038–48.
27. Seif E, Niu M, Kleiman L. In vitro SHAPE analysis of tRNA(Lys3) annealing to HIV-1 genomic RNA in wild type and protease-deficient virus. Retrovirology. 2015;12:40.
28. McCauley MJ, Rouzina I, Manthei KA, Gorelick RJ, Musier-Forsyth K, Williams MC. Targeted binding of nucleocapsid protein transforms the folding landscape of HIV-1 TAR RNA. Proc Natl Acad Sci USA. 2015;112(44):13555–60.
29. L'Hernault A, Weiss EU, Greatorex JS, Lever AM. HIV-2 genome dimerization is required for the correct processing of Gag: a second-site reversion in matrix can restore both processes in dimerization-impaired mutant viruses. J Virol. 2012;86(10):5867–76.
30. Comas-Garcia M, Davis SR, Rein A. On the selective packaging of genomic RNA by HIV-1. Viruses. 2016;8(9):246.
31. Post K, Olson ED, Naufer MN, Gorelick RJ, Rouzina I, Williams MC, et al. Mechanistic differences between HIV-1 and SIV nucleocapsid proteins and cross-species HIV-1 genomic RNA recognition. Retrovirology. 2016;13(1):89.
32. De Guzman RN, Wu ZR, Stalling CC, Pappalardo L, Borer PN, Summers MF. Structure of the HIV-1 nucleocapsid protein bound to the SL3 psi-RNA recognition element. Science. 1998;279(5349):384–8.
33. Lu K, Heng X, Garyu L, Monti S, Garcia EL, Kharytonchyk S, et al. NMR detection of structures in the HIV-1 5′-leader RNA that regulate genome packaging. Science. 2011;334(6053):242–5.
34. Clever JL, Parslow TG. Mutant human immunodeficiency virus type 1 genomes with defects in RNA dimerization or encapsidation. J Virol. 1997;71(5):3407–14.
35. Wang SW, Noonan K, Aldovini A. Nucleocapsid–RNA interactions are essential to structural stability but not to assembly of retroviruses. J Virol. 2004;78(2):716–23.
36. Potempa M, Nalivaika E, Ragland D, Lee SK, Schiffer CA, Swanstrom R. A direct interaction with RNA dramatically enhances the catalytic activity of the HIV-1 protease in vitro. J Mol Biol. 2015;427(14):2360–78.
37. Mortuza GB, Haire LF, Stevens A, Smerdon SJ, Stoye JP, Taylor IA. High-resolution structure of a retroviral capsid hexameric amino-terminal domain. Nature. 2004;431(7007):481–5.
38. von Schwedler UK, Stemmler TL, Klishko VY, Li S, Albertine KH, Davis DR, et al. Proteolytic refolding of the HIV-1 capsid protein amino-terminus facilitates viral core assembly. EMBO J. 1998;17(6):1555–68.
39. Datta SA, Clark PK, Fan L, Ma B, Harvin DP, Sowder RC 2nd, et al. Dimerization of the SP1 region of HIV-1 Gag induces a helical conformation and association into helical bundles: implications for particle assembly. J Virol. 2015;90(4):1773–87.
40. Muller B, Anders M, Akiyama H, Welsch S, Glass B, Nikovics K, et al. HIV-1 Gag processing intermediates trans-dominantly interfere with HIV-1 infectivity. J Biol Chem. 2009;284(43):29692–703.
41. Lee SK, Harris J, Swanstrom R. A strongly transdominant mutation in the human immunodeficiency virus type 1 gag gene defines an Achilles heel in the virus life cycle. J Virol. 2009;83(17):8536–43.
42. van Bel N, Ghabri A, Das AT, Berkhout B. The HIV-1 leader RNA is exquisitely sensitive to structural changes. Virology. 2015;483:236–52.
43. Lu K, Heng X, Summers MF. Structural determinants and mechanism of HIV-1 genome packaging. J Mol Biol. 2011;410(4):609–33.
44. Abbink TE, Berkhout B. A novel long distance base-pairing interaction in human immunodeficiency virus type 1 RNA occludes the Gag start codon. J Biol Chem. 2003;278(13):11601–11.
45. Mlynsky V, Bussi G. Molecular simulations reveal an interplay between SHAPE reagent binding and RNA flexibilty. J Phys Chem Lett. 2017;9:313–8.
46. Sztuba-Solinska J, Shenoy SR, Gareiss P, Krumpe LR, Le Grice SF, O'Keefe BR, et al. Identification of biologically active, HIV TAR RNA-binding small molecules using small molecule microarrays. J Am Chem Soc. 2014;136(23):8402–10.
47. Abulwerdi FA, Shortridge MD, Sztuba-Solinska J, Wilson R, Le Grice SF, Varani G, et al. Development of small molecules with a noncanonical binding mode to HIV-1 trans activation response (TAR) RNA. J Med Chem. 2016;59(24):11148–60.
48. Wilkinson KA, Gorelick RJ, Vasa SM, Guex N, Rein A, Mathews DH, et al. High-throughput SHAPE analysis reveals structures in HIV-1 genomic RNA strongly conserved across distinct biological states. PLoS Biol. 2008;6(4):e96.
49. Keane SC, Heng X, Lu K, Kharytonchyk S, Ramakrishnan V, Carter G, et al. RNA structure. Structure of the HIV-1 RNA packaging signal. Science. 2015;348(6237):917–21.
50. Stephenson JD, Li H, Kenyon JC, Symmons M, Klenerman D, Lever AM. Three-dimensional RNA structure of the major HIV-1 packaging signal region. Structure. 2013;21(6):951–62.
51. Abd El-Wahab EW, Smyth RP, Mailler E, Bernacchi S, Vivet-Boudou V, Hijnen M, et al. Specific recognition of the HIV-1 genomic RNA by the Gag precursor. Nat Commun. 2014;2(5):4304.
52. Keane SC, Van V, Frank HM, Sciandra CA, McCowin S, Santos J, et al. NMR detection of intermolecular interaction sites in the dimeric 5′-leader of the HIV-1 genome. Proc Natl Acad Sci USA. 2016;113(46):13033–8.

5

New World feline APOBEC3 potently controls inter-genus lentiviral transmission

Yoriyuki Konno[1,2], Shumpei Nagaoka[1,2], Izumi Kimura[1,3], Keisuke Yamamoto[1,4], Yumiko Kagawa[1,5], Ryuichi Kumata[1,6], Hirofumi Aso[1,3,7], Mahoko Takahashi Ueda[8], So Nakagawa[8,9], Tomoko Kobayashi[10], Yoshio Koyanagi[1] and Kei Sato[1,11,12*]

Abstract

Background: The apolipoprotein B mRNA-editing enzyme catalytic polypeptide-like 3 (APOBEC3; A3) gene family appears only in mammalian genomes. Some A3 proteins can be incorporated into progeny virions and inhibit lentiviral replication. In turn, the lentiviral viral infectivity factor (Vif) counteracts the A3-mediated antiviral effect by degrading A3 proteins. Recent investigations have suggested that lentiviral *vif* genes evolved to combat mammalian APOBEC3 proteins, and have further proposed that the Vif-A3 interaction may help determine the co-evolutionary history of cross-species lentiviral transmission in mammals.

Results: Here we address the co-evolutionary relationship between two New World felids, the puma (*Puma concolor*) and the bobcat (*Lynx rufus*), and their lentiviruses, which are designated puma lentiviruses (PLVs). We demonstrate that PLV-A Vif counteracts the antiviral action of APOBEC3Z3 (A3Z3) of both puma and bobcat, whereas PLV-B Vif counteracts only puma A3Z3. The species specificity of PLV-B Vif is irrespective of the phylogenic relationships of feline species in the genera *Puma, Lynx* and *Acinonyx*. We reveal that the amino acid at position 178 in the puma and bobcat A3Z3 is exposed on the protein surface and determines the sensitivity to PLV-B Vif-mediated degradation. Moreover, although both the puma and bobcat *A3Z3* genes are polymorphic, their sensitivity/resistance to PLV Vif-mediated degradation is conserved.

Conclusions: To the best of our knowledge, this is the first study suggesting that the host A3 protein potently controls inter-genus lentiviral transmission. Our findings provide the first evidence suggesting that the co-evolutionary arms race between lentiviruses and mammals has occurred in the New World.

Keywords: Lentivirus, FIV, APOBEC3, Vif, Evolutionary arms race, Puma, Bobcat, PLV, New World

Background

The apolipoprotein B mRNA editing enzyme catalytic polypeptide-like 3 (APOBEC3; A3) proteins are cellular cytidine deaminases that are specifically encoded in mammals but not in other vertebrates [1, 2]. Mammalian A3 proteins, particularly primate A3 proteins, are considered cellular intrinsic immune factors that potently restrict lentiviral replication. To exhibit antiviral activity, some A3 proteins are incorporated into the released progeny virions and enzymatically insert guanine-to-adenine hypermutations into the viral genome, thereby halting viral replication. In turn, the lentiviral protein, viral infectivity factor (Vif), antagonizes the A3-mediated antiviral action by degrading A3 proteins via the ubiquitin/proteasome-dependent pathway (reviewed in [3]).

Elucidating co-evolutionary relationships between hosts and their viruses is an intriguing topic in virology and is crucial to understanding how viruses influence their hosts' evolution and vice versa. Cell-based virological experiments are essential for a better understanding of the evolutionary conflict between mammals and their viruses, including lentiviruses. For instance, since the

*Correspondence: ksato@ims.u-tokyo.ac.jp
[12] Division of Systems Virology, Department of Infectious Disease Control, International Research Center for Infectious Diseases, Institute of Medical Science, The University of Tokyo, 4-6-1 Shirokanedai, Minato-ku, Tokyo 1088639, Japan
Full list of author information is available at the end of the article

interaction between host A3 and lentiviral Vif is species specific, various investigations focusing on the functional relationship between A3 and Vif have recently been conducted in combination with molecular phylogenetic approaches (reviewed in [4]). This strategy stems from the "Red Queen hypothesis [5]", which proposes that host and viral proteins have competed with one another for survival over time [6–8]. However, since most previous observations are based the Old World evolutionary events [4], whether the evolutionary arms race between mammals and lentiviruses has occurred in the New World is unclear.

Feline immunodeficiency virus (FIV) is a lentivirus that was first isolated in 1987 from domestic cats (*Felis catus*) with chronic AIDS-like disorders [9]. Domestic cats encode multiple *A3* genes, and the feline A3 protein (designated A3Z3) potently impairs FIV replication by incorporating into nascent virions [10–15]. In response, FIV Vif antagonizes the antiviral activity of feline A3Z3 by degrading this protein [11–15]. To elucidate the evolutionary relationship between FIV and felids, we have recently reported that the domestic cat *A3Z3* is polymorphic and that a haplotype of the domestic cat A3Z3 is resistant to Vif-mediated degradation by FIVfca, which is the FIV that infects domestic cats [12]. Our findings suggest that the domestic cat *A3Z3* is under positive selection due to evolutionary selective pressure caused by FIVfca or related ancestral viruses [12, 14].

In addition to FIVfca in domestic cats, various FIV types have been identified in other felids, such as FIVple in lions (*Panthera leo*), FIVpco in pumas (*Puma concolor*), and FIVlru in bobcats (*Lynx rufus*) (reviewed in [16]). Interestingly, although these felids become infected with a specific FIV species, Lee et al. [17] recently reported that a subcluster of FIVpco co-circulated in both pumas and bobcats in North America. The authors designated these viruses (FIVpco and FIVlru) puma lentiviruses (PLVs) and re-classified them as follows: PLV-A includes the FIVpco and FIVlru strains co-circulating in both pumas and bobcats, and PLV-B includes FIVpco strains circulating only in pumas (Fig. 1a) [17–19]. This scenario is the first known case indicating cross-species transmission (CST) of a lentivirus between hosts of different genera (*Puma* and *Lynx*) at present. Since *A3* genes are under positive selection [12, 20–22], the A3 sequences are highly variable among hosts of different genera. Furthermore, we can reasonably hypothesize that the specificity in which Vif counteracts the host A3 also differs in each host genus. These findings and insights raise the hypothesis that Vif-A3 interplay between FIV and the two New World felids (puma and bobcat) is closely associated with the mode of CST among lentiviruses, which may illustrate the history of the co-evolutionary arms race between lentiviruses and felids in the New World.

In this study, we perform cell-based virological experiments and demonstrate that the species specificity of the Vif proteins of PLV-A and PLV-B is different between each other. Through the combinational investigations of experimental virology and molecular phylogenetics, we also provide evidence suggesting that bobcat A3Z3 plays a role as the species barrier specifically against PLV-B. Moreover, we determine the amino acid residue responsible for the sensitivity to PLV-B Vif-dependent degradation. This is the first report addressing the co-evolutionary interplay between antiviral A3 of the two New World felids and feline lentiviral Vif.

Results

Bobcat A3Z3 is resistant to PLV-B Vif-mediated degradation

To elucidate the co-evolutionary relationship between PLVs and their host felids, particularly between viral Vif and host A3Z3, first we constructed a phylogenetic tree of FIV *vif* gene. As shown in Fig. 1a, FIVfca and FIVple formed clusters based on their hosts. In sharp contrast, the FIVlru sequences were co-mingled with some FIVpco sequences, whereas the other FIVpco sequences formed a cluster (Fig. 1a). These results are consistent with the phylogeny of the full-length FIV sequences shown in a previous study [17]. According to this previous report, here we designated the FIVlru and co-mingled FIVpco sequences PLV-A and the other FIVpco sequences PLV-B. The sequence identities of PLV-A Vif and PLV-B Vif were 90.1 ± 4.8 and $85.7 \pm 7.1\%$, respectively, whereas the Vif sequence identity between PLV-A and PLV-B was $40.6 \pm 1.1\%$. Thus, the PLV-A Vif and PLV-B Vif sequences clearly differed.

Next, we focused on the host felid evolutionary relationships. Based on comprehensive genetic information, a previous study [23] described the evolutionary history of the felids of interest in this study (Fig. 1b). The common ancestor of the puma (*Puma concolor*) and the bobcat (*Lynx rufus*) is estimated to have crossed the Bering Isthmus from Eurasia to the New World approximately 8.0–8.5 million years ago (Mya) and to have diverged approximately 7.2 Mya [23]. The puma diverged from the cheetah (*Acinonyx jubatus*) approximately 4.9 Mya, whereas the bobcat was branched from the lynx (*Lynx lynx*) approximately 3.2 Mya (Fig. 1b) [23]. Puma and bobcat resided in the New World, whereas lynx and cheetah returned to Eurasia approximately 1.2–1.6 Mya (Fig. 1b) [23]. To assess the phylogenetic relationships of the *A3Z3* genes of these felids, we collected body hairs from a bobcat and a cheetah from Japanese zoos and determined their *A3Z3* sequences (see Additional file 1: Fig. S1 and Additional file 2: Table S1). As shown in

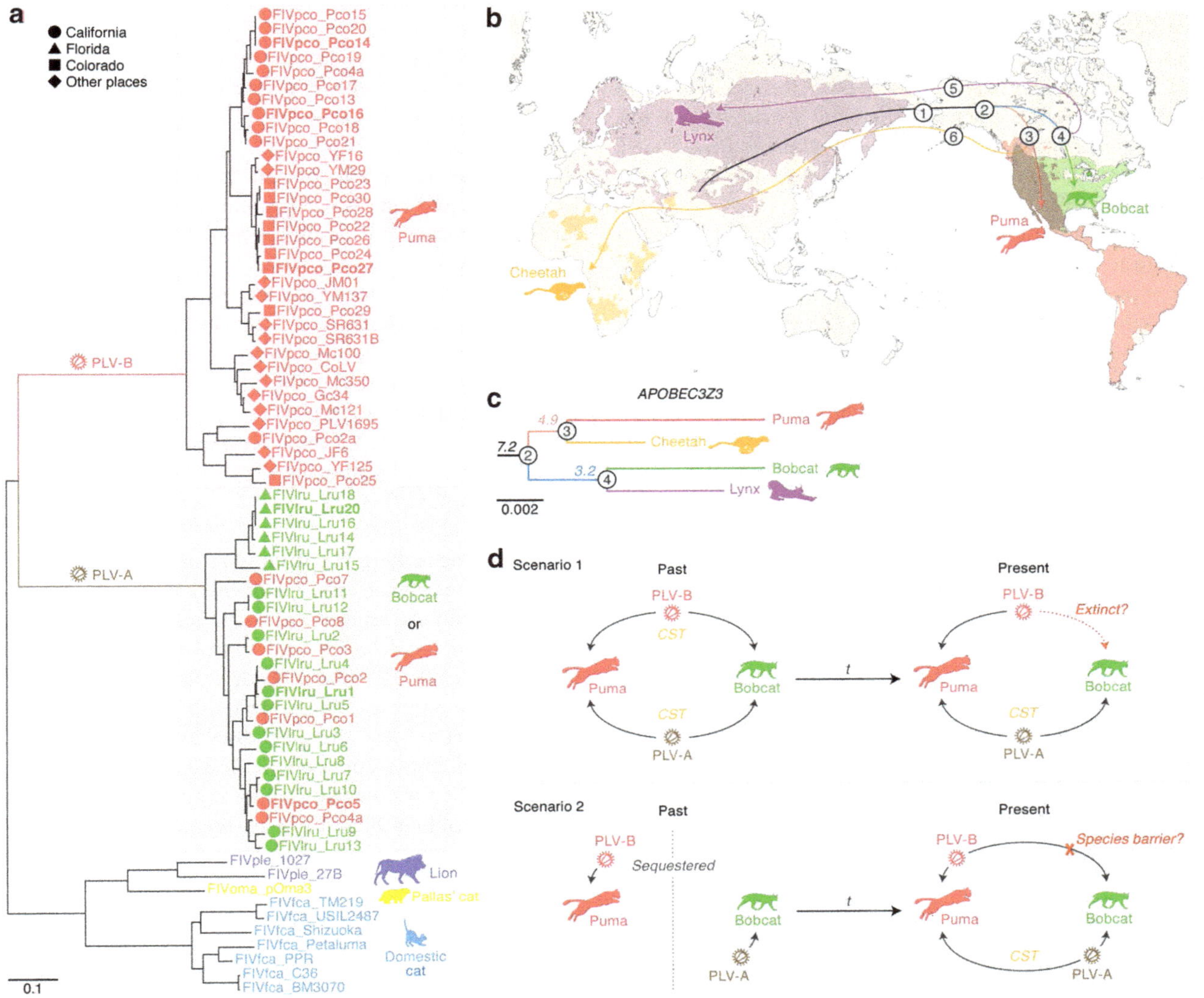

Fig. 1 Evolutionary relationship of FIV *vif* and feline *A3Z3*. **a** Phylogenetic tree of FIV *vif*. This phylogenetic tree was constructed using the ML method and displays the evolutionary relationships among the FIV sequences used in this study. The scale bar indicates 0.1 nucleotide substitutions per site. The bootstrap values are indicated on each node as follows: *, > 50 and **, > 80. The FIV *vif* genes used in this study are indicated in bold. The PLV-A and PLV-B sampling locations are available from previous studies [17, 19, 59] and are indicated by symbols. **b, c** Evolutionary history of felids in the puma and bobcat lineages. **b** The evolutionary events of the four felid species (puma, bobcat, cheetah and lynx) are summarized according to a previous report [23]. The numbers in circles indicate the events as follows: 1, migration of the common ancestor of the puma (*Puma concolor*) and bobcat (*Lynx rufus*) through the Bering Isthmus from Eurasia to the New World (ca. 8.0–8.5 Mya); 2, divergence into the two lineages (ca. 7.2 Mya); 3, divergence of the puma and cheetah (*Acinonyx jubatus*) (ca. 4.9 Mya); 4, divergence of the bobcat and lynx (*Lynx lynx*) (ca. 3.2 Mya); 5, migration of the lynx from the New World to Eurasia (ca. 1.2–1.6 Mya); and 6, migration of the cheetah from the New World to Eurasia (ca. 1.2–1.6 Mya). The current habitats of the puma (red), bobcat (green), cheetah (yellow) and lynx (purple) are indicated by each color and are referred from the IUCN Red List of Threatened Species website (http://www.iucnredlist.org/). **c** Phylogenetic tree of feline *A3Z3*. The bobcat and cheetah *A3Z3* sequences, which were newly identified in this study, were aligned with those of the puma and lynx, and the tree was reconstructed using the ML method. The branch colors correspond to those of the lines in **b**, and the circled numbers on the nodes correspond to those in **b**. The numbers under nodes in italics indicate the age of divergence (Mya) estimated in a previous study [23]. The scale bar indicates 0.002 nucleotide substitutions per site. **d** Two possible scenarios leading to the inter-species PLV transmission between the puma and bobcat. Each scenario is explained in the main text. *CST* cross-species transmission; *t* time

Fig. 1c, the topology of the phylogenetic tree of the *A3Z3* genes form these four felids corresponded to their evolutionary relationships.

Based on these observations, two scenarios are assumed. The first scenario was that both PLV-A and PLV-B co-circulated in pumas and bobcats in the New

World but PLV-B became extinct only in the bobcat population ("scenario 1" in Fig. 1d). The other possibility was that PLV-A and PLV-B were specific viruses in bobcats and pumas, respectively, in the past. After co-habitation of these two feline species in the New World, PLV-A was transferred from bobcats to pumas, whereas PLV-B CST from pumas to bobcats was hampered for unknown reasons ("scenario 2" in Fig. 1d).

To experimentally investigate the interplay between PLVs and their hosts, we analyzed the antiviral activity of the puma and bobcat A3Z3 proteins. In the absence of PLV Vif, both puma and bobcat A3Z3 were expressed at comparable levels and are incorporated into the released particles of *vif*-deficient FIV (strain Petaluma) in a dose-dependent manner (Fig. 2a). In addition, the infectivity of *vif*-deficient FIV was suppressed by both puma and bobcat A3Z3 proteins in a dose-dependent manner (Fig. 2b). Interestingly, although the protein expression (Fig. 2a, top) and incorporation levels in the released virions (Fig. 2a, bottom) were similar between the puma and bobcat A3Z3 s, the antiviral effect was significantly higher for the puma A3Z3 than for the bobcat A3Z3 (Fig. 2b).

Next, we analyzed the sensitivity of puma and bobcat A3Z3 to PLV Vif proteins. We used expression plasmids for three PLV-A Vif strains (Lru1, Lru20 and Pco5) and three PLV-B Vif strains (Pco14, Pco16 and Pco27). The sampling years and locations for each virus are summarized in Table 1. As shown in Fig. 2c, the puma A3Z3 was degraded by all of the PLV Vif proteins tested in this study, and the antiviral effect of puma A3Z3 was counteracted by these PLV Vif proteins (Fig. 2d). In sharp contrast, the bobcat A3Z3 was degraded by the PLV-A Vif proteins but was resistant to degradation mediated by the PLV-B Vif proteins (Fig. 2e, top). Bobcat A3Z3 was incorporated into nascent virions even in the presence of PLV-B Vif (Fig. 2e, bottom), which significantly decreased viral infectivity (Fig. 2f). Taken together, these findings suggest that bobcat A3Z3 is resistant to PLV-B Vif-mediated degradation.

Puma A3Z3 is uniquely sensitive to PLV-B Vif-mediated degradation

As shown in Fig. 1b, the puma and bobcat are evolutionarily similar to the cheetah (*Acinonyx jubatus*) and the lynx (*Lynx lynx*), respectively. To address whether the sensitivity of host A3Z3 to PLV-B Vif is associated with the host phylogeny, we constructed expression plasmids for cheetah and lynx A3Z3 proteins and performed similar experiments. Our results revealed that both cheetah and lynx A3Z3 were degraded by PLV-A Vif but were resistant to PLV-B Vif-mediated degradation (Fig. 3a, b, top). Similar to bobcat A3Z3, the cheetah and lynx A3Z3 proteins were incorporated into the released virions even in the presence of PLV-B Vif (Fig. 3a, b, bottom) resulting in an antiviral effect (Fig. 3c, d). These findings suggest that the PLV-A Vif antagonizes all four feline A3Z3 tested in this study, whereas PLV-B Vif antagonizes only the puma A3Z3.

The threonine residue at position 178 of bobcat A3Z3 confers the resistance to PLV-B Vif-mediated degradation

We compared the amino acid sequence of puma, bobcat, cheetah and lynx A3Z3s and found three amino acids at positions 90, 131 and 178 that consistently differed between the puma and the other felids (Fig. 4a). To assess the positions of these three residues in the tertiary structures of the puma and bobcat A3Z3 proteins, we constructed A3Z3 protein homology models for these felids (Fig. 4b). Residues 90 and 178 are located in alpha-helices 3 and 6, respectively, whereas residue 131 is positioned in the loop (Fig. 4c). Additionally, residues 90 and 178 are localized on the protein surface (Fig. 4d).

To determine the amino acid residue(s) responsible for the resistance to PLV-B Vif-mediated degradation, we constructed lines of bobcat A3Z3 mutants and performed cell-based loss-of-function experiments. Although the bobcat A3Z3 E90K and I131T mutants were resistant to PLV-B Vif-mediated degradation (strain Pco27), the bobcat A3Z3 T178M mutant was degraded by PLV-B Vif (Fig. 5a). We also used combination mutants and

(See figure on next page.)

Fig. 2 Resistance of bobcat A3Z3 to PLV-B Vif-dependent degradation. **a, b** Antiviral effects of the puma and bobcat A3Z3 proteins. Different amounts of HA-tagged expression plasmids for puma or bobcat A3Z3 (0, 50, 100, 200 and 400 ng) and the three plasmids used to produce the *vif*-deficient FIV-based reporter virus (FIV plasmids; pFP93 [200 ng], pTiger-luc [150 ng] and pMD.G [50 ng]) were co-transfected into HEK293T cells. **a** Western blotting. Representative results are shown. **b** FIV reporter assay. FIV infectivity is shown as the percentage of the value of "no A3Z3". **c–f** Puma and bobcat A3Z3 sensitivity to PLV Vif. HA-tagged expression plasmids for puma (**c, d**) and bobcat (**e, f**) A3Z3 (200 ng) and the three plasmids used to produce the *vif*-deficient FIV-based reporter virus (FIV plasmids; pFP93 [200 ng], pTiger-luc [150 ng] and pMD.G [50 ng]) were co-transfected with or without His-tagged PLV Vif expression plasmids (400 ng) into HEK293T cells. **c, e** Western blotting. Representative results are shown. **d, f** FIV reporter assay. FIV infectivity is shown as the percentage of the value of "no A3Z3". In **b**, asterisks indicate significant differences ($P < 0.05$ by Student's t test) between puma A3Z3 and bobcat A3Z3. In **d** and **f**, asterisks indicate significant differences ($P < 0.05$ by Student's t test) versus "no Vif". The assays were independently performed in triplicate. Data represent averages with SDs

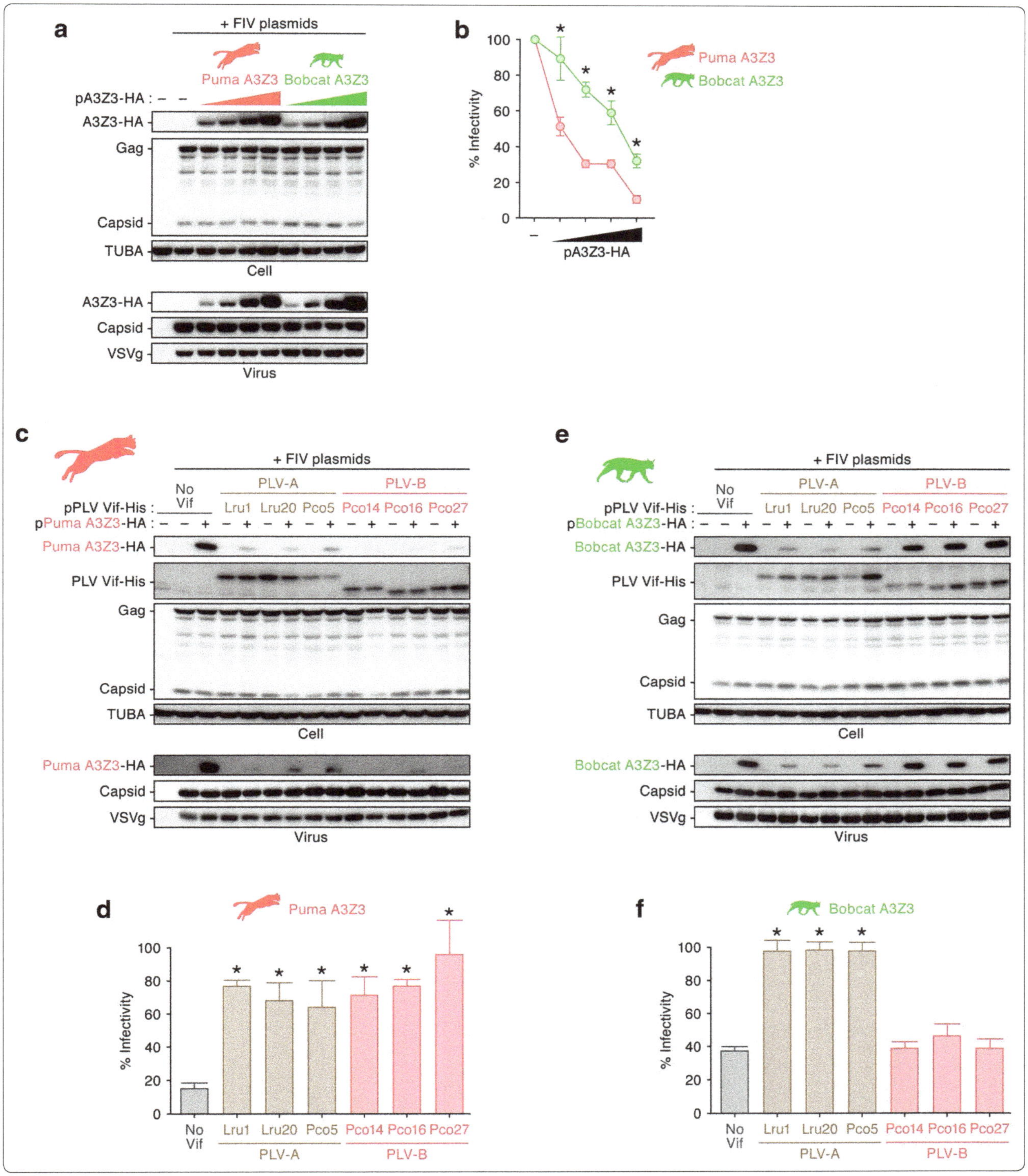
a
+ FIV plasmids
Puma A3Z3
Bobcat A3Z3
pA3Z3-HA : – –
A3Z3-HA
Gag
Capsid
TUBA
Cell
A3Z3-HA
Capsid
VSVg
Virus
b
% Infectivity
100
80
60
40
20
0
–
pA3Z3-HA
Puma A3Z3
Bobcat A3Z3
c
+ FIV plasmids
No Vif
PLV-A
PLV-B
pPLV Vif-His :
Lru1 Lru20 Pco5 Pco14 Pco16 Pco27
pPuma A3Z3-HA :
Puma A3Z3-HA
PLV Vif-His
Gag
Capsid
TUBA
Cell
Puma A3Z3-HA
Capsid
VSVg
Virus
e
+ FIV plasmids
No Vif
PLV-A
PLV-B
pPLV Vif-His :
Lru1 Lru20 Pco5 Pco14 Pco16 Pco27
pBobcat A3Z3-HA :
Bobcat A3Z3-HA
PLV Vif-His
Gag
Capsid
TUBA
Cell
Bobcat A3Z3-HA
Capsid
VSVg
Virus
d
Puma A3Z3
% Infectivity
No Vif
Lru1 Lru20 Pco5 Pco14 Pco16 Pco27
PLV-A
PLV-B
f
Bobcat A3Z3
% Infectivity
No Vif
Lru1 Lru20 Pco5 Pco14 Pco16 Pco27
PLV-A
PLV-B

Table 1 PLV Vif used in this study

Group	Strain	Sampling location	Sampling year	Accession nos.
PLV-A	Lru1	California	1996	KF906143
PLV-A	Lru20	Florida	2010	KF906162
PLV-A	Pco5	California	2004	KF906167
PLV-B	Pco14	California	2002	KF906182
PLV-B	Pco16	California	2011	KF906193
PLV-B	Pco27	Colorado	2008	KF906194

found those harboring the T178M substitution lost the ability to resist PLV-B Vif-mediated degradation (strain Pco27) (Fig. 5a). In the presence of PLV-B Vif, incorporation of bobcat A3Z3 derivatives possessing the T178M mutation into the released viruses was impaired (Fig. 5a, bottom), and the derivatives' antiviral effects were abrogated (Fig. 5b). To determine whether this observation was strain-specific, we performed similar experiments using the other PLV-B strains (Pco14 and Pco16). Similar to strain Pco27 (Fig. 5a, b), the Vif proteins of the other PLV-B strains degraded the bobcat A3Z3 T178M mutant (Fig. 5c) and significantly recovered viral infectivity (Fig. 5d). These findings suggest that the resistance of bobcat A3Z3 to PLV-B Vif-mediated degradation is determined by the amino acid residue at position 178.

To validate the importance of the amino acid at position 178 to sensitivity to PLV-B Vif-mediated degradation, we performed gain-of-function experiments based on the puma A3Z3. By substituting the methionine

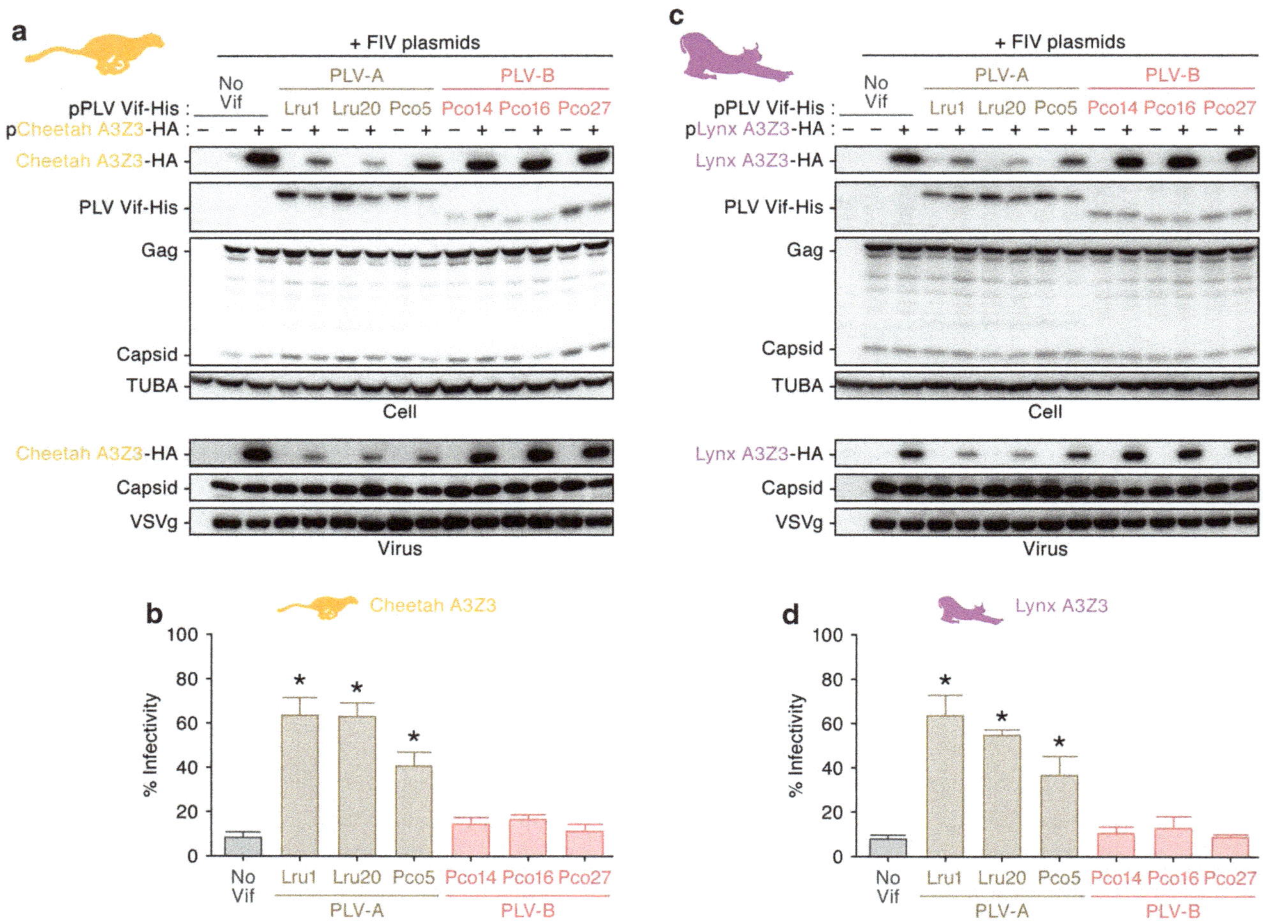

Fig. 3 A3Z3 sensitivity of the felids related to pumas and bobcats to PLV Vif-mediated degradation. HA-tagged expression plasmids for cheetah (**a**, **b**) and lynx (**c**, **d**) A3Z3 (200 ng) and the three plasmids used to produce the *vif*-deficient FIV-based reporter virus (FIV plasmids; pFP93 [200 ng], pTiger-luc [150 ng] and pMD.G [50 ng]) were co-transfected with or without His-tagged PLV Vif expression plasmids (400 ng) into HEK293T cells. **a**, **c** Western blotting. Representative results are shown. **b**, **d** FIV reporter assay. FIV infectivity is shown as the percentage of the value of "no A3Z3". In **b** and **d**, asterisks indicate significant differences ($P < 0.05$ by Student's t test) versus "no Vif". The assays were independently performed in triplicate. Data represent averages with SDs

a

	Degraded by PLV-B Vif?	Amino acid position		
		90	131	178
Puma	Yes	K	T	M
Bobcat Cheetah Lynx	No	E	I	T

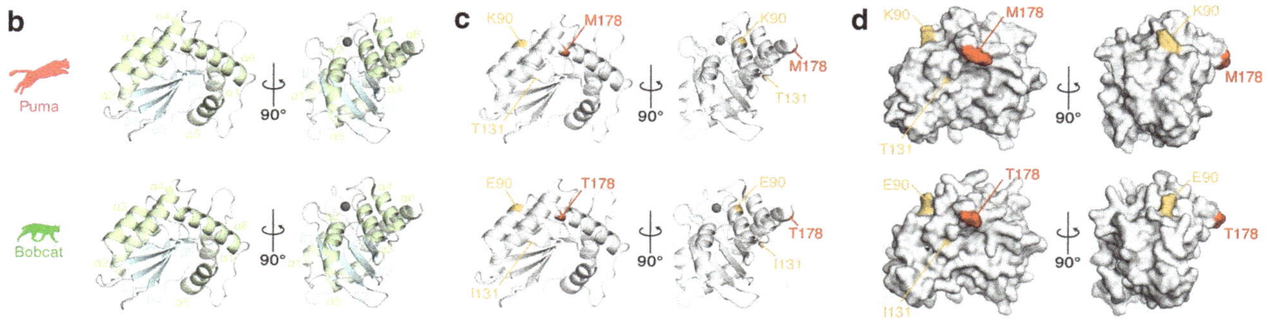

Fig. 4 Structure modeling of puma and bobcat A3Z3. **a** Comparison of amino acid residues. The feline species used for the comparison (puma, bobcat, cheetah and lynx; left), the sensitivity of A3Z3 to PLV-B Vif (middle) and the amino acid residues at positions 90, 131 and 178 (right) are summarized. **b–d** Structural homology model of puma and bobcat A3Z3. Cartoon (**b**, **c**) and surface (**d**) models of the A3Z3 protein structures of the puma (top) and bobcat (bottom) are shown. In **b**, alpha-helices and beta-sheets are shown in green and pale blue, respectively. In the right panels of **a** and **b**, Zn^{2+} is represented as a gray sphere. In **c** and **d**, the three amino acids that differed between the puma and the other felids shown in **a** (bobcat, cheetah and lynx; residues 90, 131 and 178) are represented in orange or red

residue at position 178 of the puma A3Z3 to threonine, the puma A3Z3 mutant became resistant to PLV-B Vif-mediated degradation (Fig. 6a, top). The puma A3Z3 M178T mutant was efficiently incorporated into the released virions (Fig. 6a, bottom) and exhibited significant antiviral activity (Fig. 6b). Taken together, these findings suggest that the amino acid at position 178 in the bobcat/puma A3Z3 plays a pivotal role in conferring resistance to PLV-B Vif-mediated degradation.

Puma and bobcat A3Z3 are polymorphic

Previous studies have revealed that mammalian *A3* genes, including feline *A3Z3* genes, are under positive selection and are highly diversified due to evolutionary selective pressures presumably caused by the ancestral FIV Vif [23]. For instance, the domestic cat *A3Z3* is polymorphic [24], and a haplotype of the domestic cat A3Z3 renders resistance to FIVfca Vif-mediated degradation [12]. These previous findings raise the possibility that the puma and bobcat *A3Z3* genes are also polymorphic and that the PLV Vif sensitivity may differ among haplotypes. To address this possibility, we additionally determined the *A3Z3* sequences of five pumas, one bobcat as well as one cheetah. As shown in Fig. 7a, we detected additional puma and bobcat *A3Z3* haplotypes and designated them haplotype II (hap II). One nonsynonymous mutation was detected in the puma A3Z3 hap II (a c392t mutation resulting in T131I amino acid substitution), and a heterozygous sequence was detected in the bobcat A3Z3 hap II (g347r; a g347a mutation resulting in a R116H amino acid substitution) (Fig. 7a). Next, we mapped these residues on the protein homology model. Both residue 131 of the puma A3Z3 (Fig. 4b) and residue 116 of the bobcat A3Z3 (Additional file 3: Fig. S2) are positioned in the loop structure. Then, we prepared expression plasmids for these haplotypes and assessed their antiviral activity and sensitivity to PLV Vif. As shown in Fig. 7b, puma and bobcat A3Z3 hap II were expressed at similar levels to their hap I derivatives. Notably, in the absence of PLV Vif, the antiviral effects of the puma and bobcat A3Z3 hap II were significantly higher than those of the hap I derivatives (Fig. 7c). In particular, the bobcat A3Z3 hap II exhibited stronger antiviral activity in a dose-dependent manner (Fig. 7c). Moreover, puma A3Z3 hap II was sensitive to both the PLV-A and PLV-B Vifs (Fig. 7d, e), whereas the bobcat A3Z3 hap II was sensitive to PLV-A Vif but was resistant to PLV-B Vif-mediated degradation (Fig. 7e, f). Taken together, these findings suggest that the puma and bobcat *A3Z3* genes are polymorphic and that the bobcat A3Z3 hap II exhibits a higher antiviral effect than the hap I derivative, whereas the PLV Vif sensitivity phenotype is conserved within each species.

Discussion

In this study, we identified the *A3Z3* sequences of two felids, the bobcat and cheetah, and demonstrated that PLV-A Vif counteracted the A3Z3 proteins of both the puma and bobcat lineages, whereas PLV-B Vif

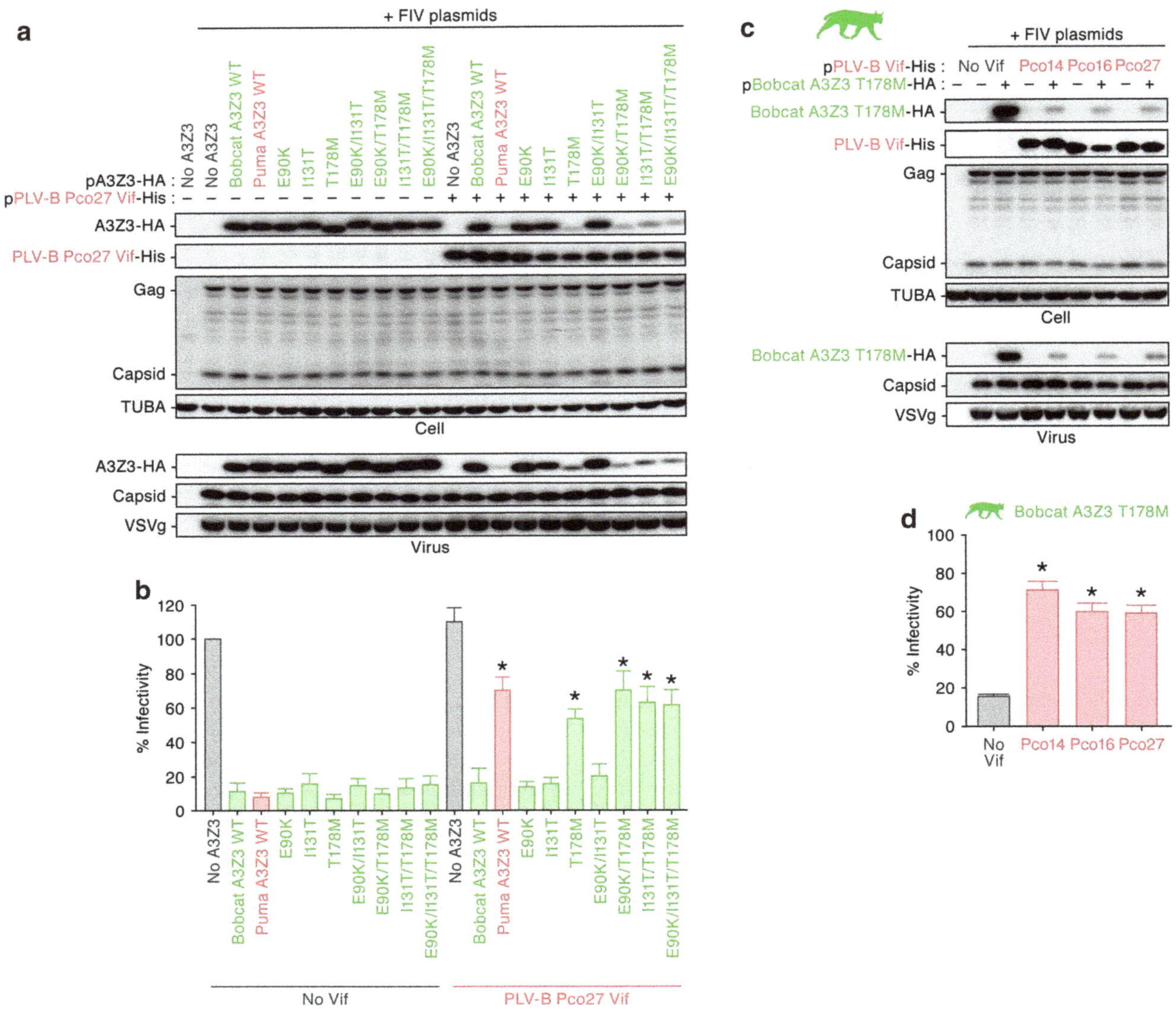

Fig. 5 Loss-of-function screening of the amino acid residues responsible for PLV-B Vif sensitivity. HA-tagged expression plasmids for the A3Z3 derivatives (indicated in the figure) of the bobcat (200 ng) and the three plasmids used to produce the *vif*-deficient FIV-based reporter virus (FIV plasmids; pFP93 [200 ng], pTiger-luc [150 ng] and pMD.G [50 ng]) were co-transfected with or without His-tagged PLV-B Vif expression plasmids (400 ng) into HEK293T cells. **a**, **c** Western blotting. Representative results are shown. **b**, **d** FIV reporter assay. FIV infectivity is shown as the percentage of the value of "no A3Z3". In **b** and **d**, asterisks indicate significant differences ($P < 0.05$ by Student's *t* test) versus "no Vif". The assays were independently performed in triplicate. Data represent averages with SDs

counteracted only the puma A3Z3 (Figs. 1, 2, 3). Through loss-of-function (Fig. 5) and gain-of-function (Fig. 6) experiments, we also determined that the amino acid at position 178 of the puma and bobcat A3Z3 was responsible for the sensitivity to PLV-B Vif. In addition, structural modeling suggested that this residue was exposed to the protein surface (Fig. 4). Furthermore, although the puma and bobcat *A3Z3* genes were polymorphic, the sensitivity to PLV-B Vif was conserved in each species (Fig. 7). Some previous studies have addressed the potential of OWM A3G to restrict HIV-1 infection [25, 26]. However, OWMs do not infect with HIV-1 in nature. Therefore, to the best of our knowledge, this study is the first to suggest that inter-species lentiviral transmission in nature is controlled by a host A3 protein.

As summarized in Fig. 1a, d, PLV-A is shared by both pumas and bobcats in North America, whereas PLV-B is detected only in pumas. Through cell-based virological experiments, here we demonstrated that PLV-A Vif degraded the A3Z3 protein of feline lineages including

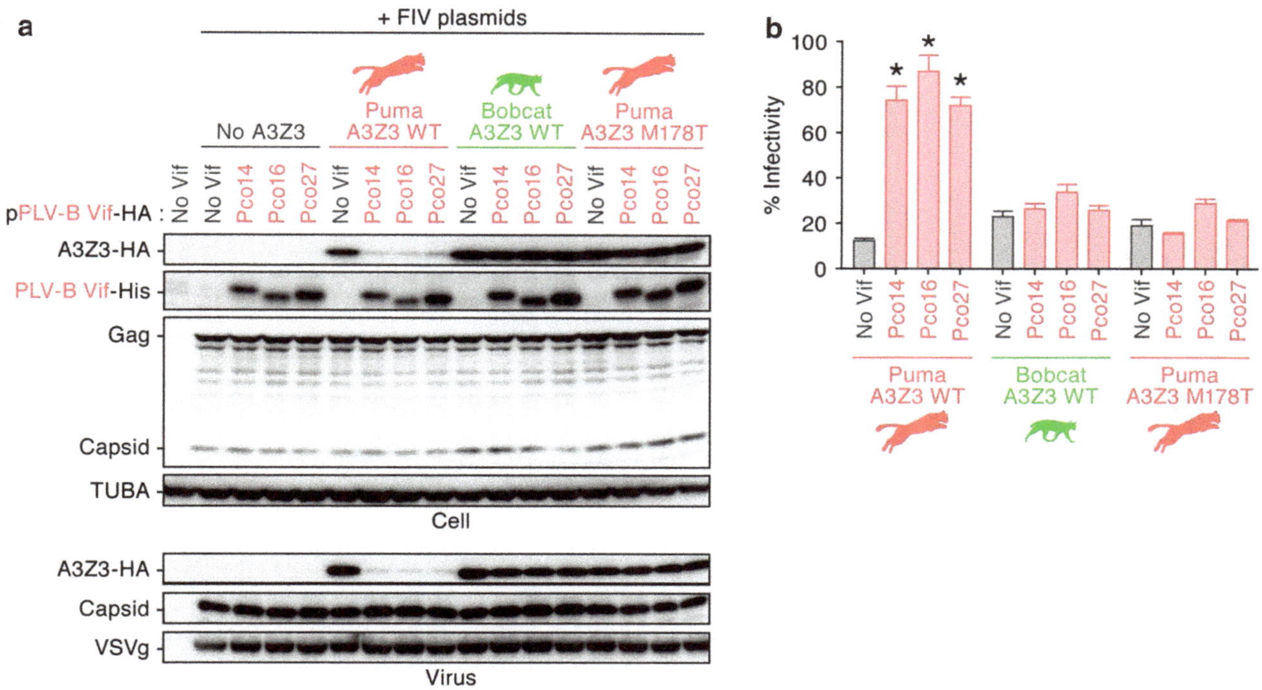

Fig. 6 Gain-of-function validation of the amino acid residues responsible for PLV-B Vif sensitivity. HA-tagged expression plasmids for puma A3Z3, bobcat A3Z3 and the puma A3Z3 M178T derivative (200 ng) and the three plasmids used to produce *vif*-deficient FIV-based reporter virus (FIV plasmids; pFP93 [200 ng], pTiger-luc [150 ng] and pMD.G [50 ng]) were co-transfected with or without His-tagged PLV-B Vif expression plasmids (400 ng) into HEK293T cells. **a** Western blotting. Representative results are shown. **b** FIV reporter assay. FIV infectivity is shown as the percentage of the value of "no A3Z3". In **b**, asterisks indicate significant differences ($P < 0.05$ by Student's *t* test) versus "no Vif". The assays were independently performed in triplicate. Data represent averages with SDs

puma and bobcat, whereas PLV-B Vif counteracted only the puma A3Z3. Additionally, the bobcat A3Z3 exhibited a significant antiviral effect even in the presence of PLV-B Vif. Because the functionality of PLV Vif is independent of the sampling location and year (Fig. 2 and Table 1), our results suggest that the functional relationship between PLV Vif and feline A3Z3 is conserved. Moreover, we found a polymorphism in the puma *A3Z3*, but the sensitivity of the puma A3Z3 variants to PLV-B Vif-mediated degradation was also conserved (Fig. 6). These findings suggest that "scenario 2" reasonably explains the difference in species tropism between PLV-A and PLV-B and that bobcat A3Z3 acts as a species barrier to restrict PLV-B CST of PLV-B from pumas (Fig. 1d). To the best of our knowledge, this is the first report demonstrating that the host A3 protein is the factor that restricts non-primate lentiviral CST.

Consistent with our results, PLV-A and PLV-B form clusters by sampling location [17–19]. Moreover, the PLV-As detected in pumas (FIVpco) and bobcats (FIVlru), particularly those isolated in California, were highly co-mingled in a cluster (Fig. 1a). These observations suggest that PLV-As have been transmitted relatively recently from bobcats to pumas after geological sequestration, and we can assume that this bobcat-to-puma CST event occurs frequently at least in California.

Our findings in New World felids were reminiscent of the fact that simian immunodeficiency viruses (SIVs) in Old World monkeys (OWMs) were transferred to chimpanzees, which originated in the emergence of a novel SIV in chimpanzees (SIVcpz) (reviewed in [27, 28]). SIV CST from OWMs to chimpanzees in Africa is explained phylogenetically and is assumed to be due to the prey-predator relationship between OWMs and chimpanzees. Small OWMs are the prey of chimpanzees in the wild; thus, chimpanzees are frequently exposed to various SIVs that infect their OWM prey species [29, 30]. Similarly, pumas are a top carnivore in the New World [31] and bobcats are their prey [32, 33]. Therefore, pumas may be frequently exposed to PLV-A in bobcats via prey-predator interactions, leading to bobcat-to-puma CST; this possibility is consistent with a previous assumption based on phylogenetic analyses [17].

Although PLV-A Vif degraded all of the feline A3Z3 proteins tested in this study, PLV-B Vif counteracted only the puma A3Z3, suggesting the evolutionary convergence

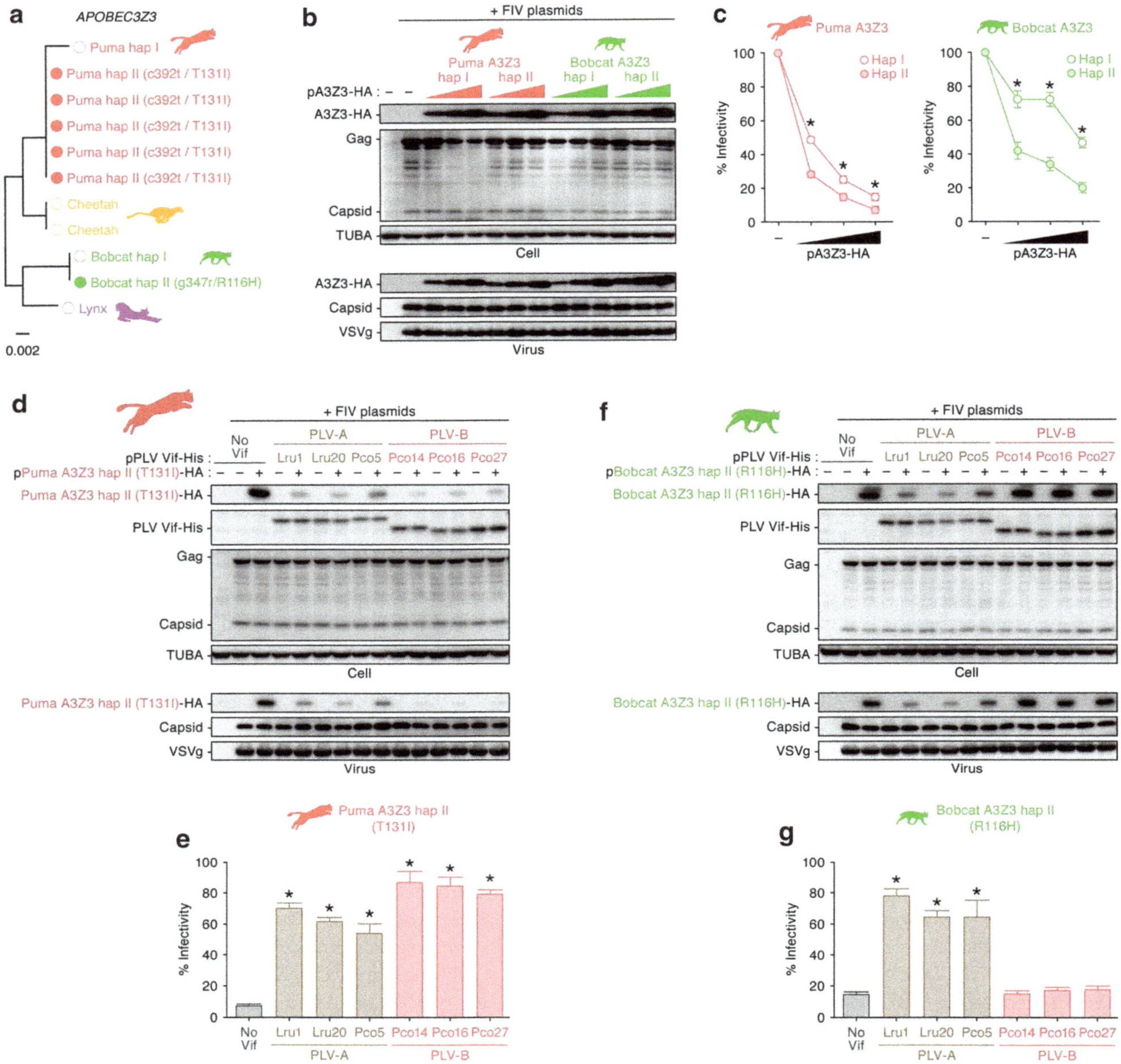

Fig. 7 Polymorphisms of puma and bobcat *A3Z3* and their association with PLV Vif sensitivity. **a** Feline *A3Z3* polymorphisms. Additional *A3Z3* sequences from five pumas, one cheetah and one bobcat were determined and the phylogenetic tree was reconstructed using the ML method. The *A3Z3* sequences used in the experiments above (summarized in Fig. 1c) are indicated with open circles and designated hap I. Additional *A3Z3* haplotypes detected in the puma and bobcat are indicated with filled circles and designated hap II. The nucleotide/amino acid substitutions detected in puma and bobcat A3Z3 hap II are indicated in parentheses. Note that the new cheetah *A3Z3* sequence was identical to the previously determined sequence (Fig. 1c). The scale bar indicates 0.002 nucleotide substitutions per site. **b–f** Antiviral activity of the puma and bobcat A3Z3 hap II proteins and their sensitivity to PLV Vif. **b**, **c** Different amounts of HA-tagged expression plasmids for A3Z3 hap I and II of puma or bobcat (0, 100, 200 and 400 ng) and the three plasmids used to produce the *vif*-deficient FIV-based reporter virus (FIV plasmids; pFP93 [200 ng], pTiger-luc [150 ng] and pMD.G [50 ng]) were co-transfected into HEK293T cells. **b** Western blotting. Representative results are shown. **c** FIV reporter assay. FIV infectivity is shown as the percentage of the value of "no A3Z3". Note that "hap I" is identical to those used in the other experiments shown in Figs. 1, 2, 4, 5 and 6. **d–g** HA-tagged expression plasmids for puma A3Z3 hap II (**d**, **e**) or bobcat A3Z3 hap II (**f**, **g**) (200 ng) and the three plasmids used to produce the *vif*-deficient FIV-based reporter virus (FIV plasmids; pFP93 [200 ng], pTiger-luc [150 ng] and pMD.G [50 ng]) were co-transfected with or without His-tagged PLV Vif expression plasmids (400 ng) into HEK293T cells. **d**, **f** Western blotting. Representative results are shown. **e**, **g** FIV reporter assay. FIV infectivity is shown as the percentage of the value of "no A3Z3". In **b**, asterisks indicate significant differences ($P < 0.05$ by Student's *t* test) between hap I and hap II. In **e** and **g**, asterisks indicate significant differences ($P < 0.05$ by Student's *t* test) versus "no Vif". The assays were independently performed in triplicate. Data represent averages with SDs

of the puma and PLV-B. Compton and Emerman reported that the Vif protein of an SIV infecting colobus monkey (*Colobus guereza*), which is an OWM in Africa, counteracted only colobus A3G [34]. Based on findings for the A3G-Vif interaction in colobus monkeys, the authors proposed convergent evolution of colobus monkeys and SIVcor and further suggested that the convergent co-evolution of these monkeys and viruses occurred at least 12 Mya [34]. Similarly, the puma diversified from the cheetah approximately 4.9 Mya (Fig. 1c) [23]. Puma and cheetah habitats were separated approximately 1.2–1.6 Mya after cheetahs immigrated from the New World to Eurasia (Fig. 1b) [23]. Since the cheetah A3Z3 was not counteracted by the PLV-B Vif (Fig. 3), our findings suggested that the co-evolutionary convergence of the puma and PLV-B arose at least 1.2–1.6 Mya. Moreover, OWMs exhibit higher viral RNA loads without any disorders [35]. Since SIVs circumvent the immune pressure from host monkeys because they are naturally infected, SIVs are less diversified in African OWMs, which is assumed to be due to the co-existence of SIVs in OWMs over a long period [35]. Lee et al. [17] recently showed that the PLV-B plasma viral RNA load in pumas was significantly higher than PLV-A loads in both pumas and bobcats, whereas PLV-B was less diversified than PLV-A. This is the first study suggesting the convergent co-evolution of a mammal (puma) and a lentivirus (PLV-B) in the New World, and PLV-B can be concluded to have co-evolved with pumas over a long period, similar to apathogenic SIV infections in OWMs.

Through mutagenesis experiments, we demonstrated that the amino acid at position 178 determined the sensitivity to PLV-B Vif-dependent degradation (Fig. 5). Since both the gain- and loss-of-function experiments indicated the importance of this residue, the species specificity of PLV-B Vif against puma and bobcat A3Z3 was determined by this single amino acid. In addition, the structural biology investigation revealed that this amino acid residue was exposed on the A3Z3 protein surface (Fig. 4). In our previous study [12], we revealed that there are at least seven haplotypes in the domestic cat *A3Z3* and that a domestic cat A3Z3 haplotype (designated haplotype V in our previous study) was resistant to FIVfca Vif-mediated degradation. Furthermore, we demonstrated that amino acid residue 65 of the domestic cat A3Z3 determines the sensitivity to FIVfca Vif [12]. Similar to this study, residue 65 in the domestic cat A3Z3 was also exposed on the protein surface, which suggested that this residue was associated with the interaction with FIV Vif [12]. However, the amino acid residue at position 65 was positioned at alpha-helix 2 in the domestic cat A3Z3 [12], whereas the residue 178 was located at alpha-helix 6 in the puma and bobcat A3Z3 (Fig. 5). Therefore, our findings suggest that these residues render resistance to FIV Vif in a different manner.

In addition to A3, other cellular proteins such as tripartite motif containing protein 5 (TRIM5) [36] and tetherin [37, 38] are known as restriction factors that inhibit lentiviral replication in primates. Therefore, it might be possible that these restriction factors in pumas and/or bobcats may also play roles in preventing FIV CST. However, feline TRIM5 is not functional because of the insertion of premature stop codon [39], and feline tetherin is unable to restrict spreading FIV infection [40]. Furthermore, although FIV Vif is a functional antagonist against feline A3, FIV lacks functional proteins that potently counteract TRIM5 and tetherin [39, 40]. Primates encode several restriction factors against lentiviruses; however, only A3 is convincing to be a restriction factor in felids. Therefore, feline A3 may play a pivotal role in controlling CST in felids.

In summary, we demonstrated that bobcat A3Z3 was resistant to PLV-B Vif-dependent degradation, whereas PLV-A Vif overcame the antiviral action mediated by both the puma and bobcat A3Z3. The co-evolutionary relationship between primate A3 proteins and their lentiviral Vifs has been rigorously investigated [20, 22, 34, 41–44]; however, few studies, including ours, have addressed the evolutionary dynamics of non-primate A3 and non-primate lentiviruses [4, 12, 14]. Here, we provided evidence suggesting that lentiviral CST between different genera (*Puma* and *Lynx*) was controlled by the Vif-A3 interaction. To the best of our knowledge, this is the first report providing evidence of the co-evolutionary arms race between mammals and lentiviruses in the New World.

Methods

Ethics statement

To determine the feline *A3Z3* sequences, blood, body hair or cryopreserved muscle tissue of pumas, bobcats and cheetahs were kindly provided by the following facilities: Tennoji zoo, Osaka, Japan; Kobe City Oji zoo, Hyogo, Japan; Omoriyama zoo, Akita, Japan; Tama Zoological Park, Tokyo, Japan; Izu Animal Kingdom, Shizuoka, Japan; and Shizuoka Municipal Nihondaira zoo, Shizuoka, Japan. Sampling was performed in accordance with the guideline of Tokyo University of Agriculture, Japan. All experimental protocols were approved by a committee at Tokyo University of Agriculture, Japan.

Sequencing PCR of the feline *A3Z3* genes

Sequencing PCR of feline *A3Z3* was performed as previously described [45, 46]. Briefly, genomic DNA was extracted from the samples described above using the DNA Extractor FM kit (Wako) or the DNeasy Blood &

Tissue it (Qiagen). PCR was performed using the PrimeSTAR GXL DNA polymerase (Takara) and the primers are listed in Additional file 2: Table S1. The obtained PCR products were purified by gel extraction using the QIAquick gel extraction kit (Qiagen). The nucleotide sequences were determined by a DNA sequencing service (Fasmac, Kanagawa, Japan) and the data were analyzed using Sequencher v5.1 software (Gene Codes Corporation).

Molecular phylogenetic analysis of the FIV *vif* and feline *A3Z3* genes

Molecular phylogenetic analyses were performed as previously described [12, 14, 45–47]. Briefly, the sequences of PLV Vif and feline *A3Z3*, some of which were newly identified in this study (Figs. 1a, c, 7a), were aligned using ClustalW implemented in MEGA7 [48]. The alignments were verified manually at the amino acid level. Phylogenetic trees (Figs. 1a, c, 7a) were reconstructed using the maximum likelihood (ML) method with PhyML [49]. To calculate the amino acid sequence diversity of PLV Vif, 26 and 34 amino acid sequences from PLV-A and PLV-B Vif (summarized in Fig. 1a) are used, respectively. A multiple alignment was generated using L-INS-i in MAFFT [50]. The gapped regions were removed using trimAl with the nogaps option [51], and 227 amino acid sites were used for the analysis. Then, we performed pairwise comparisons of 60 amino acid sequences to calculate the amino acid sequence identity using MEGA7 [48].

Plasmid construction

The expressing plasmids for HA-tagged puma and lynx A3Z3 were kindly provided by Dr. Carsten Münk [15]. The expressing plasmids for HA-tagged bobcat and cheetah A3Z3 were constructed by using the genomic DNA fragments and the primers listed in Additional file 2: Table S1. The point mutants of HA-tagged puma and bobcat A3Z3 were constructed by using a GeneArt site-directed mutagenesis system (Thermo Fisher Scientific). Each wild-type plasmid was used as the template, and the primers used are listed in Additional file 2: Table S1. The His-tagged PLV-A Vif (strains Lru1, Lru20 and Pco5) and PLV-B (strains Pco14, Pco16 and Pco27) were obtained from GeneArt gene synthesis service (Thermo Fisher Scientific). The obtained DNA fragments were digested with BamHI and SalI and inserted into the BamHI-SalI site of pDON-AI plasmid (Takara). The nucleotide sequences were determined by a DNA sequencing service (Fasmac, Kanagawa, Japan) and the data were analyzed by Sequencher v5.1 software (Gene Codes Corporation).

Cell culture and transfection

HEK293T cells (CRL-11268; ATCC) were cultured in Dulbecco's modified Eagle medium (Sigma-Aldrich) supplemented with 10% heat-inactivated fetal calf serum and antibiotics (Thermo Fisher Scientific). Transfection was performed by using PEI Max (GE Healthcare) in accordance with the manufacturers procedures and described previously [12, 14, 45–47, 52]. To analyze the dose-dependent anti-FIV activity of feline A3Z3, pFP93 (pFIVgagpolΔvif; a replication incompetent *vif*-deficient FIV packaging construct derived from clone FIV-34TF10 [GenBank accession number M25381]; kindly provided by Dr. Eric M. Poeschla) (200 ng), pTiger-luc (pFIVΨ-luc) (150 ng), and pMD.G (pVSVg; a vesicular stomatitis virus G [VSVg] expression plasmid) (50 ng) were co-transfected into HEK293T cells (1×10^5 cells) with feline A3Z3 expression plasmid (50, 100, 200, or 400 ng). To analyze the functional relationship between feline A3Z3 and PLV-Vif, feline A3Z3 expression plasmid (200 ng), pFP93 (200 ng), pTiger-luc (150 ng), and pMD.G (50 ng) were co-transfected into HEK293T cells with or without His-tagged PLV Vif expression plasmid (400 ng). After 48 h post-transfection, the transfected cells and culture supernatants were harvested as previously described [12–14, 45–47, 52].

Western blotting and virus reporter assay

Western blotting and reporter assay were performed as previously described [12–14, 45–47, 52]. For the Western blotting of virus particles, 340 µl of the culture supernatant was ultracentrifuged at 100,000×*g* for 1 h at 4 °C using a TL-100 instrument (Beckman), and the pellet was lysed with 1 × SDS buffer. For the Western blotting of transfected cells, the cells were lysed with RIPA buffer (50 mM Tris–HCl buffer [pH 7.6], 150 mM NaCl, 1% Nonidet P-40, 0.5% sodium deoxycholate, 0.1% SDS) with protease inhibitor cocktail (Roche). The following antibodies for Western blotting: anti-His polyclonal antibody (OGHis; Medical and Biological Laboratories), anti-HA antibody (3F10; Roche), anti-FIV p24 Capsid antibody (PAK3-2C1; Santa Cruz Biotechnology); anti-alpha-tubulin (TUBA) antibody (DM1A; Sigma), and anti-VSVg antibody (P5DA; Roche). For FIV reporter assays, HEK293T cells were used for the target of infection. Ten microliter of the culture supernatant of transfected cells was inoculated into HEK293T cells in a 96-well plate (Nunc), and the firefly luciferase activity was measured by using the BrillianStar-LT assay system (Toyo-b-net) and the 2030 ARVO X multilabel counter instrument (PerkinElmer) according to the manufacturers' procedures.

Protein homology modeling

Homology modeling was performed using the SWISS-MODEL server [53–56]. After BLAST searches [57] of the bobcat and puma A3Z3 amino acid sequences against protein data bank sequence entries (http://www.rcsb.org/pdb/), the crystal structure of human APOBEC3A (PDB: 5KEG) [58] was selected as the best template for homology modeling per the Global Model Quality Estimations, QMEAN statistical parameters, and modeled sequence length. Each generated model was minimized and refined using Discovery Studio (Dassault Systèmes BIOVIA, Discovery Studio Modeling Environment, Release 4.1, San Diego: Dassault Systèmes, 2007). Mutant models and 3D images were generated with PyMOL (The PyMOL Molecular Graphics System, version 1.8 Schrödinger, LLC).

Statistical analyses

The data are expressed as averages with the standard deviations (SDs), and statistically significant differences were determined using Student's *t* test.

Accession numbers

The *A3Z3* sequences of five pumas, two bobcats and two cheetahs were submitted to DDBJ (accession numbers LC376039-LC376042).

Additional files

Additional file 1: Figure S1. Scheme of the feline genome encoding *APOBEC3Z3* and the position of the primers used in this study. The scheme used for *Felis catus* chromosome B4, including the exons of feline *APOBEC3Z3,* is shown. The primers used for PCR/sequencing are shown as red arrowheads, and the names are identical to those in Table S1.

Additional file 2: Table S1. Primers used in this study. A full list of the primers used in this study.

Additional file 3: Figure S2. Structure homology model of bobcat A3Z3 hap II. Cartoon (top and middle) and surface (bottom) models of the A3Z3 protein structures of bobcat hap I (A) and hap II (B) are shown. In the top panel, alpha-helices and beta-sheets are shown in green and pale blue, respectively. Zn^{2+} is represented as a gray sphere. In the middle and bottom panels, the amino acid that differed between hap I (R116) and hap II (H116) is represented in orange.

Abbreviations

A3: apolipoprotein B mRNA-editing enzyme catalytic polypeptide-like 3 (APOBEC3); A3Z3: APOBEC3Z3; CST: cross-species transmission; FIV: feline immunodeficiency virus; ML: maximum likelihood; Mya: million years ago; OWM: Old World monkey; PLV: puma lentivirus; SIV: simian immunodeficiency virus; SD: standard deviation; TRIM5: tripartite motif containing protein 5; TUBA: alpha-tubulin; Vif: viral infectivity factor; VSVg: vesicular stomatitis virus G.

Authors' contributions

KS conceived the study and designed experiments; Y.Konno, S.Nagaoka, IK, KY, Y.Kagawa, RK, HA and TK performed experiments; Y.Konno, S.Nagaoka, IK, KY and HA analyzed data; MTU and S.Nakagawa constructed protein homology models; S.Nakagawa and KS performed phylogenetic analyses; KS wrote the paper; and Y.Konno, S.Nagaoka, IK, MTU, S.Nakagawa, Y.Koyanagi and KS edited the manuscript. All authors read and approved the final manuscript.

Author details

[1] Laboratory of Systems Virology, Institute for Frontier Life and Medical Sciences, Kyoto University, Kyoto, Japan. [2] Graduate School of Biostudies, Kyoto University, Kyoto, Japan. [3] Graduate School of Pharmaceutical Sciences, Kyoto University, Kyoto, Japan. [4] Graduate School of Medicine, Kyoto University, Kyoto, Japan. [5] Faculty of Medicine, Kyoto University, Kyoto, Japan. [6] Faculty of Science, Kyoto University, Kyoto, Japan. [7] Faculty of Pharmaceutical Sciences, Kyoto University, Kyoto, Japan. [8] Micro/Nano Technology Center, Tokai University, Kanagawa, Japan. [9] Department of Molecular Life Science, Tokai University School of Medicine, Tokai University, Kanagawa, Japan. [10] Department of Animal Science, Faculty of Agriculture, Tokyo University of Agriculture, Kanagawa, Japan. [11] CREST, Japan Science and Technology Agency, Saitama, Japan. [12] Division of Systems Virology, Department of Infectious Disease Control, International Research Center for Infectious Diseases, Institute of Medical Science, The University of Tokyo, 4-6-1 Shirokanedai, Minato-ku, Tokyo 1088639, Japan.

Acknowledgements

We thank Carsten Münk (Heinrich-Heine-Universität, Düsseldorf, Germany) and Eric M. Poeschla (University of Colorado, USA) for providing experimental materials and Tennoji zoo (Osaka, Japan), Kobe City Oji zoo (Hyogo, Japan), Omoriyama zoo (Akita, Japan), Tama Zoological Park (Tokyo, Japan), Izu Animal Kingdom (Shizuoka, Japan), and Shizuoka Municipal Nihondaira zoo (Shizuoka, Japan) for kindly providing feline samples.

Competing interests

The authors declare that no competing interests exist.

Funding

This study was supported in part by AMED J-PRIDE 17fm0208006h0001 (to KS); JST CREST (to KS); JSPS KAKENHI Grants-in-Aid for Scientific Research C 15K07166 (to KS), Scientific Research B (Generative Research Fields) 16KT0111 (to KS), and Scientific Research on Innovative Areas 16H06429 (to S. Nakagawa and KS), 16K21723 (to S.Nakagawa and KS), 17H05823 (to S. Nakagawa) and 17H05813 (to KS); Takeda Science Foundation (to KS); Salt Science Research Foundation (to KS); Smoking Research Foundation (to KS); Chube Ito Foundation (to KS); Fordays Self-Reliance Support in Japan (to KS); Mishima Kaiun Memorial Foundation (to KS); Tobemaki Foundation (to KS); ONO Medical Research Foundation (to KS); MEXT-Supported Program for the Strategic Research Foundation at Private Universities (to S.Nakagawa and MTU); Joint Usage/Research Center program of Institute for Frontier Life and Medical Sciences Kyoto University (to S.Nakagawa); JSPS Core-to-Core program, A. Advanced Research Networks (to Y.Koyanagi); and AMED Research on HIV/AIDS 16fk0410203h002 (to Y.Koyanagi).

References

1. Koito A, Ikeda T. Intrinsic restriction activity by AID/APOBEC family of enzymes against the mobility of retroelements. Mob Genet Elem. 2011;1:197–202.
2. LaRue RS, Andresdottir V, Blanchard Y, Conticello SG, Derse D, Emerman M, et al. Guidelines for naming nonprimate APOBEC3 genes and proteins. J Virol. 2009;83:494–7.
3. Harris RS, Dudley JP. APOBECs and virus restriction. Virology. 2015;479–480:131–45.
4. Nakano Y, Aso H, Soper A, Yamada E, Moriwaki M, Juarez-Fernandez G, et al. A conflict of interest: the evolutionary arms race between mammalian APOBEC3 and lentiviral Vif. Retrovirology. 2017;14:31.

5. Dawkins R, Krebs JR. Arms races between and within species. Proc R Soc Lond B Biol Sci. 1979;205:489–511.
6. Duggal NK, Emerman M. Evolutionary conflicts between viruses and restriction factors shape immunity. Nat Rev Immunol. 2012;12:687–95.
7. Gifford RJ. Viral evolution in deep time: lentiviruses and mammals. Trends Genet. 2012;28:89–100.
8. Kirchhoff F. Immune evasion and counteraction of restriction factors by HIV-1 and other primate lentiviruses. Cell Host Microbe. 2010;8:55–67.
9. Pedersen NC, Ho EW, Brown ML, Yamamoto JK. Isolation of a T-lymphotropic virus from domestic cats with an immunodeficiency-like syndrome. Science. 1987;235:790–3.
10. Münk C, Beck T, Zielonka J, Hotz-Wagenblatt A, Chareza S, Battenberg M, et al. Functions, structure, and read-through alternative splicing of feline APOBEC3 genes. Genome Biol. 2008;9:R48.
11. Wang J, Zhang W, Lv M, Zuo T, Kong W, Yu X. Identification of a Cullin5-ElonginB-ElonginC E3 complex in degradation of feline immunodeficiency virus Vif-mediated feline APOBEC3 proteins. J Virol. 2011;85:12482–91.
12. Yoshikawa R, Izumi T, Yamada E, Nakano Y, Misawa N, Ren F, et al. A naturally occurring domestic cat APOBEC3 variant confers resistance to FIV infection. J Virol. 2015;90:474–85.
13. Yoshikawa R, Nakano Y, Yamada E, Izumi T, Misawa N, Koyanagi Y, et al. Species-specific differences in the ability of feline lentiviral Vif to degrade feline APOBEC3 proteins. Microbiol Immunol. 2016;60:272–9.
14. Yoshikawa R, Takeuchi JS, Yamada E, Nakano Y, Misawa N, Kimura Y, et al. Feline immunodeficiency virus evolutionarily acquires two proteins, vif and protease, capable of antagonizing Feline APOBEC3. J Virol. 2017;91:e00250-17.
15. Zielonka J, Marino D, Hofmann H, Yuhki N, Löchelt M, Münk C. Vif of feline immunodeficiency virus from domestic cats protects against APOBEC3 restriction factors from many felids. J Virol. 2010;84:7312–24.
16. Pecon-Slattery J, Troyer JL, Johnson WE, O'Brien SJ. Evolution of feline immunodeficiency virus in Felidae: implications for human health and wildlife ecology. Vet Immunol Immunopathol. 2008;123:32–44.
17. Lee J, Malmberg JL, Wood BA, Hladky S, Troyer R, Roelke M, et al. Feline immunodeficiency virus cross-species transmission: implications for emergence of new lentiviral infections. J Virol. 2017;91:e02134-16.
18. Franklin SP, Troyer JL, Terwee JA, Lyren LM, Boyce WM, Riley SP, et al. Frequent transmission of immunodeficiency viruses among bobcats and pumas. J Virol. 2007;81:10961–9.
19. Lee JS, Bevins SN, Serieys LE, Vickers W, Logan KA, Aldredge M, et al. Evolution of puma lentivirus in bobcats (*Lynx rufus*) and mountain lions (*Puma concolor*) in North America. J Virol. 2014;88:7727–37.
20. Sawyer SL, Emerman M, Malik HS. Ancient adaptive evolution of the primate antiviral DNA-editing enzyme APOBEC3G. PLoS Biol. 2004;2:E275.
21. Vallender EJ, Lahn BT. Positive selection on the human genome. Hum Mol Genet. 2004;13(2):R245–54.
22. Krupp A, McCarthy KR, Ooms M, Letko M, Morgan JS, Simon V, et al. APOBEC3G polymorphism as a selective barrier to cross-species transmission and emergence of pathogenic SIV and AIDS in a primate host. PLoS Pathog. 2013;9:e1003641.
23. Johnson WE, Eizirik E, Pecon-Slattery J, Murphy WJ, Antunes A, Teeling E, et al. The late Miocene radiation of modern Felidae: a genetic assessment. Science. 2006;311:73–7.
24. de Castro FL, Junqueira DM, de Medeiros RM, da Silva TR, Costenaro JG, Knak MB, et al. Analysis of single-nucleotide polymorphisms in the *APOBEC3H* gene of domestic cats (*Felis catus*) and their association with the susceptibility to feline immunodeficiency virus and feline leukemia virus infections. Infect Genet Evol. 2014;27:389–94.
25. Hatziioannou T, Princiotta M, Piatak M Jr, Yuan F, Zhang F, Lifson JD, et al. Generation of simian-tropic HIV-1 by restriction factor evasion. Science. 2006;314:95.
26. Schrofelbauer B, Senger T, Manning G, Landau NR. Mutational alteration of human immunodeficiency virus type 1 Vif allows for functional interaction with nonhuman primate APOBEC3G. J Virol. 2006;80:5984–91.
27. Hahn BH, Shaw GM, De Cock KM, Sharp PM. AIDS as a zoonosis: scientific and public health implications. Science. 2000;287:607–14.
28. Sharp PM, Hahn BH. The evolution of HIV-1 and the origin of AIDS. Philos Trans R Soc Lond B Biol Sci. 2010;365:2487–94.
29. Gogarten JF, Akoua-Koffi C, Calvignac-Spencer S, Leendertz SA, Weiss S, Couacy-Hymann E, et al. The ecology of primate retroviruses: an assessment of 12 years of retroviral studies in the Tai national park area, Cote d Ivoire. Virology. 2014;460–461:147–53.
30. Leendertz SA, Locatelli S, Boesch C, Kucherer C, Formenty P, Liegeois F, et al. No evidence for transmission of SIVwrc from western red colobus monkeys (*Piliocolobus badius badius*) to wild West African chimpanzees (*Pan troglodytes verus*) despite high exposure through hunting. BMC Microbiol. 2011;11:24.
31. Lomolino MV, Riddle BR, Whittaker RJ, Brown JH: Distributions of Communities. In: Biogeography. 4th ed. Sunderland: Oxford University Press; 2010. pp. 121–164.
32. Wynne EM, Mathew WA, Jonathan NP. Quantifying risk and resource use for a large carnivore in an expanding urban-wildland interface. J Appl Ecol. 2015;53:371–8.
33. Smith JA, Wang Y, Wilmers CC. Top carnivores increase their kill rates on prey as a response to human-induced fear. Proc Biol Sci. 2015;282(1802):20142711.
34. Compton AA, Emerman M. Convergence and divergence in the evolution of the APOBEC3G-Vif interaction reveal ancient origins of simian immunodeficiency viruses. PLoS Pathog. 2013;9:e1003135.
35. Coffin JM. HIV population dynamics in vivo: implications for genetic variation, pathogenesis, and therapy. Science. 1995;267:483–9.
36. Stremlau M, Owens CM, Perron MJ, Kiessling M, Autissier P, Sodroski J. The cytoplasmic body component TRIM5alpha restricts HIV-1 infection in Old World monkeys. Nature. 2004;427:848–53.
37. Neil SJ, Zang T, Bieniasz PD. Tetherin inhibits retrovirus release and is antagonized by HIV-1 Vpu. Nature. 2008;451:425–30.
38. Van Damme N, Goff D, Katsura C, Jorgenson RL, Mitchell R, Johnson MC, et al. The interferon-induced protein BST-2 restricts HIV-1 release and is downregulated from the cell surface by the viral Vpu protein. Cell Host Microbe. 2008;3:245–52.
39. McEwan WA, Schaller T, Ylinen LM, Hosie MJ, Towers GJ, Willett BJ. Truncation of TRIM5 in the Feliformia explains the absence of retroviral restriction in cells of the domestic cat. J Virol. 2009;83:8270–5.
40. Dietrich I, McMonagle EL, Petit SJ, Vijayakrishnan S, Logan N, Chan CN, et al. Feline tetherin efficiently restricts release of feline immunodeficiency virus but not spreading of infection. J Virol. 2011;85:5840–52.
41. Compton AA, Hirsch VM, Emerman M. The host restriction factor APOBEC3G and retroviral Vif protein coevolve due to ongoing genetic conflict. Cell Host Microbe. 2012;11:91–8.
42. Compton AA, Malik HS, Emerman M. Host gene evolution traces the evolutionary history of ancient primate lentiviruses. Philos Trans R Soc Lond B Biol Sci. 2013;368:20120496.
43. Jern P, Coffin JM. Host-retrovirus arms race: trimming the budget. Cell Host Microbe. 2008;4:196–7.
44. OhAinle M, Kerns JA, Li MM, Malik HS, Emerman M. Antiretroelement activity of APOBEC3H was lost twice in recent human evolution. Cell Host Microbe. 2008;4:249–59.
45. Kobayashi T, Takeuchi JS, Ren F, Matsuda K, Sato K, Kimura Y, et al. Characterization of red-capped mangabey tetherin: implication for the co-evolution of primates and their lentiviruses. Sci Rep. 2014;4:5529.
46. Yamada E, Yoshikawa R, Nakano Y, Misawa N, Kobayashi T, Ren F, et al. A naturally occurring bovine APOBEC3 confers resistance to bovine lentiviruses: implication for the co-evolution of bovids and their lentiviruses. Sci Rep. 2016;6:33988.
47. Takeuchi JS, Ren F, Yoshikawa R, Yamada E, Nakano Y, Kobayashi T, et al. Coevolutionary dynamics between tribe Cercopithecini tetherins and their lentiviruses. Sci Rep. 2015;5:16021.
48. Kumar S, Stecher G, Tamura K. MEGA7: molecular evolutionary genetics analysis version 7.0 for bigger datasets. Mol Biol Evol. 2016;33:1870–4.
49. Guindon S, Dufayard JF, Lefort V, Anisimova M, Hordijk W, Gascuel O. New algorithms and methods to estimate maximum-likelihood phylogenies: assessing the performance of PhyML 3.0. Syst Biol. 2010;59:307–21.
50. Katoh K, Standley DM. MAFFT multiple sequence alignment software version 7: improvements in performance and usability. Mol Biol Evol. 2013;30:772–80.
51. Capella-Gutierrez S, Silla-Martinez JM, Gabaldon T. trimAl: a tool for automated alignment trimming in large-scale phylogenetic analyses. Bioinformatics. 2009;25:1972–3.
52. Nakano Y, Misawa N, Juarez-Fernandez G, Moriwaki M, Nakaoka S, Funo T, et al. HIV-1 competition experiments in humanized mice show that

APOBEC3H imposes selective pressure and promotes virus adaptation. PLoS Pathog. 2017;13:e1006348.
53. Biasini M, Bienert S, Waterhouse A, Arnold K, Studer G, Schmidt T, et al. SWISS-MODEL: modelling protein tertiary and quaternary structure using evolutionary information. Nucleic Acids Res. 2014;42:W252–8.
54. Guex N, Peitsch MC, Schwede T. Automated comparative protein structure modeling with SWISS-MODEL and Swiss-PdbViewer: a historical perspective. Electrophoresis. 2009;30(Suppl 1):S162–73.
55. Bordoli L, Kiefer F, Arnold K, Benkert P, Battey J, Schwede T. Protein structure homology modeling using SWISS-MODEL workspace. Nat Protoc. 2009;4:1–13.
56. Altschul SF, Madden TL, Schaffer AA, Zhang J, Zhang Z, Miller W, et al. Gapped BLAST and PSI-BLAST: a new generation of protein database search programs. Nucleic Acids Res. 1997;25:3389–402.
57. Arnold K, Bordoli L, Kopp J, Schwede T. The SWISS-MODEL workspace: a web-based environment for protein structure homology modelling. Bioinformatics. 2006;22:195–201.
58. Kouno T, Silvas TV, Hilbert BJ, Shandilya SMD, Bohn MF, Kelch BA, et al. Crystal structure of APOBEC3A bound to single-stranded DNA reveals structural basis for cytidine deamination and specificity. Nat Commun. 2017;8:15024.
59. Bruen TC, Poss M. Recombination in feline immunodeficiency virus genomes from naturally infected cougars. Virology. 2007;364:362–70.

6

Reduced antiretroviral drug efficacy and concentration in HIV-infected microglia contributes to viral persistence in brain

Eugene L. Asahchop[1], Oussama Meziane[8], Manmeet K. Mamik[1], Wing F. Chan[1], William G. Branton[1], Lothar Resch[3], M. John Gill[4], Elie Haddad[5,6], Jean V. Guimond[9], Mark A. Wainberg[10], Glen B. Baker[2], Eric A. Cohen[7,8] and Christopher Power[1,2,4*]

Abstract

Background: In patients with HIV/AIDS receiving antiretroviral therapy (ART), HIV-1 persistence in brain tissue is a vital and unanswered question. HIV-1 infects and replicates in resident microglia and trafficking macrophages within the brain although the impact of individual ART drugs on viral infection within these brain myeloid cells is unknown. Herein, the effects of contemporary ART drugs were investigated using in vitro and in vivo models of HIV-1 brain infection.

Results: The EC_{50} values for specific ART drugs in HIV-infected human microglia were significantly higher compared to bone marrow-derived macrophages and peripheral blood mononuclear cells. Intracellular ART drug concentrations in microglia were significantly lower than in human lymphocytes. In vivo brain concentrations of ART drugs in mice were 10 to 100-fold less in brain tissues compared with plasma and liver levels. In brain tissues from untreated HIV-infected BLT mice, HIV-encoded RNA, DNA and p24 were present in human leukocytes while ART eradicated viral RNA and DNA in both brain and plasma. Interruption of ART resulted in detectable viral RNA and DNA and increased human CD68 expression in brains of HIV-infected BLT mice. In aviremic HIV/AIDS patients receiving effective ART, brain tissues that were collected within hours of last ART dosing showed HIV-encoded RNA and DNA with associated neuroinflammatory responses.

Conclusions: ART drugs show variable concentrations and efficacies in brain myeloid cells and tissues in drug-specific manner. Despite low drug concentrations in brain, experimental ART suppressed HIV-1 infection in brain although HIV/AIDS patients receiving effective ART had detectable HIV-1 in brain. These findings suggest that viral suppression in brain is feasible but new approaches to enhancing ART efficacy and concentrations in brain are required for sustained HIV-1 eradication from brain.

Keywords: HIV-1, Nervous system, Antiretroviral therapy, BLT mouse, Macrophages, Microglia

Background

From the outset of the HIV/AIDS epidemic, early entry and expression of HIV-1 in the brain and cerebrospinal fluid (CSF) has been consistently observed [1, 2]. With the availability of increasingly efficacious antiretroviral therapy (ART), plasma and CSF viral burden is usually undetectable by conventional assays [3]. The impact of ART on viral burden in brain tissue has been less certain because of the relative inaccessibility of tissue for measurements except at autopsy; indeed, patients with HIV/AIDS rarely die while receiving effective ART. Nonetheless, with ART-associated improvement in immune status, the severity of degenerative brain disease associated with HIV/AIDS, termed HIV-associated neurocognitive disorders (HAND), has declined although it remains a

*Correspondence: chris.power@ualberta.ca
[1] Department of Medicine (Neurology), University of Alberta, Edmonton, AB, Canada
Full list of author information is available at the end of the article

serious challenge in ~ 25% of treated patients that often precludes employment and reduces overall survival [4–7]. All immunosuppressive lentiviruses (e.g., Human, Simian, Feline and Bovine Immunodeficiency Viruses) infect the brain and display a preference for macrophage tropism as well as infection and depletion of T lymphocyte populations. In the human brain, productive HIV-1 infection is evident in trafficking macrophages (bone marrow-derived macrophages) and resident microglia, but few if any infected T cells are observed [8–11]. These myeloid cell populations in the brain constitute a unique reservoir in which HIV-1 replicates because of the paucity of resident lymphoid cells within the brain and the blood–brain barrier's impervious structure, which limits drug penetration. Of note, microglia and macrophages are less permissive to HIV-1 than CD4+ T-cells, possibly due to reduced CD4 expression and variable expression of specific host restriction factors [12–14]. Furthermore, macrophage-lineage cells are long lived (half-life of months–years) and comparatively resistant to cell death upon HIV-1 infection compared to CD4+ T-cells, although some groups have reported very short half-lives in HIV-infected macrophage-lineage cells [15, 16]. Likewise, recent studies of cerebrospinal fluid (CSF) show higher than expected rates of viral escape in CSF that were associated with brain injury in the setting of undetectable virus in plasma [17].

The impact of contemporaneous ART drugs on viral expression and replication in the brain remains unclear despite their potent viral suppressive properties in blood. Likewise, the actual ART drug concentrations in brain tissue are also uncertain. Studies of brain tissues from HIV-infected humans or SIV-infected nonhuman primates show that both viral RNA and DNA are detected in brain tissues despite receiving ART [15, 18, 19]. Of note, a very recent study reported that in a macrophage-only mouse model of HIV-1 infection, HIV-infected monocyte-derived macrophages were highly responsive to ART although the brain myeloid cells and tissues were not examined [20]. In these patients and animal studies, the efficacy of ART was assessed in combination but not as individual drugs. These circumstances raise several questions regarding the efficacy of ART in HIV-infected brain myeloid cells, ART's capacity to modulate HIV-1 entry of the brain and the impact of ART on established HIV-1 infection of the brain. It is important to evaluate the efficacy of individual ART ex vivo and in animal studies to select ART drugs with a high propensity to inhibit replication in the brain.

Given the questions concerning the impact of ART on HIV-1 brain infection, we investigated ART drugs' efficacies and concentrations in different experimental platforms. The in vitro and in vivo efficacies and concentrations of different ART drugs were examined in a primary cell culture model. HIV-1 DNA and RNA quantities in brain and plasma from HIV-infected immunosuppressed mice reconstituted with human leukocytes (BLT) mice and human actively receiving ART was also quantified. These studies showed that several ART drugs exhibit reduced concentrations and efficacies in the brain myeloid cells (and in brain tissues) compared to PBMCs, but nonetheless inhibited in vivo HIV-1 brain infection in BLT mice. Investigation of brain tissues from HIV/AIDS patients indicated there was persistent viral detection in the setting of longstanding aviremia with concurrent ART.

Results

HIV-1 infection of microglia and macrophages

In the brain, HIV-1 infects both resident microglia and circulation-derived macrophages originating from the bone marrow although viral production tends to be higher in macrophages [1, 21–23]. To model HIV-1 infection of these brain myeloid cells, we established primary cultures of human fetal microglia (HFM) and bone marrow-derived macrophages (BMDMs). Following differentiation in vitro to enhance microglial and macrophage phenotypes, both cell types expressed Iba-1 and MHC Class II (Fig. 1a). HFM showed higher transcript levels of *CD68* and *CD163* but conversely BMDMs showed greater expression of *CD4* and *CCR5* (Fig. 1b). After HIV-1_{YU2} infection at matched input titers (MOI = 0.1), BMDMs showed increased release of HIV-1 p24 in supernatants over time compared to HFM, with peak p24 release at Day 12 post-infection (Fig. 1c). Thus, both BMDMs and HFM supported HIV-1 replication but BMDM exhibited higher viral production levels, possibly due to increased CD4 and CCR5 expression in keeping with previous observations suggesting there is higher viral replication in macrophages compared to microglia [23].

ART efficacy in HIV-infected myeloid cells and PBMCs

Although the efficacies of different ART drugs in T cell lines and primary CD4 T cells is well known, little is known about their effects in HIV-infected brain myeloid cells although they suppress viral replication while their relative actions in different brain myeloid cells (such as microglia) versus activated PBMCs is unknown. To assess the comparative antiviral activity of different ART drugs in HFM, BMDMs and activated peripheral blood mononuclear cells (PBMCs), representative drugs from several different classes were investigated. These included zidovudine (AZT), etravirine (ETR), raltegravir (RAL) and darunavir (DRV), maraviroc (MVC) and dolutegravir (DTG). Each drug was tested in a concentration-dependent manner by measuring p24 levels in supernatants

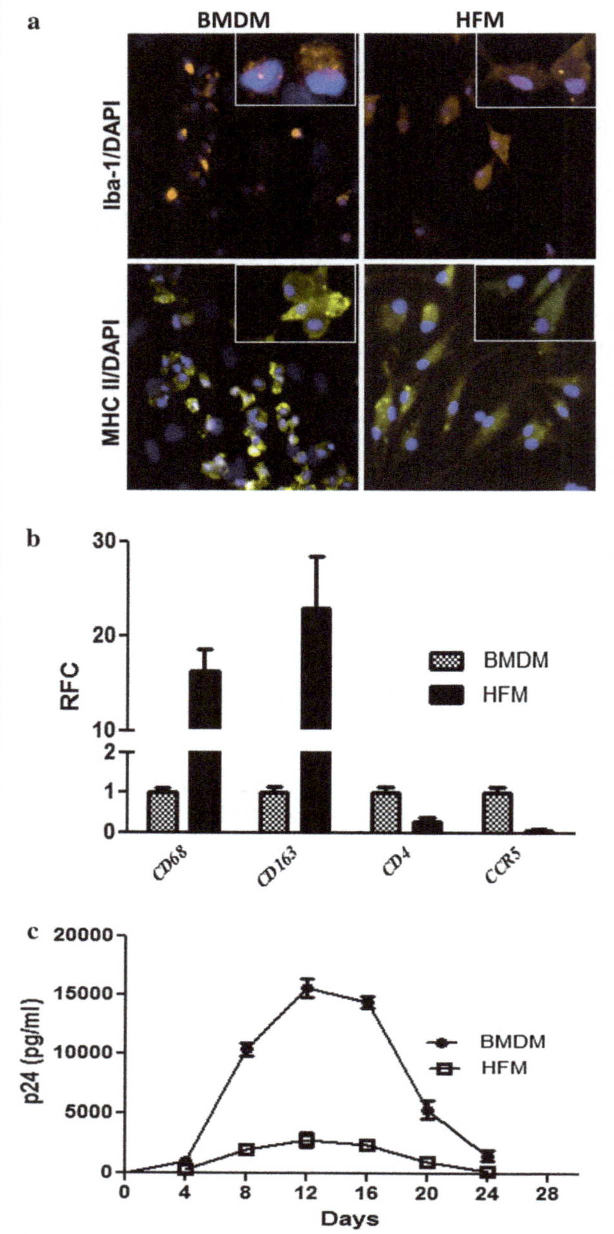

Fig. 1 Human myeloid cells and HIV-1 infection. **a** Immunolabeling of cultured human fetal microglia (HFM) and bone marrow-derived macrophages (BMDMs) with antibodies to Iba-1 (HFM: orange, BMDMs: orange) and MHC Class II (HFM: yellow, BMDMs: yellow) expression. **b** Relative *CD68, CD163, CD4* and *CCR5* transcript expression levels in HFM and BMDMs were quantified by RT-PCR. **c** HFM and BMDMs were infected at a multiplicity of infection (MOI) of 0.1 with HIV-1 YU-2 (derived by transfection in 293 T-cells) and supernatant was collected after every 4 days for HIV-1 p24 quantification by ELISA

from HIV-infected HFM, BMDMs and activated PBMCs (Additional file 1) from which EC_{50} values were determined (Table 1). Comparison of ART drug efficacies

Table 1 ART EC_{50} (ng/ml) values in different myeloid cells and PBMCs compared to published CSF concentrations

ART drug	PBMC EC_{50}	BMDM EC_{50}	HFM EC_{50}	CSF[a]
AZT	2.1 ± 0.24	14.7 ± 1.2	27.0 ± 3.1	44.0 ± 6.0
RAL	1.1 ± 0.1	1.8 ± 0.2	3.3 ± 0.6	21.3 ± 5.0
ETR	0.64 ± 0.08	1.1 ± 0.25	3.6 ± 0.2	8.4 ± 1.1
DRV	1.2 ± 0.15	1.9 ± 0.3	8.1 ± 0.78	47.0 ± 8.8
MVC	2.67 ± 0.2	0.1 ± 0.02	1.0 ± 0.15	2.87 ± 0.4
DTG	0.77 ± 0.04	0.3 ± 0.13	0.59 ± 0.05	18

[a] All CSF concentrations are published data

in each infected cell culture type revealed that DRV (Fig. 2a), ETR (Fig. 2b), AZT (Fig. 2c) and RAL (Fig. 2d) had significantly higher EC_{50} values in HFM compared to activated PBMCs and in BMDMs for DRV (Fig. 2a), ETR (Fig. 2b) and AZT (Fig. 2c). In contrast, MVC and DTG were both highly effective at inhibiting viral replication in HIV-infected BMDMs and HFM compared to activated PBMCs (Fig. 2e, f). These findings underscored the differential actions of individual ART drugs on viral replication in brain myeloid cells.

As ART concentrations within cells (and tissue) are critical determinants of their antiviral capacity, we measured DRV and RAL concentrations in cultured cells. In these analyses both RAL and DRV exhibited similar intracellular and extracellular concentrations in HFM apart from RAL, which at 44.4 ng/ml showed higher extracellular compared to intracellular concentrations (Fig. 3a). However, the intracellular concentrations for both drugs were significantly higher in activated PBMCs compared to HFM cultures (~ 100 fold) and matched extracellular activated PBMCs concentrations (Fig. 3b). Of note, exposure of differentiated human myeloid (THP-1) cells to endotoxin or HIV-1 infection did not influence RAL or DRV extra- or intracellular concentrations (Additional file 2). However, published antiretroviral (ARV) drug concentrations in cerebrospinal fluid (CSF) [24–36] were higher than the EC_{50} values for all drugs tested herein (Table 1). These data indicated that ART efficacy in productively infected brain (myeloid) cells was substantially less than in infected lymphoid cells for specific ART drugs.

In vivo tissue ART drug concentrations

Although ART drug concentrations in CSF are known in treated HIV/AIDS patients [24–36], brain tissue levels are largely unknown for most ART drugs, prompting investigation of RAL and DRV in vivo concentrations. Following intraperitoneal administration of each drug in adult mice (1.2 mg/animal), drug concentrations were measured at different time points in serum, liver and the

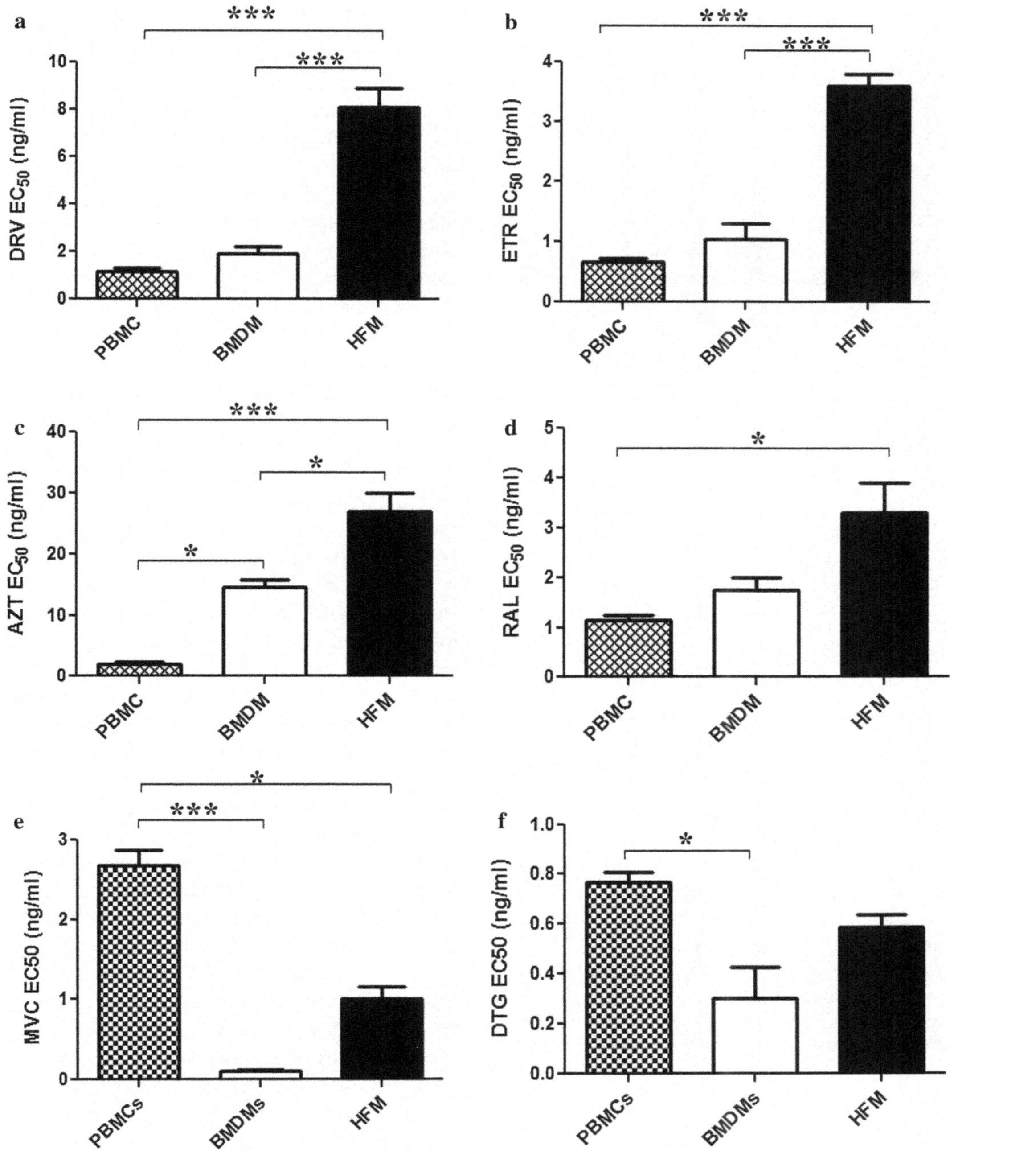

Fig. 2 ART drug EC_{50} values in HIV-infected HFM, BMDMs and PBMCs. HFM, BMDMs and PBMCs were infected with HIV-1 YU-2 (MOI of 0.1), followed by drug treatment. Mean EC_{50} values for each cell type (HFM, BMDM and PBMC) are shown during treatment with **a** DRV, **b** ETR, **c** AZT, **d** RAL, **e** MVC and **f** DTG. * $p < 0.05$; ** $p < 0.01$; *** $p < 0.001$

brain (cerebral cortex, striatum, cerebellum, medulla and hypothalamus). When RAL was administered, the mean concentrations in serum and liver were approximately 15,000 ng/ml and 27,000 ng/g tissue, respectively, at 1 h post-administration while the concentration in different regions of the brain ranged between 150 and 200 ng/g of tissue (Fig. 4a). At 4 h post-administration the serum concentrations for RAL in serum and liver were 300 ng/ml and 1200 ng/g, respectively, with only traces of RAL in brain (Fig. 4b). For DRV, there was a similar trend with

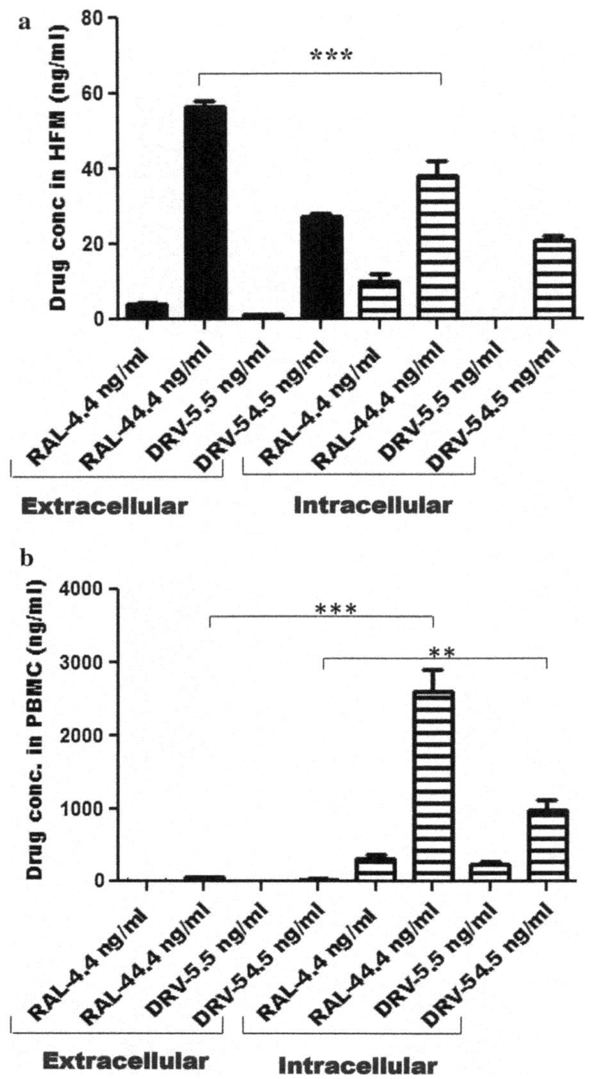

Fig. 3 Intracellular and extracellular ART drug concentrations in HFM and PBMCs. HFM (**a**) and PBMCs (**b**) were exposed to DRV and RAL at different concentrations for 24 h. Intracellular and extracellular concentrations were measured by HPLC–MS. * $p < 0.05$; ** $p < 0.01$; *** $p < 0.001$

lower drug concentrations in brain; in the serum and liver, the mean DRV concentrations were 850 ng/ml and 3800 ng/g tissue, respectively, at 1 h post drug administration (Fig. 4c). However, in different brain regions, DRV concentrations ranged between 0 and 150 ng/g (tissue) after 1 h (Fig. 4c). Again, only trace amounts of DRV were detected at 4 h post-administration (Fig. 4d). These studies showed that while ART drug levels were similar in different anatomic regions of the brain, they were substantially lower than their corresponding concentrations in serum or liver.

In vivo efficacy of ART in HIV-infected humanized mice

As contemporary ART regimens are highly effective in controlling viral burden in blood, we wanted to examine if there was a corresponding inhibition of HIV-1 infection in the brain during viral suppression in blood. To address this issue, we evaluated uninfected and HIV-infected humanized (BLT) mice with and without an ART regimen comprised of RAL, tenofovir and emtricabine or PBS (Fig. 5). The neuropathological features associated with HIV-1 infection of the brains of BLT mice were assessed in brain sections from animals in each group using antibodies to conventional neural cell markers that immunolabeled human CD68 (macrophage/microglia), GFAP (astrocytes), CD3ε (lymphocytes) and HIV-1 p24. Human CD3ε immunopositive T cells were detected in HIV[−], HIV[+] and HIV[+]/ART (Fig. 5ai–iii) animals with p24 immunoreactivity evident within CD3ε^+ cells (Fig. 5aii inset). Similarly, human CD68 immunopositive macrophages were detected in brains from each experimental group, i.e. HIV[−], HIV[+] and HIV[+]/ART (Fig. 5bi–iii) animals, also with p24 immunodetection in CD68$^+$ cells (Fig. 5bii, inset). GFAP-immunolabeled (mouse only) astrocytes were apparent in brains from all experimental groups with similar morphology and abundance in all groups (Fig. 5ci–iii). Human cells engrafted mouse brains in all experimental groups without substantial differences in host (mouse) glial responses despite the presence of HIV-1 p24 in the untreated/infected experimental group. Thus, HIV-1 infection of BLT resulted in viral neuroinvasion mediated by infected macrophages and T cells but ART treatment of infected animals lead to undetectable virus in the brain without preventing human cell trafficking into the brain.

By way of verifying the above neuropathological findings, we examined viral, human and mouse gene expression levels in brains from each experimental BLT mouse group together with examining the impact of interrupting ART for several weeks to recapitulate clinical circumstances of missed ART doses. To determine the relative number of human cells engrafting each brain, human *CD45* genomic DNA copy numbers were analyzed in brain tissues and were found to be similar across all experimental groups (Fig. 6a). Human *CD68* transcript levels in brain varied across experimental groups, with increased *CD68* expression in the brains of the HIV[+]/ART-interrupt group (Fig. 6b) while *CD3E* transcript levels were similar across groups (Fig. 6c). Analyses of human immune gene expression in brains showed that *CD163* (Additional file 3A) was detectable but did not differ across groups although IP10 (*CXCL10*) and *MX2* (Additional file 3B and C) were significantly increased in the HIV[+]/ART-interrupt group. Of note, the mouse macrophage gene, *f4/80,* was significantly

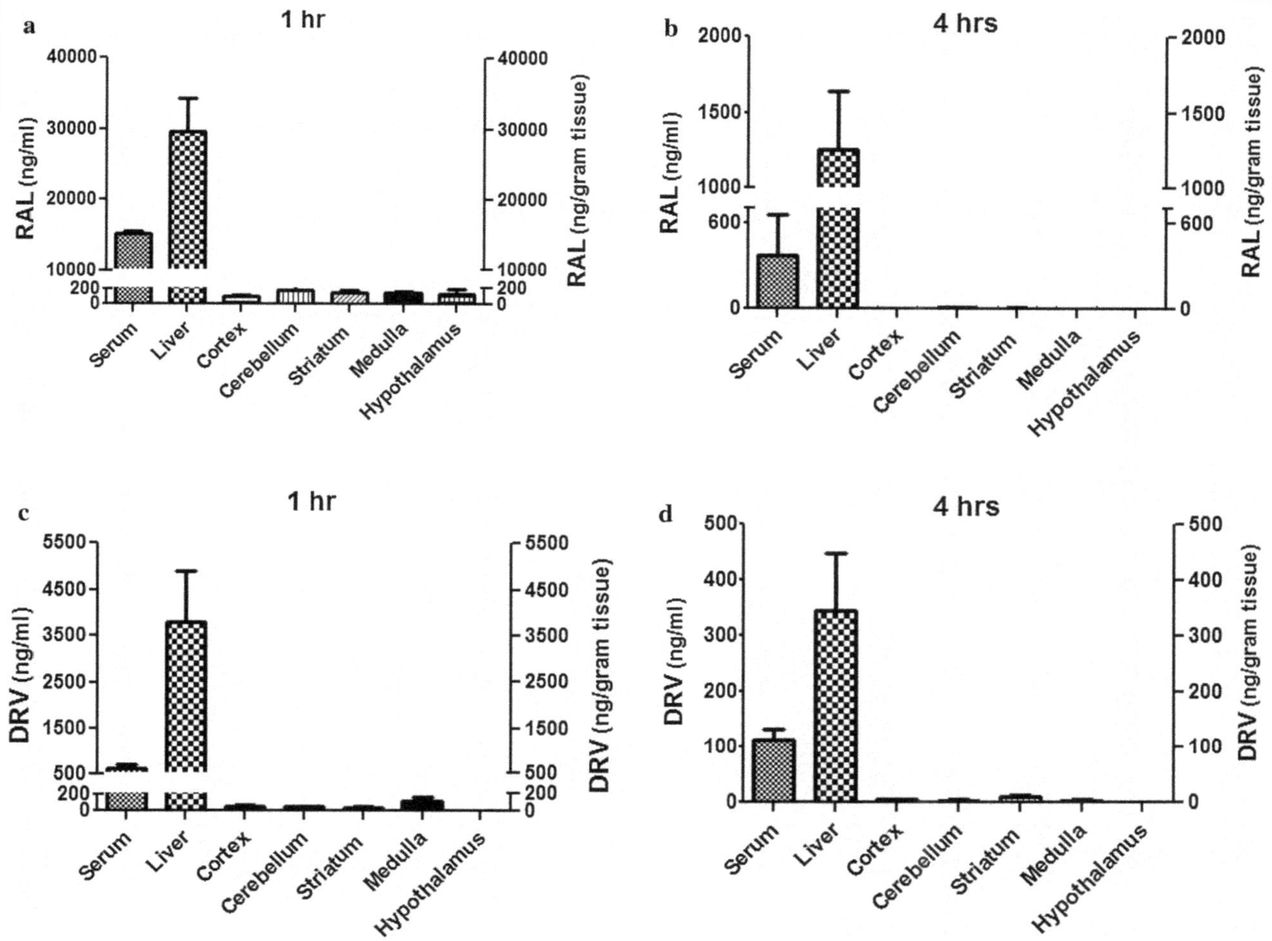

Fig. 4 In vivo ART drug concentrations in brain, blood and liver. RAL and DRV were injected intraperitoneally into adult mice (1.2 mg/animal). ART drug concentrations were measured at 1 and 4 h post-drug administration in serum, liver and brain (cortex, striatum, cerebellum, medulla and hypothalamus). RAL was detected in all tissues at 1 h (**a**). At 4 h, **b** RAL levels were markedly reduced in serum and liver and not detected in brain. **c** DRV was detected in all tissues at 1 h but levels of DRV were also markedly reduced at 4 h (**d**). Results represent mean ± SEM (n = 4)

decreased in the HIV[+]/ART-interrupt group (Additional file 3D).

Viral burden was examined next, disclosing that plasma mean HIV-1 RNA copy number in HIV-infected BLT mice ranged widely and was detected in the HIV[+] group (520,000 copy numbers/ml) and in the HIV[+]/ART-interrupt group (1,150,000 copy numbers/ml) (Fig. 6d) but was not detected in the HIV[+]/ART or HIV[-] groups. In brains from BLT mice, mean viral RNA levels ranged from 500 to 700,000 copies/g (tissue) in the HIV[+]/PBS group but HIV-1 RNA was not detectable in the HIV[−] or the HIV[+]/ART groups (Fig. 6e). In contrast, mean HIV-1 RNA copy numbers in the HIV[+]/ART-interrupt group were detectable in all animals but at low copy numbers (100–500 copies/g) (Fig. 6e). Total mean HIV-1 DNA copies were also measured in all groups, revealing that the HIV[+] group showed 50–40,000 copies/g but HIV-1 DNA was not detected in the HIV[−] or the HIV[+]/ART groups (Fig. 6f). As with HIV-1 RNA brain burden, mean HIV-1 DNA copy numbers in the HIV[+]/ART-interrupt group were detectable in all animals at 100–20,000 copies/g (Fig. 6f) although integrated proviral DNA in brain was not detected in all groups. Of note, RAL was not detected in the brains of the ART-treated animals. These findings highlighted the capacity for HIV-1 to infect brain myeloid cells but also emphasized that in HIV[+]/ART group viral RNA was undetectable in the brain.

HIV-1 detection and neuroimmune responses in human brain tissue during effective ART

To date, most, if not all, reported studies of HIV-1 quantities in brain were performed using tissues from patients who had died days to weeks after last ART dosing,

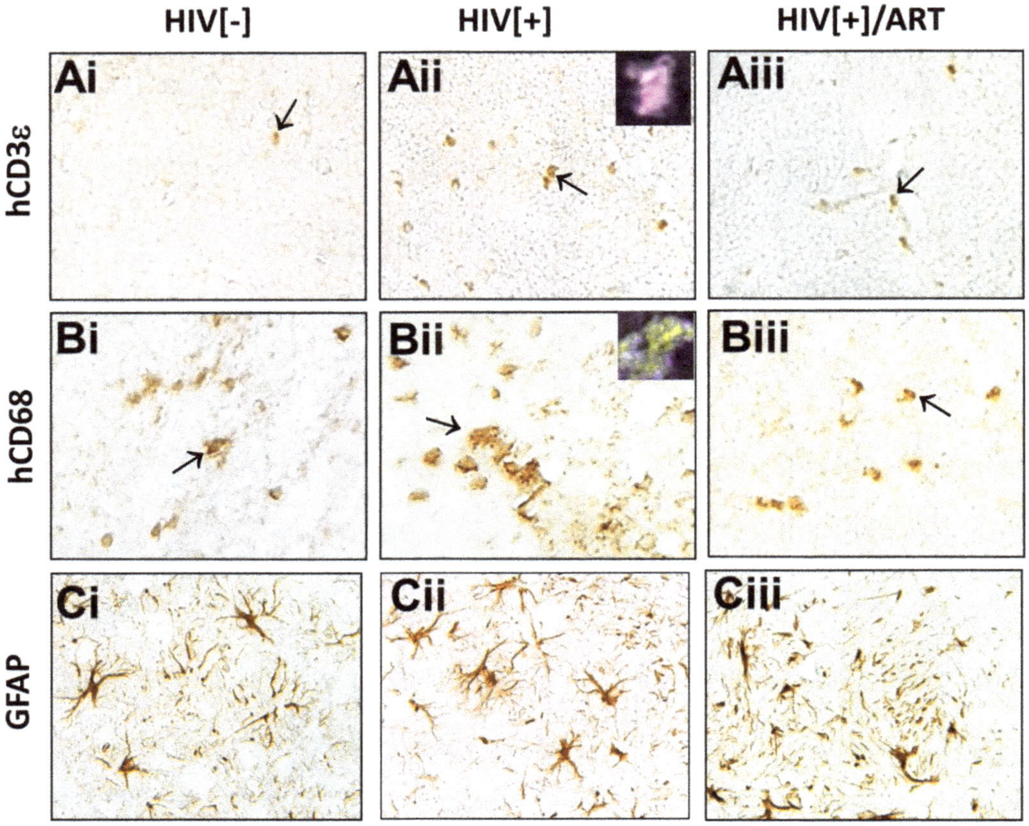

Fig. 5 Neuropathology in HIV-infected BLT mice: Human CD3ε immunolabeled T cells were observed in brains from uninfected (ai), HIV-infected (aii) and HIV-infected/ART-exposed animals (aiii). Co-labeled cells exhibiting CD3ε (yellow), p24 (purple) with DAPI nuclear labeling (inset) were evident in HIV-infected animals. Similarly, human CD68 immunolabeling of macrophages was evident in HIV[−] (bi), HIV[+] (bii) and HIV[+]/ART (biii) and was localized showing CD68 (yellow) and HIV-1 p24 (purple) (inset). GFAP-immunopositive (murine) astrocytes were apparent in brains from HIV[−] (ci), HIV[+] (cii) and HIV[+]/ART (ciii) animals

precluding a definitive comparison between virus burden in brain tissue and plasma during effective ART. We encountered two patients with HIV/AIDS from whom brain tissue was collected within hours of last ART dosing. Patient 1 (Pt-1) had a premortem diagnosis of HAND but died unexpectedly following a myocardial infarction while Patient 2 (Pt-2) had longstanding complex partial epilepsy (CPS_Z) and underwent surgical resection of the left temporal lobe to remove the epileptic lesion (see "Methods" section for clinical and demographic details). Pt-1 (Fig. 7a) and Pt-2 (Fig. 7b) were receiving effective ART regimens associated with high CNS penetration effectiveness (CPE) scores within 8 and 12 h, respectively, of brain tissue collection. Both patients had CD4 T cell levels of > 400 cells/µl and plasma RNA viral loads below the detection limit (< 40 copies/ml) for at least three previous years of clinical follow up. In multiple discrete brain samples (n = 6) from HIV-infected and non-infected (control) patients including cortex and adjacent white matter, HIV-1 RNA and DNA as well as host immune responses were measured (Fig. 7c–f and Additional file 4). All 6 brain samples from Pt-1 contained HIV-1 RNA while two of 6 samples from Pt-2 had detectable HIV-1 RNA. The brain samples from uninfected patients did not show HIV-1 RNA (Fig. 7c). Total HIV-1 DNA was also quantified from matched samples, revealing that in Pt-1 viral DNA was detected in all 6 samples but in Pt-2 only one brain sample showed viral DNA, while in uninfected controls (n = 6), viral DNA was not detected (Fig. 7d). Using an established assay that detected virus in HIV-infected human macrophages (Additional file 4A), integrated proviral DNA was measured, revealing that integrated provirus was detected in 2 of 6 brain samples from Pt-1 but was not detected in brain samples from Pt-2 (Fig. 7e). To determine the corresponding expression of cellular and immune genes in ART-suppressed HIV/AIDS patients' brains compared with uninfected patients, we examined *CD68* (macrophage marker), *CD3E* (T cell marker), *CD163*, *TNFA*, *IL6* and *IL1B* transcript levels; *CD68* (Fig. 7f), *CD3E* (Fig. 7g) and *CD163* (Additional

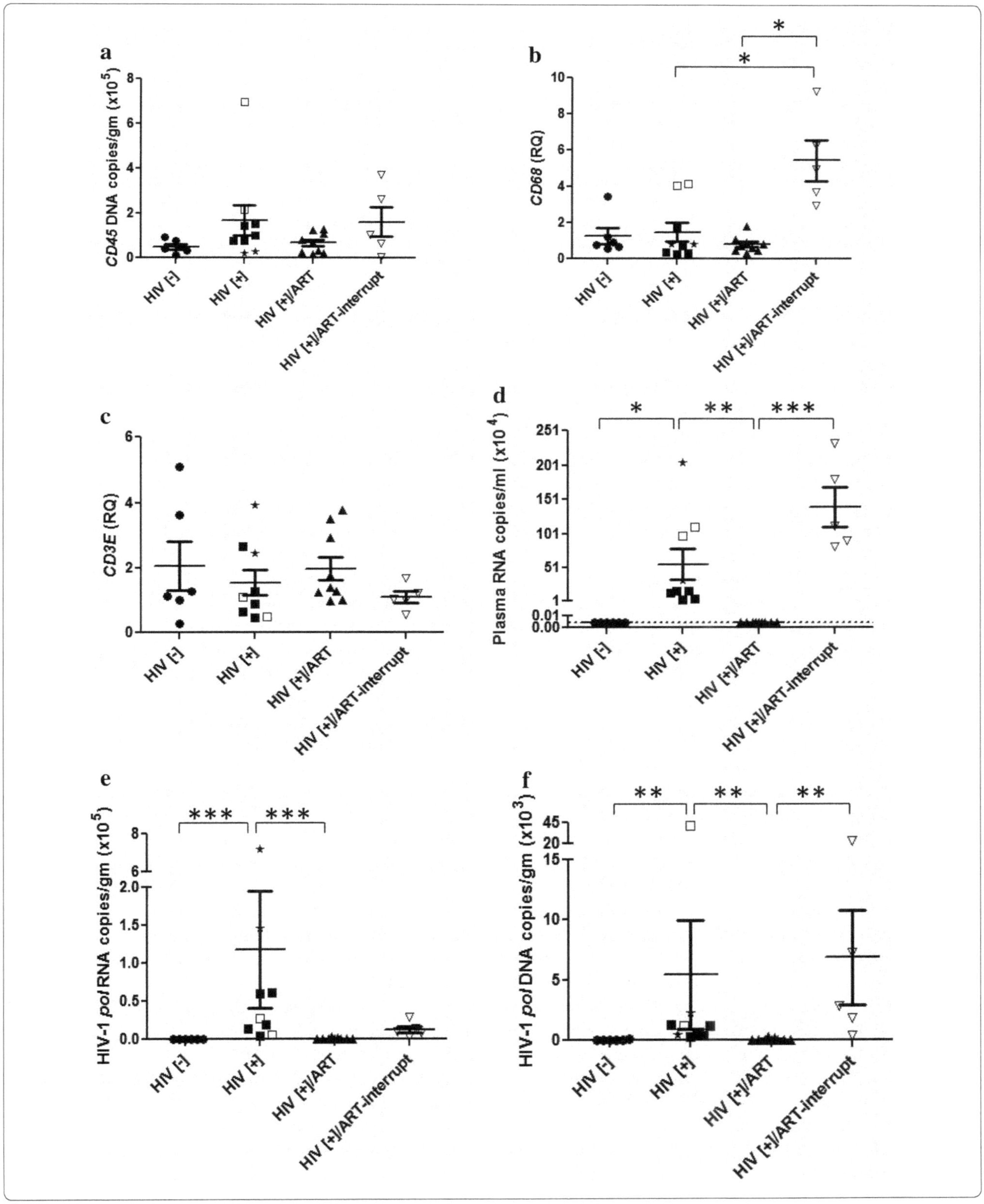
a
CD45 DNA copies/gm (x10^5)
HIV [-]
HIV [+]
HIV [+]/ART
HIV [+]/ART-interrupt
b
CD68 (RQ)
*
*
c
CD3E (RQ)
d
Plasma RNA copies/ml (x10^4)
*
**
***
e
HIV-1 pol RNA copies/gm (x10^5)
***
***
f
HIV-1 pol DNA copies/gm (x10^3)
**
**
**

(See figure on previous page.)

Fig. 6 HIV-1 and human RNA and DNA in brains from BLT mice. Four experimental groups of BLT mice were examined including uninfected (HIV[−]), HIV-infected/PBS-treated (HIV[+]), HIV-infected/ART-treated (HIV[+]/ART) and HIV-infected/ART-interrupted (HIV[+]/ART-interrupt). The mice in the HIV[+] group were infected at different $TCID_{50}$/animal as well as for different durations [two mice were infected at 50,000 $TCID_{50}$/animal for 8 weeks without viral detection and then re-infected at 50,000 $TCID_{50}$/animal for another 4 weeks (asterisk); two other mice were infected at 50,000 $TCID_{50}$/animal for 13 weeks (open square); five mice were infected at 100,000 $TCID_{50}$/animal for 7 weeks (filled square)]. For the ART-interrupt group (HIV[+]/ART-interrupt), animals were infected and treated for 2 weeks (between weeks 4–6 or 8–10), after which ART was stopped and animals were sacrificed at 13 weeks post-infection. **a** Human *CD45* DNA (copy numbers/g of tissue) was detected in all animals' brains and did not differ across experimental groups. **b** Human *CD68* transcript was also detected in all animals but increased in the HIV[+]/ART-interrupt group. **c** Human *CD3E* was detected in all experimental groups. **d** HIV-1 RNA was detected in plasma in the HIV[+] and HIV[+]/ART-interrupt groups. **e** All brains from the HIV[+] and HIV[+]/ART-interrupt groups showed detectable HIV-1 RNA. **f** HIV-1 DNA was detected in brains from the HIV[+] and HIV[+]/ART-interrupt groups. * $p < 0.05$; ** $p < 0.01$; *** $p < 0.001$

file 4B) were increased in samples from Pt-2 compared to Pt-1 and uninfected controls. In contrast, *IL6* and *TNFA* were each increased in Pt-1 and Pt-2, respectively (Additional file 4C and D), while *IL1B* expression was increased in both HIV/AIDS patients' brains compared to uninfected controls (Fig. 7h). Similarly, *APOBEC3G, MX2, MX1* and *BST2,* were upregulated in Pt-2's brain but not induced in Pt-1's brain samples (Additional file 4E–H). Using tissue from six different regions of the brain from Pt-1, the mean RAL concentrations (18 ng/g tissue) were low (Additional file 5). These data indicated that HIV-1 RNA and DNA with associated neuroinflammatory responses were present in brain tissues from patients who had achieved long-term suppression of HIV-1 in plasma although concurrent ART drug levels were low.

Discussion

The present study addresses several fundamental questions regarding the impact of ART on HIV-1 infection of the brain. To the best of our knowledge, this is the first report showing that contemporary ART drugs exert differential actions in HIV-infected brain myeloid cells depending on the infected cell type and individual drug, with overall higher EC_{50} values in HIV-infected microglia compared to macrophages and activated PBMCs. Additionally, this is the initial report comparing ART drug concentrations in human brain myeloid and activated PBMCs as well as in brain; the intracellular drug levels were substantially lower in microglia than in activated PBMCs and very low (and transient) in brain tissues from mice and humans. These studies also indicate that in ART-treated BLT mice, HIV-1 infection was undetectable within the brain (and plasma) but ART interruption resulted in a viral rebound. Our clinical observations supported the experimental findings of reduced efficacy in brain myeloid cells by disclosing that HIV-1 RNA and DNA were detected in brain tissues from stable aviremic HIV/AIDS patients receiving ART within hours of tissue collection. These findings underline the limited ART drug efficacies and concentrations in the brain together with the precarious vulnerability of the brain to HIV-1 infection derived from blood.

The human brain is a unique HIV-1 reservoir because it is established early in infection, harbors CCR5-dependent HIV-1 viruses, and the cell populations supporting HIV-1 replication are comprised solely of myeloid cells including trafficking macrophages and resident microglia [37]. Although in vitro data show that HIV-1 infects and replicates in both macrophages and microglia [21, 38, 39], data reporting replication of HIV-1 in human brain myeloid cells in the setting of ART-mediated viral suppression in blood are limited and usually complicated by interruption of ART before tissue collection [40, 41]. In SIV-infected macaques, infected macrophages and perhaps microglia contribute to viral persistence in the brain [18, 42]. Effective ART resulted in complete suppression of plasma SIV levels associated with undetectable SIV RNA in brain while viral DNA levels remain unchanged [18, 42]. A quantitative outgrowth assay was reported that assessed the contribution of productively infected microglia/macrophages in ART-suppressed SIV-infected macaques and found productively infected macrophage-lineage cells in different tissues including the brain [8].

Although integrated proviruses might exhibit defective replication [43], their capacity to exert pathogenic effects could be maintained through expression of subgenomic mRNAs encoding pathogenic viral proteins including Vpr, Tat and Nef. To date, there are few studies quantifying integrated proviral (HIV-1) DNA in human brain in the presence or absence of ART; the ratio of integrated:total HIV-1 DNA in brain ranged from 1:6 to 1–81 [11] in an early study, similar to that reported (1:86) for infected lymphocytes [44]. Another study reported matched integrated and unintegrated HIV-1 LTR sequences from the same brain were similar, but LTR activity was not assessed [45]; nonetheless, LTRs from replication-competent brain-derived HIV-1 exhibit substantial variation in transcriptional activity [46, 47]. Herein, the ratio of integrated:total viral DNA

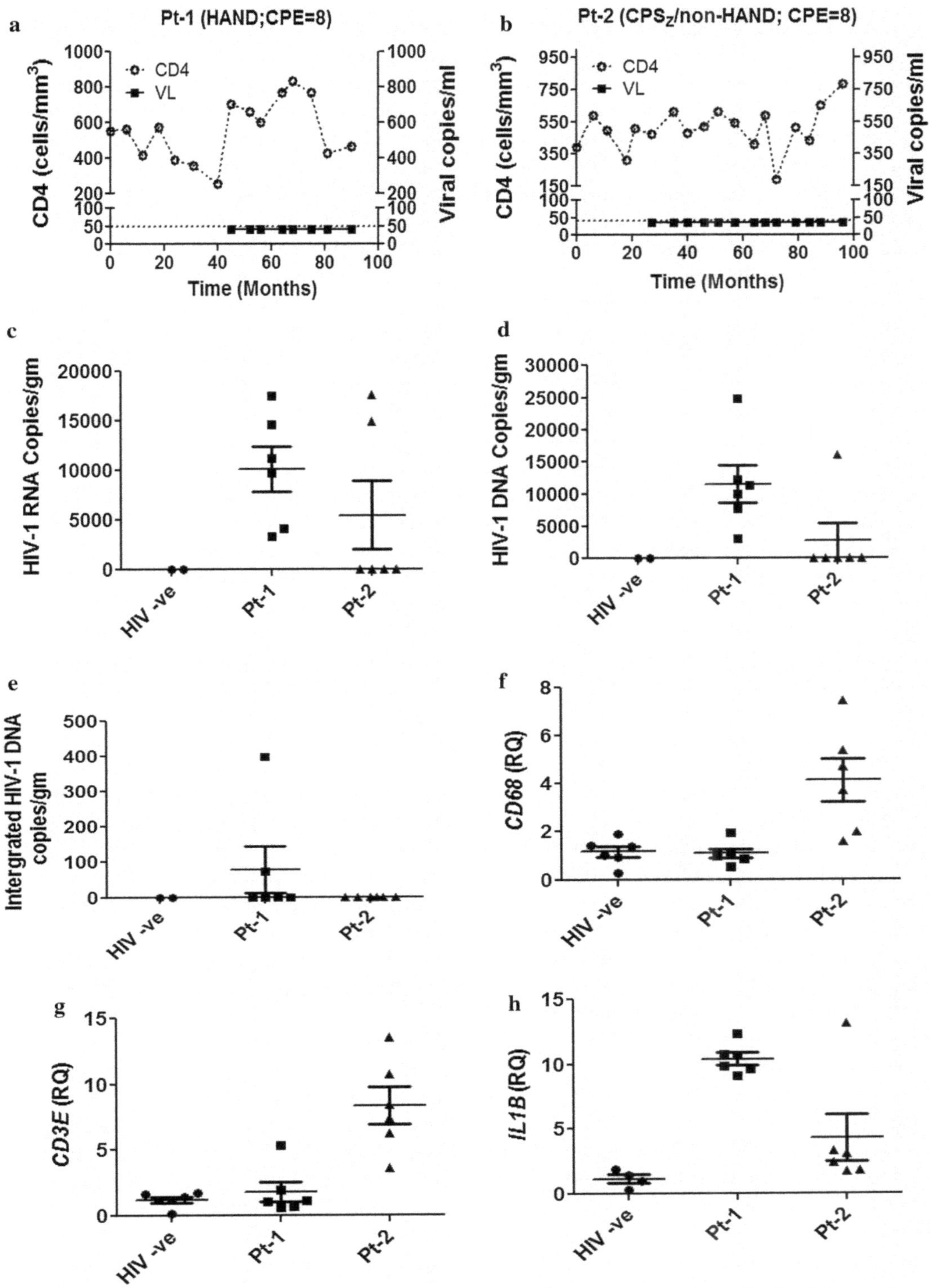

Fig. 7 HIV-1 RNA, DNA and host gene expression in human brain during ART. HIV-infected patients receiving effective ART were followed for > 90 months before brain tissue was obtained by autopsy (Pt-1) (**a**) or by surgical resection for epilepsy (Pt-2) (**b**). **c** Viral RNA and **d** total viral DNA were quantified by droplet digital PCR in brain tissues from uninfected patients (HIV − ve) and from Pt-1 and Pt-2. **e** Integrated proviral DNA was measured in brain by real-time PCR using specific primers and probe to HIV-1 *LTR* following pre-amplification with an *alu-gag* primer pair. **f** *CD68*, **g** *CD3E* and **h** *IL1B* transcript levels were measured in brain tissue by RT-PCR. Results represent mean of ± SEM

in ART-treated HIV/AIDS patients was 1:147 for Pt-1 but integrated provirus was not detected in brain samples from Pt-2 (Fig. 7e). While brain tissue from Pt-1 was collected from frontal cortex and proximal white matter, Pt-2's brain samples were from temporal lobe, an anatomic site that typically shows less viral burden [48], which may have contributed to the relative paucity of virus in Pt-2. Another differentiating aspect between the two patients that could explain high viral burden in Pt-1 compared to Pt-2 is the composition of the ART regimen at the time of tissue collection. Pt-1 was receiving abacavir, lamivudine and raltegravir while Pt-2 was on abacavir, lavimudine and dolutegravir. We observed increased antiviral efficacy of dolutegravir in brain myeloid cells compared to activated PBMCs, implying a potentially higher intracellular concentration and perhaps greater antiviral activity for longer periods, as suggested by others [49]. As our in vitro data showed comparable efficacy in HFM and BMDMs with dolutegravir, its use in Pt-2 might explain the reduced viral burden and the absence of integrated virus.

While the ART drugs we tested (DRV, RAL, ETR, AZT, MVC, DTG) are highly effective in controlling HIV-1 replication in blood, the extent to which they regulate viral replication in HIV-infected brain myeloid cells is largely unknown. Indeed, their levels in tissues are also uncertain although they are known to enter CSF which has prompted assignment of individual CPE values. Previous studies have reported that several ART drugs are efficacious in HIV-infected monocyte-derived macrophages [50] and some also inhibit HIV-transduced astrocytes [51]. However, the current data indicate that, at matched input viral titers and drug concentrations, most of the ART drugs were significantly less efficient in controlling viral replication in HIV-infected microglia compared to BMDMs or activated PBMCs (Fig. 2a–d), although both MVC and DTG were highly efficacious. The increased efficacy of MVC is supported by studies indicating that intensification of ART in both HIV-1 and SIV infections with MVC improves neurocognition and further reduces latent viral reservoir in the brain [52–55]. It is probable that this feature of MVC likely stems from its capacity to block CCR5 as well as induce PPAR-gamma expression [56]. It has previously been shown that PPAR-gamma up regulation can inhibit HIV-1 replication [57]. Notably, activated PBMCs also appeared to concentrate RAL and DRV at higher intracellular levels compared to intracellular concentrations in microglia (Fig. 3a, b). Increased intracellular concentrations would enhance antiretroviral drug efficacy over time, leading to restricted viral integration and production. Our results also suggest that brain tissue is less accessible to intraperitoneally delivered drugs; ART drug concentrations were similar in different brain regions but substantially lower than in liver or blood (Fig. 4a–d), with minimal drug detection by 4 h post-delivery. Two independent studies have shown in lymphoid tissues of HIV/AIDS patients receiving ART with undetectable plasma viremia that there is continuous evolution and trafficking of HIV-1 persistent virus as well as lower tissue ART drug concentrations than in the peripheral blood [58, 59]. These findings also illustrate that viral persistence is associated with lower ART drug concentrations in lymphoid tissues. Collectively, the present data underline the limited efficacy of present-day ART drugs in the brain and perhaps other tissues/organs harboring HIV-infected tissue macrophages. The use of sub-optimal ART regimens might lead to low drug concentrations in infected brain cells, resulting in a viral reservoir in these long-lived cells together with the potential emergence of drug resistance.

We used a BLT murine model of HIV-1 infection [60] to examine the brain viral burden in animals infected with high titer HIV-1 and their responses to an established ART regimen. The present data are similar to a recent study [60] indicating that viral RNA (Fig. 6c) and total DNA (Fig. 6d) were detected in brains from BLT mice, although our data show higher viral levels, perhaps due to a higher input viral titer or efficiency of human cell engraftment. Nonetheless, viral RNA was undetectable in brain by the ART regimen implemented in this study in all animals treated until sacrifice while in animals in which ART was interrupted before termination of the study, both viral RNA and DNA were detected in brain, which was associated with an increase in human *CD68* expression. Remarkably, other than *CD68* expression, human (and mouse) host cell abundance and immune responses in BLT brains (Fig. 4) were minimally affected despite variable infectious input titers and durations of infection, unlike in human brains (Fig. 5).

Conclusions

These findings highlight the complex circumstances involved in HIV-1 infection of the brain while also emphasizing the challenges in establishing brain-specific contemporary ART regimens. Our in vitro data are supported by data from human brain tissue showing that the efficacy of contemporary ART drugs is less in the brain compared to peripheral blood. This reduced efficacy might contribute to HIV-1 viral persistence in the brain. The higher in vitro efficacies of MVC and DTG in brain myeloid cells suggest that inclusion of these drugs in first line ART regimens might prevent or diminish the development of HIV-associated neurological impairments as well as viral persistence in the brain.

Methods

Patients

Brain tissue from two HIV-1 seropositive patients receiving ART without detectable viremia (Patients 1 and 2) and uninfected (n = 6) seronegative patients were obtained by autopsy or biopsy. HIV-infected patients were receiving ART as part of the active care program at Southern Alberta Clinic, Calgary, Alberta, Canada [4–7] while uninfected patients died of other causes. Brain tissue was collected from HIV-infected patients (Patient 1: autopsy following death from a myocardial infarction; Patient 2; surgical resection of left temporal lobe for chronic complex partial epilepsy and from adult uninfected Other Disease Control patients (n = 6; causes of death included sepsis [n = 1], cancer [n = 2], Parkinson's Disease [n = 1], pneumonia [n = 2]) at autopsy after receipt of consent. Patient 1 (Pt-1, Caucasian male, 57 yr, HIV-1 B clade seropositive for 27 yr, lifetime CD4 T cell nadir = 78 cells/μl, disease course complicated by distal sensory neuropathy and HIV-associated neurocognitive disorders [HAND, MND subtype for 10 years]) at the time of death was receiving RAL, abacavir (ABA) and lamivudine (3TC) while Patient 2 (Pt-2, Caucasian male, 38 yr, HIV-1 B clade seropositive for 12 yr, lifetime CD4 T cell nadir = 386 cells/μl, complicated by complex partial epilepsy since childhood with normal cognition and no HIV-associated complications) at the time of surgery was receiving ABA, 3TC and DTG.

CNS penetration effectiveness values of ART drugs

We used the updated CNS penetration effectiveness (CPE) values as reported previously [61]. The CPE value for DTG has not been established. However, based on a study on the ratio CSF to plasma concentration of DTG and as well as the CSF and plasma HIV-1 RNA viral load [36], we assumed a CPE value of 3 for this study.

Generation of BLT mice, infection, ART and collection of brain tissue

BLT mice were generated and infected using sub-lethally irradiated NOD/SCID/IL2Rγ$^{-/-}$ mice [obtained from Jackson Laboratory (Bal Harbor ME), 6-10 weeks old], by implantation of human fetal thymus under the kidney capsule and IV injection of fetal liver CD34+ cells [62]. After transplantation mice were monitored for engraftment of human cells by flow cytometry. The HIV-1 NL4-3 *env*$_{ADA}$-IRES-GFP construct was used for virus production. Brain tissue (forebrain) and plasma were harvested from four groups of animals including uninfected (HIV[−]; n = 5), HIV-infected with PBS treatment (HIV[+]; n = 9), HIV-infected with ART treatment (HIV[+]/ART; n = 9) and HIV-infected that were treated with ART but interrupted before sacrifice (HIV[+]/ART-interrupt; n = 5). This latter group of mice was infected and treated for 2 weeks post-infection (between 4 and 6 or 8 and 10) and after 2 weeks of treatment, ART was stopped and the virus allowed to rebound and mice were sacrificed at week 13 post-infection. Animals in the (HIV[+]; n = 9) were infected for different durations or at different input titers (two mice that did not show infection after 8 weeks and were re-infected at a $TCID_{50}$ 50,000/animal for another 4 weeks, another two mice were infected at a $TCID_{50}$ 50,000/animal for 13 weeks and the other five mice were infected at a $TCID_{50}$ 100,000/animal for 7 weeks, by intraperitoneal infection. HIV-infected animals were treated with ART (RAL, emtricitabine and tenofovir) or PBS for 3 weeks and sacrificed at week 7 post infection. In another group, mice were infected for 8 weeks, treated with the same ART regimen or PBS for weeks and sacrificed at 12 weeks post infection. Brains were harvested at the time of necropsy and a half hemisphere was fixed in 4% paraformadehyde (PFA) and the other half stored at − 80 °C. Plasma HIV RNA was quantified in 25 μl of plasma. In brief, 50 μl of plasma was diluted to 2 ml with normal human plasma; 1 ml of diluted plasma was used to quantify RNA copy numbers using the Abbott real time HIV-1 assay (sensitivity, < 40 copies/ml).

RNA and DNA preparation

Total RNA and DNA were extracted from brain of BLT mice and HIV-infected and non-infected patients using the RNeasy and DNeasy kits (Qiagen Germantown MD, USA), respectively, according to manufacturer's instructions and as previously described [18, 63].

Quantitative real time PCR

First strand cDNA synthesis was performed using DNase-treated RNA, random primers and Superscript II reverse transcriptase (Invitrogen, Carlsbad, CA, USA) at 42 °C for 90 min according to the manufacturer's instructions. cDNA was diluted 1:3 (adding 100 μl of ultrapure water to 50 μl of cDNA). Host human genes were quantified as previously described [63]. The sequences of primers used for quantification of host immune genes are included (Table 2).

Droplet digital PCR for viral RNA and DNA quantification

HIV-1 viral RNA was quantified using 5 μl of diluted cDNA as template. For genomic DNA (gDNA) quantification, 5 μl (200 ng) of DNA was incubated with the restriction enzyme Alu I for 5 min at room temperature together with the reaction mixture. The Bio-Rad QX200 droplet digital PCR system was used for quantification [57, 64]. For both RNA and DNA quantification the probe: 5′-/56-FAM/AAGCCAGGA/ZEN/AT

Table 2 List of primers used for quantitative real time PCR

Primer	Sequence
hu_CD3E-F	5′-GCCTCCGCCATCTTAGTAAAG-3′
hu_CD3E-R	5′-TCTTCATTACCATCTTGCCCC-3′
hu_CD163-F	5′-GAGTCCCTTCACCATTACTGTG-3′
hu_CD163-R	5′-GACTTTCACTTCCACTCTCCC-3′
hu_IL1B-F	5′-CCAAAGAAGAAGATGGAAAAGC-3′
hu_IL1B-R	5′-GGTGCTGATGTACCAGTTGGG-3′
hu_IL6-F	5′-ACCCCTGACCCAACCACAAAT-3′
hu_IL6-R	5′-AGCTGCGCAGAATGAGATGAG-3′
hu_APOBEC3G-F	5′-CGAAGTGAAAACAAAGGGTCC-3′
hu_APOBEC3G-R	5′-CATACTCCTGGTCACGATGC-3′
hu_MX2-F	5′-AGCAGGAGATCACAAACAGG-3′
hu_MX2-R	5′-GGTAAGTCTTTCTGCCAGTCG-3′
hu_CD68-F	5′-CATCTCTGTACTGAACCCCAAC-3′
hu_CD68-R	5′-CCATGTAGCTCAGGTAGACAAC-3′
hu_TNF alpha-F	5′-GTTTGAATTCTTAGTGGTTGCC-3′
hu_TNF alpha-R	5′-ATTCAGGAATGTGTGGCCTGC-3′
hu_GAPDH-F	5′-ACCAGGGCTGCTTTTAACTC-3′
hu_GAPDH-R	5′-TTGATTTTGGAGGGATCTCG-3′
mouse_f4/80-F	5′-GCTGTGAGATTGAAGCA-3′
mouse_f4/80-R	5′-AGTTTGCCATCCGGTTACAG-3′
mouse_GAPDH-F	5′-AATGGTGAAGGTCGGTGTG-3′
mouse_GAPDH-R	5′-GTGGAGTCATACTGGAACATGTAG-3′

GGATGGCC/3IABkFQ/-3′, primer 1: 5′-CAAATTTCTA CTAATGCTTTTATTTTTTC-3′ and primer 2: 5′-GCA CTTTAAATTTTCCCATTAGTCCTA-3′ were used. Human CD45 copy number in the brain of BLT mice was also quantified by droplet digital PCR. The probe: 5′-/5HEX/AGAATGTTC/ZEN/TGGCCCCTCAGT GC/3IABkFQ/-3′ and primer 1: 5′-GAAAAGCTCCCTG AAGCAAAG-3′, primer 2: 5′-GGTTTGGAGTTTCCTC ATTTATGTC-3′. Droplets were generated from each reaction and consisted of 10 µl droplet digital PCR probe Supermix, 900 nM primers, 250 nM probe and template cDNA or gDNA in a final volume of 20 µl using a droplet generator. PCR was performed using an S1000 thermal cycler (Bio-Rad). The cycling conditions were as follows: 10 min at 95 °C, 30 s denaturation at 94 °C, 58 °C extension for 60 s, and 10 min at 98 °C for a total of 40 cycles. After the cycling, droplets were analyzed immediately or stored at 4 °C overnight and until analysis. Raw data were analyzed with the specific software (QX200) by setting a common fixed fluorescence threshold intensity based on non-template control (NTC). The number of template copies was then estimated based on the number of positives detected in a corresponding channel and the number of total accepted droplets. Samples were quantified in duplicate and the analysis repeated two or three times. Template copies per sample were calculated averaging the overall replicate wells per sample. We estimated the total number of cells from the concentration of gDNA input as described previously [64].

Quantification of HIV-1 integration by real time PCR

A two step PCR amplification was performed using DNA extracted from tissues as template. The primers, probes and cycling conditions described in a previous study were used [65]. A serial dilution of U1 cells with a known concentration of DNA was also included and used to generate standard curve for quantification.

Cell cultures, virus infections and ART drugs

For in vitro ART studies, HFM, BMDMs or adult PBMCs were infected and allowed to replicate in the presence or absence of ART drugs [38, 57]. Primary HFM were isolated from 16 to 20 week fetal brain tissues, as previously described [38]. Briefly, fetal brain tissues were dissected, meninges were removed, and a single cell suspension was prepared through enzymatic digestion with 0.25% trypsin and 0.2 mg/ml DNase I for 60 min at 37 °C in a water bath. This was followed by passage through a 70-µm cell strainer. The cells were washed twice and plated in vented Cell+ T-75 tissue culture flasks (Sarstedt) and cultures were incubated at 5% CO_2 for two weeks. Cultures were maintained in MEM supplemented with 10% FBS, 2 mM L-glutamine, 1 mM sodium pyruvate, 1 × MEM nonessential amino acids, 0.1% dextrose, 100 U/ml Penicillin, 100 µg/ml streptomycin. After 2 weeks, cultures were gently rocked for 10 min to suspend the weakly adhering microglia in medium, which were then decanted, washed and plated. Purity of cultures was verified by immunofluorescence [38]. BMDMs were isolated from 16 to 20 week fetal femurs. Briefly, using a 5 ml syringe each femur was drilled and flushed with DMEM into a tube. The cell suspension was then incubated with ACK lysis buffer to remove red blood cells, washed and centrifuged twice. The cell pellets were re-suspended in DMEM (10% FBS) containing 25 ng/ml of macrophage colony stimulating factor (M-CSF), transferred into a six well culture plate and incubated at 37 °C and 10% CO_2. Cell culture medium was replaced every 3 days and macrophages were harvested on Day 7. Human THP-1 cells were cultured in RPMI (10% FBS) and differentiated for 72 h with PMA (50 nM) [66]. After exposure to PMA, cells were washed with PBS and fresh medium without PMA was added and incubation continued for another 48 h before use. PBMCs were isolated from healthy donors using the Ficoll-Paque (Sigma) density gradient separation, stimulated with phytohemagglutinin-P (Sigma) for 3 days at 5 µg/ml and re-suspended at 3.5×10^6 cells/ml in RPMI 1640 supplemented with 10% heat-inactivated FBS, 1%

penicillin/streptomycin, 1% glutamine and 1% recombinant human IL-2 at which time cells were infected. Viruses were produced by transfection of 16 μg of HIV-1 YU-2 plasmid into HEK293T cells using Lipofectamine 2000 (Invitrogen, Carlsbad, CA). Two days after transfection, supernatants from transfected cells were harvested and centrifuged at 1500 rpm for 10 min to remove cellular debris, filtered through a 0.22-μm-pore-size filter, and stored in aliquots at − 80 °C. Virus production was confirmed by HIV-1 p24 ELISA [38]. For in vitro ART studies, HFM or BMDMs (5×10^4) were seeded into 96-well plates. Plates were incubated at 37 °C and 5% CO_2 overnight to allow cells to adhere to the bottom of the plate and infected the following day for 4 h at 37° C and 5% CO_2 at a multiplicity of infection (MOI) of 0.1 for 4 h and washed (×3). For HIV-1 replication, supernatant was collected at pre-defined time points post-infection. A total of 250 μl of diluted drug at different concentrations was added to each well post-infection and incubated for 5 days. On Day 5 post-infection, supernatants were collected for HIV-1 p24 quantification by ELISA. For PBMCs infections, cells were infected with HIV-1 YU-2 for 2 h (MOI 0.05), washed and re-suspended with fresh medium and 125 μl of cell suspension containing 250,000 cells were seeded into each well of a 96-well plate. Diluted drugs (125 μl) at different concentrations were added to each well post-infection. On day 4 post-infection supernatant was collected for HIV-1 p24 quantification.

AZT, ETR, RAL and DRV were obtained from the NIH AIDS Research and Reference Reagent Program. Additional RAL and DRV used for animal studies were purchased (Cedarlane or Sigma respectively).

Immunofluorescence and immunocytochemistry studies

Cultured fetal HFM, BMDMs and fixed brains were immunolabeled with specific antibodies as described previously [57, 67]. Briefly, fetal HFM and BMDMs were seeded into an 8-well chamber μ-slide (Ibidi, DE). Cells were fixed (4% PFA) for 20 min. and immunolabeled with antibodies to Iba-1 (rabbit pAb 1:250, Wako) or MHC II (Mouse mAb, 1:300, Dako) over night. After washes with PBS, cells were incubated with anti-rabbit (AlexaFluor 647 nm, red) and anti-mouse (AlexaFluor 488 nm, green) (1:250, Life Technologies) antibodies. Cell nuclei were stained with DAPI (1:100, ImmunoChemistry Technologies) for 15 min. An Olympus IX-81 confocal microscope (Quorum Technologies) was used to capture the images. Fixed brains from uninfected [−] and HIV-infected [+] BLT mice were paraffin-embedded followed by sectioning and mounting. Slides were rehydrated and subjected to antigen retrieval by boiling in 10 mM sodium citrate, pH6 before immunostaining. Tissues were labeled with primary antibodies [anti-hCD68 (Abcam), −hCD3ε (Millipore), anti-HIV-1 p24 (NIH AIDS reagents #530) and GFAP (Dako)] followed by application of appropriate secondary antibodies [57].

ART concentration measurements

RAL and DRV were injected intraperitoneally into adult mice (8–12 months old) at a concentration of 1.2 mg per animal in a total of 4 animals per group and the subsequent drug concentrations were measured in blood, liver and brain. Animals were sacrificed at 1, 4, 8 or 24 h post-drug administration and whole blood, liver and brain (cortex, striatum, cerebellum, medulla and hypothalamus) were harvested and the concentrations of RAL and DRV were measured using HPLC–MS after homogenizing in 80% methanol, centrifuging, and running the supernatants through Oasis HLB Prime uelution plates followed by elution of the drugs. A similar procedure was used for analysis of RAL concentrations in human brain.

Statistical analyses

One-way ANOVA (Kruskal–Wallis test) with multiple comparisons was used to analyze in vitro and in vivo experiments. Statistical tests were applied using GraphPad Prism version 5 (GraphPad Software, San Diego California USA, www.graphpad.com). Results are expressed as mean ± SEM. p values < 0.05 were considered significant.

Additional files

Additional file 1. Representative dose-response curves for ART drugs (RAL, AZT and DRV measured by p24 ELISA in supernatant from PBMCs (**A**, **D** and **G**), BMDM (**B**, **E** and **H**) and HFM (**C**, **F** and **I**).

Additional file 2. Intracellular and extracellular ARV drug concentrations in differentiated human THP-1 cells that were stimulated with LPS or infected with HIV-1 and treated with DRV or RAL at different concentrations for 24 hours. Drug concentrations were measured by HPLC-MS.

Additional file 3. Brain tissue (70–100 mg) from all BLT HIV-infected and uninfected animals was used to extract RNA. The RNA was used to quantify host genes (**A**) human *CD163* (**B**) human *IP10* (**C**) human *MX2* and (**D**) mouse *f4/80*.

Additional file 4. Differentiated THP-1 cells were infected with HIV-1 YU-2 and SF162. DNA was extracted from infected cells and used to establish the integration assay (**A**). Host genes expression from patients 1 and 2 compared to uninfected controls measured by quantitative RT-PCR for, (**B**) *CD163*, (**C**) *IL6*, (**D**) *TNFA*, (**E**) *APOBEC3G*, (**F**) *MX2*, (**G**) *MX1* and (H) *BST2*.

Additional file 5. Brain tissue (70–100 mg) from different anatomic regions from patient 1 and uninfected controls were used to measure RAL concentration by HPLC-MS.

Authors' contributions

ELA and CP designed the experiments and wrote the paper; ELA, OM, MKM, WFC, WGB, and GBB performed all experiments and analyzed results; LR provided and analysed brain tissues from Pt-1 and Pt-2; MAW provided Dolutegravir; MJG and CP administered ART and clinical follow up of patients; EH and JVG generated the BLT mice; EAC supervised the BLT mice experiments. All authors read and approved the final manuscript.

Author details
[1] Department of Medicine (Neurology), University of Alberta, Edmonton, AB, Canada. [2] Department of Psychiatry, University of Alberta, Edmonton, AB, Canada. [3] Department of Pathology, University of Calgary, Calgary, AB, Canada. [4] Department of Medicine, University of Calgary, Calgary, AB, Canada. [5] CHU Sainte-Justine, Montréal, Canada. [6] Department of Pediatrics, Université de Montréal, Montréal, Canada. [7] Department of Microbiology, Infectiology and Immunology, Université de Montréal, Montréal, Canada. [8] Montreal Clinical Research Institute, Montréal, Canada. [9] CIUSSS du Centre-Sud-de-l'île-Montréal, CLSC des Faubourgs, Montréal, QC, Canada. [10] McGill University AIDS Centre, Lady Davis Institute for Medical Research, Jewish General Hospital, Montréal, QC, Canada.

Acknowledgements
The expert technical assistance of Ms. Gail Rauw is gratefully acknowledged. The authors also thank Dr. Walter Hader for providing surgical brain samples as well as the patients and their families for their willingness to participate in the present studies. The authors thank Merck and Gilead for supplying antiretroviral drugs for the in vivo animal studies. The authors dedicate this paper to our colleague and co-author, Dr. Mark Wainberg, who died unexpectedly while this manuscript was being prepared.

Competing interest
The following co-authors have no conflict of interest; Eugene L. Asahchop, Oussama Meziane, Manmeet K. Mamik, Wing F. Chan, William G. Branton, Lothar Resch, Jean V. Guimond, Mark A. Wainberg, Glen B Baker, Eric A. Cohen, Christopher Power. However, Dr. Elie Haddad is a consultant for SigmaTau, and has received investigator initiated grants from CSL-Behring. Dr. M John Gill has served as an ad hoc member of National HIV advisory boards to Merck, Gilead and ViiV in last 3 years.

Consent for publication
The use of autopsied brain tissues for research has been approved (Pro00002291) by the University of Alberta Human Research Ethics Board. The University of Alberta Ethics Board has also approved the use of animals in research.

Sources of funding
The Canadian Institutes of Health Research (CIHR), Canada Foundation for Innovation (CFI), Alberta Innovates-Health Solution (AIHS), University of Alberta Hospital Foundation, University of Alberta Vice President of Research Office.

References

1. Koenig S, Gendelman HE, Orenstein JM, Dal Canto MC, Pezeshkpour GH, Yungbluth M, et al. Detection of AIDS virus in macrophages in brain tissue from AIDS patients with encephalopathy. Science. 1986;233:1089–93.
2. Elovaara I, Iivanainen M, Valle SL, Suni J, Tervo T, Lahdevirta J. CSF protein and cellular profiles in various stages of HIV infection related to neurological manifestations. J Neurol Sci. 1987;78:331–42.
3. Antinori A, Giancola ML, Grisetti S, Soldani F, Alba L, Liuzzi G, et al. Factors influencing virological response to antiretroviral drugs in cerebrospinal fluid of advanced HIV-1-infected patients. AIDS. 2002;16:1867–76.
4. Vivithanaporn P, Heo G, Gamble J, Krentz HB, Hoke A, Gill MJ, et al. Neurologic disease burden in treated HIV/AIDS predicts survival: a population-based study. Neurology. 2010;75:1150–8.
5. Van Marle G, Rourke SB, Zhang K, Silva C, Ethier J, Gill MJ, et al. HIV dementia patients exhibit reduced viral neutralization and increased envelope sequence diversity in blood and brain. AIDS. 2002;16:1905–14.
6. McCombe JA, Vivithanaporn P, Gill MJ, Power C. Predictors of symptomatic HIV-associated neurocognitive disorders in universal health care. HIV Med. 2013;14:99–107.
7. Maingat F, Halloran B, Acharjee S, van Marle G, Church D, Gill MJ, et al. Inflammation and epithelial cell injury in AIDS enteropathy: involvement of endoplasmic reticulum stress. FASEB J. 2011;25:2211–20.
8. Avalos CR, Price SL, Forsyth ER, Pin JN, Shirk EN, Bullock BT, et al. Quantitation of productively infected monocytes and macrophages of simian immunodeficiency virus-infected macaques. J Virol. 2016;90:5643–56.
9. Honeycutt JB, Wahl A, Baker C, Spagnuolo RA, Foster J, Zakharova O, et al. Macrophages sustain HIV replication in vivo independently of T cells. J Clin Investig. 2016;126:1353–66.
10. Koyanagi Y, Itoyama Y, Nakamura N, Takamatsu K, Kira J, Iwamasa T, et al. In vivo infection of human T-cell leukemia virus type I in non-T cells. Virology. 1993;196:25–33.
11. Pang S, Koyanagi Y, Miles S, Wiley C, Vinters HV, Chen IS. High levels of unintegrated HIV-1 DNA in brain tissue of AIDS dementia patients. Nature. 1990;343:85–9.
12. Cobos Jimenez V, Booiman T, de Taeye SW, van Dort KA, Rits MA, Hamann J, et al. Differential expression of HIV-1 interfering factors in monocyte-derived macrophages stimulated with polarizing cytokines or interferons. Sci Rep. 2012;2:763.
13. Dimitrov DS, Norwood D, Stantchev TS, Feng Y, Xiao X, Broder CC. A mechanism of resistance to HIV-1 entry: inefficient interactions of CXCR4 with CD4 and gp120 in macrophages. Virology. 1999;259:1–6.
14. Pauls E, Jimenez E, Ruiz A, Permanyer M, Ballana E, Costa H, et al. Restriction of HIV-1 replication in primary macrophages by IL-12 and IL-18 through the upregulation of SAMHD1. J Immunol. 2013;190:4736–41.
15. Micci L, Alvarez X, Iriele RI, Ortiz AM, Ryan ES, McGary CS, et al. CD4 depletion in SIV-infected macaques results in macrophage and microglia infection with rapid turnover of infected cells. PLoS Pathog. 2014;10:e1004467.
16. Askew K, Li K, Olmos-Alonso A, Garcia-Moreno F, Liang Y, Richardson P, et al. Coupled proliferation and apoptosis maintain the rapid turnover of microglia in the adult brain. Cell Rep. 2017;18:391–405.
17. Kugathasan R, Collier DA, Haddow LJ, El Bouzidi K, Edwards SG, Cartledge JD, et al. Diffuse white matter signal abnormalities on magnetic resonance imaging are associated with human immunodeficiency virus type 1 viral escape in the central nervous system among patients with neurological symptoms. Clin Infect Dis. 2017;64:1059–65.
18. Zink MC, Brice AK, Kelly KM, Queen SE, Gama L, Li M, et al. Simian immunodeficiency virus-infected macaques treated with highly active antiretroviral therapy have reduced central nervous system viral replication and inflammation but persistence of viral DNA. J Infect Dis. 2010;202:161–70.
19. Lamers SL, Rose R, Maidji E, Agsalda-Garcia M, Nolan DJ, Fogel GB, et al. HIV DNA is frequently present within pathologic tissues evaluated at autopsy from combined antiretroviral therapy-treated patients with undetectable viral loads. J Virol. 2016;90:8968–83.
20. Honeycutt JB, Thayer WO, Baker CE, Ribeiro RM, Lada SM, Cao Y, et al. HIV persistence in tissue macrophages of humanized myeloid-only mice during antiretroviral therapy. Nat Med. 2017;23:638–43.
21. Watkins BA, Dorn HH, Kelly WB, Armstrong RC, Potts BJ, Michaels F, et al. Specific tropism of HIV-1 for microglial cells in primary human brain cultures. Science. 1990;249:549–53.
22. Brinkmann R, Schwinn A, Narayan O, Zink C, Kreth HW, Roggendorf W, et al. Human immunodeficiency virus infection in microglia: correlation between cells infected in the brain and cells cultured from infectious brain tissue. Ann Neurol. 1992;31:361–5.
23. Ghorpade A, Nukuna A, Che M, Haggerty S, Persidsky Y, Carter E, et al. Human immunodeficiency virus neurotropism: an analysis of viral replication and cytopathicity for divergent strains in monocytes and microglia. J Virol. 1998;72:3340–50.
24. Calcagno A, Cusato J, Simiele M, Motta I, Audagnotto S, Bracchi M, et al. High interpatient variability of raltegravir CSF concentrations in HIV-positive patients: a pharmacogenetic analysis. J Antimicrob Chemother. 2014;69:241–5.
25. Calcagno A, Yilmaz A, Cusato J, Simiele M, Bertucci R, Siccardi M, et al. Determinants of darunavir cerebrospinal fluid concentrations: impact of once-daily dosing and pharmacogenetics. AIDS. 2012;26:1529–33.

26. Croteau D, Letendre S, Best BM, Ellis RJ, Breidinger S, Clifford D, et al. Total raltegravir concentrations in cerebrospinal fluid exceed the 50-percent inhibitory concentration for wild-type HIV-1. Antimicrob Agents Chemother. 2010;54:5156–60.
27. Croteau D, Rossi SS, Best BM, Capparelli E, Ellis RJ, Clifford DB, et al. Darunavir is predominantly unbound to protein in cerebrospinal fluid and concentrations exceed the wild-type HIV-1 median 90% inhibitory concentration. J Antimicrob Chemother. 2013;68:684–9.
28. Yilmaz A, Watson V, Else L, Gisslen M. Cerebrospinal fluid maraviroc concentrations in HIV-1 infected patients. AIDS. 2009;23:2537–40.
29. Yilmaz A, Gisslen M, Spudich S, Lee E, Jayewardene A, Aweeka F, et al. Raltegravir cerebrospinal fluid concentrations in HIV-1 infection. PLoS ONE. 2009;4:e6877.
30. van Praag RM, van Weert EC, van Heeswijk RP, Zhou XJ, Sommadossi JP, Jurriaans S, et al. Stable concentrations of zidovudine, stavudine, lamivudine, abacavir, and nevirapine in serum and cerebrospinal fluid during 2 years of therapy. Antimicrob Agents Chemother. 2002;46:896–9.
31. Tiraboschi JM, Niubo J, Vila A, Perez-Pujol S, Podzamczer D. Etravirine concentrations in CSF in HIV-infected patients. J Antimicrob Chemother. 2012;67:1446–8.
32. Tiraboschi JM, Niubo J, Curto J, Podzamczer D. Maraviroc concentrations in cerebrospinal fluid in HIV-infected patients. J Acquir Immune Defic Syndr. 2010;55:606–9.
33. Foudraine NA, Hoetelmans RM, Lange JM, de Wolf F, van Benthem BH, Maas JJ, et al. Cerebrospinal-fluid HIV-1 RNA and drug concentrations after treatment with lamivudine plus zidovudine or stavudine. Lancet. 1998;351:1547–51.
34. Nguyen A, Rossi S, Croteau D, Best BM, Clifford D, Collier AC, et al. Etravirine in CSF is highly protein bound. J Antimicrob Chemother. 2013;68:1161–8.
35. Croteau D, Best BM, Letendre S, Rossi SS, Ellis RJ, Clifford DB, et al. Lower than expected maraviroc concentrations in cerebrospinal fluid exceed the wild-type CC chemokine receptor 5-tropic HIV-1 50% inhibitory concentration. AIDS. 2012;26:890–3.
36. Letendre SL, Mills AM, Tashima KT, Thomas DA, Min SS, Chen S, et al. ING116070: a study of the pharmacokinetics and antiviral activity of dolutegravir in cerebrospinal fluid in HIV-1-infected, antiretroviral therapy-naive subjects. Clin Infect Dis. 2014;59:1032–7.
37. Sturdevant CB, Joseph SB, Schnell G, Price RW, Swanstrom R, Spudich S. Compartmentalized replication of R5 T cell-tropic HIV-1 in the central nervous system early in the course of infection. PLoS Pathog. 2015;11:e1004720.
38. Walsh JG, Reinke SN, Mamik MK, McKenzie BA, Maingat F, Branton WG, et al. Rapid inflammasome activation in microglia contributes to brain disease in HIV/AIDS. Retrovirology. 2014;11:35.
39. Gunyat F, Rainho JN, West B, Swainson L, McCune JM, Stevenson M. CSF-1R antagonists sensitize HIV-1-infected macrophages to TRAIL-mediated killing. J Virol. 2016;90(14):6255–62.
40. Lamers SL, Rose R, Maidji E, Agsalda-Garcia M, Nolan DJ, Fogel GB, et al. HIV DNA is frequently present within pathologic tissues evaluated at autopsy from cART-treated patients with undetectable viral load. J Virol. 2016;90(20):8968–83.
41. Gelman BB, Lisinicchia JG, Morgello S, Masliah E, Commins D, Achim CL, et al. Neurovirological correlation with HIV-associated neurocognitive disorders and encephalitis in a HAART-era cohort. J Acquir Immune Defic Syndr. 2013;62(5):487–95.
42. Damouche A, Lazure T, Avettand-Fenoel V, Huot N, Dejucq-Rainsford N, Satie AP, et al. Adipose tissue is a neglected viral reservoir and an inflammatory site during chronic HIV and SIV infection. PLoS Pathog. 2015;11:e1005153.
43. Bruner KM, Murray AJ, Pollack RA, Soliman MG, Laskey SB, Capoferri AA, et al. Defective proviruses rapidly accumulate during acute HIV-1 infection. Nat Med. 2016;22:1043–9.
44. Suspene R, Meyerhans A. Quantification of unintegrated HIV-1 DNA at the single cell level in vivo. PLoS ONE. 2012;7:e36246.
45. Teo I, Veryard C, Barnes H, An SF, Jones M, Lantos PL, et al. Circular forms of unintegrated human immunodeficiency virus type 1 DNA and high levels of viral protein expression: association with dementia and multinucleated giant cells in the brains of patients with AIDS. J Virol. 1997;71:2928–33.
46. Burdo TH, Gartner S, Mauger D, Wigdahl B. Region-specific distribution of human immunodeficiency virus type 1 long terminal repeats containing specific configurations of CCAAT/enhancer-binding protein site II in brains derived from demented and nondemented patients. J Neurovirol. 2004;10(Suppl 1):7–14.
47. Gray LR, Cowley D, Welsh C, Lu HK, Brew BJ, Lewin SR, et al. CNS-specific regulatory elements in brain-derived HIV-1 strains affect responses to latency-reversing agents with implications for cure strategies. Mol Psychiatry. 2016;21:574–84.
48. Kure K, Weidenheim KM, Lyman WD, Dickson DW. Morphology and distribution of HIV-1 gp41-positive microglia in subacute AIDS encephalitis. Pattern of involvement resembling a multisystem degeneration. Acta Neuropathol. 1990;80:393–400.
49. Osman N, Mesplede T, Quashie PK, Oliveira M, Zanichelli V, Wainberg MA. Dolutegravir maintains a durable effect against HIV replication in tissue culture even after drug washout. J Antimicrob Chemother. 2015;70:2810–5.
50. Scopelliti F, Pollicita M, Ceccherini-Silberstein F, Di Santo F, Surdo M, Aquaro S, et al. Comparative antiviral activity of integrase inhibitors in human monocyte-derived macrophages and lymphocytes. Antivir Res. 2011;92:255–61.
51. Gray LR, Tachedjian G, Ellett AM, Roche MJ, Cheng WJ, Guillemin GJ, et al. The NRTIs lamivudine, stavudine and zidovudine have reduced HIV-1 inhibitory activity in astrocytes. PLoS ONE. 2013;8:e62196.
52. Gates TM, Cysique LA, Siefried KJ, Chaganti J, Moffat KJ, Brew BJ. Maraviroc-intensified combined antiretroviral therapy improves cognition in virally suppressed HIV-associated neurocognitive disorder. AIDS. 2016;30:591–600.
53. Martin-Blondel G, Brassat D, Bauer J, Lassmann H, Liblau RS. CCR5 blockade for neuroinflammatory diseases—beyond control of HIV. Nat Rev Neurol. 2016;12:95–105.
54. Kelly KM, Beck SE, Pate KAM, Queen SE, Dorsey JL, Adams RJ, et al. Neuroprotective maraviroc monotherapy in simian immunodeficiency virus-infected macaques: reduced replicating and latent SIV in the brain. AIDS. 2013;27:F21–8.
55. Tiraboschi JM, Munoz-Moreno JA, Puertas MC, Alonso-Villaverde C, Prats A, Ferrer E, et al. Viral and inflammatory markers in cerebrospinal fluid of patients with HIV-1-associated neurocognitive impairment during antiretroviral treatment switch. HIV Med. 2015;16:388–92.
56. Liu FC, Tsai YF, Yu HP. Maraviroc attenuates trauma-hemorrhage-induced hepatic injury through PPAR gamma-dependent pathway in rats. PLoS ONE. 2013;8:e78861.
57. Mamik MK, Asahchop EL, Chan WF, Zhu Y, Branton WG, McKenzie BA, et al. Insulin treatment prevents neuroinflammation and neuronal injury with restored neurobehavioral function in models of HIV/AIDS neurodegeneration. J Neurosci. 2016;36:10683–95.
58. Fletcher CV, Staskus K, Wietgrefe SW, Rothenberger M, Reilly C, Chipman JG, et al. Persistent HIV-1 replication is associated with lower antiretroviral drug concentrations in lymphatic tissues. Proc Natl Acad Sci USA. 2014;111:2307–12.
59. Lorenzo-Redondo R, Fryer HR, Bedford T, Kim EY, Archer J, Kosakovsky Pond SL, et al. Persistent HIV-1 replication maintains the tissue reservoir during therapy. Nature. 2016;530:51–6.
60. Arainga M, Su H, Poluektova LY, Gorantla S, Gendelman HE. HIV-1 cellular and tissue replication patterns in infected humanized mice. Sci Rep. 2016;6:23513.
61. Letendre SL, Ellis RJ, Ances BM, McCutchan JA. Neurologic complications of HIV disease and their treatment. Top HIV Med. 2010;18:45–55.
62. Dave VP, Hajjar F, Dieng MM, Haddad E, Cohen EA. Efficient BST2 antagonism by Vpu is critical for early HIV-1 dissemination in humanized mice. Retrovirology. 2013;10:128.
63. Branton WG, Ellestad KK, Maingat F, Wheatley BM, Rud E, Warren RL, et al. Brain microbial populations in HIV/AIDS: alpha-proteobacteria predominate independent of host immune status. PLoS ONE. 2013;8:e54673.

64. Bosman KJ, Nijhuis M, van Ham PM, Wensing AM, Vervisch K, Vandekerckhove L, et al. Comparison of digital PCR platforms and semi-nested qPCR as a tool to determine the size of the HIV reservoir. Sci Rep. 2015;5:13811.
65. Yu JJ, Wu TL, Liszewski MK, Dai J, Swiggard WJ, Baytop C, et al. A more precise HIV integration assay designed to detect small differences finds lower levels of integrated DNA in HAART treated patients. Virology. 2008;379:78–86.
66. Mamik MK, Hui E, Branton WG, McKenzie BA, Chisholm J, Cohen EA, et al. HIV-1 Viral Protein R Activates NLRP3 Inflammasome in Microglia: implications for HIV-1 Associated Neuroinflammation. J Neuroimmune Pharmacol. 2017;12(2):233–48.
67. Branton WG, Lu JQ, Surette MG, Holt RA, Lind J, Laman JD, et al. Brain microbiota disruption within inflammatory demyelinating lesions in multiple sclerosis. Sci Rep. 2016;6:37344.

7

The RNA surveillance proteins UPF1, UPF2 and SMG6 affect HIV-1 reactivation at a post-transcriptional level

Shringar Rao[1,2], Raquel Amorim[1,3], Meijuan Niu[1], Abdelkrim Temzi[1] and Andrew J. Mouland[1,2,3*]

Abstract

Background: The ability of human immunodeficiency virus type 1 (HIV-1) to form a stable viral reservoir is the major obstacle to an HIV-1 cure and post-transcriptional events contribute to the maintenance of viral latency. RNA surveillance proteins such as UPF1, UPF2 and SMG6 affect RNA stability and metabolism. In our previous work, we demonstrated that UPF1 stabilises HIV-1 genomic RNA (vRNA) and enhances its translatability in the cytoplasm. Thus, in this work we evaluated the influence of RNA surveillance proteins on vRNA expression and, as a consequence, viral reactivation in cells of the lymphoid lineage.

Methods: Quantitative fluorescence in situ hybridisation—flow cytometry (FISH-flow), si/shRNA-mediated depletions and Western blotting were used to characterise the roles of RNA surveillance proteins on HIV-1 reactivation in a latently infected model T cell line and primary CD4+ T cells.

Results: UPF1 was found to be a positive regulator of viral reactivation, with a depletion of UPF1 resulting in impaired vRNA expression and viral reactivation. UPF1 overexpression also modestly enhanced vRNA expression and its ATPase activity and N-terminal domain were necessary for this effect. UPF2 and SMG6 were found to negatively influence viral reactivation, both via an interaction with UPF1. UPF1 knockdown also resulted in reduced vRNA levels and viral gene expression in HIV-1-infected primary CD4+ T cells.

Conclusion: Overall, these data suggest that RNA surveillance proteins affect HIV-1 gene expression at a post-transcriptional level. An elucidation of the role of vRNA metabolism on the maintenance of HIV-1 persistence can lead to the development of novel curative strategies.

Keywords: HIV-1 latency, RNA surveillance proteins, HIV-1 genomic RNA stability, Post-transcriptional regulation, Nonsense-mediated mRNA decay, NMD, UPF1, UPF2, SMG6

Background

The implementation of combination antiretroviral therapy (cART) to treat human immunodeficiency virus type 1 (HIV-1) has led the infection to be likened to a chronic condition, with patients on cART having near-normal life expectancy [1]. However, this therapy is not without its drawbacks, such as adverse side effects that lower the adherence rates, the development of drug resistance and its economic repercussions [2–4]. But one of the biggest disadvantages of this therapy is that it is not curative and an infected individual needs to be on cART for the entire duration of their lifetime to effectively suppress viremia. The major hurdle towards an HIV-1 cure is the property of virus to form a stable latent reservoir upon infection that is responsible for the rapid rebound of plasma viral loads when cART is discontinued [5]. This reservoir is primarily composed of resting memory CD4+ T cells along with monocytes and macrophages [6] in peripheral blood and other anatomical compartments such as the gut, lymph nodes and central nervous system. Latency in HIV-1 infection is defined as a reversibly non-productive state of infection which is characterised by the presence

*Correspondence: andrew.mouland@mcgill.ca
[1] HIV-1 RNA Trafficking Laboratory, Lady Davis Institute at the Jewish General Hospital, Montreal, QC H3T 1E2, Canada
Full list of author information is available at the end of the article

of infected cells that do not actively produce viral particles, but retain the ability to do so [7]. Latent cells harbour a replication competent proviral DNA integrated in their genomes [8]. Many research groups have studied the functional aspects of the maintenance of latency in cells by investigating the molecular mechanisms leading to a block at the level of transcription (reviewed in [6, 9, 10]). However, certain studies also highlight that co and post-transcriptional events can also contribute to the maintenance of latency in HIV-1 infected cells [11–13]. These include defective splicing of the genomic viral RNA (vRNA) [14], inhibition of nucleocytoplasmic export of vRNA [13, 15, 16] or an impediment to vRNA translation [17, 18]. Thus, in this work, we investigate the role of the RNA surveillance proteins on the post-transcriptional events that are involved in the maintenance of HIV-1 latency.

RNA surveillance is a host quality control mechanism that identifies and degrades unspliced, aberrantly spliced, intron-containing, upstream open reading frame-containing and premature termination codon (PTC)-containing mRNAs to prevent the accumulation of potentially toxic truncated proteins within the cell (reviewed in [19]). A central player in this mechanism is the Up Frameshift Protein 1 (UPF1), an RNA binding protein that has ATPase and RNA helicase activity [20]. It is a multifunctional protein that has defined roles in DNA repair and replication [21, 22], RNA stability [23–25], telomere metabolism [21] and cell cycle progression [22] (reviewed in [26]). Its most characterised function, however, is its role in nonsense-mediated mRNA decay (NMD) during which UPF1 interacts with a family of proteins such as UPF2, UPF3A and UPF3B, a kinase SMG1 and an endonuclease SMG6 resulting in the degradation of aberrant mRNAs (reviewed in [19, 27]). Although NMD was previously implicated only in the degradation of aberrant mRNA, it is now widely accepted that NMD also targets up to 10% of other physiological mRNAs for degradation in response to cellular needs [19, 28–30], including transcripts that contain long 3′UTRs [31].

In order to promote their survival, viruses have evolved numerous strategies to either evade or manipulate the RNA surveillance pathways (reviewed in [32]). Retroviruses, despite containing long 3′UTRs that are recognised by UPF1, are capable of evading NMD by virtue of the presence of RNA stability elements in their genome [33] (reviewed in [34, 35]). In previous studies, our group has demonstrated that HIV-1 not only evades NMD, it also hijacks UPF1 to form an RNP that promotes vRNA stability and nucleocytoplasmic export [36, 37]. This effect may be exerted during the rapid, co-transcriptional association of UPF1 with vRNA during transcription [38]. UPF2, another protein involved in NMD, has been shown to block nucleocytoplasmic export of the vRNA by binding to UPF1 and preventing its nucleocytoplasmic shuttle [37]. Once in the cytoplasm, UPF1 assembles in another distinct RNP on the vRNA resulting in not only the increased stability of the vRNA, but also in its enhanced translation leading to increased levels of the HIV-1 structural protein pr55^{Gag} viral production [36]. Additionally, UPF1 interacts with vRNA in an RNA length-dependent manner and this could contribute to its incorporation into progeny HIV-1 virions [38–41]. Therefore, there is substantial evidence to show that UPF1 can affect vRNA metabolism at different levels.

In this study, we investigated the ability of UPF1 and its associated proteins UPF2 and SMG6 to influence the HIV-1 gene expression and, as a consequence, viral reactivation at a post-transcriptional level by overexpression and siRNA-mediated knockdown studies in cells of the lymphoid lineage. We employed a fluorescence in situ hybridisation/flow cytometry (FISH-Flow) to monitor vRNA expression levels and viral protein production in a latently-infected T cell line. We observed that these proteins can modulate the HIV-1 gene expression and thus the post-transcriptional maintenance of HIV-1 latency. We have also identified the domains responsible for these effects on viral reactivation by mutational studies. Importantly, we also demonstrate a direct effect of UPF1 on vRNA expression in primary HIV-1 infected CD4+ T cells.

Results

FISH-Flow can be used to monitor vRNA levels and viral reactivation in J-Lat cells

UPF1 has previously been demonstrated to affect vRNA metabolism at three distinct stages: overall vRNA stability, the nucleocytoplasmic export of the vRNA, and vRNA translation in the cytoplasm [36, 37]. Therefore, we employed the FISH-Flow technique using probes against the GagPol region of the vRNA in latently infected J-Lat 10.6 cells to monitor both the transcriptional as well as translational products of the HIV-1 provirus. This technique has previously been employed to assess ongoing HIV-1 replication, to quantify the size of the inducible latent reservoir in HIV-infected individuals, to determine the kinetics of latency reversal and to characterize the specific cell subpopulations of CD4+ T cells that transcribe HIV-1 RNA [17, 42–44] (reviewed in [45, 46]). Using this technique, it is possible to distinguish between cells that contain both vRNA and viral proteins, and cells that only contain untranslated vRNA, thus differentiating between the transcription-competent and translation-competent viral reservoir [45, 46]. Cells can then also be seeded on a coverslip to determine the sub-cellular localisation of the vRNA using laser scanning confocal microscopy (LCSM). This comprehensive analysis enables us to investigate how UPF1 influences viral reactivation and to distinguish between an effect on vRNA expression, export or translation. J-Lat 10.6 cells, a well-established model of studying HIV-1 latency and reactivation [88, 47, 48], and primary CD4+ T cells are used in this study. The J-Lat cells have a GFP reporter in the *nef* open reading frame of the virus to monitor viral gene expression and, thus, viral reactivation. The cells can be reactivated by treatment with phorbol myristate acetate (PMA) or TNFα (Additional file 1: Figure S1A). To assess whether the FISH-Flow technique can be used in the J-Lat cell model to measure reactivation, cells were either mock treated with dimethyl sulfoxide (DMSO) or treated with PMA to reactivate the cells. PMA is a protein kinase C agonist and is a strong activator of cellular transcription and was the latency reversing agent of choice because it leads to maximal reactivation of the J-Lat 10.6 cells [49]. We also validated the PMA treatment did not affect the baseline expression levels of our proteins of interest: UPF1, UPF2 and SMG6 (Additional file 1: Figure S1B–D). Jurkat cells were used as a negative, uninfected control to determine the specificity of the FISH-Flow technique. Upon treatment with PMA, 60.89 (±11.35)% of J-Lat cells produced GFP indicating viral protein production and reactivation (Fig. 1a, b). Efficient GagPol mRNA staining was also observed in 63.78 (±15.16)% of PMA-treated cells. (PE channel, Fig. 1a, b). It is also important to note that 4.79 (±2.44)% of PMA-treated cells contained vRNA but not GFP, representing the transcription-competent viral reservoir as previously described [45, 46]. The 2.48 (±1.17) of PMA-treated cells that were GFP+ but did not contain vRNA represent the cells that are generating multiply-transcripts but not full length transcripts, since the GFP codon is present on the *nef* open reading frame [88]. The uninduced J-Lat cells contained some residual vRNA and GFP production, with 2.59 (±1.76)% of cells expressing GFP and 0.27 (±0.11)% of cells expressing vRNA (Fig. 1a, b). Although the vRNA is the unspliced genomic viral RNA whereas GFP is generated from the multiply spliced viral RNA, GFP was used as a marker for viral reactivation rather than intracellular p24 due to the efficiency of measuring viral reactivation at a single cell level by Flow cytometry due to the stability of GFP. The levels of $pr55^{Gag}$, coded for by the vRNA, can be measured by Western blot to further correlate effects vRNA transcription and translation, if necessary. Jurkat cells did not show any vRNA+ cells, indicating that this technique is highly specific (Fig. 1a). Cells from each of these conditions were seeded onto coverslips and observed by laser scanning confocal microscopy (Fig. 1c) to view the subcellular localisation of the vRNA. Therefore, the FISH-Flow technique is an efficient method to monitor viral reactivation at the transcriptional and translational levels in J-Lat cells.

UPF1 knockdown attenuates HIV-1 proviral reactivation

In previous studies conducted by our group, we observed that UPF1 knockdown lead to reduced vRNA stability in the nucleus and in the cytoplasm of cell [36]. Thus, we hypothesised that the depletion of UPF1 can reduce vRNA expression at a post-transcriptional level and thereby inhibit viral reactivation. To evaluate the effect of UPF1 levels on proviral reactivation, J-Lat cells were either transfected with a non-silencing siRNA (siNS) or with siRNA against UPF1 (siUPF1). In each of these conditions, cells were either left uninduced (DMSO) or treated with PMA to reactivate the cells. The percentage of reactivation in the form of GFP production was monitored by flow cytometry and the cell lysates were subjected to Western blotting to validate UPF1 knockdown using antibodies against UPF1, $pr55^{Gag}$ and actin. Treatment of cells with siUPF1 resulted in a 68.9 (±29.9)% decrease in UPF1 protein levels as measured by Western blot, demonstrating the efficiency of siUPF1 treatment (Additional file 1: Figure S2A). UPF1 knockdown had no significant effect on viral reactivation in the uninduced condition (Fig. 2a). However, upon reactivation with PMA, UPF1 knockdown lead to a 35.3 (±8.4)% decrease in viral reactivation as compared to the siNS condition (Fig. 2a), which correlated with reduced $pr55^{Gag}$ levels observed by Western blots (Fig. 2b). In order to determine if this decrease in viral reactivation was due to an effect on the vRNA levels or due to inefficient nucleocytoplasmic export or translation of the vRNA, we also conducted FISH-Flow analyses in each of the above reactions. The levels of vRNA were also quantified by RT-qPCR. Upon treatment with PMA, UPF1 knockdown lead to a 23.5 (±4.8)% decrease in the number of vRNA expressing cells as compared to the siNS treated cells (Fig. 2c, d) as well as a 72.6 (±0.1)% decrease in the levels of vRNA as quantified by RT-qPCR (Fig. 2e). Of these vRNA expressing cells, a knockdown of UPF1 also led to a 28.0 (±11.8)% decrease in per cell vRNA levels as measured median fluorescence intensity (MFI) of the vRNA channel (PE) as compared the vRNA in the siNS treated cells (Additional file 1: Figure S2B). This is in accordance with our previous work where we demonstrated that a knockdown of UPF1 resulted in a decrease in vRNA stability [36]. The reduction in vRNA levels as quantified by RT-qPCR in the siUPF1 condition is more dramatic than the reduction of GFP production in the same condition, possibly due to increased stability of GFP as compared to the vRNA. It is also important to note that these detrimental effects of UPF1 knockdown on vRNA levels is specific to the vRNA, since no significant differences were observed in the % of cell expressing a housekeeping mRNA RPL13A and the MFI of the RPL13A mRNA channel measured by FISH-Flow, or

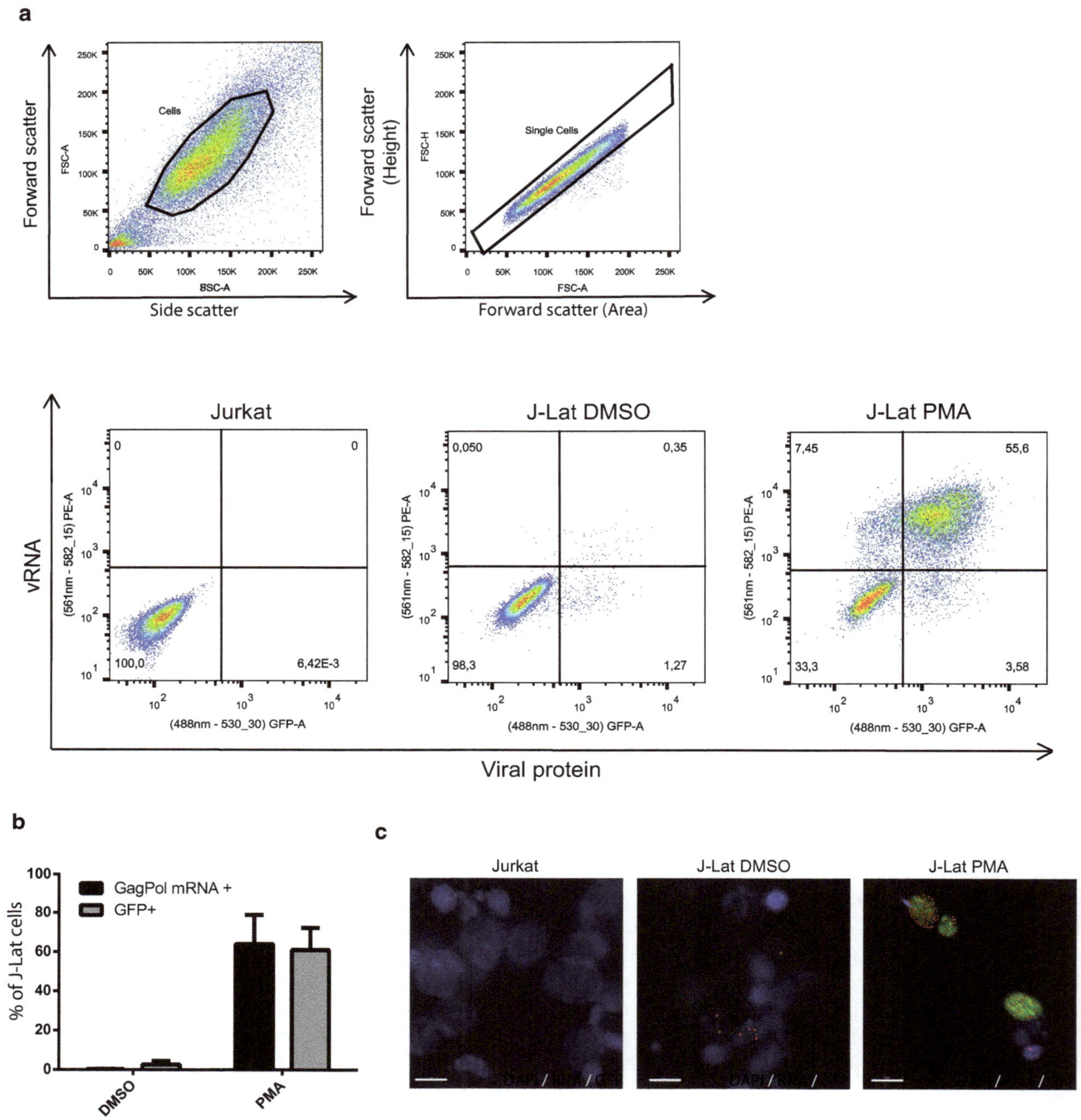

Fig. 1 Characterisation of FISH-Flow technique in J-Lat cells. J-Lat cells were either treated with DMSO or with PMA to reactivate the provirus. Jurkat cells were used as an uninfected negative control. **a** Dot plots representing cells gates for size by forward and side scatter, for singlets by forward scatter height versus area and finally for GFP expression and vRNA staining. **b** The % of GFP+ and the % of vRNA-expressing cells were quantified. Error bars represent the standard deviation from three independent experiments. **c** Representative images of cells in each of the above conditions imaged by confocal microscopy. In example images from sorted populations, DAPI is in blue, vRNA in red, and cells making viral protein produce GFP in green. Scale bars represent 10 µm

in the relative levels of housekeeping mRNA GAPDH measured by RT-qPCR (Additional file 1: Figure S2C–E). However, in these experimental conditions, we can not differentiate between cells that have successful knockdown of UPF1 and non-transfected cells. Therefore, to partially overcome this caveat, we also stained the cells with a UPF1 mRNA probe and, using FISH-Flow analysis, we delineated between UPF1 high vs.

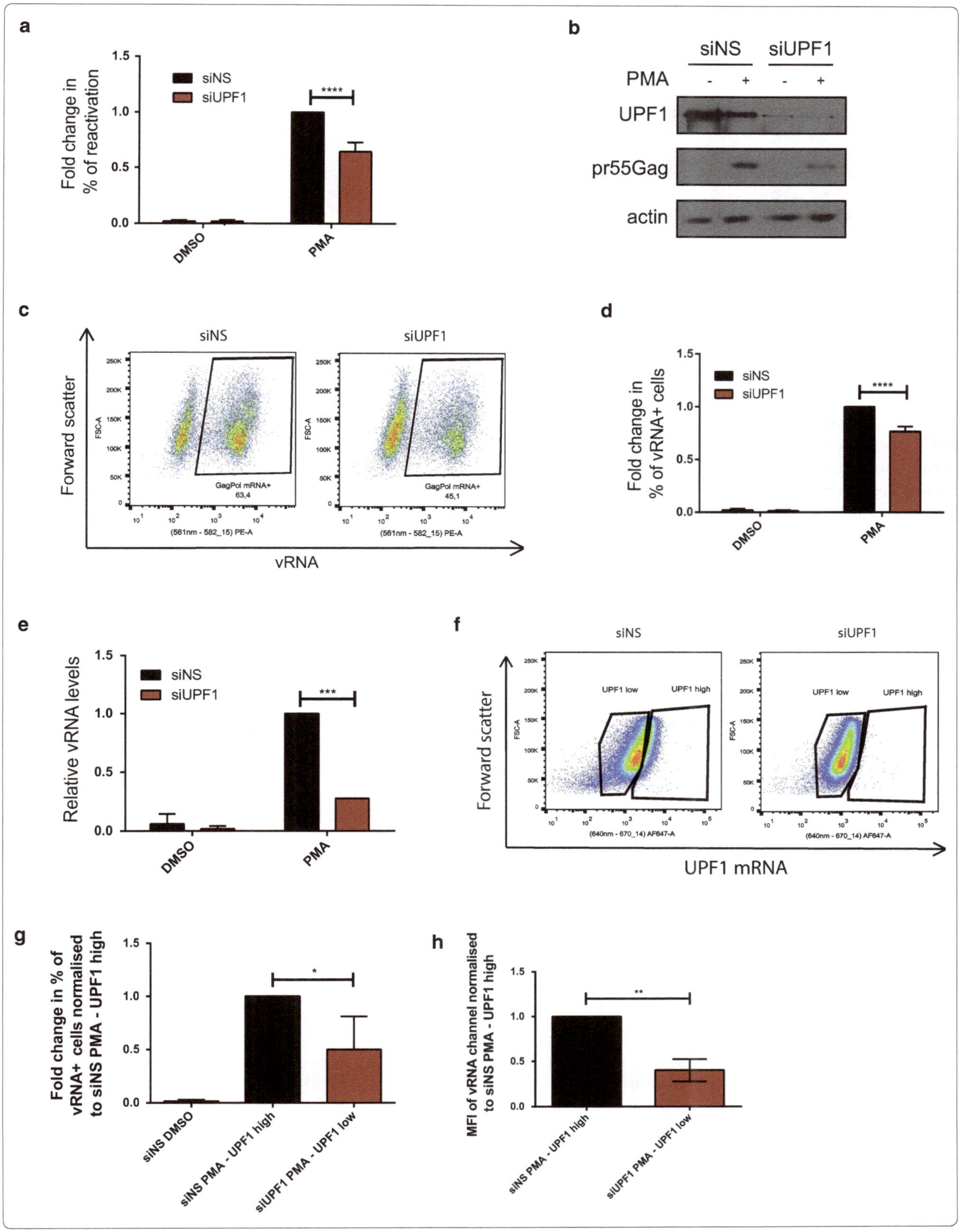
a
Fold change in % of reactivation
siNS
siUPF1
****
DMSO
PMA
b
siNS
siUPF1
PMA
UPF1
pr55Gag
actin
c
siNS
siUPF1
Forward scatter
GagPol mRNA+
63,4
GagPol mRNA+
45,1
(561nm - 582_15) PE-A
vRNA
d
Fold change in % of vRNA+ cells
****
e
Relative vRNA levels
***
f
UPF1 low
UPF1 high
(640nm - 670_14) AF647-A
UPF1 mRNA
g
Fold change in % of vRNA+ cells normalised to siNS PMA - UPF1 high
*
siNS DMSO
siNS PMA - UPF1 high
siUPF1 PMA - UPF1 low
h
MFI of vRNA channel normalised to siNS PMA - UPF1 high
**

(See figure on previous page.)

Fig. 2 UPF1 knockdown attenuates reactivation of HIV-1 in J-Lat cells. J-Lat 10.6 cells were either transfected with siNS or siUPF1 and were uninduced (DMSO) or reactivated (PMA). **a** Reactivation, monitored by GFP production, was quantified by Flow cytometry and the percentages of reactivation were normalised to the siNS-PMA reactivated condition. Error bars represent the standard deviation from three independent experiments with at least 10,000 cells counted per treatment. Asterisks represent statistically significant difference between groups (Two-way ANOVA; $p<0.0001$). **b** Cell lysates were run on SDS-PAGE gels and UPF1 and $pr55^{Gag}$ protein levels were detected by Western Blotting. **c** Example dot plot depicting vRNA expression in siNS-PMA and siUPF1 PMA conditions using FISH-Flow technique and, **d** The % of vRNA expressing cells were quantified and normalised to the siNS-PMA condition. Error bars represent the standard deviation from three independent experiments with at least 10,000 cells counted per treatment. Asterisks represent statistically significant difference between groups (Two-way ANOVA; $p<0.0001$). **e** Levels of vRNA were quantified using RT-qPCR and normalised to the siNS-PMA condition. Error bars represent the standard deviation from two independent experiments, each done in triplicate. Asterisks represent statistically significant difference between groups (Two-way ANOVA; $p<0.001$). **f** Gating strategy of cells separated into UPF1 low or high by detecting UPF1 mRNA levels by FISH-Flow. **g** The % of vRNA expressing cells in each condition normalised to the siNS-PMA/UPF1-high condition. Error bars represent the standard deviation from three independent experiments. Asterisks represent statistically significant difference between groups (One-way ANOVA; $p<0.05$). **h** MFI of the vRNA signal were quantified. Asterisks represent statistically significant difference between groups (student's t-test; $p<0.01$)

UPF1 low cells (Fig. 2f). Using this gating strategy, it was observed that the UPF1 low population of the siUPF1-PMA treated cells showed a 50.5 ($\pm$31.07) reduction in the % of vRNA-expressing cells as compared to the UPF1 high population of the siNS-PMA condition (Fig. 2g). Of these vRNA expressing cells, a knockdown of UPF1 also led to a 1.66 fold reduction in the median fluorescence intensity (MFI) of the vRNA channel (PE) as compared the vRNA in the siNS treated cells (Fig. 2h). Since UPF1 has previously characterised roles in nuclear export [37], we determined if a knockdown of UPF1 resulted in increased nuclear retention of the vRNA. Cellular fractionation was performed and the vRNA present in whole cell, cytoplasmic and nuclear fractions were quantified by RT-PCR (Additional file 1: Figure S3A, B). A decrease is vRNA levels was observed in all fractions, thus implying that in these experimental conditions, UPF1 is acting on vRNA expression rather than on nuclear export (Additional file 1: Figure S3A, B). Taken together, these data suggest that a knockdown of UPF1 leads to attenuated HIV-1 proviral reactivation in J-Lat cells at a post-transcriptional level, by reducing vRNA levels and thus, viral reactivation and protein production.

UPF1 overexpression enhances HIV-1 proviral reactivation by enhancing vRNA levels

UPF1 overexpression has been shown to enhance vRNA stability, nucleocytoplasmic export and translation in previous studies [36, 37]. Therefore, we hypothesised that UPF1 overexpression could enhance proviral reactivation. J-Lat cells were either mock transfected or transfected with FLAG-UPF1. They were then either left uninduced (DMSO) or reactivated with PMA. We employed the FISH-Flow technique using probes against the vRNA as well as UPF1 mRNA to gate for UPF1-overexpressing populations (Fig. 3a). The percentage of reactivation was monitored by flow cytometry and the cell lysates were subjected to Western blotting to validate UPF1 overexpression using antibodies against UPF1, $pr55^{Gag}$ and actin (Fig. 3b, c). UPF1 overexpression resulted in a 21.3

(See figure on next page.)

Fig. 3 UPF1 overexpression leads to enhanced reactivation of HIV-1 in J-Lat cells. J-Lat 10.6 cells were either mock transfected or transfected with Flag-UPF1 and were uninduced (DMSO) or reactivated (PMA). **a** Gating strategy to detect UPF1 overexpressing cells by detecting UPF1 mRNA levels by FISH-Flow. **b** Of the UPF1 overexpressing cells gated for in **a**, reactivation, monitored by GFP production, was quantified by flow cytometry and the percentages of reactivation were normalised to the mock-PMA reactivated condition. Error bars represent the standard deviation from three independent experiments with at least 10,000 cells counted per treatment. Asterisks represent statistically significant difference between groups (Two-way ANOVA; $p<0.05$). **c** Cell lysates were run on acrylamide gels and UPF1 and $pr55^{Gag}$ protein levels were detected by Western Blotting. **d** Example dot plot depicting vRNA expression in mock transfected and UPF1 overexpressing populations using FISH-Flow technique. **e** The % of vRNA expressing cells were quantified and normalised to the mock-PMA condition. Error bars represent the standard deviation from three independent experiments. Asterisks represent statistically significant difference between groups (One-way ANOVA; $p<0.01$). **f** J-Lat cells were either mock transfected and uninduced (Mock DMSO), mock transfected and reactivated with PMA (Mock PMA) or transfected with FLAG-UPF1 and left uninduced (FLAG-UPF1 DMSO). The % of vRNA expressing cells were quantified. Error bars represent the standard deviation from three independent experiments. Asterisks represent statistically significant difference between groups (One-way ANOVA; $p>0.05$). **g** J-Lat cells were mock transfected or transfected with FLAG-UPF1, FLAG-UPF1-Δ20-150 or FLAG-UPF1-DE and reactivated using PMA. Cell lysates were run on SDS-PAGE gels and UPF1 and $pr55^{Gag}$ protein levels were detected by Western Blotting. **h** Reactivation was quantified in FLAG-UPF1-Δ20-150 and FLAG-UPF1-DE expressing cells and the percentages of reactivation were normalised to the mock-PMA reactivated condition. Error bars represent the standard deviation from three independent experiments with at least 10,000 cells counted per treatment. Asterisks represent statistically significant difference between groups (One-way ANOVA; $p<0.0001$ and $p<0.05$ respectively)

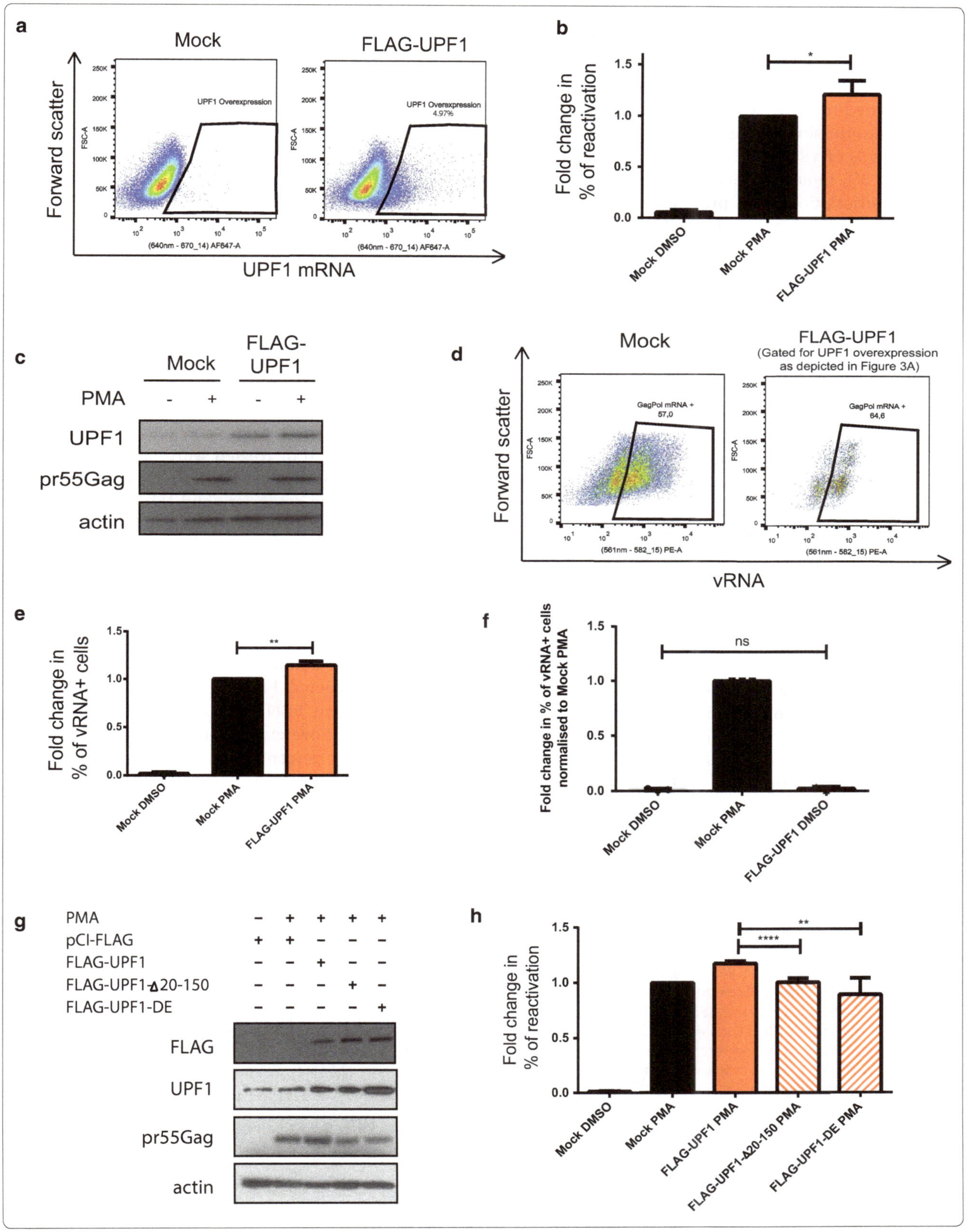
a
Mock
FLAG-UPF1
Forward scatter
UPF1 Overexpression
UPF1 Overexpression
4.97%
FSC-A
(640nm - 670_14) AF647-A
UPF1 mRNA
b
Fold change in
% of reactivation
1.5
1.0
0.5
0.0
*
Mock DMSO
Mock PMA
FLAG-UPF1 PMA
c
Mock
FLAG-
UPF1
PMA
- + - +
UPF1
pr55Gag
actin
d
Mock
FLAG-UPF1
(Gated for UPF1 overexpression
as depicted in Figure 3A)
Forward scatter
GagPol mRNA +
57,0
GagPol mRNA +
64,6
FSC-A
(561nm - 582_15) PE-A
vRNA
e
Fold change in
% of vRNA+ cells
**
Mock DMSO
Mock PMA
FLAG-UPF1 PMA
f
Fold change in % of vRNA+ cells
normalised to Mock PMA
ns
Mock DMSO
Mock PMA
FLAG-UPF1 DMSO
g
PMA − + + + +
pCI-FLAG + + − − −
FLAG-UPF1 − − + − −
FLAG-UPF1-Δ20-150 − − − + −
FLAG-UPF1-DE − − − − +
FLAG
UPF1
pr55Gag
actin
h
Fold change in
% of reactivation
**
****
Mock DMSO
Mock PMA
FLAG-UPF1 PMA
FLAG-UPF1-Δ20-150 PMA
FLAG-UPF1-DE PMA

(±13.5)% increase in viral reactivation upon PMA treatment as compared to the mock-transfected condition (Fig. 3b). UPF1 overexpression also led to a 14.4 (±4.2)% increase in vRNA levels in the UPF1 overexpressing cells and compared to the mock transfected cells (Fig. 3d, e). UPF1 overexpression in uninduced condition shows no increase in % of vRNA cells as demonstrated by FISH-Flow (Fig. 3f), indicating that UPF1 alone is unable to activate transcription of the provirus and PMA is necessary for transcription to take place. UPF1 overexpression also does not result in a change in the % of vRNA+/GFP- cells as compared to mock treated cells (Additional file 1: Figure S3C). This implies that enhanced viral reactivation upon UPF1 overexpression is due to an effect on vRNA levels rather than an increase in the translation of the transcriptional-competent reservoir. Hence, UPF1 overexpression enhances proviral reactivation at a post-transcriptional level by modestly increasing the expression of the vRNA, thereby resulting in enhanced viral reactivation. This is consistent with our previous work where we demonstrated that an overexpression of UPF1 results in enhanced vRNA stability [36].

In order to determine which domain of UPF1 is responsible for enhancing vRNA expression, we either mock transfected cells, or transfected them with FLAG-UPF1 or other constructs of UPF1 that contain deletions in the N-terminal region (FLAG-UPF1-Δ20-150), deletions in the C-terminal (FLAG-UPF1-1-1074), mutations in the RNA helicase domain of UPF1 (FLAG-UPF1-RR857AA), mutations leading to a deficiency in UPF2 binding ability (FLAG-UPF1-LECY) or mutations in the ATPase region of UPF1 (FLAG-UPF1-DE). These cells were then treated with PMA and the % of reactivation was monitored by flow cytometry (Additional file 1: Figure S4A). The ability of UPF1 overexpression to enhance viral reactivation was lost when the FLAG-UPF1-Δ20-150 construct, which contains an N-terminal deletion or the FLAG-UPF1-DE, that has impaired ATPase activity, were used (Fig. 3g, h). The overexpression of these UPF1 mutants resulted in reactivation at levels comparable to the mock transfected cells treated with PMA. These results indicate that the N-terminal domain and ATPase activity of UPF1 are necessary for its mild effect on enhancing vRNA expression and are consistent with our previous work [36].

UPF2 overexpression attenuates HIV-1 reactivation via an interaction with UPF1

Previous work from our lab has demonstrated that UPF2 is excluded from the HIV-1 RNP and that its overexpression can block UPF1-mediated nucleocytoplasmic export of vRNA [37]. UPF2 is also known to bind UPF1 with a high affinity [50]. For these reasons, we hypothesised that when UPF2 is present in excess it can sequester UPF1 in the cytoplasm resulting in reduced UPF1 being bound to vRNA. J-Lat cells were either mock transfected or transfected with FLAG-UPF2 and cells were either left uninduced (DMSO) or treated with PMA. The percentage of reactivation in the form of GFP production was monitored by flow cytometry and the cell lysates were subjected to Western blotting to validate UPF2 overexpression using antibodies against UPF2, pr55^{Gag} and actin. Upon reactivation with PMA, UPF2 overexpression resulted in a 25.95 (±16.8)% decrease in viral reactivation (Fig. 4a) and viral protein production (Fig. 4b). To differentiate between UPF2 overexpressing cells from the whole population and to see if it has any effect on vRNA levels, we conducted FISH-Flow using probes against UPF2 mRNA and vRNA (Fig. 4c). Upon reactivation with PMA, UPF2 overexpression led to a 57.36 (±27.83) decrease in the percentage of vRNA expressing cells as compared to the mock transfected cells (Fig. 4d, e). Therefore, an overexpression of UPF2 resulted in a modest, albeit statistically significant ($p<0.05$) decrease in viral reactivation due to a reduction in vRNA expression.

(See figure on next page.)

Fig. 4 UPF2 overexpression inhibits the reactivation of HIV-1 in J-Lat cells. J-Lat 10.6 cells were either mock transfected or transfected with Flag-UPF2 and were uninduced (DMSO) or reactivated (PMA). **a** Reactivation, monitored by GFP production, was quantified by flow cytometry and the percentages of reactivation were normalised to the mock-PMA reactivated condition. Error bars represent the standard deviation from three independent experiments with at least 10,000 cells counted per treatment. Asterisks represent statistically significant difference between groups (Two-way ANOVA; $p<0.05$). **b** Cell lysates were run on SDS-PAGE gels and UPF2 and pr55^{Gag} protein levels were detected by Western Blotting. **c** Gating strategy to detect UPF2 overexpressing cells by detecting UPF2 mRNA levels by FISH-Flow. **d** Of the UPF2-mRNA expressing cells gated for in (**c**), the % of vRNA expressing cells were quantified. Error bars represent the standard deviation from three independent experiments. Asterisks represent statistically significant difference between groups (One-way ANOVA; $p<0.05$). **e** Example dot plot depicting vRNA expression in mock transfected and UPF2 overexpressing populations using FISH-Flow technique. **f** J-Lat cells were mock transfected or transfected with FLAG-UPF2 or FLAG-UPF2-1-1096. Cell lysates were run on acrylamide gels and UPF2 and pr55^{Gag} protein levels were detected by Western Blotting. **g** J-Lat cells were mock transfected, transfected with FLAG-UPF2 or co-transfected with FLAG-UPF1 or FLAG-UPF1-LECY. Cell lysates were run on acrylamide gels and UPF2, UPF1 and pr55^{Gag} protein levels were detected by Western Blotting. **h** Reactivation in the form of GFP expression was quantified in cells transfected as in (**f**) and (**g**) and the percentages of reactivation were normalised to the mock-PMA reactivated condition. Error bars represent the standard deviation from three independent experiments with at least 10,000 cells counted per treatment. Asterisks represent statistically significant difference between groups (One-way ANOVA; $p<0.0001$)

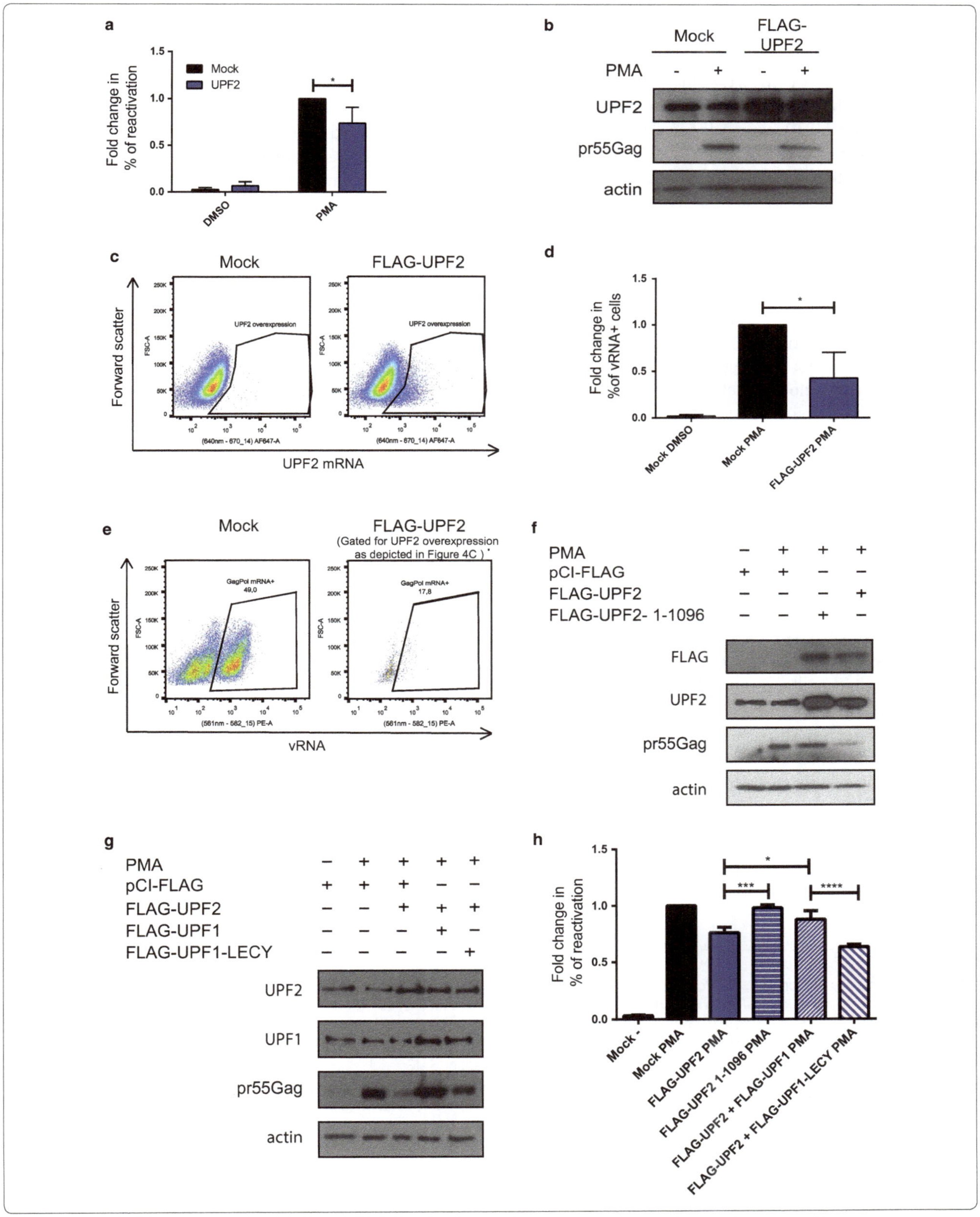

a
Fold change in % of reactivation
1.5
1.0
0.5
0.0
Mock
UPF2
*
DMSO
PMA
b
Mock
FLAG-UPF2
PMA - + - +
UPF2
pr55Gag
actin
c
Mock
FLAG-UPF2
Forward scatter
UPF2 overexpression
UPF2 mRNA
d
Fold change in %of vRNA+ cells
1.5
1.0
0.5
0.0
*
Mock DMSO
Mock PMA
FLAG-UPF2 PMA
e
Mock
FLAG-UPF2
(Gated for UPF2 overexpression as depicted in Figure 4C)
GagPol mRNA+ 49,0
GagPol mRNA+ 17,8
Forward scatter
vRNA
f
PMA − + + +
pCI-FLAG + + − −
FLAG-UPF2 − − − +
FLAG-UPF2- 1-1096 − − + −
FLAG
UPF2
pr55Gag
actin
g
PMA − + + + +
pCI-FLAG + + + − −
FLAG-UPF2 − − + + +
FLAG-UPF1 − − − + −
FLAG-UPF1-LECY − − − − +
UPF2
UPF1
pr55Gag
actin
h
Fold change in % of reactivation
1.5
1.0
0.5
0.0
*
***
****
Mock -
Mock PMA
FLAG-UPF2 PMA
FLAG-UPF2 1-1096 PMA
FLAG-UPF2 + FLAG-UPF1 PMA
FLAG-UPF2 + FLAG-UPF1-LECY PMA

In order to determine if this detrimental effect of UPF2 on vRNA levels is an indirect effect due to its binding to UPF1, we transfected cells with a mutant of UPF2 that does not bind to UPF1 [37, 51, 89] (FLAG-UPF2-1-1096) and compared the % of reactivation in the mock transfected cells, the UPF2 expressing cells and the UPF2-1-1096-expressing cells. It was observed that when UPF2 loses the ability to bind UPF1, there is a loss of its inhibitory effect on reactivation, with reactivation at levels comparable to the mock treated cells (Fig. 4f, h). We also co-transfected FLAG-UPF2 with either FLAG-UPF1 or with FLAG-UPF1-LECY that contains a mutation in the UPF2 binding site and monitored the % of reactivation. UPF1 coexpression is able to rescue the deleterious effect of UPF2 on viral reactivation, but not when in contains a mutation to the UPF2-binding site (Fig. 4g, h). This indicates that the deleterious effect of UPF2 on viral reactivation is a result of its binding to UPF1 which is sequestered and unable to exert a positive effect on vRNA expression, consistent with previous reports [37].

SMG6 overexpression is detrimental to HIV-1 proviral reactivation

UPF1 is an integral member of a network of proteins involved in NMD, including UPF2, UPF3A, UPF3B, SMG6, SMG5, SMG7 and SMG1. SMG6 is the endonuclease involved in the final step of the degradation of aberrant RNA in NMD [52, 53] and has a direct influence on RNA levels. Thus, to evaluate the roles of SMG6 in proviral reactivation, we either mock transfected J-Lat cells or transfected them with HA-SMG6 and either left them uninduced or reactivated them with PMA. The percentage of reactivation in the form of GFP production was monitored by flow cytometry (Fig. 5a) and the cell lysates were subjected to Western blotting to validate SMG6 overexpression using antibodies against SMG6, pr55^{Gag} and actin (Fig. 5b). Overexpression of SMG6 resulted in a 21.2 ($\pm$9.1)% decrease in reactivation (Fig. 5a). Furthermore, upon reactivation with PMA, FISH-Flow analyses revealed a small but significant decrease (7.6$\pm$4.1%) in the percentage of vRNA expressing cells upon SMG6 overexpression as compared to the mock-transfected cells (Fig. 5c, d). Of the vRNA present upon SMG6 overexpression, there was a 1.25-fold decrease in the median fluorescence intensity (Fig. 5e). Thus, SMG6 is detrimental to vRNA expression and attenuates PMA-induced proviral reactivation.

SMG6 contains an exon junction binding domain (EBM) [54], a 14-3-3-like domain that binds to phosphorylated UPF1 [55] and a PilT N-terminus (PIN) domain [56] that possesses the endonuclease activity [56–58]. In order to determine which of these domains are responsible for the negative effect on vRNA levels, we transfected J-Lat cells with plasmids that express SMG6 with mutations in each of the aforementioned domains; HA-SMG6-mEBM, HA-SMG6-m14-3-3 and HA-SMG6-mPIN respectively. These cells were reactivated with PMA and the percentage of reactivation was monitored using flow cytometry. While the overexpression of HA-SMG6 and the exon junction binding mutant HA-SMG6-mEBM attenuated proviral reactivation, the overexpression of HA-SMG6-m14-3-3 and HA-SMG6-mPIN displayed reactivation levels similar to the mock transfected cells (Fig. 5f, g). Thus, these results demonstrate that both, the binding of SMG6 to phosphorylated UPF1 and its endonuclease activity are necessary for its inhibitory effect on vRNA levels (Fig. 5f, g).

SMG6 knockdown increases vRNA expression, but does not affect viral reactivation

To determine the effect of SMG6 depletion on HIV-1 proviral reactivation, we conducted siRNA mediated knockdown studies. J-Lat cells were either transfected with a non-silencing siRNA (siNS) or with siRNA against SMG6 (siSMG6) and cells were either left uninduced

(See figure on next page.)

Fig. 5 SMG6 overexpression leads to attenuated reactivation of HIV-1. **a** J-Lat 10.6 cells were either mock transfected or transfected with HA-SMG6 and were uninduced (DMSO) or reactivated (PMA). Reactivation, monitored by GFP production, was quantified by flow cytometry and the percentages of reactivation were normalised to the mock-PMA reactivated condition. Error bars represent the standard deviation from three independent experiments with at least 10,000 cells counted per treatment. Asterisks represent statistically significant difference between groups (Two-way ANOVA; $p < 0.01$). **b** Cell lysates were run on acrylamide gels and SMG6 and pr55^{Gag} protein levels were detected by Western Blotting. **c** Example dot plot depicting vRNA expression in mock PMA and SMG6 PMA conditions using FISH-Flow technique. **d** The % of vRNA expressing cells were quantified. Error bars represent the standard deviation from three independent experiments. Asterisks represent statistically significant difference between groups (One-way ANOVA; $p < 0.05$). **e** MFI of the vRNA signal were quantified. Asterisks represent statistically significant difference between groups (student's t-test; $p < 0.05$). **f** J-Lat cells were mock transfected or transfected with HA-SMG6, HA-SMG6-mEBM, HA-SMG6-m14-3-3 or HA-SMG6-mPIN and reactivated with PMA. Cell lysates were run on acrylamide gels and SMG6 and pr55^{Gag} protein levels were detected by SDS-PAGE followed by Western Blotting. **g** Reactivation in the above conditions was quantified and the percentages of reactivation were normalised to the mock PMA reactivated condition. Error bars represent the standard deviation from three independent experiments with at least 10,000 cells counted per treatment. Asterisks represent statistically significant difference between groups (One-way ANOVA; $p < 0.05$)

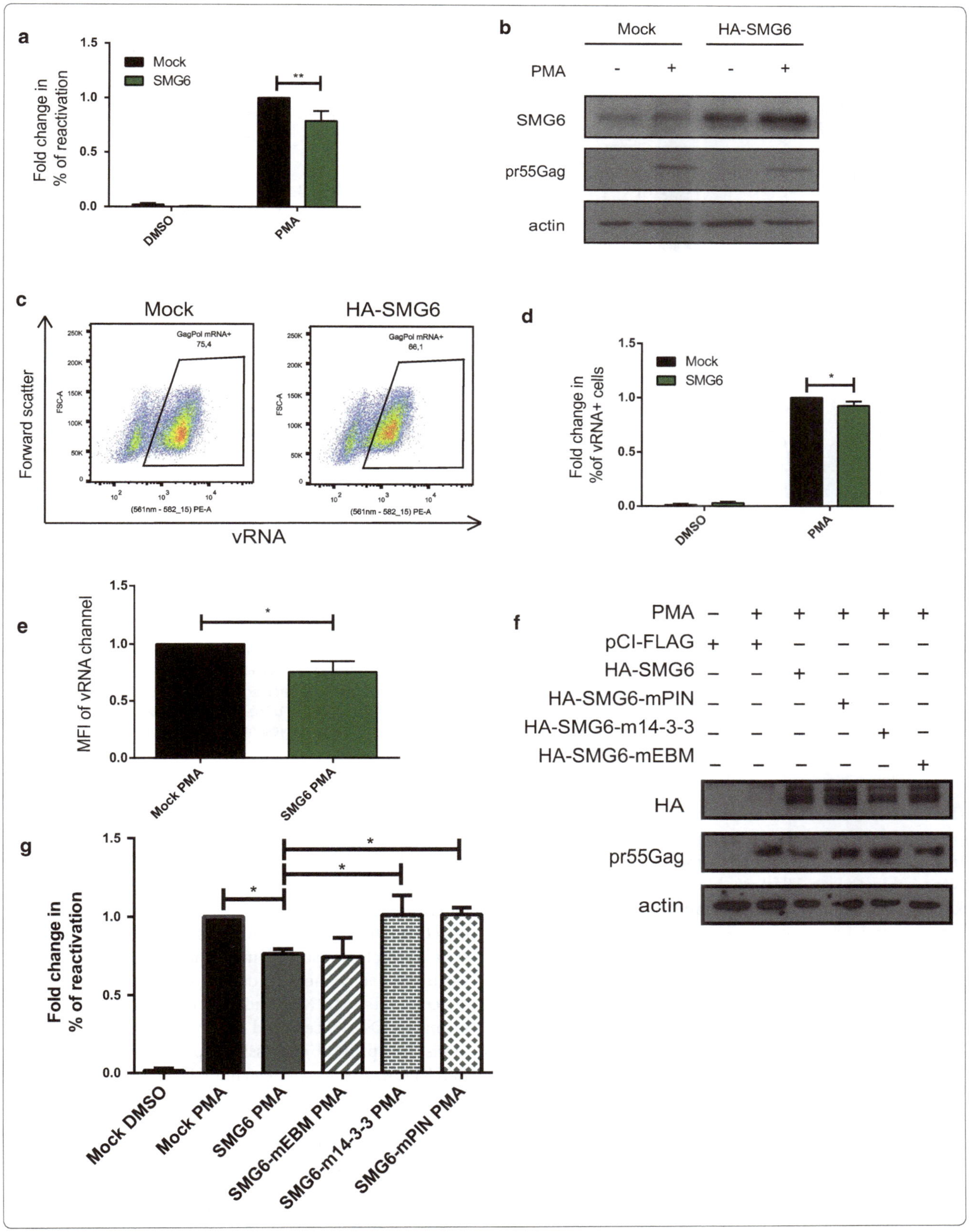
a
Mock
SMG6
Fold change in % of reactivation
DMSO
PMA
b
Mock
HA-SMG6
PMA
SMG6
pr55Gag
actin
c
Mock
HA-SMG6
GagPol mRNA+ 75,4
GagPol mRNA+ 66,1
Forward scatter
FSC-A
(561nm - 582_15) PE-A
vRNA
d
Mock
SMG6
Fold change in %of vRNA+ cells
DMSO
PMA
e
MFI of vRNA channel
Mock PMA
SMG6 PMA
f
PMA
pCI-FLAG
HA-SMG6
HA-SMG6-mPIN
HA-SMG6-m14-3-3
HA-SMG6-mEBM
HA
pr55Gag
actin
g
Fold change in % of reactivation
Mock DMSO
Mock PMA
SMG6 PMA
SMG6-mEBM PMA
SMG6-m14-3-3 PMA
SMG6-mPIN PMA

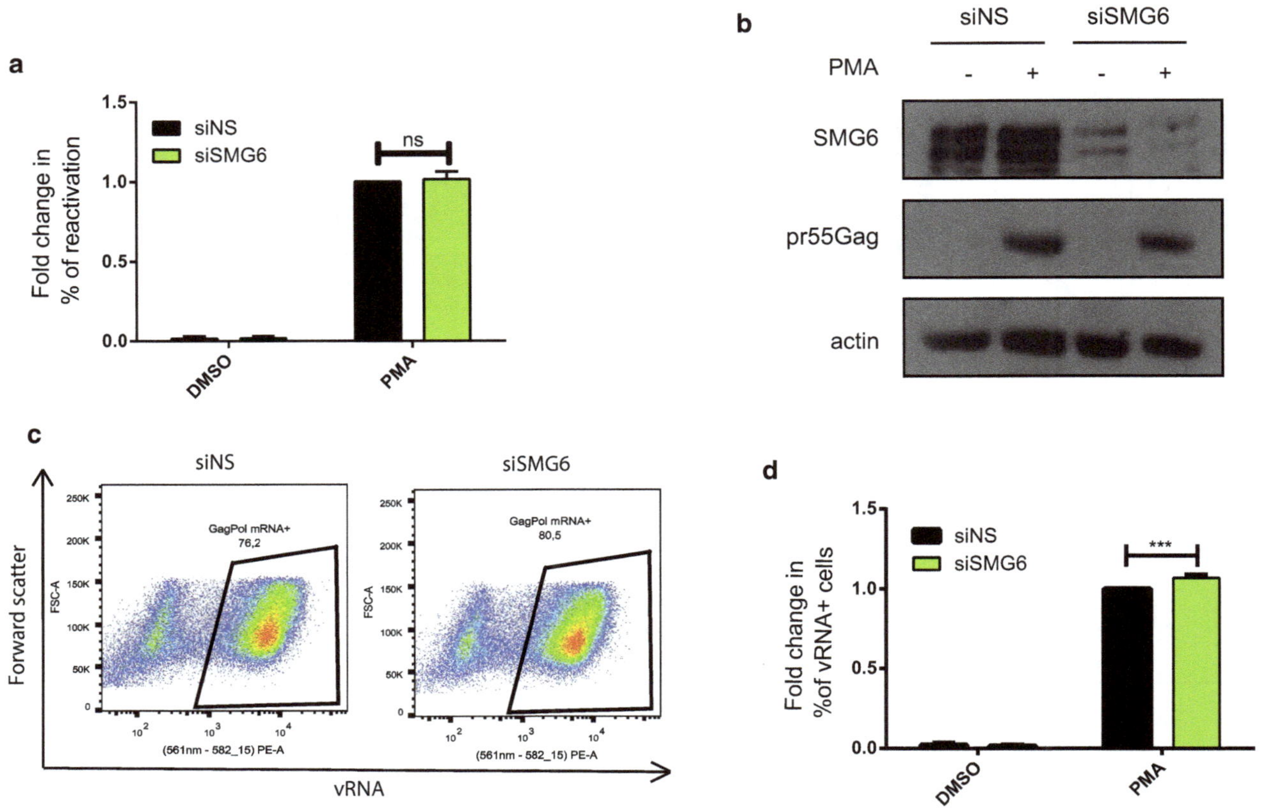

Fig. 6 SMG6 knockdown leads to increased vRNA levels, but not reactivation in J-Lat cells. J-Lat 10.6 cells were either transfected either siNS or siSMG6 and were either uninduced (DMSO) or reactivated (PMA). **a** Reactivation monitored by GFP production was measured by flow cytometry. Error bars represent the standard deviation from three independent experiments with at least 10,000 cells counted per treatment. (Two-way ANOVA; $p > 0.05$). **b** Cell lysates were run on acrylamide gels and SMG6 and pr55^{Gag} protein levels were detected by Western Blotting. **c** Example dot plot depicting vRNA expression in siNS PMA and siSMG6 PMA conditions using FISH-Flow technique and, **d** The % of vRNA expressing cells were quantified. Error bars represent the standard deviation from three independent experiments. Asterisks represent statistically significant difference between groups (One-way ANOVA; $p < 0.001$)

(DMSO) or treated with PMA to reactivate the cells. The percentage of reactivation in the form of GFP production was monitored by flow cytometry and the cell lysates were subjected to Western blotting to validate SMG6 knockdown using antibodies against SMG6, pr55^{Gag} and actin. A knockdown of SMG6 did not have a significant effect on viral reactivation at the level of protein production (Fig. 6a, b). However, upon reactivation with PMA and using FISH-Flow using probes against vRNA, SMG6 knockdown resulted in a small but significant increase ($6.9 \pm 1.8\%$) in the total number of vRNA expressing cells as compared to the siNS condition (Fig. 6c, d). This further illustrates that SMG6 is detrimental to vRNA expression.

UPF1 knockdown impairs vRNA expression in primary HIV-1 infected CD4+ T cells

UPF1 enhances vRNA expression and, as a consequence, viral reactivation in J-Lat cells. UPF2 and SMG6 are detrimental to vRNA expression, both, via interactions with UPF1. We also assessed the effects of UPF1, UPF2 and SMG6 overexpression on TNFα-induced reactivation of J-Lat cells and observed comparable results (Additional file 1: Figure S4B). However, whether these effects of UPF1 on vRNA expression and pr55^{Gag} expression were also observed in primary CD4+ T cells was yet to be determined. In order to address this question, we conducted shRNA-mediated knockdown of UPF1 in primary CD4+ T cells and observed the effects on vRNA levels and pr55^{Gag} expression upon HIV-1 infection by FISH-Flow. Negatively selected CD4+ T cells from three donors were activated with phytohemagglutinin (PHA). They were then transduced with shUPF1-containing lentiviral particles. Lentiviral particles containing a scrambled sequence were used as a negative control (shNS). The cells were infected with HIV-1 24 h post transduction by spinoculation. Cells were collected 6 days post infection and FISH-Flow was conducted to monitor vRNA

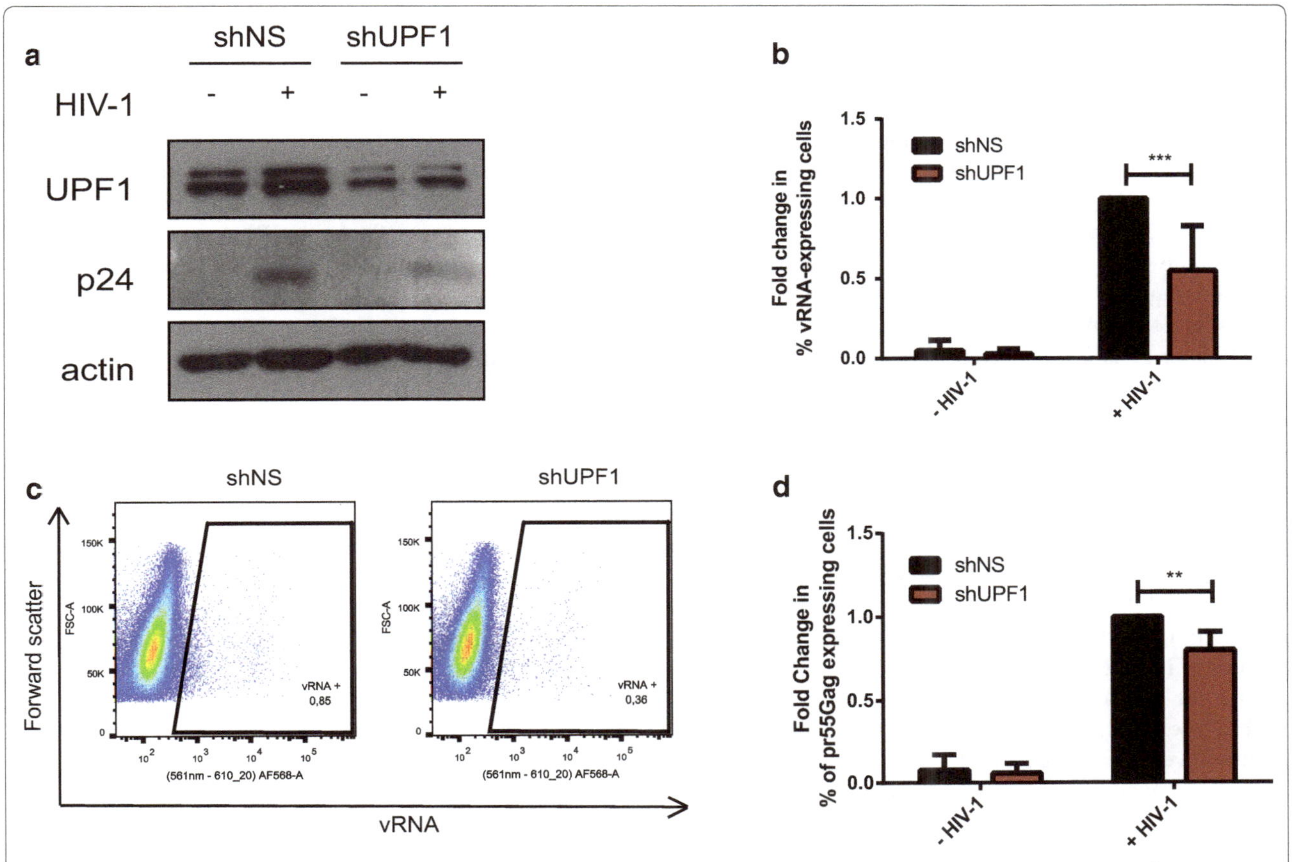

Fig. 7 UPF1 knockdown leads to reduced vRNA levels and Gag expression in primary HIV-1 infected CD4+ T cells. Primary CD4+ T cells were either transduced with shNS or shUPF1-containing lentiviral particles and either left uninfected or infected with HIV-1. **a** Cell lysates were run on SDS-PAGE gels and UPF1 and pr55^{Gag} protein levels were detected by Western Blotting. **b** The % of vRNA expressing cells were quantified and normalised to shNS HIV-1-infected condition. Error bars represent the standard deviation from six independent experiments (three donors in duplicate) with at least 5,000,000 cells counted per experiment. Asterisks represent statistically significant difference between groups (Two-way ANOVA; $p<0.001$). **c** Example dot plot depicting vRNA expression in HIV-1 infected shNS and shUPF1 conditions using FISH-Flow technique and, **d** The % of Gag expressing cells were quantified and normalised to shNS HIV-1-infected condition. Error bars represent the standard deviation from nine independent experiments (three donors in duplicate) with at least 5,000,000 cells counted per experiment. Asterisks represent statistically significant difference between groups (Two-way ANOVA; $p<0.01$)

and intracellular pr55^{Gag} levels. Cell lysates were also subjected to Western blotting to validate UPF1 knockdown (Fig. 7a). In humans, UPF1 has two isoforms and both isoforms are detected in primary CD4+ T cells [90] (Additional file 1: Figure S5A). However, in J-Lat cells, only the larger one is expressed at high enough levels to be detected by the UPF1 antibody (Additional file 1: Figure S5A). shUPF1 treatment in primary T cells resulted in a 53.8 (±4.5)% decrease in UPF1 protein levels as compared to the shNS-treated cells (Additional file 1: Figure S5B). Results from three independent donors demonstrated that a knockdown of UPF1 resulted in a 45.16 (±27.9)% decrease in vRNA levels as compared to the mock treated cells (Fig. 7b, c). This also corresponded with 20.1 (±10.9)% reduced intracellular pr55^{Gag} staining (Fig. 7d). Therefore, UPF1 also enhances vRNA levels and promotes viral gene expression in primary CD4+ T cells.

Discussion

The 'active viral reservoir' has been defined as the HIV-1 infected cells that contain viral RNA species but do not produce infectious viral particles [59, 60] and this highlights the post-transcriptional maintenance of HIV-1 latency. Latently-infected resting CD4+ cells T cells have been demonstrated to contain cell-associated unspliced and multiply spliced HIV-1 RNA [11, 61]. In these cells, the vRNA was sequestered within the nucleus and could be efficiently rescued through the overexpression of the host protein polypyrimidine tract binding protein (PTB), suggesting that latency can be reversed at a post-transcriptional level [61]. Two characterised primary T cell models of latency have also demonstrated a post-transcriptional block to HIV-1 reactivation, either by sequestration of the vRNA in the nucleus or splicing defects [14, 16, 62]. In addition, microRNAs have been implicated in the maintenance of HIV-1 latency (reviewed in [18]), providing another example of how post-transcriptional events can affect proviral reactivation. In the quest for an HIV-1 cure, the importance of investigating the contribution of post-transcriptional events and vRNA metabolism in the maintenance of HIV-1 latency is being recognised [63–65]. One HIV-1 cure strategy is the 'shock and kill' approach which involves reactivating the latent provirus by small molecules (shock) and then to eliminating the virus (kill) using intensive cART and/or immunomodulators [66]. Numerous compounds are under investigation as candidates for latency-reversing agents (LRAs) which promote the transcription of the provirus (reviewed in [67, 68]). So far, the use of LRAs have limited ability to decrease the size of the viral reservoir, with only two reports of successful reduction in reservoir size [7, 69, 70]. The shortcomings of current LRAs is highlighted in a recent study using FISH-Flow in which CD4+ T cells from HIV-1 infected patients were reactivated with the LRAs romidepsin or PMA/ionomycin and only 2–10% of cells that expressed vRNA produced viral proteins [17]. Therefore, the LRAs might be more effective if used in combination with drugs that affect vRNA metabolism at a post-transcriptional level. By modulating the activities of the RNA surveillance proteins or creating small molecules that mimic their activity, we can increase the stability of the vRNA to facilitate reactivation of these latent cells so that they are visible to the immune system and can be targeted by host immune responses and antiretrovirals. Alternatively, we can also apply this study to create novel long-lasting antiretrovirals by designing small molecules to inhibit the binding of UPF1 to vRNA thereby decreasing vRNA stability and reducing viral production.

Using FISH-Flow, this study demonstrates that the RNA surveillance proteins UPF1, SMG6 and UPF2 can affect HIV-1 gene expression, and thus viral reactivation at a post-transcriptional level. Although the effects of UPF1, UPF2 and SMG6 overexpression on modulating viral latency are modest (Figs. 3b, 4a and 5a), these effects nevertheless provide novel evidence of the contribution of post-transcriptional events in viral reactivation from latency. UPF1 was demonstrated to be a positive regulator of viral reactivation in the J-Lat 10.6 latent T cell model. Notably, we also demonstrate a direct effect of UPF1 on enhancing vRNA levels and viral gene expression in primary CD4+ T cells. The overexpression of the ATPase mutant of UPF1 (FLAG-DE-UPF1) did not lead to enhanced reactivation of HIV-1 in J-Lat cells (Fig. 3g, h), indicating that the ATPase activity is responsible for enhanced vRNA expression and viral reactivation. This is in concordance with our previous work where we showed that this UPF1 construct was unable to upregulate vRNA levels and enhance vRNA stability [36]. This ATPase mutant has impaired RNA-binding capacity [71]. To exert its positive effects on vRNA metabolism, UPF1 needs to be able to bind to the vRNA and subsequently lead to the assembly of distinct RNPs that promote vRNA stability, export and translation [37]. An impairment of RNA binding capability could lead to a dissociation of UPF1 from the vRNA, thereby providing another possible explanation why no enhanced viral reactivation was observed when the ATPase mutant of UPF1 was used.

The HIV-1 vRNA metabolism is controlled by numerous cis-acting RNA sequences [72], such as the cis-repressive sequences or instability sequences (INS) [73]. UPF1 contains two zinc fingers that have been implicated to bind to INSs [74] and thus, could promote vRNA stability. The FLAG-UPF1-Δ20-150 construct contains a deletion in the zinc finger motif [36] that could lead to impaired binding to the HIV-1 INS. In agreement with our previous studies where we demonstrate that an overexpression of FLAG-UPF1-Δ20-150 does not lead to enhanced vRNA expression levels [36]; here we demonstrated that, in the context of reversal from viral latency, an overexpression of FLAG-UPF1-Δ20-150 does not lead to enhanced proviral reactivation (Fig. 3g, h), most likely due to impaired binding of UPF1 to the vRNA due to the loss of a zinc finger motif.

We have also previously shown that UPF2 is excluded from HIV-1 RNPs through antagonistic interactions with the viral or host proteins such as Rev or Staufen1 [37]. The binding of UPF2 to UPF1 has been reported to induce a conformational change in UPF1 that stimulates its RNA helicase activity and dampens its RNA binding capability, thereby hampering its binding to the vRNA [75, 76]. UPF2 also binds to UPF1 with high affinity [77] and this could limit the availability of UPF1 to bind to the vRNA. Our data reinforce the hypothesis that UPF2 is detrimental to vRNA metabolism, as we observed that overexpression of UPF2 resulted in reduced vRNA expression and viral reactivation (Fig. 4a–e). This deleterious effect is likely a result UPF2 binding to UPF1 and its sequestration, since viral reactivation was restored to levels similar to control cells when the UPF2 mutant deficient in UPF1 binding was used (Fig. 4f–h). In accordance with our work, a previous report using an shRNA library in J-Lat 5A8 cells showed that shRNAs against UPF1 were disenriched in the reactivated population as compared to the latent population, indicating that it exerts a positive effect on the reactivation of the HIV-1 provirus

[78]; whereas shRNAs against UPF2 were enriched in the reactivated population, indicating that UPF2 promotes that maintenance of latency in J-Lat cells [78].

SMG6 is the endonuclease responsible for cleaving mRNAs that are targeted for NMD [52, 53]. Both SMG6 and UPF1 have been reported to be present at transcription sites [79] and SMG6 interacts with UPF1 in a phospho-dependent [55] and a phospho-independent manner [90]. Furthermore, because of its endonuclease activity, SMG6 could have a direct effect on UPF1-bound mRNA levels, such as the vRNA. Our observation that an overexpression of SMG6 results in a decrease of vRNA expression and, consequently, decreased viral reactivation, suggests that SMG6 is detrimental to vRNA stability (Fig. 5a–g). Using mutational studies, we identified that the binding of SMG6 via its 14-3-3 like domain to phosphorylated UPF1 as well its endonuclease activity via its PIN region is necessary to downregulate the viral reactivation (Fig. 5f, g).

Recent transcriptome analyses have demonstrated that UPF1 binds promiscuously to all cellular RNAs; both, canonically identified NMD targets as well as to non-NMD targets and long non-coding RNAs [39, 80–83]. The marker for a cellular NMD target has been revealed to be the RNA's binding to phosphorylated UPF1 [19, 84]. UPF1 interacts with the PIK-related protein kinase SMG1, SMG8, SMG9, and the two translation termination factors eRF1 and eRF3 to form a decay inducing complex called the SURF [85, 86]. The phosphorylation of UPF1 by SMG1 is necessary for mRNA decay and creates an N-terminal binding platform for SMG6 that cleaves the targeted mRNAs [52, 53, 55]. Hyperphosphorylated UPF1 has been also shown to attract downstream NMD machinery with higher affinity [87]. Therefore, we can speculate that in the context of the interaction between UPF1 and the vRNA, the hyperphosphorylation of UPF1 would be detrimental to vRNA stability due to increased recruitment of SMG6 and other mRNA decay factors. The ATP deficient UPF1 mutant FLAG-UPF1-DE has also been demonstrated to be hyperphosphorylated and assembles complexes with SMG6 on both target and non-target mRNAs [83]. This could provide another possible explanation why the overexpression of the ATPase defective UPF1 did not result in enhanced viral reactivation (Fig. 3g, h). Further investigation is required to elucidate the roles of the phosphorylation status of UPF1 on proviral reactivation.

Conclusion

In this manuscript, we provide evidence that the RNA surveillance proteins UPF1, UPF2 and SMG6 can affect vRNA expression and thus, the maintenance of HIV-1 latency. These findings can be applied to bolster the reactivation of the HIV-1 provirus to effectively decrease the size of the viral reservoir using a shock and kill approach or can be harnessed to create a novel set of antiretrovirals.

Methods

Cell culture

J-Lat 10.6 cells (J-Lat full-length clone 10.6; NIH AIDS Reagent Program) are a Jurkat derived T cell line that is latently infected with HIV-1 in which the *nef* sequence was replaced with a green fluorescent protein (GFP) coding sequence [88]. J-Lat latent proviruses were reactivated by adding 20 ng/mL of phorbol 12-myristate 13-acetate (PMA) (Sigma-Aldrich) to the culture media for 24 h. In case of reactivation with TNFα, 10 ng/ml TNFα (Sigma-Aldrich) was added to the culture media for 24 h. Reactivation of cells was quantified by measuring GFP expression by flow cytometry. All cell cultures were maintained in RPMI 1640 (Life Technologies) supplemented with 10% fetal bovine serum (Hyclone) and 1% penicillin/streptomycin (Life Technologies) at 37 °C and 5% CO_2. HEK293T cells were purchased from the American Type Culture Collection (ATCC). TZM-bl HeLa cell line was obtained from NIH AIDS Reference and Reagent Program. Both of these cells lines were grown in Dulbecco's modified Eagle medium (DMEM, Invitrogen) containing 10% fetal bovine serum (HyClone) and 1% penicillin–streptomycin (Invitrogen). PBMCs were isolated from leukophoresed blood collected from healthy donors. All subjects provided informed consent for participating in this study. The research ethics boards of the recruiting sites, the Centre Hospitalier de l'Universite de Montreal and McGill University Health Centre approved this study. PBMCs were isolated by density-gradient centrifugation using lymphocyte separation medium (Corning). CD4+ T cells were negatively selected using the EasySep human T cell enrichment kit according to manufacturer's protocol (StemCell). Negatively selected CD4+ T cells were maintained in RPMI 1640 (Life Technologies) supplemented with 10% fetal bovine serum (Hyclone) and IL-2 (Sigma-Aldrich). CD4+ T cells were activated by treating them with 10ug/ml PHA (Sigma-Aldrich) for 72 h.

Antibodies

Mouse anti-p24 was obtained from NIH AIDS Reagents Program; rabbit antisera to UPF1 and UPF2 were generously supplied by Jens Lykke-Andersen (University of California, San Diego, CA, USA); rabbit anti-EST1A (SMG6) and mouse anti-actin were purchased from Abcam; rabbit anti-FLAG was purchased from

Sigma-Aldrich; mouse anti-HA was purchased from Roche; mouse anti-GAPDH was purchased from Techniscience; mouse anti-nucleolin was purchased from Santa-Cruz Biochemistry; KC57-FITC was purchased from Beckman Coulter; horseradish peroxidase-conjugated secondary antibodies were purchased from Rockland Immunochemicals.

Plasmids

The plasmids pCI-FLAG, FLAG-UPF1, FLAG-UPF1-Δ20-150, FLAG-UPF1-1-1074, FLAG-UPF1-RR857AA, FLAG-UPF1-LECY, FLAG-UPF1-DE, FLAG-UPF2 and FLAG-UPF2-1-1096 were described previously [36, 37, 89]. HA-SMG6, HA-SMG6-mEBM, HA-SMG6-m14-3-3 and HA-SMG6-mPIN were a kind gift from Dr. Oliver Muhlemann and are previously described [90]. pNL4.3 was obtained from NIH AIDS Reagents Program.

Gene silencing

Custom siRNA duplexes were synthesised by Qiagen. The target sequence for UPF1 was 5′-AAGATGCAGTTC CGCTCCATT-3′ and for SMG6 was 5′-GCTGCAGGT TACTTACAAG-3′. The siNS used in this study is a commercially available non-silencing control duplex with target sequence 5′-AATTCTCCGAACGTGTCACGT′-3′.

Transfections

J-Lat or Jurkat T cells were transfected with either 1 μg of plasmid DNA or 20 nM of siRNA per 1×10^6 cells using the Neon Transfection System (Thermo Fisher Scientific) according to manufacturer's protocols using the following electroporation parameters: three pulses of 1350 V and 10 ms at a cell density of 1×10^7/mL. J-Lat cells were reactivated 24 h after transfection. HEK293T cells were transfected using JetPrime transfection reagent according to manufacturer's protocol using 1ul of Jetprime (Polyplus) for 1ug of plasmid DNA.

Viral transduction

psPAX2, pMD2.G and the pLKO-shNS lentiviral control plasmid containing scrambled non-target shRNA used as a negative control were kind gifts from Dr. Marc Fabian (McGill University). pLKO-shUPF1 (TRCN0000022254) expression vector containing shRNA to UPF1 was obtained from the McGill genetic perturbation service. HEK293T cells were plated in 10 cm-dishes plates and were co-transfected with either shNS or shUPF1 expressing lentivirus, psPAX2 and pMD2.G. Supernatants were collected 48 h post-transfection, passed through a 0.45-μm filter (Pall) and supplemented with 5 μg/ml polybrene (Sigma-Aldrich). The viral particles were added to the primary CD4+ T cells (1 ml of supernatant per 10,000,000 cells) and incubated for 16 h, following which they were infected with HIV-1.

HIV-1 virus production and infection

NL4.3 virus particles were prepared by transfection of HEK293T cells with HIV-1 NL4-3 provirus-encoding plasmid pNL4.3 using the JetPrime transfection reagent. The supernatants were collected 48 h post transfection, filtered through a 0.45-μm filter (Pall) and centrifuged at 20,000 r.p.m. for 1 h at 4 °C to pellet the virus. Viruses were resuspended in RPMI and stored at −80 °C. The multiplicity of infection (MOI) of viruses were quantified using the X-gal staining assay in TZM-bl cells as described in [91]. CD4+ T cells in RPMI were infected with an MOI of 0.5 NL4.3 viruses by spinoculation at 1800 r.p.m. for 45 min. Following spinoculation, the cells were washed and replenished with complete culture media. Cells were collected 6 days post infection.

Western blotting

Cells were lysed in NP40 lysis buffer (50 mM Tris pH 7.4, 150 mM NaCl, 0.5 mM EDTA, 0.5% NP40). Protein concentration on each cell lysate was quantified by Bradford assay. Equal amounts of protein (20 μg) were separated by SDS-PAGE and transferred to a nitrocellulose membrane (Bio-Rad). Blocking was performed using 5% non-fat milk in Tris-buffered saline (pH 7.4) with 0.1% Tween 20 (TBST) for 1 h at room temperature. Membranes were probed with the indicated primary and corresponding horseradish peroxidase-conjugated secondary antibodies. Proteins were detected using Western Lightning Plus-ECL (PerkinElmer). Signal intensities were scanned by densitometry using ImageJ software (NIH, Bethseda, USA).

FISH-flow

Cells were collected, fixed, permeabilized and subjected to the PrimeFlow RNA assay (Thermo Fisher Scientific) following the manufacturer's instructions and as described in [42, 92]. For intracellular $pr55^{Gag}$ staining in primary CD4+ T cells, KC57-FITC antibody (Beckman Coulter) was used in permeabilisation buffer from the kit at a dilution of 1:50 for 30 min at room temperature, followed by 30 min at 4 °C. For all samples, mRNA was labelled with a set of 40 probe pairs diluted 1:20 in diluent provided in the kit and hybridized to the target mRNA for 2 h at 40 °C. The probes for GagPol, UPF1, UPF2 and SMG6 used had the following catalog numbers: GagPol HIV-1 VF10-10884, UPF1 VA1-3004200, UPF2 VA1-3007897 and SMG6 VA1-3001031. Positive control probes against the house-keeping gene RPL13A (VA1-13100) were included in each experiment. Samples were washed to remove excess probes and stored overnight in the presence of RNAsin. Signal amplification was then performed by sequential 1.5 h, 40 °C incubations with the pre-amplification and amplification mix. Amplified mRNA was labelled with fluorescently-tagged probes for 1 h at 40 °C. Gates were set on the uninfected Jurkat cells, unstimulated J-Lat control or uninfected primary CD4+ T cells where appropriate. Samples were acquired on a BD LSR Fortessa Analyzer. Analysis was performed using the FlowJo V10 software (Treestar).

Confocal microscopy following FISH-flow

Cells that underwent the FISH-Flow assay described above were seeded on 18 mm diameter coverslips and air dried. Coverslips were mounted in ProLong Gold Antifade Reagent with DAPI (Life Technologies). Laser scanning confocal microscopy was performed on a Leica DM16000B microscope equipped with a WaveFX spinning disk confocal head (Quorum Technologies) using a 63X objective lens. Images were acquired with a Hamamatsu ImageEM EM-charges coupled device (CCD) camera and image reconstruction was performed with the Imaris software (v. 8.4.1, Bitplane, Inc.).

RT-qPCR

For data presented in Fig. 2e, total RNA was extracted from cells using Aurum Total RNA Mini kits (Bio-Rad). RT-qPCR analysis of HIV-1 RNA levels was performed as previously described [93, 94]. For data presented in Additional file 1: Figure S2E and Additional file 1: Figure S3B, cellular fractionation was performed as described in [95]. RNA extraction from each fraction were performed using Trizol Reagent (Thermo Fisher Scientific) following manufacturer's instructions. cDNA was obtained using the High-Capacity cDNA Reverse Transcription Kit (Applied Biosystems). cDNA and primers were then added to GoTaq Green Master Mix (Promega). GAPDH was amplified using the primers GAPDH_1 forward 5′-TGACCACAGTCCATGCCATC-3′ and GAPDH_1 reverse 5′-ATGATGTTCTGGAGAGCCCC-3′ and HIV-1 vRNA using the primers pNL4-3_1 forward 5′-GGGAGCTAGAACGATTCGCA-3′ and pNL4-3_1 reverse 5′-GGATGGTTGTAGCTGTCCCA-3′. The PCR products were visualised in a 1% agarose gel by staining the DNA with RedSafe Nucleic Acid Staining Solution (iNtRON). Signals were captured using a Gel Doc System and intensities were normalised to the GAPDH signal.

Statistical analysis

All experiments were performed in triplicate, and the data are presented as the mean ± standard deviation (SD). A p value of < 0.05 in a student's t-test, one-way or two-way ANOVA test was considered statistically significant. GraphPad Prism 6 (Graphpad Software Inc.) was used to conduct statistical analyses and create graphs.

Additional file

Additional file 1: Figure S1. A J-Lat cells were treated with TNF-alpha or different concentrations of PMA and the % of GFP positive cells were measured. Example dot plot depicting **B** UPF1 mRNA, **C** UPF2 mRNA and **D** SMG6 mRNA expression in mock transfected cells with and without PMA addition using FISH-Flow technique. **Figure S2.** J-Lat 10.6 cells were either transfected with siNS or siUPF1 and were uninduced (DMSO) or reactivated (PMA). **A** Quantification of UPF1 protein expression by densitometry analysis of Western blots. **B** MFI of the vRNA signal were quantified. Asterisks represent statistically significant difference between groups (student's t-test; $p < 0.05$). **C** The % of RPL13A mRNA expressing cells were quantified. **D** MFI of the PRL13A signal were quantified. **E** Relative GAPDH mRNA levels as measured by RT-PCR. For all graphs, error bars represent the standard deviation from three independent experiments. **Figure S3. A** Cellular fractionation was performed in siNS or siUPF1 treated conditions, with and without PMA treatment. The fractions were run on SDS-PAGE gels and GAPDH and nucleolin protein levels were detected by Western Blotting to confirm fractionation. **B** The relative amounts of vRNA in each fraction were quantified by RT-PCR and normalised to levels of GAPDH mRNA. Error bars represent the standard deviation from three independent experiments. **C** J-Lat cells were mock transfected, transfected with FLAG-UPF1, FLAG-UPF2 or HA-SMG6 and reactivated with PMA. The % of vRNA+/GFP-cells was quantified. Error bars represent the standard deviation from three independent experiments. **Figure S4. A** J-Lat cells were mock transfected, transfected with FLAG-UPF1 or with FLAG-UPF1 mutants and reactivated with PMA. Reactivation in the above conditions was quantified and the percentages of reactivation were normalised to the Mock PMA reactivated condition. Error bars represent the standard deviation from three independent experiments with at least 10000 cells counted per treatment. Asterisks represent statistically significant difference between groups. **B** J-Lat cells were mock transfected, transfected with FLAG-UPF1, FLAG-UPF2 or HA-SMG6 and reactivated with TNF-alpha. Reactivation in the above conditions was quantified and the percentages of reactivation were normalised to the Mock PMA reactivated condition. Error bars represent the standard deviation from three independent experiments with at least 10000 cells counted per treatment. **Figure S5. A** Equal amounts of cell lysates from J-Lat 10.6 and primary CD4+ T cells were subjected to Western blotting and probed for UPF1 and actin. **B** Primary CD4+ T cells were either transduced with shNS or shUPF1-containing lentiviral particles and either left uninfected or infected with HIV-1. Quantification of UPF1 protein expression by densitometry analysis of Western blots.

Authors' contributions

SR, RA and AJM conceived of the study and designed experiments. SR conducted most of the experiments with significant contribution from RA. MN and AT provided technical help and comments on experimental design. SR drafted the manuscript with the support and comments from RA and AJM. All authors read and approved the final manuscript.

Author details

[1] HIV-1 RNA Trafficking Laboratory, Lady Davis Institute at the Jewish General Hospital, Montreal, QC H3T 1E2, Canada. [2] Department of Microbiology and Immunology, McGill University, Montreal, QC H3A 2B4, Canada. [3] Department of Medicine, McGill University, Montreal, QC H3A 0G4, Canada.

Acknowledgements

We thank the late Mark Wainberg, Andreas Kulozik, Niels Gehring, Jens Lykke-Andersen, Oliver Muehlemann, Mark Fabian, blood donors, Jean-Pierre Routy, Mario Legault and the Réseau SIDA et maladie infectieuses of the Fonds de recherche Santé-Québec for generous provision of cells and reagents;

Niels Gehring and Nada Hafez for helpful discussions; Daniel Kaufmann, Alan Cochrane, Amy Baxter and Lewis Liu for assay development; Christian Young for technical assistance; and Alessandro Cinti for critical reading of the manuscript.

Competing interests
The authors declare that they have no competing interests.

Funding
This study, S.R. and R.A. were supported by The Canadian HIV Cure Enterprise Team Grant HIG-133050 (to A.J.M.) from the Canadian Institutes of Health Research (CIHR) in partnership with Canadian Foundation for HIV-1/AIDS Research and International AIDS Society and from the Lady Davis Research Institute/Jewish General Hospital. R.A. was funded in part by a Conselho Nacional de Desenvolvimento Científico e Tecnológico Fellowship (Brazil). M.N. was supported by a generous contribution to this project from the Lady Davis Institute/Jewish General Hospital. The funders had no role in study design, data collection and interpretation, or the decision to submit the work for publication.

References

1. Antiretroviral Therapy Cohort C. Life expectancy of individuals on combination antiretroviral therapy in high-income countries: a collaborative analysis of 14 cohort studies. Lancet. 2008;372:293–9.
2. Hoffmann C, Rockstroh JK. HIV 2015/16. Hamburg: Medizin Fokus Verlag; 2015.
3. Al-Dakkak I, Patel S, McCann E, Gadkari A, Prajapati G, Maiese EM. The impact of specific HIV treatment-related adverse events on adherence to antiretroviral therapy: a systematic review and meta-analysis. AIDS Care. 2013;25:400–14.
4. Nakagawa F, Miners A, Smith CJ, Simmons R, Lodwick RK, Cambiano V, Lundgren JD, Delpech V, Phillips AN. Projected lifetime healthcare costs associated with HIV infection. PLoS ONE. 2015;10:e0125018.
5. Chun TW, Moir S, Fauci AS. HIV reservoirs as obstacles and opportunities for an HIV cure. Nat Immunol. 2015;16:584–9.
6. Ruelas DS, Greene WC. An integrated overview of HIV-1 latency. Cell. 2013;155:519–29.
7. Martin AR, Siliciano RF. Progress toward HIV eradication: case reports, current efforts, and the challenges associated with cure. Annu Rev Med. 2016;67:215–28.
8. Chun TW, Finzi D, Margolick J, Chadwick K, Schwartz D, Siliciano RF. In vivo fate of HIV-1-infected T cells: quantitative analysis of the transition to stable latency. Nat Med. 1995;1:1284–90.
9. Mbonye U, Karn J. Transcriptional control of HIV latency: cellular signaling pathways, epigenetics, happenstance and the hope for a cure. Virology. 2014;454–455:328–39.
10. Cary DC, Fujinaga K, Peterlin BM. Molecular mechanisms of HIV latency. J Clin Invest. 2016;126:448–54.
11. Chun TW, Justement JS, Lempicki RA, Yang J, Dennis G Jr, Hallahan CW, Sanford C, Pandya P, Liu S, McLaughlin M, et al. Gene expression and viral prodution in latently infected, resting CD4+ T cells in viremic versus aviremic HIV-infected individuals. Proc Natl Acad Sci USA. 2003;100:1908–13.
12. Sarracino A, Marcello A. The relevance of post-transcriptional mechanisms in HIV latency reversal. Curr Pharm Des. 2017;23:4103–11.
13. Lassen KG, Ramyar KX, Bailey JR, Zhou Y, Siliciano RF. Nuclear retention of multiply spliced HIV-1 RNA in resting CD4+ T cells. PLoS Pathog. 2006;2:e68.
14. Swiggard WJ, Baytop C, Yu JJ, Dai J, Li C, Schretzenmair R, Theodosopoulos T, O'Doherty U. Human immunodeficiency virus type 1 can establish latent infection in resting CD4+ T cells in the absence of activating stimuli. J Virol. 2005;79:14179–88.
15. Saleh S, Solomon A, Wightman F, Xhilaga M, Cameron PU, Lewin SR. CCR7 ligands CCL19 and CCL21 increase permissiveness of resting memory CD4+ T cells to HIV-1 infection: a novel model of HIV-1 latency. Blood. 2007;110:4161–4.
16. Saleh S, Wightman F, Ramanayake S, Alexander M, Kumar N, Khoury G, Pereira C, Purcell D, Cameron PU, Lewin SR. Expression and reactivation of HIV in a chemokine induced model of HIV latency in primary resting CD4+ T cells. Retrovirology. 2011;8:80.
17. Grau-Expósito J, Serra-Peinado C, Miguel L, Navarro J, Curran A, Burgos J, Ocaña I, Ribera E, Torrella A, Planas B, et al. A novel single-cell fish-flow assay identifies effector memory CD4+ T cells as a major niche for HIV-1 transcription in HIV-infected patients. MBio. 2017. https://doi.org/10.1128/mBio.00876-17.
18. Sun B, Yang R, Mallardo M. Roles of microRNAs in HIV-1 replication and latency. Microrna. 2016;5:120–3.
19. Kurosaki T, Maquat LE. Nonsense-mediated mRNA decay in humans at a glance. J Cell Sci. 2016;129:461–7.
20. Bhattacharya A, Czaplinski K, Trifillis P, He F, Jacobson A, Peltz SW. Characterization of the biochemical properties of the human Upf1 gene product that is involved in nonsense-mediated mRNA decay. RNA. 2000;6:1226–35.
21. Chawla R, Redon S, Raftopoulou C, Wischnewski H, Gagos S, Azzalin CM. Human UPF1 interacts with TPP1 and telomerase and sustains telomere leading-strand replication. EMBO J. 2011;30:4047–58.
22. Azzalin CM, Lingner J. The human RNA surveillance factor UPF1 is required for S phase progression and genome stability. Curr Biol. 2006;16:433–9.
23. Kim YK, Furic L, Desgroseillers L, Maquat LE. Mammalian Staufen1 recruits Upf1 to specific mRNA 3′UTRs so as to elicit mRNA decay. Cell. 2005;120:195–208.
24. Ciaudo C, Bourdet A, Cohen-Tannoudji M, Dietz HC, Rougeulle C, Avner P. Nuclear mRNA degradation pathway(s) are implicated in Xist regulation and X chromosome inactivation. PLoS Genet. 2006;2:e94.
25. Maekawa S, Imamachi N, Irie T, Tani H, Matsumoto K, Mizutani R, Imamura K, Kakeda M, Yada T, Sugano S, et al. Analysis of RNA decay factor mediated RNA stability contributions on RNA abundance. BMC Genom. 2015;16:154.
26. Isken O, Maquat LE. The multiple lives of NMD factors: balancing roles in gene and genome regulation. Nat Rev Genet. 2008;9:699–712.
27. Fatscher T, Boehm V, Gehring NH. Mechanism, factors, and physiological role of nonsense-mediated mRNA decay. Cell Mol Life Sci. 2015;72:4523–44.
28. Nickless A, Bailis JM, You Z. Control of gene expression through the nonsense-mediated RNA decay pathway. Cell Biosci. 2017;7:26.
29. McIlwain DR, Pan Q, Reilly PT, Elia AJ, McCracken S, Wakeham AC, Itie-Youten A, Blencowe BJ, Mak TW. Smg1 is required for embryogenesis and regulates diverse genes via alternative splicing coupled to nonsense-mediated mRNA decay. Proc Natl Acad Sci USA. 2010;107:12186–91.
30. Lykke-Andersen S, Jensen TH. Nonsense-mediated mRNA decay: an intricate machinery that shapes transcriptomes. Nat Rev Mol Cell Biol. 2015;16:665–77.
31. Kebaara BW, Atkin AL. Long 3′-UTRs target wild-type mRNAs for nonsense-mediated mRNA decay in Saccharomyces cerevisiae. Nucleic Acids Res. 2009;37:2771–8.
32. Balistreri G, Bognanni C, Muhlemann O. Virus escape and manipulation of cellular nonsense-mediated mRNA decay. Viruses. 2017;9:24.
33. Withers JB, Beemon KL. The structure and function of the rous sarcoma virus RNA stability element. J Cell Biochem. 2011;112:3085–92.
34. Mocquet V, Durand S, Jalinot P. How retroviruses escape the nonsense-mediated mRNA decay. AIDS Res Hum Retrovir. 2015;31:948–58.
35. Toro-Ascuy D, Rojas-Araya B, Valiente-Echeverria F, Soto-Rifo R. Interactions between the HIV-1 unspliced mRNA and host mRNA decay machineries. Viruses. 2016;8:320.
36. Ajamian L, Abrahamyan L, Milev M, Ivanov PV, Kulozik AE, Gehring NH, Mouland AJ. Unexpected roles for UPF1 in HIV-1 RNA metabolism and translation. RNA. 2008;14:914–27.
37. Ajamian L, Abel K, Rao S, Vyboh K, Garcia-de-Gracia F, Soto-Rifo R, Kulozik AE, Gehring NH, Mouland AJ. HIV-1 recruits UPF1 but excludes UPF2 to promote nucleocytoplasmic export of the genomic RNA. Biomolecules. 2015;5:2808–39.

38. Kula A, Guerra J, Knezevich A, Kleva D, Myers MP, Marcello A. Characterization of the HIV-1 RNA associated proteome identifies Matrin 3 as a nuclear cofactor of Rev function. Retrovirology. 2011;8:60.
39. Hogg JR, Goff SP. Upf1 senses 3′UTR length to potentiate mRNA decay. Cell. 2010;143:379–89.
40. Serquina AK, Das SR, Popova E, Ojelabi OA, Roy CK, Gottlinger HG. UPF1 is crucial for the infectivity of human immunodeficiency virus type 1 progeny virions. J Virol. 2013;87:8853–61.
41. Abrahamyan LG, Chatel-Chaix L, Ajamian L, Milev MP, Monette A, Clement JF, Song R, Lehmann M, DesGroseillers L, Laughrea M, et al. Novel Staufen1 ribonucleoproteins prevent formation of stress granules but favour encapsidation of HIV-1 genomic RNA. J Cell Sci. 2010;123:369–83.
42. Baxter AE, Niessl J, Fromentin R, Richard J, Porichis F, Charlebois R, Massanella M, Brassard N, Alsahafi N, Delgado GG, et al. Single-cell characterization of viral translation-competent reservoirs in HIV-infected individuals. Cell Host Microbe. 2016;20:368–80.
43. Martrus G, Niehrs A, Cornelis R, Rechtien A, Garcia-Beltran W, Lutgehetmann M, Hoffmann C, Altfeld M. Kinetics of HIV-1 latency reversal quantified on the single-cell level using a novel flow-based technique. J Virol. 2016;90:9018–28.
44. Baxter AE, Niessl J, Morou A, Kaufmann DE. RNA flow cytometric FISH for investigations into HIV immunology, vaccination and cure strategies. AIDS Res Ther. 2017;14:40.
45. Prasad VR, Kalpana GV. FISHing out the hidden enemy: advances in detecting and measuring latent HIV-infected cells. MBio. 2017;8:e01433.
46. Baxter AE, O'Doherty U, Kaufmann DE. Beyond the replication-competent HIV reservoir: transcription and translation-competent reservoirs. Retrovirology. 2018;15:18.
47. Planelles V, Wolschendorf F, Kutsch O. Facts and fiction: cellular models for high throughput screening for HIV-1 reactivating drugs. Curr HIV Res. 2011;9:568–78.
48. Spina CA, Anderson J, Archin NM, Bosque A, Chan J, Famiglietti M, Greene WC, Kashuba A, Lewin SR, Margolis DM, et al. An in-depth comparison of latent HIV-1 reactivation in multiple cell model systems and resting CD4+ T cells from aviremic patients. PLoS Pathog. 2013;9:e1003834.
49. Brogdon J, Ziani W, Wang X, Veazey RS, Xu H. In vitro effects of the small-molecule protein kinase C agonists on HIV latency reactivation. Sci Rep. 2016;6:39032.
50. Kadlec J, Guilligay D, Ravelli RB, Cusack S. Crystal structure of the UPF2-interacting domain of nonsense-mediated mRNA decay factor UPF1. RNA (New York, NY). 2006;12:1817–24.
51. Clerici M, Mourao A, Gutsche I, Gehring NH, Hentze MW, Kulozik A, Kadlec J, Sattler M, Cusack S. Unusual bipartite mode of interaction between the nonsense-mediated decay factors, UPF1 and UPF2. EMBO J. 2009;28:2293–306.
52. Eberle AB, Lykke-Andersen S, Muhlemann O, Jensen TH. SMG6 promotes endonucleolytic cleavage of nonsense mRNA in human cells. Nat Struct Mol Biol. 2009;16:49–55.
53. Huntzinger E, Kashima I, Fauser M, Sauliere J, Izaurralde E. SMG6 is the catalytic endonuclease that cleaves mRNAs containing nonsense codons in metazoan. RNA. 2008;14:2609–17.
54. Kashima I, Jonas S, Jayachandran U, Buchwald G, Conti E, Lupas AN, Izaurralde E. SMG6 interacts with the exon junction complex via two conserved EJC-binding motifs (EBMs) required for nonsense-mediated mRNA decay. Genes Dev. 2010;24:2440–50.
55. Okada-Katsuhata Y, Yamashita A, Kutsuzawa K, Izumi N, Hirahara F, Ohno S. N- and C-terminal Upf1 phosphorylations create binding platforms for SMG-6 and SMG-5:SMG-7 during NMD. Nucleic Acids Res. 2012;40:1251–66.
56. Glavan F, Behm-Ansmant I, Izaurralde E, Conti E. Structures of the PIN domains of SMG6 and SMG5 reveal a nuclease within the mRNA surveillance complex. EMBO J. 2006;25:5117–25.
57. Takeshita D, Zenno S, Lee WC, Saigo K, Tanokura M. Crystallization and preliminary X-ray analysis of the PIN domain of human EST1A. Acta Crystallogr, Sect F: Struct Biol Cryst Commun. 2006;62:656–8.
58. Takeshita D, Zenno S, Lee WC, Saigo K, Tanokura M. Crystal structure of the PIN domain of human telomerase-associated protein EST1A. Proteins. 2007;68:980–9.
59. Pasternak AO, Lukashov VV, Berkhout B. Cell-associated HIV RNA: a dynamic biomarker of viral persistence. Retrovirology. 2013;10:41.
60. Pasternak AO, Berkhout B. What do we measure when we measure cell-associated HIV RNA. Retrovirology. 2018;15:13.
61. Lassen KG, Bailey JR, Siliciano RF. Analysis of human immunodeficiency virus type 1 transcriptional elongation in resting CD4+ T cells in vivo. J Virol. 2004;78:9105–14.
62. Pace MJ, Graf EH, Agosto LM, Mexas AM, Male F, Brady T, Bushman FD, O'Doherty U. Directly infected resting CD4+ T cells can produce HIV Gag without spreading infection in a model of HIV latency. PLoS Pathog. 2012;8:e1002818.
63. Le Douce V, Janossy A, Hallay H, Ali S, Riclet R, Rohr O, Schwartz C. Achieving a cure for HIV infection: Do we have reasons to be optimistic? J Antimicrob Chemother. 2012;67:1063–74.
64. Deeks SG, Lewin SR, Ross AL, Ananworanich J, Benkirane M, Cannon P, Chomont N, Douek D, Lifson JD, Lo YR, et al. International AIDS society global scientific strategy: towards an HIV cure 2016. Nat Med. 2016;22:839–50.
65. Schwartz C, Bouchat S, Marban C, Gautier V, Van Lint C, Rohr O, Le Douce V. On the way to find a cure: purging latent HIV-1 reservoirs. BCP Biochem Pharmacol. 2017;146:10–22.
66. Deeks SG. HIV: shock and kill. Nature. 2012;487:439–40.
67. Delagreverie HM, Delaugerre C, Lewin SR, Deeks SG, Li JZ. Ongoing clinical trials of human immunodeficiency virus latency-reversing and immunomodulatory agents. Open Forum Infect Dis. 2016;3:ofw189.
68. Darcis G, Van Driessche B, Van Lint C. Preclinical shock strategies to reactivate latent HIV-1: an update. Curr Opin HIV AIDS. 2016;11:388–93.
69. Leth S, Schleimann MH, Nissen SK, Hojen JF, Olesen R, Graversen ME, Jorgensen S, Kjaer AS, Denton PW, Mork A, et al. Combined effect of Vacc-4x, recombinant human granulocyte macrophage colony-stimulating factor vaccination, and romidepsin on the HIV-1 reservoir (REDUC): a single-arm, phase 1B/2A trial. Lancet HIV. 2016;3:e463–72.
70. Guihot A, Marcelin AG, Massiani MA, Samri A, Soulie C, Autran B, Spano JP. Drastic decrease of the HIV reservoir in a patient treated with nivolumab for lung cancer. Ann Oncol. 2018;29:517–8.
71. Cheng Z, Muhlrad D, Lim MK, Parker R, Song H. Structural and functional insights into the human Upf1 helicase core. EMBO J. 2007;26:253–64.
72. Andrew JM, Eric AC, Luc D. Trafficking of HIV-1 RNA: recent progress involving host cell RNABinding proteins. Curr Genomics. 2003;4:237–51.
73. Schwartz S, Felber BK, Pavlakis GN. Distinct RNA sequences in the gag region of human immunodeficiency virus type 1 decrease RNA stability and inhibit expression in the absence of Rev protein. J Virol. 1992;66:150–9.
74. Applequist SE, Selg M, Raman C, Jack HM. Cloning and characterization of HUPF1, a human homolog of the Saccharomyces cerevisiae nonsense mRNA-reducing UPF1 protein. Nucleic Acids Res. 1997;25:814–21.
75. Chakrabarti S, Jayachandran U, Bonneau F, Fiorini F, Basquin C, Domcke S, Le Hir H, Conti E. Molecular mechanisms for the RNA-dependent ATPase activity of Upf1 and its regulation by Upf2. Mol Cell. 2011;41:693–703.
76. Chamieh H, Ballut L, Bonneau F, Le Hir H. NMD factors UPF2 and UPF3 bridge UPF1 to the exon junction complex and stimulate its RNA helicase activity. Nat Struct Mol Biol. 2008;15:85–93.
77. Ohnishi T, Yamashita A, Kashima I, Schell T, Anders KR, Grimson A, Hachiya T, Hentze MW, Anderson P, Ohno S. Phosphorylation of hUPF1 induces formation of mRNA surveillance complexes containing hSMG-5 and hSMG-7. Mol Cell. 2003;12:1187–200.
78. Besnard E, Hakre S, Kampmann M, Lim HW, Hosmane NN, Martin A, Bassik MC, Verschueren E, Battivelli E, Chan J, et al. The mTOR complex controls HIV latency. Cell Host Microbe. 2016;20:785–97.
79. de Turris V, Nicholson P, Orozco RZ, Singer RH, Muhlemann O. Cotranscriptional effect of a premature termination codon revealed by live-cell imaging. RNA. 2011;17:2094–107.
80. Hurt JA, Robertson AD, Burge CB. Global analyses of UPF1 binding and function reveal expanded scope of nonsense-mediated mRNA decay. Genome Res. 2013;23:1636–50.
81. Kurosaki T, Maquat LE. Rules that govern UPF1 binding to mRNA 3′ UTRs. Proc Natl Acad Sci USA. 2013;110:3357–62.
82. Zund D, Gruber AR, Zavolan M, Muhlemann O. Translation-dependent displacement of UPF1 from coding sequences causes its enrichment in 3′ UTRs. Nat Struct Mol Biol. 2013;20:936–43.
83. Lee SR, Pratt GA, Martinez FJ, Yeo GW, Lykke-Andersen J. Target discrimination in nonsense-mediated mRNA decay requires Upf1 ATPase activity. Mol Cell. 2015;59:413–25.
84. Kurosaki T, Li W, Hoque M, Popp MW, Ermolenko DN, Tian B, Maquat LE. A post-translational regulatory switch on UPF1 controls targeted mRNA degradation. Genes Dev. 2014;28:1900–16.

85. Kashima I, Yamashita A, Izumi N, Kataoka N, Morishita R, Hoshino S, Ohno M, Dreyfuss G, Ohno S. Binding of a novel SMG-1-Upf1-eRF1-eRF3 complex (SURF) to the exon junction complex triggers Upf1 phosphorylation and nonsense-mediated mRNA decay. Genes Dev. 2006;20:355–67.
86. Yamashita A, Izumi N, Kashima I, Ohnishi T, Saari B, Katsuhata Y, Muramatsu R, Morita T, Iwamatsu A, Hachiya T, et al. SMG-8 and SMG-9, two novel subunits of the SMG-1 complex, regulate remodeling of the mRNA surveillance complex during nonsense-mediated mRNA decay. Genes Dev. 2009;23:1091–105.
87. Durand S, Franks TM, Lykke-Andersen J. Hyperphosphorylation amplifies UPF1 activity to resolve stalls in nonsense-mediated mRNA decay. Nat Commun. 2016;7:12434.
88. Jordan A, Bisgrove D, Verdin E. HIV reproducibly establishes a latent infection after acute infection of T cells in vitro. EMBO J. 2003;22:1868–77.
89. Serin G, Gersappe A, Black JD, Aronoff R, Maquat LE. Identification and characterization of human orthologues to *Saccharomyces cerevisiae* Upf2 protein and Upf3 protein (*Caenorhabditis elegans* SMG-4). Mol Cell Biol. 2001;21:209–23.
90. Nicholson P, Josi C, Kurosawa H, Yamashita A, Muhlemann O. A novel phosphorylation-independent interaction between SMG6 and UPF1 is essential for human NMD. Nucleic Acids Res. 2014;42:9217–35.
91. Xing L, Wang S, Hu Q, Li J, Zeng Y. Comparison of three quantification methods for the TZM-bl pseudovirus assay for screening of anti-HIV-1 agents. J Virol Methods. 2016;233:56–61.
92. Baxter AE, Niessl J, Fromentin R, Richard J, Porichis F, Massanella M, Brassard N, Alsahafi N, Routy JP, Finzi A, et al. Multiparametric characterization of rare HIV-infected cells using an RNA-flow FISH technique. Nat Protoc. 2017;12:2029–49.
93. Wong R, Balachandran A, Mao AY, Dobson W, Gray-Owen S, Cochrane A. Differential effect of CLK SR Kinases on HIV-1 gene expression: potential novel targets for therapy. Retrovirology. 2011;8:47.
94. Duffy S, Cochrane A. Analysis of HIV-1 RNA Splicing. In: Alternative pre-mRNA splicing. Wiley-VCH Verlag GmbH & Co. KGaA; 2012. pp. 438–48.
95. Suzuki K, Bose P, Leong-Quong RY, Fujita DJ, Riabowol K. REAP: a two minute cell fractionation method. BMC Res Notes. 2010;3:294.

8

HIV-1 protease with leucine zipper fused at N-terminus exhibits enhanced linker amino acid-dependent activity

Fu-Hsien Yu and Chin-Tien Wang*

Abstract

Background: HIV-1 protease (PR) activation is triggered by Gag-Pol dimerization. Premature PR activation results in reduced virion yields due to enhanced Gag cleavage. A p6* transframe peptide located directly upstream of protease is believed to play a modulating role in PR activation. Previous reports indicate that the C-terminal p6* tetra-peptide prevents premature PR activation triggered by a leucine zipper (LZ) dimerization motif inserted in the deleted p6* region. To clarify the involvement of C-terminal p6* residues in mitigating enhanced LZ-incurred Gag processing, we engineered constructs containing C-terminal p6* residue substitutions with and without a mutation blocking the p6*/PR cleavage site, and created other Gag or p6* domain-removing constructs. The capabilities of these constructs to mediate virus maturation were assessed by Western blotting and single-cycle infection assays.

Results: p6*-PR cleavage blocking did not significantly reduce the LZ enhancement effect on Gag cleavage when only four amino acid residues were present between the p6* and PR. This suggests that the potent LZ dimerization motif may enhance PR activation by facilitating PR dimer formation, and that PR precursors may trigger sufficient enzymatic activity without breaking off from the PR N-terminus. Enhanced LZ-induced activation of PR embedded in Gag-Pol was found to be independent of the Gag assembly domain. In contrast, the LZ enhancement effect was markedly reduced when six amino acids were present at the p6*-PR junction, in part due to impaired PR maturation by substitution mutations. We also observed that a proline substitution at the P3 position eliminated the ability of p6*-deleted Gag-Pol to mediate virus maturation, thus emphasizing the importance of C-terminal p6* residues to modulating PR activation.

Conclusions: The ability of HIV-1 C-terminal p6* amino acid residues to modulate PR activation contributes, at least in part, to their ability to counteract enhanced Gag cleavage induced by a leucine zipper substituted for a deleted p6*. Changes in C-terminal p6* residues between LZ and PR may affect PR-mediated virus maturation, thus providing a possible method for assessing HIV-1 protease precursor activation in the context of virus assembly.

Keywords: HIV-1, Gag-Pol, P6pol, Protease maturation, Virus maturation, Gag cleavage

Background

The HIV-1 retrovirus contains three major genes (gag, pol and env) and several accessory genes [1]. HIV-1 pol encodes viral enzymes such as protease (PR), reverse transcriptase (RT) and integrase (IN), while gag encodes viral structural proteins. Both Pol and Gag are translated from the same mRNA template. Pol is translated as a Gag-Pol fusion protein associated with a ribosome shift during Gag translation that occurs at a 5% frequency, leading to a Gag-Pol versus Gag expression ratio of approximately 1:20 [2]. Pr160gag-pol and Pr55gag are transported to plasma membranes, where Pr55gag molecules assemble into virus particles [3]; Pr160gag-pol is incorporated into these particles via Pr55gag interaction [4–7]. During or after virus budding, activated PR autocleaves from Gag-Pol and mediates virus maturation

*Correspondence: chintien@ym.edu.tw
Department of Medical Research, Taipei Veterans General Hospital and Institute of Clinical Medicine, National Yang-Ming University School of Medicine, 201, Sec. 2, Shih-Pai Road, Taipei 11217, Taiwan

through the proteolytic processing of Pr55gag and Pr160gag-pol [8]. Pr55gag cleavage yields four main products: matrix (p17; MA), capsid (CA; p24), nucleocapsid (NC; p7) and C-terminal p6 [9]. Two spacer peptides—SP1 (or p2) and SP2 (or p1)—respectively separate NC from CA and p6. Pr160gag-pol cleavage generates RT and IN in addition to the Gag proteins MA, CA and NC. PR-mediated virus maturation is necessary for viral infectivity acquisition [10, 11].

It is generally believed that Gag-Pol dimerization triggers PR activation [3]. In agreement with this assumption, mutations upstream or downstream of PR may significantly reduce PR-mediated Gag cleavage efficiency due to inadequate Gag-Pol dimerization [12–16]. In contrast, the promotion of PR activation as a result of enhanced Gag-Pol dimerization or Gag-PR dimer interaction likely triggers premature or enhanced Pr55gag cleavage, resulting in markedly reduced virus production. [17–19]. Accordingly, preventing premature PR activation is central to virus assembly.

Within Gag-Pol, truncated p1-p6gag is replaced with a transframe region referred to as p6* or p6pol. Located directly adjacent to the PR N-terminus, p6* has been described as playing a role in modulating PR activation even though it lacks a specific structure. Mature HIV-1 PR has a dimeric form, and the removal of p6* from PR precursor is essential for PR to be fully functional [20–23]. Mutations that block p6*-PR cleavage markedly impede PR-mediated virus maturation, implying a suppressive effect of p6* on PR activation [22, 23]. Molecular models suggest that p6* may prevent early PR maturation by inducing instability in the folded PR dimer structure [24–27]. However, virus assembly-associated evidence in support of this assumption is limited, since deletion analysis of p6* function has the potential to compromise virus assembly due to an overlap of p6* with the p6gag budding domain.

To investigate p6* function without affecting the p6gag coding region, we engineered an HIV-1 virus-producing vector by placing the pol coding sequence at the PR-inactivated C-terminus. This construct, designated Dp6*PR, was capable of assembling and processing virus particles in a manner similar to that of wild-type (wt). Replacement of p6* with a leucine-zipper (LZ) dimerization domain has been shown to eliminate virus production as a result of enhanced Gag cleavage, but as few as four C-terminal p6 residues remaining between LZ and PR significantly counteract the LZ enhancement effect [28]. These observations provide supporting evidence that p6* may contribute to the prevention of premature PR activation, but it remains unknown whether a correlation exists between the ability of C-terminal p6* residues to counteract LZ enhancement and their ability to modulate PR maturation. Further, it is unknown whether specific amino acids must be present between LZ and PR in order to counteract the LZ enhancement effect.

To study these questions, we engineered multiple constructs to further analyze the role of C-terminal p6* residues in counteracting LZ enhancement and modulating PR activation. Our results indicate that an HIV-1 PR precursor containing a leucine zipper (LZ) motif linked at the N-terminus eliminated virus particle production associated with enhanced Gag cleavage, suggesting that the HIV-1 PR precursor is capable of exhibiting enhanced enzymatic activity. C-terminal p6* residue substitutions can subvert the Gag cleavage enhancement effect induced by a LZ substitution for p6*, likely the result of interference with PR maturation. While p6*-deleted Gag-Pol (Δp6*fs) containing the last two remaining C-terminal p6* residues was still capable of producing infectious virions following co-expression with Pr55gag, a single amino acid residue change at the deleted p6* region completely removed the ability of Δp6*fs to mediate virus maturation. Our results confirm the importance of C-terminal p6* residues for the spatiotemporal modulation of PR activity, and provide a virus assembly system for studying HIV-1 protease precursor activation by manipulating linker residues between fused peptides and PR.

Methods

Plasmid construction

The parental HIV-1 proviral sequence in this study is HXB2 [29]. The HIV-1 proviral plasmid HIVgpt is considered the backbone of all expression constructs [30]. The constructs used in this research were mostly derived from Dp6*PR, DPR, DWzPR, DWz/PR and DWz//PR. As described previously [31], Dp6*PR contains p6* domain between an inactivated and an active PR. DPR contains BamHI-linked duplicate PR pairs, with the proximal PR was inactivated. DWzPR, DWz/PR and DWz//PR contain leucine zipper replacements of p6* with two, four and six C-terminal p6* residues remaining in the p6*/PR junction, respectively [28]. PSHL and PIDL substitutions for the four C-terminal p6* residues in DWz/PR yielded DWz/PSHL/PR and DWz/PIDL/PR [28]. Primers used for engineering the designated mutations were listed in Table 1.

V/P, as described previously was created by changing amino acid residues at p6*-PR cleavage site from Phe/Pro into Val/Pro [32]. BamHI-containing forward primers were used to amplify the p6* C-terminal coding fragment containing the desired mutation DWzPRV/P, DWz/PRV/P, DWz//PRV/P or DWz/PANF/PR (Table 1). V/P or HIVgpt served as a template and the reverse primer (nt.3116-90) sequence was 5′-TACATACAAATCATCCATGTTATTGATA-3′. Amplified fragments

Table 1 Primer sequences used for plasmid construction

Constructs	Forward primer (5′–3′)[a]
DWzPR	5′ CTGTGGATCCT**AACTTC**CCTCAGGTAACGTTATGGCAA 3′-nt 2273
DWz/PR	5′ CGGGATCCT**TCCTTTAACTTC**CCTCAGGTCACGTTATGG 3′-nt 2270
DWz//PR	5′ CGGGATCCT**ACTGTATCCTTTAACTTC**CCTCAGGTCACGTTATGG 3′-nt 2270
DWzPRV/P	5′ CGGGATCCT**AACGTT**CCTCAGATCACGTTATGG 3′-nt 2270
DWz/PRV/P	5′ CGGGATCCT**TCCTTTAACGTT**CCTCAGATCACGTTATGG 3′-nt 2270
DWz//PRV/P	5′ CGGGATCCT**ACTGTATCCTTTAACGTT**CCTCAGATCACGTTATGG 3′-nt 2270
DWz/PSHL/PR	5′ CGGGATCCT**CCCTCTCACCTC**CCTCAGGTCACTCTTTGG 3′-nt 2270
DWz/PIDL/PR	5′ CGGGATCCT**CCCATTGACCTC**CCTCAGGTCACTCTTTGG 3′-nt 2270
DWz/PANF/PR	5′ CGGGATCCT**CCCGCTAACTTC**CCTCAGGTCACTCTTTGG 3′-nt 2270
p6*PSHL	nt 2221-5′ CCGATCGACAAGGAACTGTA**CCCTCTCACCTC**CCTCAG 3′-nt 2258
p6*PIDL	nt 2221-5′ CCGATCGACAAGGAACTGTA**CCCATTGACCTC**CCTCAG 3′-nt 2258
p6*PANF	nt 2221-5′ CCGATCGACAAGGAACTGTA**CCCGCTAACTT**CC 3′-nt 2254

[a] The numbers at the 3′ and/or 5′ ends denote HIV-1 proviral DNA nucleotide positions. Nucleotides corresponding to mutated amino acid residues are shown in boldface. Most of the primers contain BamHI sites (underlined) to facilitate cloning

were digested with BamHI and EcoRV, and subcloned into pBRClaSal/DPR [28]. BamHI-flanking leucine zipper coding fragments derived from PRWzPR [28] were inserted into each pBRClaSal/DPR recombinant, yielding DWzPRV/P, DWz/PRV/P, DWz//PRV/P and DWz/PANF/PR, respectively.

p6*PSHL, p6*PIDL and p6*PANF were constructed by megaprimer PCR method [33] using a forward primer containing the desired mutation (Table 1) and a reverse primer 5′-GGTACAGTCTCAATAGGGCTAATG-3. HIVgpt serves as a template. The amplified fragments were digested with ApaI and BclI, and subcloned into a plasmid cassette pBRCla-Sal that contains HIV-1 coding sequence (from ClaI-nt.831 to SalI-nt.5786). Each mutation-containing pBRCla-Sal cassette was then digested with SpeI and SalI, and ligated into HIVgpt, yielding p6*PSHL, p6*PIDL and p6*PANF.

p6*PSHL, p6*PIDL and p6*PANF were digested with BglII and EcoRV and ligated into DPR digested with BamHI and EcoRV, yielding Dp6*PSHL, Dp6*PIDL/PR and Dp6*PANF/PR, respectively.

GPfs has the Gag and Pol in the same reading frame due to a deletion of the frame shift signal [34]. DWzPR and DWz/PR were digested with BclI. The BclI-flanking fragment containing WzPR and Wz/PR mutations were ligated into PR-inactivated GPfs (fsd) digested with BclI, yielding fsdWzPR and fsdWz/PR. Recombination of fsdWzPR and a Gag-deleted fsd [34] generated construct fsdWzPRΔGag.

Cell culture and transfection

293T and HeLa cells were maintained in DMEM supplemented with 10% fetal calf serum. Confluent 293T cells were trypsinized, split 1:10 and seeded onto 10-cm plates 18–24 h before transfection. For each construct, 293T cells were transfected with 20 μg of plasmid DNA by the calcium phosphate precipitation method, with the addition of 50 μM chloroquine to enhance transfection efficiency. Culture media and cells were harvested for protein analysis at 48–72 h post-transfection. When pGAG was co-transfected with the Gag-Pol expression constructs at a DNA ratio of 1:1 or 10 to 1, 10 or 15 μg of pGAG were used with the addition of pBlueScript plasmid DNA to a final quantity of 20 μg DNA. The cells and media were harvested for protein analysis 48–72 h post-transfection.

Single-cycle infection assays

293T cells were either co-transfected with 10 μg wt or each of the mutant HIVgpt plus 5 μg of the VSV-G protein expression plasmid pHCMV-G [35], or co-transfected with 1 μg of the GPfs or the p6*-deleted GPfs plasmid with 10 μg pGAG plus 5 μg pHCMV-G. At 48 h after transfection, virus-containing supernatants were collected, filtered, diluted, and used to infect HeLa cells. Aliquots of the same filtered supernatants and cell samples were prepared and subjected to Western blot. Adsorption of virions is allowed to proceed in the presence of 4 μg/ml polybrene. Twenty-four hours after infection, cells were trypsinized, split into dishes, and refed with medium containing drug selection cocktail [36]. Selected drug resistant colonies were fixed and stained with 50% methanol containing 0.5% methylene blue. Numbers of drug-resistant colonies were converted into titers (cfu/ml). Infectivity was expressed as the ratio of the mutant titer to the titer of wt, and normalized to Gag protein levels in parallel experiments.

Western immunoblot analysis

Culture media from transfected 293T cells were filtered through 0.45-μm pore-size and then centrifuged through 2 ml 20% sucrose in TSE (10 mM Tris–HCl, pH 7.5, 100 mM NaCl, 1 mM EDTA) containing 0.1 mM phenylmethylsulfonyl fluoride (PMSF) at 4 °C for 40 min at 274,000 × *g*. Viral pellets and cell lysates mixed with sample buffer were then subjected to SDS-10% PAGE or 4–12% Bis–Tris gradient gels (NuPage Bis–Tris Mini Gels; Thermo Fisher Scientific) followed by immunoblotting analysis as previously described [37]. HIV-1 Gag proteins were probed with an anti-p24gag monoclonal antibody (mouse hybridoma clone 183-H12-5C) from ascites. For HIV-1 RT detection, the primary antibody was rabbit antiserum or a mouse anti-RT monoclonal antibody [38, 39]. Cellular β-actin was detected using a mouse anti-β-actin monoclonal antibody (Sigma). The secondary antibody was either a sheep anti-mouse or a donkey anti-rabbit horseradish peroxidase (HRP)-conjugated antibody (Jackson ImmunoResearch). An enhanced chemiluminescence (ECL) detection system (SuperSignal West Pico Chemiluminescent Substrate; Thermo Fisher Scientific) was used to detect membrane-bound proteins.

Statistical analysis

Differences between control (wt) and experimental (mutant) groups were assessed using Student's t-tests. Data are expressed as mean ± standard deviation. Significance was defined as $^{*}p < 0.05$, $^{**}p < 0.01$, $^{***}p < 0.001$.

Results

p6*-PR cleavage blocking does not significantly mitigate leucine zipper-induced Gag cleavage enhancement

In a previous study we reported that an HIV-1 mutant (DWzPR) containing a LZ dimerization motif adjacent to and upstream of PR was not capable of producing virions due to the strong enhancement of Gag cleavage [28]. DWzPR is derived from Dp6*PR by replacing a deleted p6* with LZ, but retaining the last two C-terminal p6* residues at the LZ/PR junction (Fig. 1a). We observed that this LZ replacement of p6* led to the elimination of virus assembly, likely due to premature PR activation triggered by the LZ. Based on our prior finding that the DWzPR virus assembly defect is PR activity-dependent, and since p6*-PR cleavage is required for fully active PR, we postulated that blocking p6*-PR cleavage within the LZ/PR junction might help restore DWzPR-associated virus production by reducing PR activity. To test this possibility, we substituted Val for the last C-terminal p6* residue (Phe) at the LZ/PR junction of DWzPR and designated the resulting construct as DWzPRV/P (Fig. 1a; note that p6*-PR cleavage site residues were changed from F/P to V/P). Our data indicate that DWzPR virus yields were still hardly detected following the Val substitution unless it was accompanied by treatment with an HIV-1 PR inhibitor (Fig. 1b middle panel, lane 4 vs. lane 5). These results suggest that (a) the blocking of p6*-PR cleavage exerted no major impacts on LZ-induced Gag cleavage enhancement, and (b) a HIV-1 PR precursor containing a LZ motif fused at the N-terminus exhibited enhanced enzymatic activity.

As a control, Val substitution for Phe (referred to as a V/P mutation) significantly reduced virus processing efficiency when tested in a wild-type (wt) HIV-1 Gag/Gag-Pol expression vector (Fig. 1c, middle panel, lanes 2 and 6). Constructs with either four or six remaining C-terminal p6* residues at the LZ/PR junction (DWz/PR or DWz//PR) exhibited particle assembly and processing profiles similar to that of the wt (Fig. 1c middle panel, lanes 4 and 5). While the V/P mutation exerted no major impacts on wt virus production, it significantly reduced DWz/PR and DWz//PR virus yields in addition to impairing virus maturation (Fig. 1c middle panel lanes 4–5 vs. 8–9 and panels d and e). The capacity of the C-terminal p6* tetra-peptide to counteract the LZ enhancement effect and modulate PR activation was subject to weakening by the V/P mutation. This may account, at least in part, for the decreased virus assembly and processing efficiency of DWz/PRV/P and DWz//PRV/P (Fig. 1c middle panel, lanes 8 and 9).

Specific C-terminal p6* residues are required to modulate PR maturation

Our results support the proposal that the C-terminal p6* tetra-peptide plays a central role in mitigating PR maturation, likely due to the absence of an intact C-terminal p6* tetra-peptide, which lets LZ dictate the PR maturation process and trigger premature PR activation. To test this hypothesis, we inserted substitution mutations at the C-terminal p6* tetra-peptide without affecting the p6gag amino acid residues. Given our observation of DWzPR exhibiting Gag cleavage enhancement, the last two C-terminal p6* residues (NF) remaining at the LZ/PR junction might be required for PR activation. We therefore engineered a DWz/PANF/PR construct by replacing SF with PA (Fig. 2a). Both the DWz/PSHL/PR and DWz/PIDL/PR constructs contain substitutions for all four C-terminal p6* residues. We found that DWz/PANF/PR displayed a virus assembly and processing profile similar to that of DWz/PR (Fig. 2b, lane 4 vs. lane 1), but evidence from statistical analyses suggest that its virus particle processing was not as efficient as that of DWz/PR (Fig. 2b, d). Whereas DWz/PIDL/PR exhibited readily detected virus-associated p24gag and mature PR, DWz/PSHL/PR had virus-associated Gag or PR mostly present in unprocessed or incompletely processed precursor forms

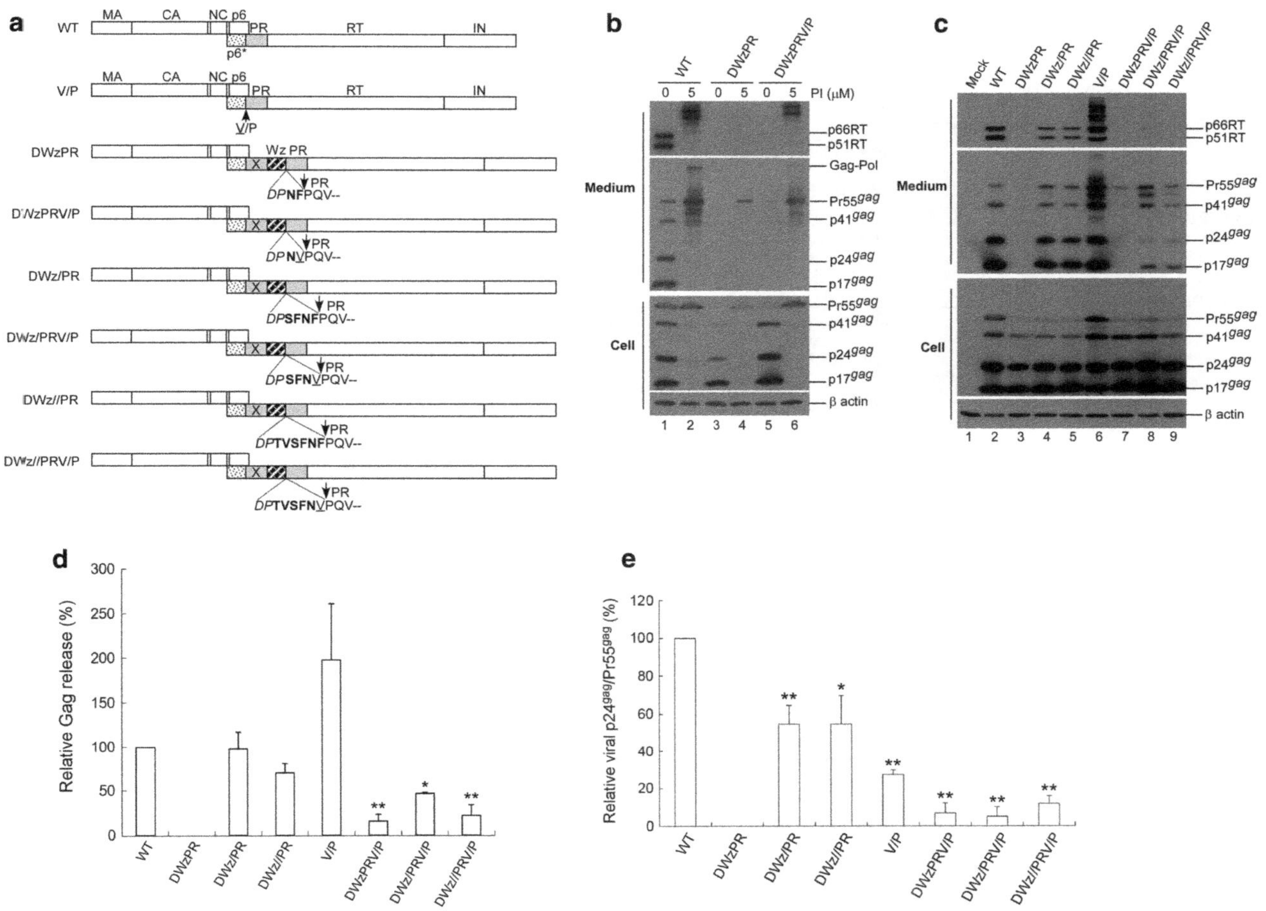

Fig. 1 Effects of C-terminal HIV-1 p6* residue substitutions on virus assembly and processing. **a** Schematic representations of HIV-1 Gag and Gag-Pol expression constructs. Indicated are the HIV-1 Gag protein domains MA (matrix), CA (capsid), NC (nucleocapsid), p6, pol-encoded p6*, PR, RT and IN. "X" denotes a PR-inactivated mutation. Arrows indicate PR cleavage sites. Underlined "V" indicates a Val residue substitution for the final C-terminal p6* residue Phe. Striped (Wz) box denotes wild-type (wt) leucine zipper (LZ). Remaining C-terminal p6* residues are in boldface. Altered or additional residues are in italics. **b** Blocking p6*-PR cleavage is insufficient for mitigating the enhanced Gag cleavage incurred by an LZ replacement for a deleted p6* domain. 293T cells were transfected with designated constructs. At 4 h post-transfection, equal amounts of cells were plated on two dishes and either left untreated or treated with saquinavir (a HIV-1 protease inhibitor) at a concentration of 5 μM. Supernatants and cells were collected 48 h post-transfection, prepared, and subjected to Western immunoblotting. **c** Blocking p6*-PR cleavage (V/P mutation) disrupted the function of a C-terminal p6* tetra-peptide for modulating PR activation. 293T cells were transfected with designated constructs. Culture supernatants and cells were collected and subjected to Western immunoblotting at 48–72 h post-transfection. **d** Relative virus assembly efficiencies of HIV-1 mutants. Gag proteins from medium or cell samples were quantified by scanning mutant and wt p24gag-associated band densities from immunoblots. Ratios of total Gag protein levels in medium to those in cells were determined for each construct and compared with wt release levels; release ratios for each mutant were divided by wt ratios in parallel experiments. Error bars indicate standard deviation. $*p < 0.05$; $**p < 0.01$. **e** Relative virus particle processing efficiency data for HIV-1 mutants. Virus-associated Pr55gag and p24gag levels were quantified by scanning immunoblot band densities. Ratios of p24gag to p55gag were determined for each mutant and normalized to those of the wt in parallel experiments. Bars indicate standard deviations. $*p < 0.05$; $**p < 0.01$

(Fig. 2b, lane 3 vs. lane 2). Similar effects on virus processing were observed when PSHL, PIDL or PANF mutations were cloned into Dp6*PR (Fig. 2c). Western blot data indicate a strong correlation between the virus processing efficiencies of the mutants and their virus-associated mature PR levels. Although Dp6*PANF/PR and DWz/PANF/PR both exhibited relatively low processing efficiency, no statistical significance was noted when compared with their Dp6*PR and DWZ/PR prototypes (Fig. 2d).

To confirm our conclusions, we tested PIDL, PSHL and PANF substitution mutations in a wt HIVgpt backbone (Fig. 3a). Results indicate that neither PIDL nor PANF mutations significantly affected virus infectivity

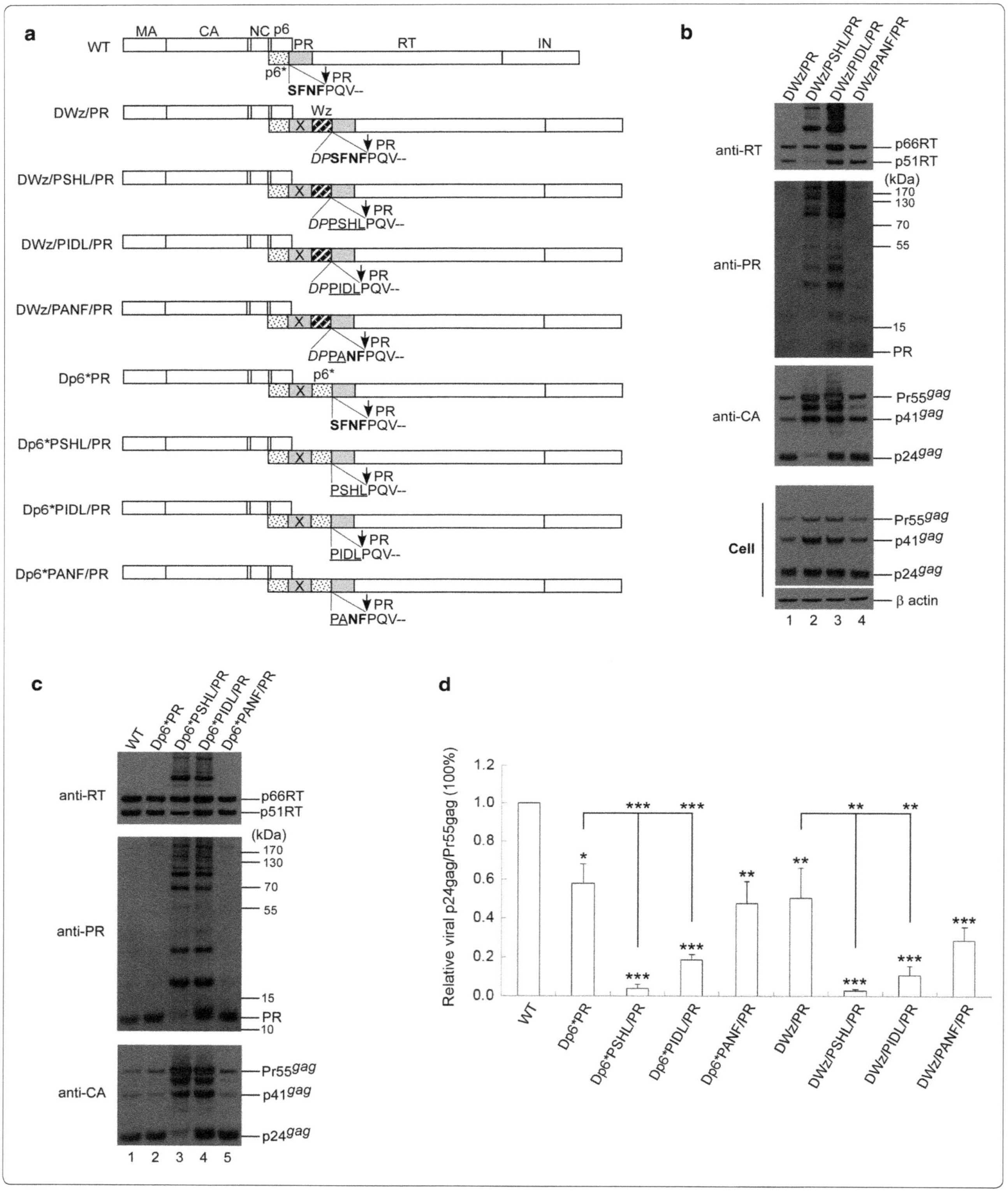
a
WT
MA
CA
NC
p6
PR
RT
IN
p6*
PR
SFNFPQV--
DWz/PR
Wz
X
PR
DPSFNFPQV--
DWz/PSHL/PR
PR
DPPSHLPQV--
DWz/PIDL/PR
PR
DPPIDLPQV--
DWz/PANF/PR
PR
DPPANFPQV--
Dp6*PR
p6*
PR
SFNFPQV--
Dp6*PSHL/PR
PR
PSHLPQV--
Dp6*PIDL/PR
PR
PIDLPQV--
Dp6*PANF/PR
PR
PANFPQV--
b
DWz/PR
DWz/PSHL/PR
DWz/PIDL/PR
DWz/PANF/PR
anti-RT
p66RT
p51RT
(kDa)
170
130
70
55
anti-PR
15
PR
anti-CA
Pr55gag
p41gag
p24gag
Cell
Pr55gag
p41gag
p24gag
β actin
1 2 3 4
c
WT
Dp6*PR
Dp6*PSHL/PR
Dp6*PIDL/PR
Dp6*PANF/PR
anti-RT
p66RT
p51RT
(kDa)
170
130
70
55
anti-PR
15
PR
10
anti-CA
Pr55gag
p41gag
p24gag
1 2 3 4 5
d
Relative viral p24gag/Pr55gag (100%)
1.2
1.0
0.8
0.6
0.4
0.2
0.0
WT
Dp6*PR
Dp6*PSHL/PR
Dp6*PIDL/PR
Dp6*PANF/PR
DWz/PR
DWz/PSHL/PR
DWz/PIDL/PR
DWz/PANF/PR

(See figure on previous page)

Fig. 2 Effects of C-terminal p6* amino acid substitutions on protease maturation and virus processing. **a** Schematic representations of HIV-1 Gag and Gag-Pol expression constructs. HIV-1 Gag protein domains and pol-encoded proteins and the leucine zipper (LZ) motif (striped box, Wz) are indicated as described in the Fig. 1 caption. Also indicated are amino acid residues in the junction area. Two or four C-terminal p6* residues remaining at the LZ/PR junction are shown in boldface. Amino acid changes at the C-terminal p6* tetra-peptide are underlined. Altered or additional residues are in italics. **b, c** 293T cells were transfected with designated constructs. Culture supernatants and cells were collected at 48–72 h post-transfection. To detect PR-associated products, aliquots of supernatant samples were separated by 4–12% Bis–Tris gradient gels. Membrane-bound proteins were initially probed with anti-PR serum prior to stripping and probing with anti-RT serum, followed by probing with an anti-p24CA monoclonal antibody. Molecular weight size markers (in kDa) are indicated on right side (middle panels). **d** Relative virus particle processing efficiency of HIV-1 mutants. Virus-associated Pr55gag and p24gag levels were quantified by scanning immunoblot band densities. Ratios of p24gag to p55gag were determined for each mutant and normalized to those of the wt in parallel experiments. Bars indicate standard deviations. $^{*}p < 0.05$; $^{**}p < 0.01$; $^{***}p < 0.001$

in single cycle infection assays, although both mutations reduced virus processing efficiency (Fig. 3b–d). The data also show that PANF exerted a weaker impact on virus processing and infectivity compared to PIDL. In contrast, the PSHL mutation significantly impaired both virus processing and infectivity (Fig. 3c, d). Combined, these results suggest that specific C-terminal p6* residues, especially the last two, are essential for modulating PR activation.

A single amino acid residue change eliminated the ability of p6*-deleted Gag-Pol to mediate virus maturation

The above conclusions agree with our past observations of HIV-1 Gag-Pol (Δp6*fs) with most p6* deleted, but with the final two C-terminal p6* residues still capable of mediating virus maturation, although less efficiently than wt Gag-Pol [40]. To determine if additional changes at C-terminal p6* residues affect the ability of Δp6*fs to mediate Gag processing, we engineered p6*-deleted constructs with two, four or six C-terminal p6* residues remaining at the p6*/PR junction (respectively designated Δp6*fsPR, Δp6*fs/PR and Δp6*fs//PR). All three contained an additional Pro residue insertion in the deleted p6* region due to cloning procedures (Fig. 4a). All of the p6*-deleted Gag-Pol mutants and wild-type GPfs were co-transfected with a Pr55gag expression vector designated pGAG. Unsurprisingly, virus production was almost completely blocked when each construct was co-expressed with equal amounts of pGAG (10 μg each), presumably due to enhanced Gag cleavage from over-expressed PR activity. The only exception was Δp6*fsPR, which produced significant amounts of mostly unprocessed or incompletely processed virus-associated Gag (Fig. 4b middle panel, lane 7). All constructs other than Δp6*fsPR were capable of producing readily detectable virus-associated p24gag and p66/51RT when co-expressed with pGAG at a plasmid DNA ratio of 1:10 (Fig. 4b, c). In contrast, Δp6*fsPR co-expression with pGAG consistently yielded virions that mostly contained incomplete or unprocessed Gag and RT-associated Gag-Pol (Fig. 4b, lane 4, and Fig. 4c, lane 6). These data suggest that Δp6*fsPR is profoundly defective in auto-processing and the *in trans* processing of virus particles.

In another test designed to determine whether the p6*-deleted Gag-Pol mutants mediated virus maturation and produced infectious virions, each Gag-Pol construct was co-expressed with pGAG plus a VSV-G envelope expression plasmid. Culture supernatants were collected for protein analysis and used to infect HeLa cells. Our data indicate that with the exception of Δp6*fsPR, all p6*-deleted Gag-Pol mutants were capable of producing infectious virions with infectivity levels of approximately 20–40% relative to wt GPfs (Fig. 4d). The third N-terminal PR residue at Δp6*fsPR (Val instead of Ile) apparently does not account for the virus processing defect, since a Val/Ile polymorphism was found at this position. Further, both Δp6*fs/PR and Δp6*fs//PR containing a Val polymorphism were capable of mediating virus particle maturation. Δp6*fsPR contains an inserted proline adjacent to the last two C-terminal p6* residues. Although Δp6*fs/PR and Δp6*fs//PR both contain the same proline insertion at the p6*-deleted region, both were still found to be capable of mediating virus maturation. The four remaining C-terminal p6* residues within Δp6*fs/PR and Δp6*fs//PR might prevent the Pro insertion from interfering with PR activation.

Enhanced PR activation due to the LZ replacement of p6* is Gag domain-independent

Given the contribution that Gag makes to PR activation by promoting Gag-Pol dimerization, we hypothesized that Gag removal from DWzPR might impair Gag-Pol dimerization, thereby mitigating the LZ enhancement effect on PR activation and reducing both Gag-Pol autocleavage and Gag cleavage efficiency. To test this idea, we engineered GPfs versions of DWzPR with a Gag deletion (fsdWzPRΔGag) and without one (fsdWzPR) (Fig. 5a). GPfs and a GPfs version of DWz/PR (designated fsdDWz/PR) served as controls. Each construct was co-expressed with D25, a PR-inactivated Gag/Gag-Pol expression plasmid. According to our results, both fsdWzPR and fsdWzPRΔGag produced barely detectable

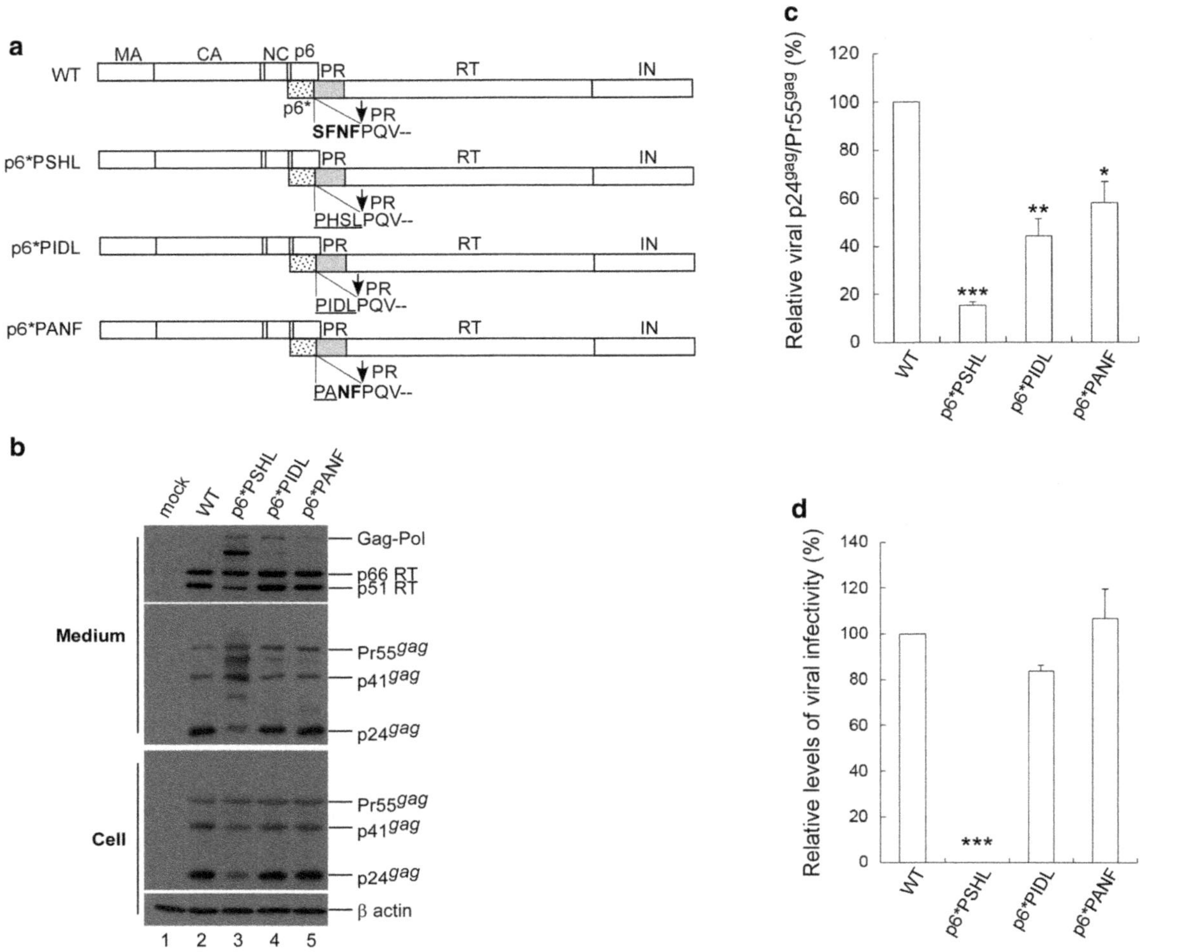

Fig. 3 Effects of C-terminal p6* tetra-peptide mutations on virus processing and infectivity. **a** Schematic representations of HIV-1 Gag and Gag-Pol expression constructs. HIV-1 Gag protein domains and pol-encoded proteins are indicated as described in the Fig. 1 caption. Native C-terminal p6* residues are shown in boldface. Altered amino acid residues are underlined. **b** 293T cells were transfected with designated constructs. Culture supernatants and cells were collected 48–72 h post-transfection and subjected to Western immunoblotting. **c** Relative virus particle processing efficiency of HIV-1 mutants. Virus-associated Pr55gag and p24gag levels were quantified by scanning immunoblot band densities. Ratios of p24gag to p55gag were determined for each mutant and normalized to those of the wt in parallel experiments. Bars indicate standard deviations. *$p < 0.05$; **$p < 0.01$; ***$p < 0.001$. **d** Infectivity of HIV-1 mutants. 293T cells were co-transfected with one of the designated constructs plus a VSV-G expression vector. At 48 h post-transfection, supernatants were collected, filtered, and used to infect HeLa cells. Infection and selection of drug-resistant colonies was performed as described in Methods. Infectivity for each mutant was determined as the ratio of mutant titers to wt titers, normalized to Gag protein levels in parallel experiments. ***$p < 0.001$

virus-associated p24gag when co-transfected with D25 at a DNA ratio of 1:10 (Fig. 5b middle panel, lanes 3 and 5). In contrast, virus-associated p24gag was readily detected in fsdWz/PR-plus-D25 co-transfection samples although at a much lower level compared to Pr55gag and p41gag (Fig. 5b, lane 7). Incorporated Gag-Pol deficiency due to premature or enhanced Gag-Pol auto-cleavage might result in insufficient virus processing. The over-expression of fsdWzPRΔGag and other GPfs mutants led to the complete blocking of virus assembly (Fig. 5b, lanes 4, 6 and 8).

Virus-associated Gag-Pol molecules detected in medium were likely from D25 (Fig. 5b, upper panel). It is possible that D25 Gag-Pol competes with other Gag-Pol mutants in terms of viral incorporation. There is also the possibility that D25 PR-defective Gag-Pol interferes with the ability of incorporated Gag-Pol mutants to mediate virus particle processing. This may partly explain the relatively lower levels of virus-associated p24gag that we observed in fsdWz/PR co-transfection samples (Fig. 5b middle panel lane 7). To study these possibilities, Gag-Pol mutants were co-expressed with pGAG. Results indicate

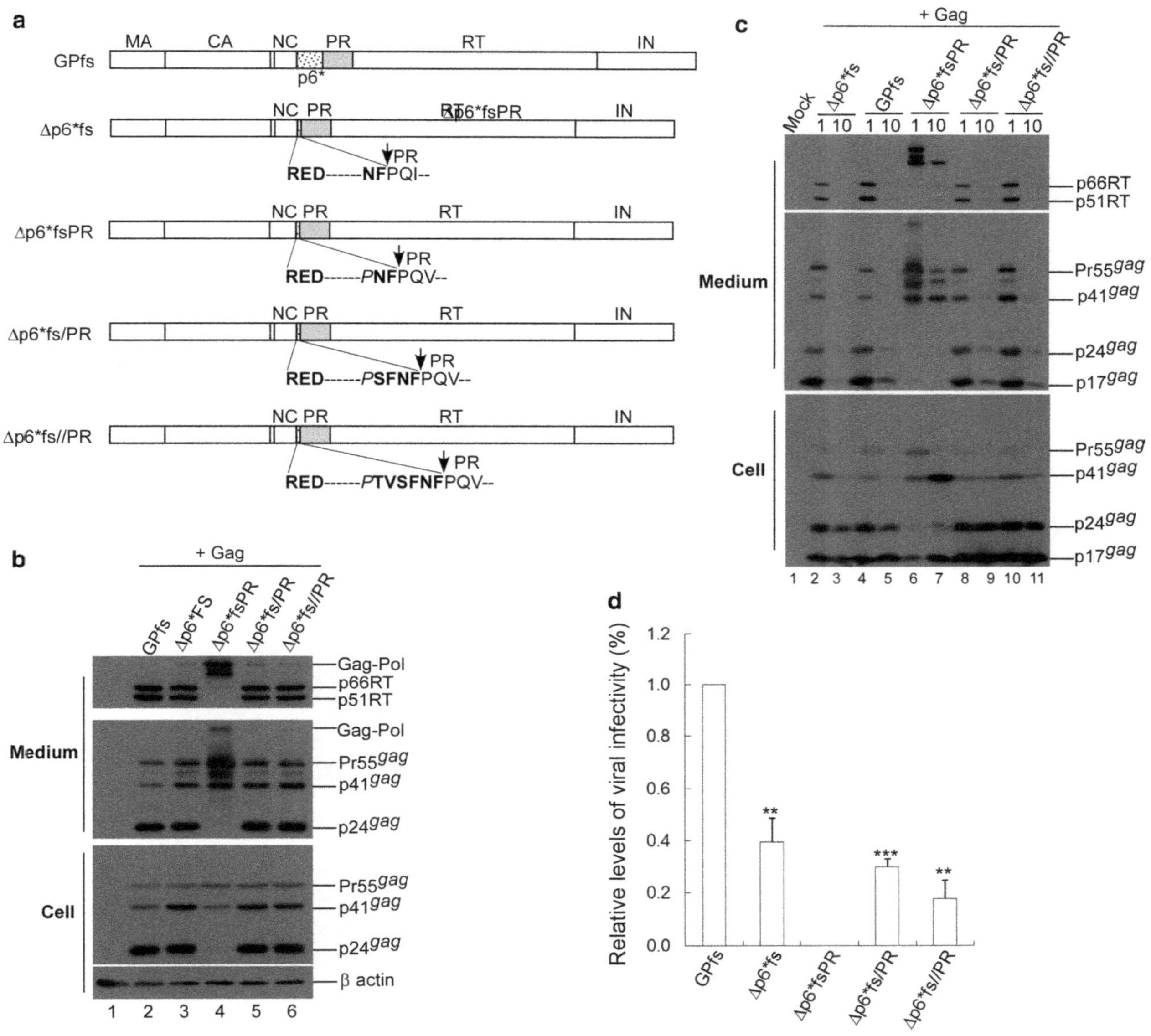

Fig. 4 Effects of C-terminal p6* residue substitutions on the capability of p6*-deleted Gag-Pol mutants to mediate virus maturation. **a** Schematic representations of HIV-1 Gag-Pol expression constructs with deletions of most p6* coding sequences. HIV-1 Gag domains, pol-encoded p6*, PR, RT and IN are indicated. All constructs contain a frame shift (fs) mutation forcing gag and pol into the same reading frame. Dashed lines denote deleted p6* regions. Remaining N-terminal and C-terminal p6* residues are indicated in boldface. Altered or foreign residues are in italics. **b**, **c** 293T cells were co-transfected with 10 µg of an HIV-1 Pr55gag expression plasmid (pGAG) and 1 or 10 µg (panel **b**) or 1 µg (panel **c**) of the designated Gag-Pol expression construct. At 48 h post-transfection, cells and supernatants were collected and analyzed by Western immunoblotting. Membrane-bound proteins were initially probed with anti-RT serum, stripped, and probed again with anti-p17MA and anti-p24CA monoclonal antibodies. Indicated are HIV-1 Gag-Pol, 66/51RT, Pr55gag, p41gag, p24gag and p17gag positions. **d** A single amino acid change blocked the capability of p6*-deleted Gag-Pol to confer virus infectivity. 293T cells were co-transfected with 10 µg of an HIV-1 Pr55gag expression vector (pGAG) and 1 µg of one of the designated constructs plus 5 µg of a VSV-G expression vector. At 48 h post-transfection, supernatants were collected, filtered, and used to infect HeLa cells. Infectivity for each Gag-Pol construct was determined as the ratio of mutant titers to wt Gag-Pol titers, normalized to virus-associated p24gag protein levels in parallel experiments. $^{**}p < 0.01$; $^{***}p < 0.001$

that virus-associated RT and p24gag were both readily detected in wt GPfs or fsdWz/PR co-transfection samples (Fig. 5c). In contrast, virus-associated RT and p24gag were both barely detectable in fsdWzPR co-transfection samples, likely due to a Gag-Pol incorporating defect (Fig. 5c, lane 3).

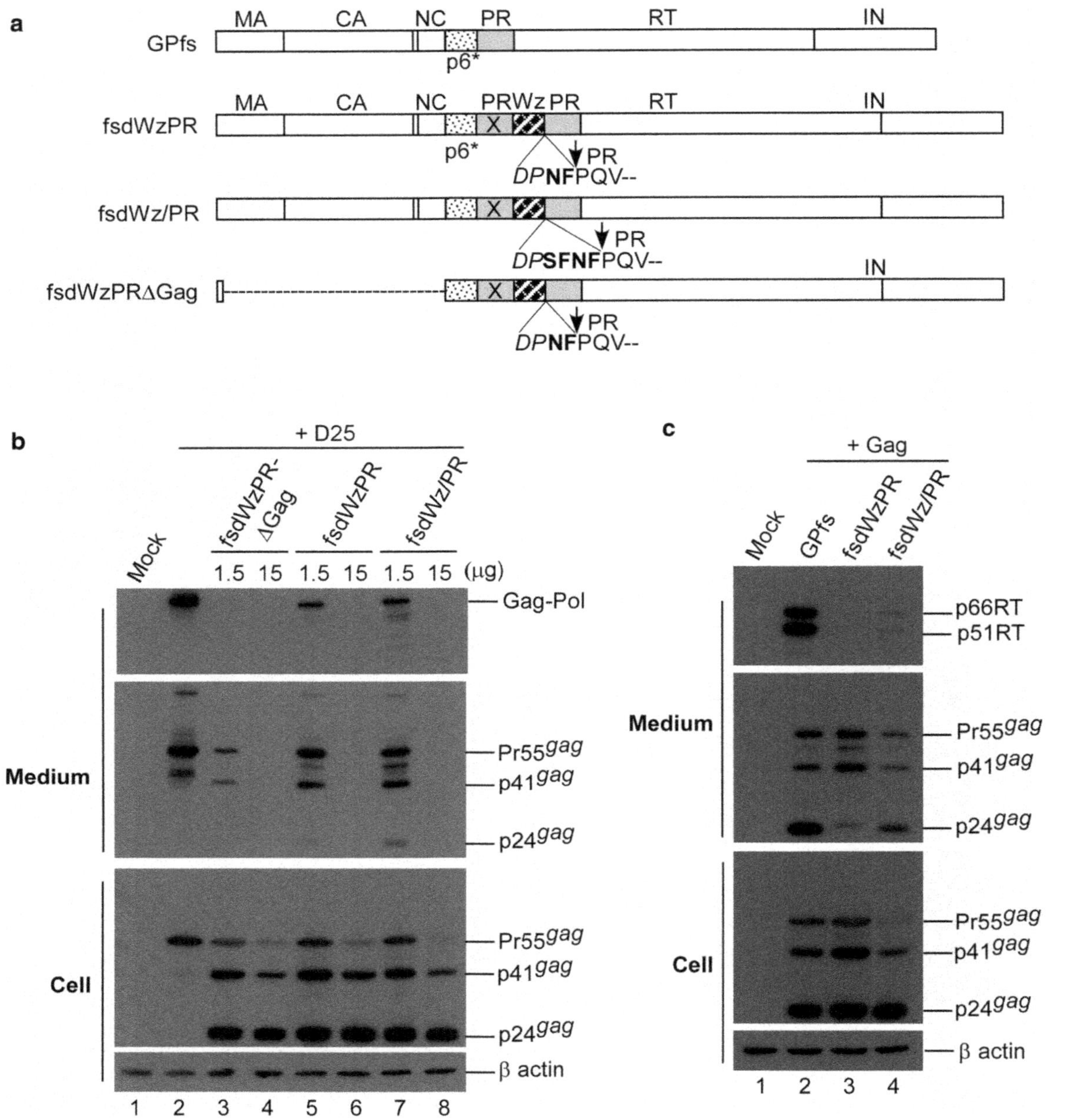

Fig. 5 Enhanced Gag-Pol auto-cleavage reduces virus yields and Gag-Pol viral incorporation. **a** Schematic representations of HIV-1 Gag-Pol expression constructs in a gag-pol frame shift (fs) mutation backbone are as described in the Fig. 4 caption. Striped (Wz) box denotes leucine zipper (LZ). Dashed line indicates deleted Gag coding sequence. "X" denotes a PR-inactivated mutation. Remaining C-terminal p6* residues are in boldface. Altered or additional residues are shown in italics. **b**, **c** Leucine zipper-induced Gag-Pol auto-cleavage enhancement leads to reductions in virion yields and Gag-Pol packaging. Indicated amounts of designated plasmids were co-expressed with 15 μg of an HIV-1 protease-defective (D25) Gag/Gag-Pol expression vector (panel b) or co-expressed with pGAG (panel c) at a DNA ratio of 1:10. Culture supernatants and cells were collected at 48–72 h post-transfection and subjected to Western immunoblotting. Indicated are HIV-1 Gag-Pol, 66/51RT, Pr55gag, p41gag and p24gag positions

Combined, these results suggest that an LZ replacement for p6* leads to enhanced Gag-Pol auto-cleavage and associated Gag-Pol incorporation deficiency, and that Gag removal does not reduce the LZ enhancement effect on PR activation.

Discussion

Even though the removal of p6* from PR precursor is necessary for PR to be fully functional, the enhancement effect of LZ on PR-mediated Gag processing was not significantly compromised by blocking p6*-PR cleavage.

Further, we noted that DWz/PRV/P, DWz//PRV/P and other constructs with substitution mutations at the remaining C-terminal p6* tetra-peptide were all capable of producing readily detectable virus-associated Gag, although with processing defects for some of the mutants (Figs. 1, 2). These results suggest that the enhancement of PR activation by LZ may be reduced when as few as six amino acid residues, whether native or foreign, are present between LZ and PR. The hexa-peptide between LZ and PR may serve as a spacer that prevents the potent LZ dimerization motif from facilitating PR dimer formation. Additionally, substitutions at the C-terminal p6* tetra-peptide might interfere with PR maturation, thereby contributing, at least in part, to reduced Gag processing efficiency.

As a result of blocked cleavage at the PR-RT site, HIV-1 PR-RT fusion is capable of mediating virus processing and supporting virus replication [41]. In contrast, p6* appears to be capable of inhibiting PR maturation, and therefore must be removed for PR to be fully active. However, DWzPR virus yields were not significantly restored when p6*-PR cleavage was blocked, suggesting that HIV-1 PR precursors containing a LZ motif fused at the N-terminus may exhibit markedly enhanced enzymatic activity even when free mature PR is not released. Theoretically, PR-associated products with LZ linked at the PR N-terminus may exist in DWzPRV/P transfectant samples due to blocked cleavage at p6*/PR. Since attempts to detect these PR-associated products were unsuccessful, we believe PR may access putative cryptic cleavage sites within the LZ in addition to being self-degrading. Regardless, our findings suggest that PR precursors containing foreign peptides fused at the N-terminus are still capable of being functionally active. Specifically, PR activity may be significantly enhanced when a potent dimerization motif is present at the PR N-terminus.

Although C-terminal p6* substitutions with either PIDL or PANF residues exerted no major impacts on HIVgpt virus infectivity according to single-round infection assays, they did reduce virus processing efficiency (Fig. 3). These data agree with an earlier report that substitutions for C-terminal p6* residues can impair PR maturation, resulting in a virus processing defect [20]. Negative effects of C-terminal p6* tetra-peptide substitution mutations on virus processing become increasingly noticeable when tested with a Dp6*PR backbone (Fig. 2c), presumably due to PR function perturbation by an upstream inactivated PR copy.

According to single-cycle infection assays, Δp6*fs generated mature infectious virions. In contrast, Δp6*fsPR (with only one amino acid difference from Δp6*fs in the linker region) was completely blocked in terms of mediating virus maturation (Fig. 4). Δp6*fs has a Leu at the P3 position, while Δp6*fsPR has a foreign Pro insertion at P3. Pro residues were barely detected at the P3 position flanking the PR cleavage site (i.e., the third amino acid residue immediately upstream from the PR substrate cleavage site). Leu was one of several amino acids found at the P3 position [42]. This raises the possibility of P3-Pro disrupting PR maturation, thereby blocking the ability of Δp6*fsPR to mediate virus processing. Even though they both contain a Pro insertion in the deleted p6* region, Δp6*fs/PR and Δp6*fs//PR were still capable of mediating virus maturation, likely due to the containment of an intact C-terminal p6* tetra-peptide SFNF at the p6*/PR junction. This agrees with our observations involving DWz/PR and DWz//PR, both of which contain an intact C-terminal p6* tetra-peptide and are capable of counteracting the LZ enhancement of PR activation.

In addition to containing only four residues between LZ and PR, DWzPR has a Pro at the P3 position, which might contribute to its failure to counteract the LZ enhancement effect on PR-mediated Gag cleavage. C-terminal p6* residues might inhibit PR maturation by destabilizing the structure of the folded PR dimer [43]. The absence of regulated PR dimer structure folding due to mutations at C-terminal p6* residues might allow the potent LZ dimerization motif to facilitate or exert a synergetic effect on Gag-Pol or PR dimer formation via PR dimer stabilization. PR is consequently prematurely activated, resulting in a significant enhancement of Gag processing as was observed for DWzPR and DWzPRV/P.

In conclusion, our findings suggest that an HIV-1 PR precursor containing an N-terminally extended peptide can function efficiently without liberating free mature PR. HIV-1 PR precursor activity might be manipulated via the altering of peptide residues adjacent to the PR N-terminus.

Authors' contributions

CTW designed this project and wrote the manuscript. FHY performed the experiment and helped with the data analysis. Both authors read and approved the final manuscript.

Acknowledgements

We thank K.J. Huang and Y.R. Lin for materials and reagents. We also thank the following individuals from the National Institutes of Health AIDS Research and Reference Reagent Program for their assistance in obtaining the following reagents: Stephen Hughes for MAb21 anti-RT monoclonal antibodies and D. Bailey and Mark Page for HIV-1 PR antiserum.

Competing interests

The authors declare that they have no competing interests.

Funding
This work was supported by Grants V104C-056 and V105C-036 from the Taipei Veterans General Hospital and by Taiwan Ministry of Science and Technology Grants 104-2320-B-010-035, 105-2320-B-010-023 and 106-2320-B-010-017-MY2.

References

1. Petropoulos C. Retroviral taxonomy, protein structures, sequences, and genetic maps. In: Coffin JM, Hughes SH, Varmus HE, editors. Retroviruses. New York: Cold Spring Harbor Laboratory Press; 1997.
2. Jacks T, Power MD, Masiarz FR, Luciw PA, Barr PJ, Varmus HE. Characterization of ribosomal frameshifting in HIV-1 gag-pol expression. Nature. 1988;331:280–3.
3. Freed EO. HIV-1 assembly, release and maturation. Nat Rev Microbiol. 2015;13:484.
4. Huang M, Martin MA. Incorporation of Pr160(gag-pol) into virus particles requires the presence of both the major homology region and adjacent C-terminal capsid sequences within the Gag-Pol polyprotein. J Virol. 1997;71:4472–8.
5. Smith AJ, Srinivasakumar N, Hammarskjöld ML, Rekosh D. Requirements for incorporation of Pr160gag-pol from human immunodeficiency virus type 1 into virus-like particles. J Virol. 1993;67:2266–75.
6. Srinivasakumar N, Hammarskjöld ML, Rekosh D. Characterization of deletion mutations in the capsid region of human immunodeficiency virus type 1 that affect particle formation and Gag-Pol precursor incorporation. J Virol. 1995;69:6106–14.
7. Park J, Morrow CD. The nonmyristylated Pr160gag-pol polyprotein of human immunodeficiency virus type 1 interacts with Pr55gag and is incorporated into viruslike particles. J Virol. 1992;66:6304–13.
8. Lee SK, Potempa M, Swanstrom R. The choreography of HIV-1 proteolytic processing and virion assembly. J Biol Chem. 2012;287:40867–74.
9. Swanstrom R, Wills JW. Synthesis, assembly, and processing of viral proteins. In: Coffin JM, Hughes SH, Varmus HE, editors. Retroviruses. New York: Cold Spring Harbor Laboratory Press; 1997.
10. Pettit SC, Moody MD, Wehbie RS, Kaplan AH, Nantermet PV, Klein CA, Swanstrom R. The p2 domain of human immunodeficiency virus type 1 Gag regulates sequential proteolytic processing and is required to produce fully infectious virions. J Virol. 1994;68:8017–27.
11. Peng C, Ho BK, Chang TW, Chang NT. Role of human immunodeficiency virus type 1-specific protease in core protein maturation and viral infectivity. J Virol. 1989;63:2550–6.
12. Liao WH, Wang CT. Characterization of human immunodeficiency virus type 1 Pr160 gag-pol mutants with truncations downstream of the protease domain. Virology. 2004;329:180–8.
13. Louis JM, Nashed NT, Parris KD, Kimmel AR, Jerina DM. Kinetics and mechanism of autoprocessing of human immunodeficiency virus type 1 protease from an analog of the Gag-Pol polyprotein. Proc Natl Acad Sci USA. 1994;91:7970–4.
14. Pettit SC, Gulnik S, Everitt L, Kaplan AH. The dimer interfaces of protease and extra-protease domains influence the activation of protease and the specificity of GagPol cleavage. J Virol. 2003;77:366–74.
15. Quillent C, Borman AM, Paulous S, Dauguet C, Clavel F. Extensive regions of pol are required for efficient human immunodeficiency virus polyprotein processing and particle maturation. Virology. 1996;219:29–36.
16. Zybarth G, Carter C. Domains upstream of the protease (PR) in human immunodeficiency virus type 1 Gag-Pol influence PR autoprocessing. J Virol. 1995;69:3878–84.
17. Figueiredo A, Moore KL, Mak J, Sluis-Cremer N, de Bethune MP, Tachedjian G. Potent nonnucleoside reverse transcriptase inhibitors target HIV-1 Gag-Pol. PLoS Pathog. 2006;2:e119.
18. Tachedjian G, Orlova M, Sarafianos SG, Arnold E, Goff SP. Nonnucleoside reverse transcriptase inhibitors are chemical enhancers of dimerization of the HIV type 1 reverse transcriptase. Proc Natl Acad Sci USA. 2001;98:7188–93.
19. Pan YY, Wang SM, Huang KJ, Chiang CC, Wang CT. Placement of leucine zipper motifs at the carboxyl terminus of HIV-1 protease significantly reduces virion production. PLoS One. 2012;7:e32845.
20. Ludwig C, Leiherer A, Wagner R. Importance of protease cleavage sites within and flanking human immunodeficiency virus type 1 transframe protein p6* for spatiotemporal regulation of protease activation. J Virol. 2008;82:4573–84.
21. Partin K, Zybarth G, Ehrlich L, DeCrombrugghe M, Wimmer E, Carter C. Deletion of sequences upstream of the proteinase improves the proteolytic processing of human immunodeficiency virus type 1. Proc Natl Acad Sci USA. 1991;88:4776–80.
22. Tessmer U, Krausslich HG. Cleavage of human immunodeficiency virus type 1 proteinase from the N-terminally adjacent p6* protein is essential for efficient Gag polyprotein processing and viral infectivity. J Virol. 1998;72:3459–63.
23. Paulus C, Ludwig C, Wagner R. Contribution of the Gag-Pol transframe domain p6* and its coding sequence to morphogenesis and replication of human immunodeficiency virus type 1. Virology. 2004;330:271–83.
24. Chatterjee A, Mridula P, Mishra RK, Mittal R, Hosur RV. Folding regulates autoprocessing of HIV-1 protease precursor. J Biol Chem. 2005;280:11369–78.
25. Ishima R, Torchia DA, Louis JM. Mutational and structural studies aimed at characterizing the monomer of HIV-1 protease and its precursor. J Biol Chem. 2007;282:17190–9.
26. Louis JM, Clore GM, Gronenborn AM. Autoprocessing of HIV-1 protease is tightly coupled to protein folding. Nat Struct Biol. 1999;6:868–75.
27. Sadiq SK, Noe F, De Fabritiis G. Kinetic characterization of the critical step in HIV-1 protease maturation. Proc Natl Acad Sci USA. 2012;109:20449–54.
28. Yu F-H, Huang K-J, Wang C-T. C-terminal HIV-1 transframe p6* tetrapeptide blocks enhanced Gag cleavage incurred by leucine zipper replacement of a deleted p6* domain. J Virol. 2017;91:e00103-17.
29. Ratner L, Haseltine W, Patarca R, Livak KJ, Starcich B, Josephs SF, Doran ER, Rafalski JA, Whitehorn EA, Baumeister K, et al. Complete nucleotide sequence of the AIDS virus. HTLV-III. Nature. 1985;313:277–84.
30. Page KA, Landau NR, Littman DR. Construction and use of a human immunodeficiency virus vector for analysis of virus infectivity. J Virol. 1990;64:5270–6.
31. Yu F-H, Chou T-A, Liao W-H, Huang K-J, Wang C-T. Gag-Pol transframe domain p6* is essential for HIV-1 protease-mediated virus maturation. PLoS ONE. 2015;10:e0127974.
32. Chen S-W, Chiu H-C, Liao W-H, Wang F-D, Chen SSL, Wang C-T. The virus-associated human immunodeficiency virus type 1 Gag-Pol carrying an active protease domain in the matrix region is severely defective both in autoprocessing and in trans processing of gag particles. Virology. 2004;318:534–41.
33. Boles E, Miosga T. A rapid and highly efficient method for PCR-based site-directed mutagenesis using only one new primer. Curr Genet. 1995;28:197–8.
34. Chiu H-C, Yao S-Y, Wang C-T. Coding sequences upstream of the human immunodeficiency virus type 1 reverse transcriptase domain in Gag-Pol are not essential for incorporation of the Pr160gag-pol into virus particles. J Virol. 2002;76:3221–31.
35. Yee JK, Friedmann T, Burns JC. Generation of high-titer pseudotyped retrovirus with very broad host range. Methods in Cell Biology. 1994;43:99–112.
36. Chen YL, Ts'ai PW, Yang CC, Wang CT. Generation of infectious virus particles by transient co-expression of human immunodeficiency virus type 1 gag mutants. J Gen Virol. 1997;78(Pt 10):2497–501.
37. Chiang C-C, Wang S-M, Tseng Y-T, Huang K-J, Wang C-T. Mutations at human immunodeficiency virus type 1 reverse transcriptase tryptophan repeat motif attenuate the inhibitory effect of efavirenz on virus production. Virology. 2009;383:261–70.
38. Ferris AL, Hizi A, Showalter SD, Pichuantes S, Babe L, Craik CS, Hughes SH. Immunologic and proteolytic analysis of HIV-1 reverse transcriptase structure. Virology. 1990;175:456–64.
39. Hizi A, McGill C, Hughes SH. Expression of soluble, enzymatically active, human immunodeficiency virus reverse transcriptase in Escherichia coli and analysis of mutants. Proc Natl Acad Sci. 1988;85:1218–22.

40. Chiu H-C, Wang F-D, Chen Y-MA, Wang C-T. Effects of human immunodeficiency virus type 1 transframe protein p6* mutations on viral protease-mediated Gag processing. J Gen Virol. 2006;87:2041–6.
41. Cherry E, Liang C, Rong L, Quan Y, Inouye P, Li X, Morin N, Kotler M, Wainberg MA. Characterization of human immunodeficiency virus type-1 (HIV-1) particles that express protease-reverse transcriptase fusion proteins11. J Mol Biol. 1998;284:43–56.
42. Pettit SC, Simsic J, Loeb DD, Everitt L, Hutchison CA, Swanstrom R. Analysis of retroviral protease cleavage sites reveals two types of cleavage sites and the structural requirements of the P1 amino acid. J Biol Chem. 1991;266:14539–47.
43. Tang C, Louis JM, Aniana A, Suh JY, Clore GM. Visualizing transient events in amino-terminal autoprocessing of HIV-1 protease. Nature. 2008;455:693–6.

9

Promoter expression of HERV-K (HML-2) provirus-derived sequences is related to LTR sequence variation and polymorphic transcription factor binding sites

Meagan Montesion[1,4], Zachary H. Williams[1], Ravi P. Subramanian[1,5], Charlotte Kuperwasser[2,3] and John M. Coffin[1*]

Abstract

Background: Increased transcription of the human endogenous retrovirus group HERV-K (HML-2) is often seen during disease. Although the mechanism of its tissue-specific activation is unclear, research shows that LTR CpG hypomethylation alone is not sufficient to induce its promoter activity and that the transcriptional milieu of a malignant cell contributes, at least partly, to differential HML-2 expression.

Results: We analyzed the relationship between LTR sequence variation and promoter expression patterns in human breast cancer cell lines, finding them to be positively correlated. In particular, two proviruses (3q12.3 and 11p15.4) displayed increased activity in almost all tumorigenic cell lines sampled. Using a transcription factor binding site prediction algorithm, we identified two unique binding sites in each 5′ LTR that appeared to be associated with inducing promoter activity during neoplasia. Genomic analysis of the homologous proviruses in several non-human primates indicated post-integration genetic drift in two transcription factor binding sites, away from the ancestral sequence and towards the active form. Based on the sequences of 2504 individuals from the 1000 Genomes Project, the active form of the 11p15.4 site was found to be polymorphic within the human population, with an allele frequency of 51%, whereas the activating mutation in the 3q12.3 provirus was fixed in humans but not present in the orthologous provirus in chimpanzees or gorillas.

Conclusions: These data suggest that stage-specific transcription factors at least partly contribute to LTR promoter activity during transformation and that, in some cases, transcription factor binding site polymorphisms may be responsible for the differential HML-2 expression often seen between individuals.

Keywords: Endogenous retrovirus, HERV-K, HML-2, LTR, Transcription, Tumorigenesis

Background

Retroviruses are unique in that they are the only virus family known to exist in both endogenous and exogenous forms [1, 2]. Their integrated DNA sequences, known as proviruses, include at least four genes (*gag, pro, pol,* and *env*) flanked by long terminal repeats (LTRs), which contain all elements necessary to initiate and terminate viral transcription [2, 3]. Genetic transmission of these sequences occurs with germline integration, producing endogenous retroviruses (ERVs). ERVs are inherited in a Mendelian fashion and are subject to natural selection; those with deleterious effects are generally either lost from the population or inactivated by mutation, whereas those with neutral or advantageous effects remain [2, 4]. As a consequence of the accumulation of these elements over time, nearly 8% of the human genome is derived from such viral sequences [5–7].

*Correspondence: john.coffin@tufts.edu
[1] Department of Molecular Biology and Microbiology, Tufts University School of Medicine, Boston, MA, USA
Full list of author information is available at the end of the article

Once classified with other "junk DNA", ERVs are now credited with providing genomic plasticity through the use of viral proteins for host functions and alternative regulation of host gene transcription. For example, proviruses contain numerous promoters, splice sites, transcription factor binding sites, and polyadenylation signals, all of which can have significant effects on neighboring host gene expression [2, 8, 9]. Syncytins, fusogenic proteins derived from ERV *env* sequences, are essential for placenta development and mediate cell fusion to form the syncytiotrophoblast layer [10, 11]. Although ERV expression is usually silenced through epigenetic and chromatin modification, primarily via CpG methylation [8, 12–14], there are a few known instances of host cell co-option of ERV expression. Recent studies show human endogenous retrovirus (HERV) expression to be increased in human embryonic stem cells (hESCs) and human preimplantation embryos and to play a critical role during embryogenesis through the maintenance of pluripotency and hESC identity [15–19]. Increased expression of HERV proteins was found to be correlated with increased IFITM1 expression, resulting in viral immunoprotection during human embryogenesis [19, 20].

Despite these exceptions, increased HERV activity is largely associated with malignancy, especially cancer. Activation of stem cell-associated retroviruses (SCARs) in human cancer is hypothesized to be associated with increased likelihood of metastasis, immune evasion of cancer cells, and a predictive marker of poor prognosis [21, 22]. Increased cancer-related expression is attributable in part to global hypomethylation, a common consequence of tumorigenesis, and LTR hypomethylation is widely documented to result in promoter activation [11, 13, 23]. However, in vitro treatment with 5-aza-2′-deoxycytidine, a DNA methyltransferase inhibitor, shows that LTR hypomethylation alone is not always sufficient to induce promoter activity, suggesting that the proper transcriptional milieu of a cell may also be required [24–26]. Ubiquitous transcription factors, such as Sp1, Sp3, and YY1, are linked with LTR activity but do not explain the cell-specific expression that is often seen [8, 25, 27].

Expression from HERV-K (HML-2), the most recently integrated and biologically active HERV group, is upregulated in up to 85% of breast cancer samples, although the mechanism of activation is still unclear [28–31]. RNA sequence analysis of cells in an in vitro mammary carcinogenesis model shows that LTR-driven transcription of HML-2 proviruses is restricted to tumorigenic human mammary epithelial cells (HMECs), suggesting that stage-specific transcription factors appearing during malignant transformation play a role in LTR activation [32]. The goal of this study was to investigate how LTR sequence variation among the various HML-2 proviruses affects activation of its promoter during HMEC transformation.

Overall, we found the most widespread increase in promoter activity during transformation in two proviruses (located at 3q12.3 and 11p15.4). Through a combination of reporter construct assays and RNA-Seq analyses, we identified two transcription factor binding sites on each 5′ LTR that were associated with promoter activity in transformed cells. Further genomic analysis of these proviruses, using data from the 1000 Genomes Project as well as comparison with homologous proviruses in several other hominoid species, showed that both of these sites had been created by mutations in the 5′ LTR that occurred post viral integration. The 3q12.3 site has become fixed in the human population whereas that at 11p15.4 is polymorphic, with the active form having an allele frequency of 51%. In both cases, these sites have evolved away from the inactive ancestral sequence and towards an active form. These results emphasize the importance of studying HERV transcription at the single provirus and single nucleotide level, as polymorphisms in critical binding sites may be responsible for the differential HML-2 expression often seen between individuals.

Results

Differential HML-2 promoter expression is correlated with 5′ LTR sequence similarity

The HML-2 5′ LTR contains all elements necessary for driving transcription. Removal of core promoter elements results in reduced promoter activity, suggesting that these sequences are critical for proper LTR-driven expression [9, 25, 33]. Each provirus has accumulated numerous unique mutations over time, suggesting that LTR sequence variation could contribute to differential HML-2 expression, particularly through the alteration of transcription factor binding sites. Through a series of dual-luciferase assays, we sought to evaluate whether LTR sequence identity is correlated with similar promoter expression patterns during breast cancer tumorigenesis.

The proviruses of interest for this study were chosen based on a preliminary investigation in which we used single-genome sequencing to detect provirus-specific transcripts from eight human breast cancer cell lines. From this analysis, we produced a list of the top ten highest expressing HML-2 proviruses within these cell lines tested (Additional file 1). 16p11.2 and K105 were excluded from our study since 16p11.2 has no 5′ LTR and the unmapped K105 exhibited cloning inconsistencies caused by its location within the unassembled centromeric region Un_g1000219 [4, 34]. The remaining eight proviruses, plus 8p23.1c, a segmental duplication of 11p15.4 [4], were chosen as our loci of interest. The alternative names and chromosomal locations of these proviruses are listed in Table 1.

Phylogenetic analysis of the LTRs from these nine proviruses shows that most of them are classified as LTR-HS, the LTR group that contains the youngest proviruses, including ~90% of the human-specific integrations (Fig. 1a) [4, 9, 35]. The 5′ LTR sequences from each provirus were cloned into pGL4.17[*luc2*/Neo], a promoterless firefly luciferase vector, directly upstream of the *luc2* gene. The relative promoter activity of these sequences was determined based on *luc2* expression and normalized against that of an internal control vector, containing a *Renilla* luciferase gene (R*luc*) driven by an SV40 promoter (Fig. 1b). A panel of eighteen human cell lines was transiently co-transfected with these vectors. The panel comprised of two immortalized HMEC cell lines, fifteen tumorigenic breast cancer cell lines (representing all three molecular subtypes), and one teratocarcinoma cell line known to produce HML-2 transcripts and retroviral-like particles (RVLPs) at high levels [9, 36]. Characterization of the cell lines used is shown in Table 2.

Although minimal promoter activity was detected in immortalized HMECs transfected with any of the HML-2 LTR reporter constructs, significant upregulation of expression driven by one or more LTRs was seen in 73% (11/15) of the tumorigenic breast cancer cell lines (Fig. 1c). This expression pattern is consistent with previous reports that suggest up to 85% of breast cancer samples have a significant increase in HML-2 activity [29, 31, 37]. Overall, each LTR was significantly expressed in at least one cell line tested but showed differential expression across the panel. Two proviruses (3q12.3 and 11p15.4) were significantly upregulated in nearly all neoplastic cell lines investigated, whereas others were only upregulated in a select few (Fig. 1d). The highest level of combined HML-2 expression in a breast cell line was exhibited by T47D (Fig. 1c), a tumorigenic breast cancer cell line known to produce RVLPs under hormonally-stimulated conditions [3, 38, 39]. However, this activity level was only about half that seen in the Tera-1 cells, consistent with our previous report that Tera-1 cells produce markedly higher numbers of HML-2 transcripts than breast cancer cell lines [32].

We next sought to determine if LTRs of similar sequence share similar patterns of promoter activity. For this purpose, we created a percent sequence identity matrix, by multiple sequence alignment using Clustal Omega [40], and an HML-2 percent expression similarity matrix, determined through pairwise comparisons of significant promoter activity within each cell line tested (Additional file 2). We found the two values to be correlated, suggesting that LTRs with high sequence similarity are more likely to exhibit significant promoter activity under the transcriptional environment of the same cell line (Fig. 2a). Overall, LTRs with ~70% sequence similarity shared promoter expression patterns ~60% of the time, whereas LTRs with ~95% sequence identity shared promoter expression patterns ~90% of the time (Fig. 2b). With the exception of the 5′ LTR of 3q12.3 (Fig. 2, red), the sequences clustered into two observable groups. The expression pattern of the 3q12.3 5′ LTR was not similar to any other LTR and instead exhibited unusually high promoter activity levels, with significant promoter expression seen in almost every transformed cell line investigated (Fig. 1d).

Table 1 HML-2 proviruses with alternative names and genomic coordinates

Provirus	Alternative names	Chromosomal location (hg19)
1q22	K102, K(C1b), K50a, ERVK-7	chr1:155,596,457–155,605,636
3q12.3	KII, ERVK-5	chr3:101,410,737–101,419,859
3q21.2	KI, ERVK-4	chr3:125,609,302–125,618,439
5p13.3	K104, K50d	chr5:30,486,760–30,496,205
7p22.1b	K108R, ERVK-6	chr7:4,630,561–4,640,031
8p23.1c		chr8:12,073,970–12,083,497
11p15.4	K7	chr11:3,468,656–3,478,209
21q21.1	K60, ERVK-23	chr21:19,933,659–19,941,962
22q11.21	K101, K(C22), ERVK-24	chr22:18,926,187–18,935,361

From Subramanian et al. [4] and Montesion et al. [32]

Identification of transcription factor binding sites critical for HML-2 promoter activity during neoplasia

The association between LTR sequence and cell line-specific expression suggests that certain sequence-specific elements, such as transcription factor binding sites, play a large role in determining differential promoter activity. Increased HML-2 expression is largely seen during tumorigenesis and our recent results indicate that LTR-driven transcription does not occur until post-transformation [32]. The following experiments were performed to further investigate the relationship between malignant transformation and expression and to elucidate the specific LTR sequences required.

For this purpose, we focused on three cell lines: HME, HMLE-Her2, and HMLE-Ras. These cells were all derived from the same HMEC population and are therefore isogenic, differing only by oncogene overexpression. HME cells are non-transformed but immortalized through hTERT (human telomerase reverse transcriptase) overexpression. The HMLE cells, in addition to being hTERT-immortalized, are transformed through the introduction of SV40 large and small T antigens. HMLE-Her2 and

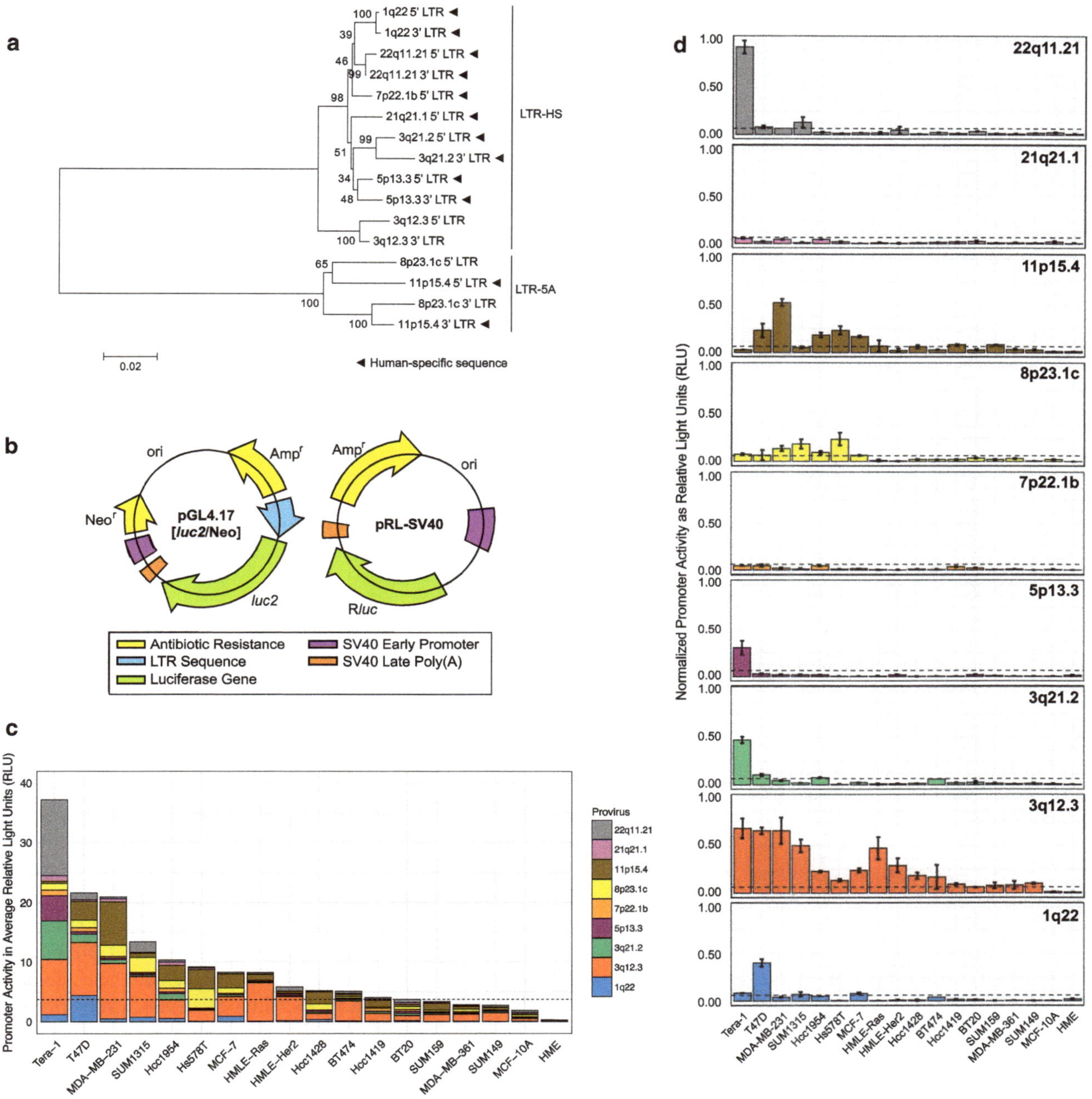

Fig. 1 HML-2 proviruses exhibit differential promoter activity in tumorigenic cells and negligible activity in immortalized HMEs. **a** Neighbor-joining tree displaying 5′ and 3′ LTR sequence relationship of the nine HML-2 proviruses used in this study. Bootstrap values are shown to the left of each node and scale is substitutions/site. LTR type (LTR-HS or LTR-5A) is shown to the right of the tree. Human-specific sequences are designated with a black triangle. **b** Schematic of the reporter constructs used in the dual-luciferase assay. Left, promoter-less firefly luciferase vector (pGL4.17[*luc2*/Neo]). Right, control *Renilla* luciferase vector (pRL-SV40). Direction of gene transcription is shown by arrows. Important gene regions are differentiated by colors and the names associated with those colors are displayed underneath. **c**, **d** Relative 5′ LTR promoter activity determined by dual luciferase assay in eighteen human cell lines. Data are organized by cell line in (**c**) and by provirus in (**d**). Promoter activity is displayed as relative light units (RLU) normalized against the internal control *Renilla* expression. Data in (**d**) are normalized against the highest expression value in the dataset. Statistical significance (dashed line, $p < 0.05$) was generated by ANOVA with Bonferroni's multiple comparisons test and is based on comparisons to HME expression. All experiments were conducted in triplicate and data displayed as mean (**c**) or mean ± standard deviation (**d**)

Table 2 Characterization of cell lines used for transfection

Breast cancer molecular subtype	Hormone receptor status	Cell lines
Luminal	ER+ and/or PR + HER2 ±	T47D, MCF-7, Hcc1428, BT474, MDA-MB-361
HER2/*neu*	ER− PR− HER2 +	SUM1315, Hcc1954, Hcc1419
Basal	ER− PR− HER2 −	MDA-MB-231, Hs578T, BT20, SUM159, SUM149
Additional cell types		**Cell lines**
Immortalized human mammary epithelial cells		HME, MCF-10A
Transformed human mammary epithelial cells		HMLE-Her2, HMLE-Ras
Human teratocarcinoma cells		Tera-1

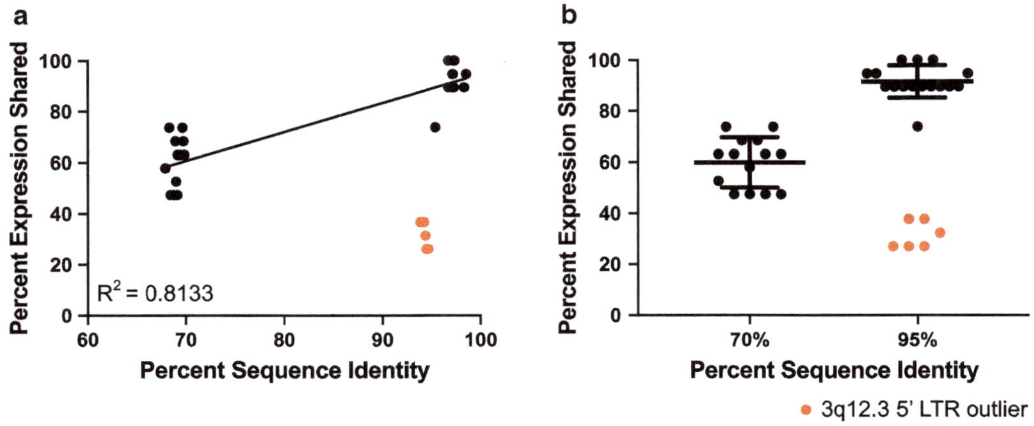

Fig. 2 LTR sequence identity is correlated with promoter expression patterns, with the exception of 3q12.3. Scatter plots displaying the correlation between percent sequence identity and shared percent expression. Raw values are shown in Additional file 2 and are based on pairwise comparisons. Best fit line and its R^2 value are shown for (**a**) (black values only). Error bars depict the mean ± standard deviation in (**b**) (black values only). Outlying 3q12.3 5′ LTR data points are shown in red for both plots

HMLE-Ras differ from one another by their oncogene overexpression, *ERBB2* (also known as HER2/*neu*) and *HRAS*, respectively. These cell lines provided the opportunity to investigate how specific differences in the transcriptional environment of the cell can affect LTR expression.

We detected increased promoter activity from 3q12.3 and 11p15.4 in HMLE-Ras cells as well as increased activity from 3q12.3 in HMLE-Her2 cells. The significance of this expression was determined as compared to the HME cell line (Fig. 3a). In effort to explain this pattern, we sought to identify transcription factor binding sites that are unique to each LTR and therefore may be responsible for the selective activation seen of one LTR over another. Using MatInspector, a transcription factor binding site prediction software by Genomatix [41], we found a total of 63 unique sites among the nine LTRs in this study. Of those, 13 were unique to 3q12.3 and 20 were unique to 11p15.4 (Table 3).

The same software was used to create a list of transcription factors predicted to bind to the unique sites on the 5′ LTRs of 3q12.3 and 11p15.4. In a previous study [32] the expressed RNAs of the HMLE-Ras, HMLE-Her2, and HME cell lines were sequenced, alongside the established human breast cancer cell line Hcc1954, using Illumina MiSeq sequencing. The transcript abundance levels, measured as FPKM, of these transcription factors were compared to assess upregulation of their expression in the tumorigenic cell lines as compared to the non-transformed HME control, and related to levels of expression of the proviruses at 3q12.3 and 11p15.4. Overall, we saw a significant increase in expression of transcription factors known to bind to the HOX-PBX and RFX3 sites on the 3q12.3 5′ LTR as well as a significant increase in those known to bind to the ATF and RORA sites on the 11p15.4 5′ LTR (Fig. 3b), implicating these sites and one or more of the upregulated factors in LTR activation during neoplasia.

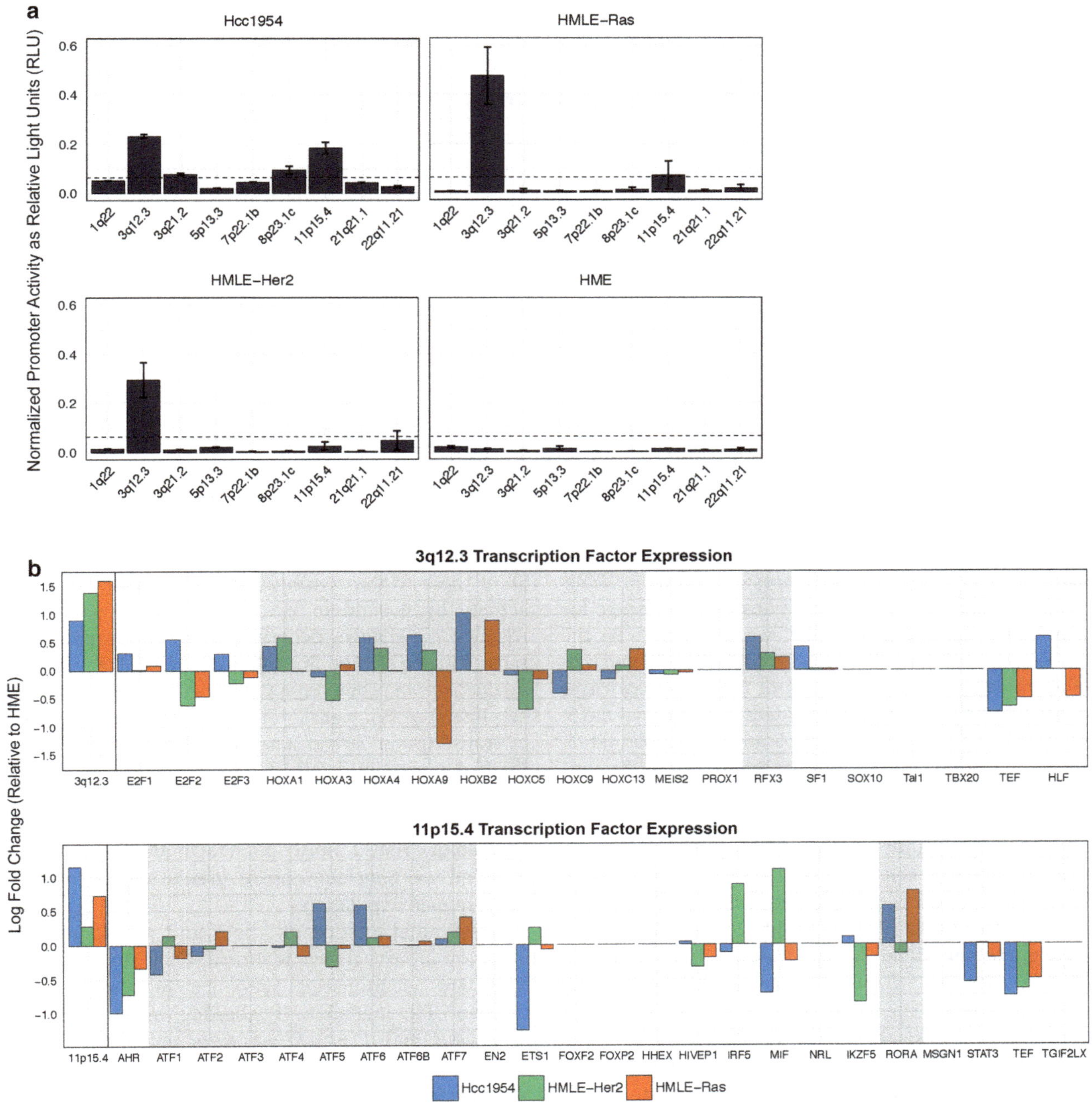

Fig. 3 Identification of transcription factor binding sites critical for tumorigenic HML-2 promoter activity. **a** Relative 5′ LTR promoter activity in the tumorigenic Hcc1954, HMLE-Ras, and HMLE-Her2 cell lines as well as the immortalized HME cell line. Promoter activity is determined as relative light units (RLU) normalized against the internal control *Renilla* luciferase expression and normalized against the highest expression value in the dataset. Statistical significance (dashed line, $p < 0.05$) was assessed by ANOVA with Bonferroni's multiple comparisons test and is based on comparisons to HME expression. All experiments were conducted in triplicate and data displayed as the mean ± standard deviation. **b** The log fold change of relative transcript abundance levels, in FPKM, of transcription factors predicted to bind to unique sites on the 5′ LTR of 3q12.3 (top) and 11p15.4 (bottom). Fold change is relative to expression in the HME cell line. The log fold change of relative promoter activity (RLU) in the 5′ LTR of 3q12.3 and 11p15.4 is shown to the left of each respective plot. Highlighted in gray are the transcription factors known to bind to the HOX-PBX and RFX3 binding sites (top) as well as the transcription factors known to bind to the ATF and RORA binding sites (bottom)

Table 3 Unique transcription factor binding sites found in HML-2 5′ LTRs of interest

Provirus	Unique binding site
1q22	NBRE‡
3q12.3	CDE, E2F, HOX-PBX, MRG1‡, PROX1, RFX3†, SF1†, SOX10, TAL1-E2A, TBX20, TEF-HLF‡, TGIF†, TR2‡
3q21.2	GLI1†, IK3, NFY‡, NKX29†, SIX2‡, STAT5
5p13.3	CARF‡, MYBL1‡
7p22.1b	EKLF‡, GAGA‡, GLI3‡
8p23.1c	AML1‡, BHLHB2‡, DMRT7‡, HMGA‡, HOX1-3‡, MAFF‡, MEF2‡, NRF1‡, PAX1‡, SOX17‡, STAT5A‡
11p15.4	AHRARNT‡, ATF, ATF6, CETS1P54, EN2‡, ETS1, FOXP2†, FREAC2‡, HDBP1-2, HHEX, HIVEP1†, IRF5, MIF1†, NRL, PEGASUS, RORA, SGN1, STAT3, TEF†, TGIF2LX
21q21.1	CHOP†, NFKAPPAB50†, USF†, ZNF300†
22q11.21	GRHL1‡, MASH1†, TAL1BETAHEB†

Only sites unique to each 5′ LTR, as compared to the other eight 5′ LTRs, are shown

† Present only in other HML-2 solo LTR(s)

‡ Present in other HML-2 full length provirus(es) and solo LTR(s)

Removal of critical binding sites decreases HML-2 promoter activity in neoplastic cell lines

The functionality of these sites was assessed by mutating each one individually. A multiple sequence alignment was performed using the sequences of all nine 5′ LTRs. From this analysis, we created a consensus sequence for each critical binding site, which we deemed to be the "non-active" version of each site. The full binding site sequence in each 5′ and 3′ LTR of the nine proviruses of interest in this study are provided in Additional file 3, Additional file 4, Additional file 5 and Additional file 6. The 3q12.3 HOX-PBX binding site differed from the consensus non-active sequence by a five base pairs, including a duplication of four nucleotides (Fig. 4a). Reversion of these sites significantly decreased LTR promoter activity in both neoplastic cell lines, with activity decreasing by twofold in HMLE-Ras cells (Fig. 4c, left) and by sevenfold in HMLE-Her2 cells (Fig. 4c, middle). The 3q12.3 RFX3 binding site only differed from the consensus sequence by one nucleotide, an A to C transversion (Fig. 4b), and yet removal of this site decreased LTR activity by fivefold in both HMLE-Ras cells (Fig. 4c, left) and HMLE-Her2 cells (Fig. 4c, middle). Activity was decreased to levels comparable to that of 1q22, a proviral LTR with no significant promoter activity in these cell lines (Fig. 4c). Mutating these sites did not significantly decrease LTR promoter activity in Hcc1954 cells (Fig. 4c, right), which also showed elevated expression of transcription factors known to bind to five unique 3q12.3 sites (E2F, HOX-PBX, RFX3, SF1, TEF-HLF) (Fig. 3b), suggesting that the other active binding sites can compensate for promoter activity when only some of them are removed.

Similar results were seen with the 11p15.4 5′ LTR. The consensus sequence differed from the ATF binding site by nine nucleotides (Fig. 5a) and back mutating the binding site to match the consensus sequenced decreased promoter activity by sixfold in HMLE-Ras cells (Fig. 5c, left). The RORA binding site differed by eleven nucleotides from the consensus sequence (Fig. 5b) and mutating all of these to the consensus bases decreased promoter activity by fivefold in HMLE-Ras cells (Fig. 5c, left). Again, these changes decreased activity to levels comparable with 1q22 (Fig. 5c, left). As in the case of 3q12.3, no decrease in promoter activity was seen in the Hcc1954 cell line (Fig. 5c, right), which had elevated expression of transcription factors known to bind to four unique 11p15.4 sites (ATF, HIVEP1, PEGASUS, RORA) (Fig. 3b, bottom).

Most unique HML-2 transcription factor binding sites were acquired over time following integration and are fixed in the human population

At the time of integration, the 5′ and 3′ LTRs of a provirus are almost always identical. Over time, as mutations are accumulated, sequence variation between the two LTRs increases. By aligning the 5′ and 3′ LTRs of 3q12.3 and 11p15.4, we were able to determine whether these critical transcription factor binding sites were present at the time of insertion (as evidenced by its presence in both LTRs) or were acquired over time (and found in only one LTR). We determined that one of the sites, RFX3 found in 3q12.3, was present at the time of insertion, but that three of the binding sites were acquired over time (Table 4). We analyzed the remaining unique binding sites in this same manner, with the exception of sites found on 7p22.1b and 21q21.1, which do not have full 3′ LTRs. Overall, only 21% (12/56) of the unique sites were present at the time of insertion (Fig. 6a, left), the majority of which (58%, 7/12) were found in the 3q12.3 5′ LTR (Fig. 6b).

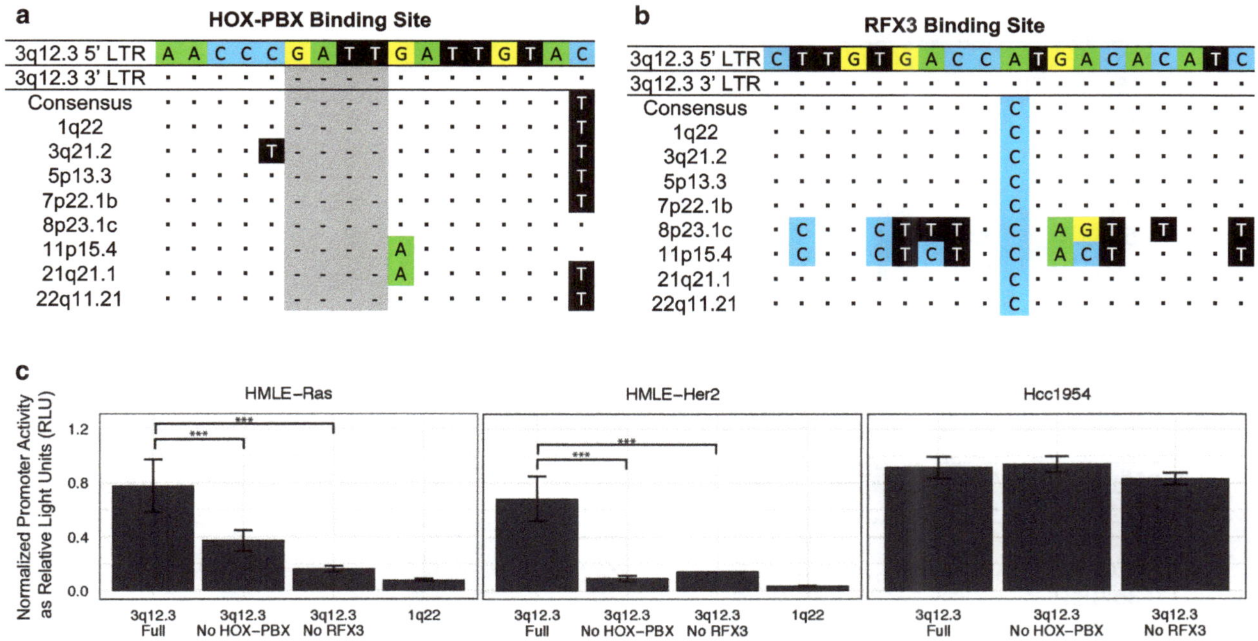

Fig. 4 Back mutation of critical transcription factor binding sites to consensus sequences on the 3q12.3 provirus. **a**, **b** Multiple sequence alignment of the **a** HOX-PBX and **b** RFX3 binding regions on the nine 5′ LTRs of interest in this study as well as a consensus sequence of the site. Sequences are compared against the 3q12.3 5′ LTR site, dots are used for shared identity, and dashes and shading indicate indels. **c** Relative 5′ LTR promoter activity in HMLE-Ras cells, HMLE-Her2 cells, and Hcc1954 cells. Constructs used either contained full HOX-PBX and RFX3 binding sites, or had a binding site removed through back mutation to the consensus sequence. Promoter activity of the 1q22 5′ LTR is shown for comparison. Promoter activity is determined as relative light units (RLU) normalized against the internal control *Renilla* luciferase expression. Statistical significance was assessed by ANOVA with Bonferroni's multiple comparisons test (***$p < 0.0005$). All experiments were conducted in triplicate and data are display as the mean ± standard deviation

To determine the distribution of these sites within the human population, we analyzed the VCF (Variant Call Format) files of 2504 individuals, as supplied by phase 3 of the 1000 Genomes Project [42]. Of the four binding sites that we found to be critical for HML-2 promoter expression during neoplasia, three had allele frequencies > 99% and are therefore fixed in the population. The RORA binding site, found in the 11p15.4 5′ LTR, was found to be polymorphic with an allele frequency of 50.76% (Table 4). Overall, only 8% (5/63) of the unique sites that we identified were polymorphic in the human population (Fig. 6a, right).

Evolution of the HML-2 HOX-PBX and RORA binding sites

Alignment of the 5′ and 3′ LTRs of the 3q12.3 provirus revealed a 4 bp insertion, found in the middle of the HOX-PBX site, resulting from duplication of a GATT sequence (Fig. 4a). This provirus is estimated to have integrated ~ 10 million years ago and is present in gorillas, chimpanzees, and bonobos, as well as humans [4]. Using the UCSC Genome Browser, we examined this LTR in several non-human primate reference genomes. We found that despite the conservation of the 3q12.3 provirus across multiple hominoid species, the 4 bp insertion, and consequently the HOX-PBX binding site, is only present in humans and Denisovans (Fig. 6c). These results suggest that this binding site was acquired sometime after the human-chimpanzee evolutionary split and has been stably integrated in the human genome ever since.

The RORA binding site on 11p15.4 was one of the only polymorphic unique binding sites that we identified. This polymorphism is due to a single nucleotide change, where 51% of alleles in the human population contains an A at the 23rd base pair in the site (and therefore an intact RORA site) and 49% of the population contains a T. This provirus is of particular interest because 11p15.4 is a segmental duplication of 8p23.1c, which is estimated to have integrated ~ 20 million years ago. Although the proviral sequence is quite old, the duplication occurred after the human-chimpanzee split, and the 11p15.4 sequence is human-specific [4]. We aligned the 5′ and 3′ LTRs of these two proviruses and compared their sequences at the RORA binding site. We found that although both of the 3′ LTRs at this site are identical, the 5′ LTRs differ by one nucleotide,

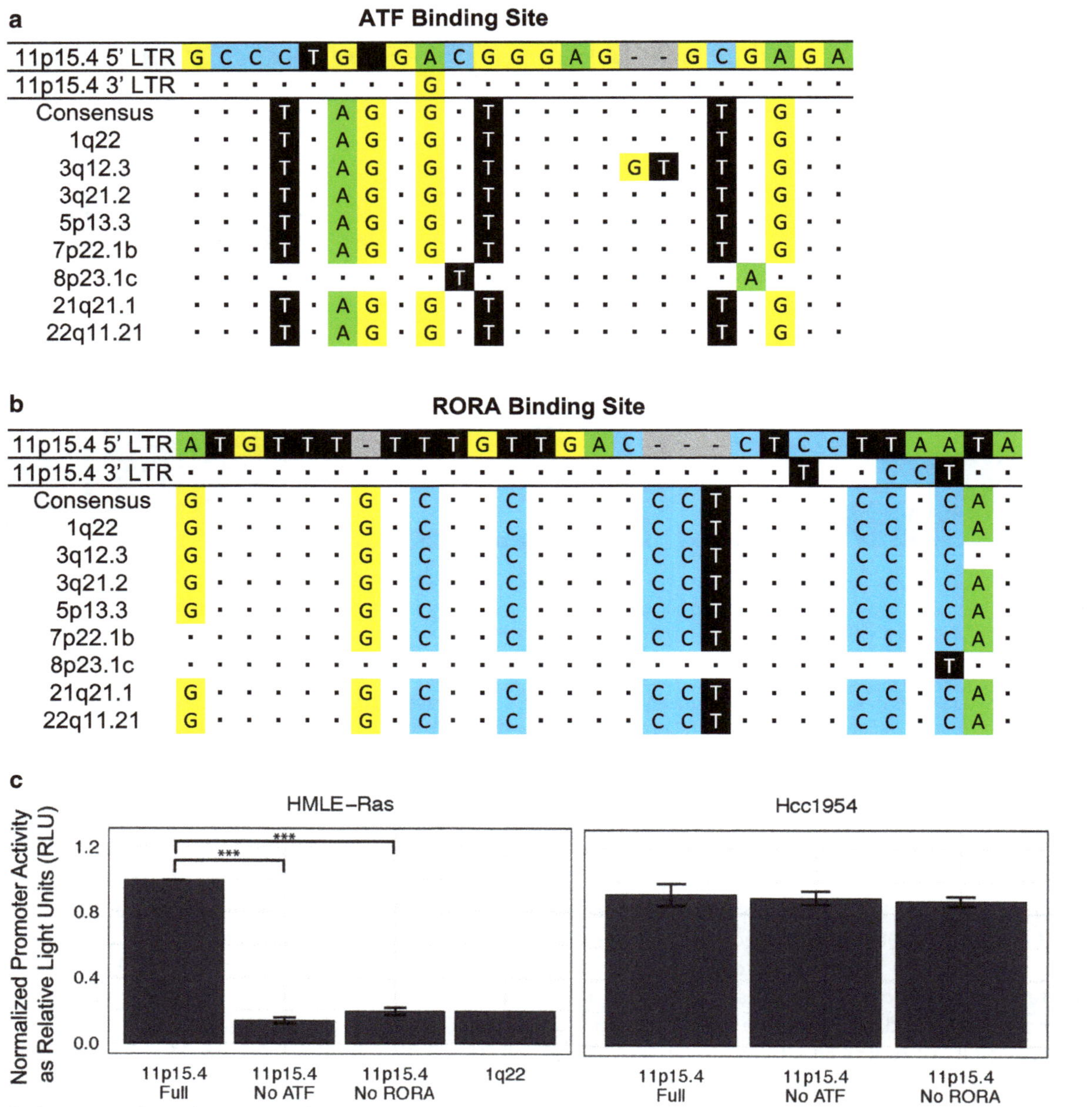

Fig. 5 Back mutation of critical transcription factor binding sites to consensus sequences on the 11p15.4 provirus. **a**, **b** Multiple sequence alignment of the **a** ATF and **b** RORA binding regions on the nine 5′ LTRs of interest in this study as well as a consensus sequence of the site. Sequences are compared against the 11p15.4 5′ LTR site, dots are used for shared identity, and dashes indicate indels. **c** Relative 5′ LTR promoter activity in HMLE-Ras cells and Hcc1954 cells. Constructs used either contained full ATF and RORA binding sites, or had a binding site removed through back mutation to the consensus sequence. Promoter activity of the 1q22 5′ LTR is shown for comparison. Promoter activity is determined as relative light units (RLU) normalized against the internal control *Renilla* expression. Statistical significance was generated by ANOVA with Bonferroni's multiple comparisons test (***p < 0.0005). All experiments were conducted in triplicate and data displayed as the mean ± standard deviation

the same 23rd nucleotide that is responsible for the RORA polymorphism (Fig. 6d). The 5′ LTR also differed from the 3′ by deletion of 3 bp, which must have predated the segmental duplication of this provirus, as it was also found in the 8p23.1c 5′ LTR. Based on these observations, it appears as though the provirus at 11p15.4 in half of the human population has evolved away from the ancestral 8p23.1c sequence, resulting in a functional RORA binding site.

Table 4 Characterization of LTR binding sites critical for 3q12.3 and 11p15.4 promoter activity in tumorigenic cells

Provirus	Binding site	LTR	Binding site allele frequency	Binding site evolution
3q12.3	HOX-PBX	5′ LTR	99.68% (fixed)	Acquired
	RFX3	5′ and 3′ LTR	99.96% (fixed)	Present at the time of insertion
11p15.4	ATF	5′ LTR	99.88% (fixed)	Acquired
	RORA	5′ LTR	50.76% (polymorphic)	Acquired

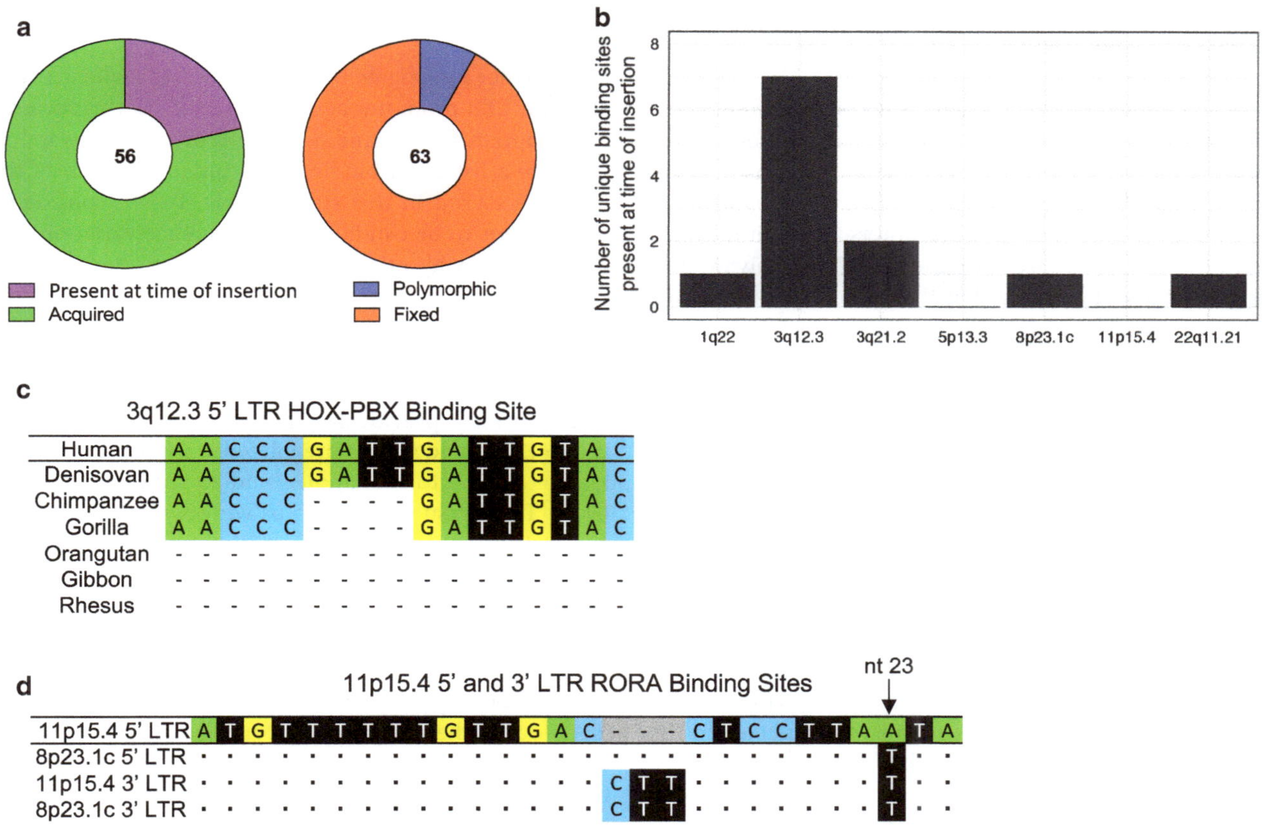

Fig. 6 Characterization of unique transcription factor binding sites. **a** Pie charts describing the unique binding sites in the 5′ and 3′ LTRs of proviruses in this study. Shown is the number of unique binding sites that were either present at the time of insertion or acquired over time (left) as well as the number of unique binding sites that are either polymorphic or fixed in the human population (right). **b** The number of unique transcription factor binding sites identified as being present at the time of insertion found on each 5′ LTR of interest in this study. **c** Multiple sequence alignment of the human HOX-PBX binding site in the 3q12.3 5′ LTR as compared to the homologous sequences in different non-human primate reference genomes. Dashes indicate indels. **d** Multiple sequence alignment of the RORA binding site in the 11p15.4 5′ and 3′ LTRs as compared to the 5′ and 3′ LTRs of the 8p23.1c provirus. Dots are used for shared identity and dashes indicate indels. The nucleotide that appears to be responsible for the RORA functional polymorphism, is shown with an arrow

Discussion

Post-integration, retroviral sequences are transcribed and translated like any other cellular gene and are subject to the same selective pressures. Germline sequences with neutral or advantageous effects can become fixed in the population, resulting in endogenization [2, 25, 43]. These sequences provide unique opportunities to study the evolutionary relationship between host and pathogen, including adaptations for assimilation within the host genome.

The full biological significance of HERVs remains to be uncovered. Repetitive mobile sequences are often credited with contributing to genome plasticity and HERVs, equipped with multiple splice junctions, promoter/

enhancer sites, and polyadenylation signals, are abundantly capable of altering host gene expression [2, 8, 43]. A number of endogenous retroviruses, including the mouse mammary tumor virus, murine leukemia virus, and Jaagsiekte sheep retrovirus, exhibit both endogenous and exogenous transmission and are capable of inducing carcinogenesis. Since the pathogenicity of these viruses is generally due to LTR activity and integration site, which can result in the alteration of expression of nearby proto-oncogenes [43, 44], endogenous viral sequences are often silenced through epigenetic and chromatin modifications such as CpG methylation [12–14].

Our group recently characterized the HML-2 transcriptome during HMEC transformation and found that the site of proviral integration is often crucial for expression, with the majority of expressed proviruses being transcribed by non-LTR-driven mechanisms such as read-through from adjacent promoters. When it was present, LTR-driven transcription was detected only in tumorigenic cells, suggesting that the altered transcriptional milieu of a transformed cell is critical for LTR promoter activation [32]. The goal of this study was to investigate the interplay between LTR sequence variation and cellular environment and to look for evidence of evolutionary adaptations that could result in increased activity during neoplasia.

LTR hypomethylation, commonly seen in malignant cells, is well documented to result in increased ERV expression [13, 29, 45]. To eliminate this issue, we decided to investigate the relationship between LTR sequence similarity and differential expression patterns using reporter construct assays, where methylation status is not a factor. We chose to study nine HML-2 proviruses, shown by single-genome sequencing to be highly transcribed across a number of breast cancer cell lines (Additional file 1). Phylogenetic analysis of these 5′ LTRs classified most of them as LTR-HS, the LTR group that includes the youngest proviruses and most human-specific integrations [4, 9]. Of these, only one provirus (3q12.3) is known to not be human-specific, as it is present in gorillas and chimpanzees as well (Fig. 1a). All proviruses in this study are fixed in the human population, although one provirus (7p22.1b) is considered to be allelically polymorphic. It is present as either a solo LTR, formed through the recombination of the 5′ and 3′ LTRs and excision of the internal proviral sequence, or a full ("2-LTR") provirus [4, 7]. However, in either case, the 5′ LTR of interest is fixed and as such, for the purpose of this study, we do not consider any of these LTRs to be insertionally polymorphic.

Overall, we found significant HML-2 promoter activity in 73% (11/15) of tumorigenic HME cell lines (Fig. 1c), consistent with previous reports of increased HML-2 expression in up to 85% of breast cancer samples [30, 31, 37]. Molecular subtype, as denoted by hormone receptor status, of the cell lines was noted (Table 2), but no significant correlation with HML-2 promoter expression was observed (Additional file 7). Pairwise comparisons of 5′ LTR sequence identity and promoter expression in our luciferase panel revealed a positive correlation between the results of the two assays (Fig. 2). These results suggest that LTRs with similar sequences share similar promoter expression patterns, most likely due to conservation of the same transcription factor binding sites and core promoter elements.

To further investigate the importance of sequence variation on LTR promoter activity, we used MatInspector, a transcription factor binding site prediction software, to generate a list of all binding sites unique to each of the nine LTRs used in this study (Table 3). We considered unique sites to be candidates for sequence variation that may explain why one LTR would be activated under a certain cellular condition instead of another. Two proviruses, 3q12.3 and 11p15.4, exhibited the highest levels of promoter activity across our luciferase panel (Fig. 1d). We used the MatInspector data, alongside RNA-Seq results from a previously published experiment by our group [32], to identify upregulated transcription factors known to bind to the unique binding sites on these two LTRs. These results provided us with two candidate sites per 5′ LTR for the promoter activation we saw during neoplasia: the HOX-PBX and RFX3 sites on 3q12.3 and the ATF and RORA sites on 11p15.4 (Fig. 3b). Removal of these sites individually decreased LTR promoter activity in HMLE-Ras and HMLE-Her2 cells by two to sevenfold (Figs. 4c, 5c).

All four of these binding sites are known to be involved with transcriptional activation, particularly during the regulation of human embryogenesis [46–49]. Interestingly, this observation is consistent with previous literature suggesting that HERVs are regulated in manners similar to stem cell genes, relying on cell-specific transcription factors and epigenetic modifications rather than TATA boxes or other canonical promoter elements [25, 33]. The evolution of the HOX-PBX and RORA binding sites were of most interest. Although the 3q12.3 provirus can be traced back through the primate lineage to gorillas, the HOX-PBX binding site is only found in the 5′ LTR in Denisovan and human genomes (Fig. 6c). Due to lack of coverage of the Neandertal reference genome at this location, it's unclear if this binding site is present in that species. This site, created by duplication of a GATT sequence, appears to have been acquired after the evolutionary split between humans and chimpanzees and has been fixed in the human population ever since. This analysis suggests that although the 3q12.3 provirus is evolutionarily conserved amongst several non-human primate species, the HOX-PBX binding site is human-specific. Although HOX proteins are widely expressed during development, aberrant expression has been documented during malignancy and increased HOX gene expression is being investigated as a potential breast cancer biomarker [50].

Alignments between the 5′ and 3′ LTR of proviruses shed light on the evolution of unique transcription factor binding sites. We were able to determine if sites were present at the time of insertion (present in both LTRs) or acquired over time (present in only one LTR). Only 21% of the unique binding sites that we identified were present at the time of insertion (Fig. 6a, left), implying that the expression patterns observed for these proviruses would not have reflected those of the ancestral virus that gave rise to them. Furthermore, the majority of unique sites were in the 3q12.3 5′ LTR (Fig. 6b). This distribution is consistent with the greater genetic distance and greater age of this provirus from the rest of the LTR-HS group (Fig. 1a). The high degree of unique sites present at the time of insertion may also explain why this particular provirus had an expression pattern widely different from the other LTRs in this study (Fig. 2).

Due to their possible role in pathogenicity, it is essential to study the genetic differences of HML-2 elements among individuals. Most often, such studies focus on whole proviruses, studying insertional polymorphism and its possible contribution to disease. Thus far, however, no polymorphic proviruses have been found to play a role in the genesis of cancer [34, 51]. To our knowledge, ours is the first study to investigate genetic differences at the single nucleotide level, by examining SNPs within LTRs. Of the 63 binding sites unique to one of the expressed LTRs that we identified, only five of them were found to be polymorphic within the 2504 genomes mined (Fig. 6a, right). These allele frequencies were further broken down by super-population, showing only slightly higher prevalence of these binding sites in the African population (Additional file 8).

The RORA binding site, harbored on the 5′ LTR of the 11p15.4 provirus, was the only site critical for HML-2 activation during neoplasia that was also polymorphic (Table 4). This provirus is of particular interest because it is a segmental duplication of 8p23.1c [4], which showed no LTR activity during tumorigenesis. After examining the RORA binding sites on both of these LTRs, we found that 51% of the population contains an active RORA site whereas the other half of the population contains an inactive RORA site, identical to the ancestral 8p23.1c 5′ LTR. Thus, more than half of the human population has evolved away from the ancestral sequence and towards a more active LTR version (Fig. 6d).

Conclusions

The role, if any, of HERV activity during tumorigenesis is unknown. It is currently unclear if HML-2 expression is an ancillary consequence of transformation or if it somehow aids in the event; although recent work shows that Env protein expression may increase the ability of tumor cells to evade immune surveillance during some cancers [52] or even participate directly in the transformation process by interacting with cellular proto-oncogenes [53]. Although no provirus of interest in our study is believed to have a viable open reading frame for any viral gene, protein production in these cell lines as well as any sample used in future investigations, should be examined. Our results show that HML-2 promoter activity is present in the majority (73%) of breast cancer cell lines tested and that LTR sequence similarity is correlated with promoter expression patterns. From there, we were able to map binding sites seemingly crucial for HML-2 promoter expression during neoplasia, many of which were acquired over evolutionary time. The polymorphism of certain sites provides another dimension in regards to what causes differential expression of ERVs between individuals. These data may shed light on adaptive co-evolution of ERVs within their host cells.

In recent years, there have been numerous reports of co-option of endogenous proviral sequences to disparate features of normal human and vertebrate biology, including protection against infection by related exogenous viruses [54], formation of the placental syncytiotrophoblast layer [10], expression of salivary amylase [55], stimulation of innate immunity [56], stimulation of neurological synapses promoting long-term memory [57], among others. It is particularly noteworthy that transcription of the two most highly expressed proviruses in our panel of ex vivo transformed cancer cell lines was facilitated through binding sites that were created by mutations in the 5′ LTRs that arose and spread in the human population following integration, implying that the expression patterns observed do not reflect those of the ancestral virus. It is tempting to speculate that responsiveness of the mutant proviruses to common, development-specific transcription factors might have given them some beneficial property along the lines of the ones listed above, thereby providing a selective advantage to the individuals carrying them and promoting their rapid fixation in the population.

Methods

Cell culture

The HME, HMLE-Her2, HMLE-Ras, MCF-10A, SUM149, SUM159, MDA-MB-361, Hcc1419, Hcc1428, and SUM1315 cell lines were grown in the Kuperwasser lab at Tufts University as previously described [32] and all other cell lines were obtained from ATCC (Manassas, VA, USA). All cell lines were grown as per ATCC's recommendations and detailed information regarding their origin and culture conditions can be found in Additional file 9.

Single-genome sequencing

ZR-75-1, MCF-7, T47D, SK-BR-3, Hcc1954, BT20, Hs578T, and MDA-MB-231 breast cancer cells were grown to 90% confluency. RNA was extracted and purified using the RNeasy Mini Kit (Qiagen, Valencia, CA, USA, Cat. No. 74104) and all DNA contamination was removed through DNase treatment (Turbo DNA-*free* Kit, Ambion, Foster City, CA, USA, Cat. No. AM1907). RT reactions were set up as recommended by the manufacturer's protocol using an oligo(dT) primer (SuperScript III One-Step RT-PCR System, Invitrogen, Carlsbad, CA, USA, Cat. No. 12574-018). The resulting cDNA was serially diluted down to an average of 1/3 genome per sample and amplified using Taq DNA polymerase (Invitrogen, Cat. No. 10342-020). Two forward primers (5′-TTCCTT TACAAAGTTGCGTAAAGC-3′, 5′-GTTGCGTAAAGC CCCCTTAT-3′) and one reverse primer (5′-CACAGA CACAGTAACAATCTG-3′), all targeting the HML-2 *env* region, were used in the reaction. The amplified products were gel extracted with the QIAquick Gel Extraction Kit (Qiagen, Cat. No. 28704) and purified samples were sent out for sequencing. The primers used for sequencing were 5′-GACTCCCAGACTATAACCTGTC-3′ and 5′-CGAAGCATCAAAAGCCCA-3′. Sequencing results were BLAT searched in the UCSC Genome Browser [58] to identify expressed proviruses.

Phylogenetic analysis

The 5′ and 3′ LTR sequence of each provirus of interest was obtained from the UCSC Genome Browser's RepeatMasker Track [58, 59] and imported as FASTA files into the Molecular Evolutionary Genetics Analysis (MEGA, v6.06) program for alignment using Multiple Sequence Comparison by Log-Expectation (MUSCLE) [60, 61]. Phylogeny of aligned sequences was determined by sequence dissimilarity and a neighbor-joining tree was constructed using a p-distance algorithm. Bootstrap values were determined using 1000 replicate tests.

Dual-luciferase assay

Primers for LTR amplification were selected using the Primer3 program [62]. Restriction enzyme cleavage sites were appended to the 5′ end of the primer sequences for proper vector ligation. The primers created are listed in Additional file 10. The LTR sequences were PCR-amplified using Taq DNA polymerase. Template DNA was purified from Tera-1 cells using the DNeasy Blood and Tissue Kit (Qiagen, Cat. No. 69504). The amplified sequences were cloned using basic molecular biology techniques and ligated into the multiple cloning region of the pGL4.17[*luc2*/Neo] promoter-less firefly luciferase vector (Promega, Madison, WI, USA, Cat. No. E6721). All constructs were sequenced to check for PCR-induced mutations before transfection. All cell cultures were seeded in triplicate at 100,000 cells/well in a 24-well plate for transfection. Cultures were co-transfected with the pGL4 vector alongside a pRL-SV40 internal control *Renilla* luciferase vector (Promega, Cat. No. E2231) at a 30:1 ratio using Opti-MEM reduced serum media (Gibco, Cat. No. 31985-070) and Lipofectamine 2000 (Thermo Fisher Technologies, Cat. No. 11668-019), as recommended by the manufacturer's protocol. Post-transfection, cells were incubated at 37 °C for 48 h before lysis and analysis. Luminescence was measured via the dual-luciferase assay system (Promega, Cat. No. E1910) and quantified as relative light units (RLU) on a BioTek Synergy HT plate reader using Gen5 Data Analysis Software (BioTek Instruments, Winooski, VT, USA). Empty vectors as well as non-transfected cells were measured as a control to determine any cell-specific background signal. LTR promoter activity was calculated as *luc2* activity normalized against that of the internal *Renilla* luciferase control signal.

HML-2 similarity matrices

The sequence of each "full length" (i.e., not solo LTR) HML-2 provirus annotated within the human reference genome (hg19 build) was obtained from the UCSC Genome Browser [58]. These sequences were input into the Clustal Omega program (The European Bioinformatics Institute (EMBL-EBI), Hinxton, Cambridge, UK) [40] to create a multiple sequence alignment using the HHalign algorithm [63] and to create a percent sequence identity matrix. The HML-2 percent expression similarity matrix was created by making pairwise comparisons of significant promoter expression in each of the eighteen cell lines used in our dual-luciferase analysis.

Transcription factor binding site analysis

The full sequence of each 5′ LTR of interest was imported into MatInspector, a transcription factor binding site prediction software provided by Genomatix [41]. Any site that was identified in more than one provirus was removed from the analysis to produce a list containing all predicted binding sites unique to each LTR. This program also provided information regarding transcription factors that are known to bind to these sites. Transcript abundance levels of these transcription factors in the Hcc1954, HMLE-Ras, HMLE-Her2, and HME cell lines were determined by Cuffdiff analysis of our previous RNA-Seq results. A full description of the study used to obtain these values is detailed in our previous publication [32] and the RNA-Seq data are deposited in the NCBI Gene Expression Omnibus database under Accession Number GSE84275.

Consensus sequences of the HOX-PBX, RFX3, ATF, and RORA binding sites were determined through a separate MEGA alignment. New reporter constructs containing the consensus (non-active) sites were created through IDT's gBlocks® Gene Fragments synthesis service (Integrated DNA Technologies, Inc., Coralville, IA, USA). These fragments were directly cloned into the pGL4[*luc2*/Neo] firefly luciferase vector and transfected into cell lines as previously described in the Dual-Luciferase Assay section of the Materials and Methods.

The 5′ and 3′ LTRs of each of the nine proviruses of interest were analyzed in an additional MEGA alignment. All unique transcription factor binding sites found in only one LTR were regarded as being "acquired" and any unique binding sites found in both LTRs were characterized as "present at time of insertion". Sites located in the 7p22.1b and 21q21.1 proviruses were excluded from the analysis since they no longer possess intact 3′ LTRs [4].

The allele frequencies of each unique binding site were calculated from the VCF (Variant Call Format) files of 2504 individuals, as supplied by phase 3 of the 1000 Genomes Project [42]. VCF files were analyzed computationally using VCFtools, by specifying the genomic coordinates (hg19 build) of each site of interest. All sites with an allele frequency of at least 89% were considered to be fixed in the human population. All sites that were classified as polymorphic within the population had allele frequencies of 52% or less. No binding site that we identified had an allele frequency intermediate of those two thresholds, i.e. calculated to be greater than 52% but less than 89%.

The HOX-PBX binding site was further analyzed in several non-human primate reference genomes as supplied by the UCSC Genome Browser [58]. The Denisovan reference genome sequence was obtained from the Denisova High-Coverage Sequence Reads of the Denisova Seq Track. The chimpanzee, gorilla, orangutan, gibbon, and rhesus reference genome sequences were obtained from the Vertebrate Multiz Alignment & Conservation Track.

Additional files

Additional file 1: Table S1. HML-2 transcript levels detected through single-genome sequencing in breast cancer cell lines of varying molecular subtype.

Additional file 2: Table S2. HML-2 similarity matrices.

Additional file 3: Table S3. HOX-PBX binding site sequences and genomic coordinates (hg19).

Additional file 4: Table S4. RFX3 binding site sequences and genomic coordinates (hg19).

Additional file 5: Table S5. ATF binding site sequences and genomic coordinates (hg19).

Additional file 6: Table S6. RORA binding site sequences and genomic coordinates (hg19).

Additional file 7: Figure S1. HML-2 promoter activity is not breast cancer subtype-specific. Total relative 5′ LTR promoter activity levels of fifteen tumorigenic breast cancer cell lines broken down by molecular subtype (luminal, HER2+, and basal-like) as compared to two immortalized HME cell lines. Hormone receptor status and cell lines identified as being each molecular subtype are shown in detail in Table 2. All experiments were conducted in triplicate and data display the mean ± standard deviation.

Additional file 8: Figure S2. Allele frequencies of polymorphic HML-2 5′ LTR transcription factor binding sites within each superpopulation. Allele frequencies were determined for the proviruses shown from 2504 individuals from the 1000 Genomes Project and broken down by superpopulation (EAS = East Asian; AMR = Ad Mixed American; AFR = African; EUR = European; SAS = South Asian). The name of the transcription factor binding site as well as the provirus of interest are shown at the top of each graph.

Additional file 9: Table S7. Culture methods for cell lines used.

Additional file 10: Table S8. Primers used to amplify 5′ LTRs of transfected HML-2 proviruses.

Authors' contributions

MM, CK, and JMC conceived and designed the experiments. MM and RPS performed the experiments. MM, ZHW, RPS, and JMC analyzed the data. MM and JMC wrote the paper and all authors read and approved the final manuscript.

Author details

[1] Department of Molecular Biology and Microbiology, Tufts University School of Medicine, Boston, MA, USA. [2] Department of Developmental, Chemical, and Molecular Biology, Tufts University School of Medicine, Boston, MA, USA. [3] Raymond and Beverly Sackler Convergence Laboratory, Tufts University School of Medicine, Boston, MA, USA. [4] Present Address: Foundation Medicine, Inc., Cambridge, MA, USA. [5] Present Address: Excerpta Medica, New York, NY, USA.

Acknowledgements

We thank the Tufts University Genomics core facility for their RNA-Seq advice and as John Yoon for helpful discussion and editorial advice.

Competing interests

The authors declare that they have no competing interests.

Funding

This project was supported by research Grants R37 CA 089441 and R35 CA 200421 from the National Cancer Institute.

References

1. Lower R, Lower J, Tondera-Koch C, Kurth R. A general method for the identification of transcribed retrovirus sequences (R-U5 PCR) reveals the expression of the human endogenous retrovirus loci HERV-H and HERV-K in teratocarcinoma cells. Virology. 1993;192:501–11. https://doi.org/10.1006/viro.1993.1066.
2. Jern P, Coffin JM. Effects of retroviruses on host genome function. Annu Rev Genet. 2008;42:709–32. https://doi.org/10.1146/annurev.genet.42.110807.091501.
3. Ono M, Yasunaga T, Miyata T, Ushikubo H. Nucleotide sequence of human endogenous retrovirus genome related to the mouse mammary tumor virus genome. J Virol. 1986;60:589–98.
4. Subramanian RP, Wildschutte JH, Russo C, Coffin JM. Identification, characterization, and comparative genomic distribution of the HERV-K (HML-2) group of human endogenous retroviruses. Retrovirology. 2011;8:90. https://doi.org/10.1186/1742-4690-8-90.
5. Armbruester V, Sauter M, Krautkraemer E, Meese E, Kleiman A, Best B, et al. A novel gene from the human endogenous retrovirus K expressed in transformed cells. Clin Cancer Res. 2002;8:1800–7.
6. Gonzalez-Hernandez MJ, Cavalcoli JD, Sartor MA, Contreras-Galindo R, Meng F, Dai M, et al. Regulation of the human endogenous retrovirus K (HML-2) transcriptome by the HIV-1 Tat protein. J Virol. 2014;88:8924–35. https://doi.org/10.1128/JVI.00556-14.
7. Wildschutte JH, Williams ZH, Montesion M, Subramanian RP, Kidd JM, Coffin JM. Discovery of unfixed endogenous retrovirus insertions in diverse human populations. Proc Natl Acad Sci U S A. 2016. https://doi.org/10.1073/pnas.1602336113.
8. Schmitt K, Reichrath J, Roesch A, Meese E, Mayer J. Transcriptional profiling of human endogenous retrovirus group HERV-K(HML-2) loci in melanoma. Genome Biol Evol. 2013;5:307–28. https://doi.org/10.1093/gbe/evt010.
9. Bhardwaj N, Montesion M, Roy F, Coffin JM. Differential expression of HERV-K (HML-2) proviruses in cells and virions of the teratocarcinoma cell line Tera-1. Viruses. 2015;7:939–68. https://doi.org/10.3390/v7030939.
10. Mi S, Lee X, Li X, Veldman GM, Finnerty H, Racie L, et al. Syncytin is a captive retroviral envelope protein involved in human placental morphogenesis. Nature. 2000;403:785–9. https://doi.org/10.1038/35001608.
11. Reiss D, Zhang Y, Mager DL. Widely variable endogenous retroviral methylation levels in human placenta. Nucleic Acids Res. 2007;35:4743–54. https://doi.org/10.1093/nar/gkm455.
12. Gotzinger N, Sauter M, Roemer K, Mueller-Lantzsch N. Regulation of human endogenous retrovirus-K Gag expression in teratocarcinoma cell lines and human tumours. J Gen Virol. 1996;77(Pt 12):2983–90. https://doi.org/10.1099/0022-1317-77-12-2983.
13. Florl AR, Lower R, Schmitz-Drager BJ, Schulz WA. DNA methylation and expression of LINE-1 and HERV-K provirus sequences in urothelial and renal cell carcinomas. Br J Cancer. 1999;80:1312–21. https://doi.org/10.1038/sj.bjc.6690524.
14. Conklin KF, Coffin JM, Robinson HL, Groudine M, Eisenman R. Role of methylation in the induced and spontaneous expression of the avian endogenous virus ev-1: DNA structure and gene products. Mol Cell Biol. 1982;2:638–52.
15. Santoni FA, Guerra J, Luban J. HERV-H RNA is abundant in human embryonic stem cells and a precise marker for pluripotency. Retrovirology. 2012;9:111. https://doi.org/10.1186/1742-4690-9-111.
16. Xie W, Schultz MD, Lister R, Hou Z, Rajagopal N, Ray P, et al. Epigenomic analysis of multilineage differentiation of human embryonic stem cells. Cell. 2013;153:1134–48. https://doi.org/10.1016/j.cell.2013.04.022.
17. Smith ZD, Chan MM, Humm KC, Karnik R, Mekhoubad S, Regev A, et al. DNA methylation dynamics of the human preimplantation embryo. Nature. 2014;511:611–5. https://doi.org/10.1038/nature13581.
18. Ohnuki M, Tanabe K, Sutou K, Teramoto I, Sawamura Y, Narita M, et al. Dynamic regulation of human endogenous retroviruses mediates factor-induced reprogramming and differentiation potential. Proc Natl Acad Sci U S A. 2014;111:12426–31. https://doi.org/10.1073/pnas.1413299111.
19. Grow EJ, Flynn RA, Chavez SL, Bayless NL, Wossidlo M, Wesche DJ, et al. Intrinsic retroviral reactivation in human preimplantation embryos and pluripotent cells. Nature. 2015. https://doi.org/10.1038/nature14308.
20. Frank JA, Feschotte C. Co-option of endogenous viral sequences for host cell function. Curr Opin Virol. 2017;25:81–9. https://doi.org/10.1016/j.coviro.2017.07.021.
21. Glinsky GV. Activation of endogenous human stem cell-associated retroviruses (SCARs) and therapy-resistant phenotypes of malignant tumors. Cancer Lett. 2016;376:347–59. https://doi.org/10.1016/j.canlet.2016.04.014.
22. Kudo-Saito C, Yura M, Yamamoto R, Kawakami Y. Induction of immunoregulatory CD271+ cells by metastatic tumor cells that express human endogenous retrovirus H. Cancer Res. 2014;74:1361–70. https://doi.org/10.1158/0008-5472.CAN-13-1349.
23. Kreimer U, Schulz WA, Koch A, Niegisch G, Goering W. HERV-K and LINE-1 DNA methylation and reexpression in urothelial carcinoma. Front Oncol. 2013;3:255. https://doi.org/10.3389/fonc.2013.00255.
24. Lavie L, Kitova M, Maldener E, Meese E, Mayer J. CpG methylation directly regulates transcriptional activity of the human endogenous retrovirus family HERV-K(HML-2). J Virol. 2005;79:876–83. https://doi.org/10.1128/JVI.79.2.876-883.2005.
25. Fuchs NV, Kraft M, Tondera C, Hanschmann KM, Lower J, Lower R. Expression of the human endogenous retrovirus (HERV) group HML-2/HERV-K does not depend on canonical promoter elements but is regulated by transcription factors Sp1 and Sp3. J Virol. 2011;85:3436–48. https://doi.org/10.1128/JVI.02539-10.
26. Stengel S, Fiebig U, Kurth R, Denner J. Regulation of human endogenous retrovirus-K expression in melanomas by CpG methylation. Genes Chromosomes Cancer. 2010;49:401–11. https://doi.org/10.1002/gcc.20751.
27. Knossl M, Lower R, Lower J. Expression of the human endogenous retrovirus HTDV/HERV-K is enhanced by cellular transcription factor YY1. J Virol. 1999;73:1254–61.
28. Ono M, Kawakami M, Ushikubo H. Stimulation of expression of the human endogenous retrovirus genome by female steroid hormones in human breast cancer cell line T47D. J Virol. 1987;61:2059–62.
29. Wang-Johanning F, Frost AR, Johanning GL, Khazaeli MB, LoBuglio AF, Shaw DR, et al. Expression of human endogenous retrovirus k envelope transcripts in human breast cancer. Clin Cancer Res. 2001;7:1553–60.
30. Wang-Johanning F, Frost AR, Jian B, Epp L, Lu DW, Johanning GL. Quantitation of HERV-K env gene expression and splicing in human breast cancer. Oncogene. 2003;22:1528–35. https://doi.org/10.1038/sj.onc.1206241.
31. Zhao J, Rycaj K, Geng S, Li M, Plummer JB, Yin B, et al. Expression of human endogenous retrovirus type K envelope protein is a novel candidate prognostic marker for human breast cancer. Genes Cancer. 2011;2:914–22. https://doi.org/10.1177/1947601911431841.
32. Montesion M, Bhardwaj N, Williams ZH, Kuperwasser C, Coffin JM. Mechanisms of HERV-K (HML-2) Transcription during human mammary epithelial cell transformation. J Virol. 2018. https://doi.org/10.1128/jvi.01258-17.
33. Manghera M, Douville RN. Endogenous retrovirus-K promoter: a landing strip for inflammatory transcription factors? Retrovirology. 2013;10:16. https://doi.org/10.1186/1742-4690-10-16.
34. Wildschutte JH, Ram D, Subramanian R, Stevens VL, Coffin JM. The distribution of insertionally polymorphic endogenous retroviruses in breast cancer patients and cancer-free controls. Retrovirology. 2014;11:62. https://doi.org/10.1186/PREACCEPT-1720768941312026.
35. Buzdin A, Ustyugova S, Khodosevich K, Mamedov I, Lebedev Y, Hunsmann G, et al. Human-specific subfamilies of HERV-K (HML-2) long terminal repeats: three master genes were active simultaneously during branching of hominoid lineages. Genomics. 2003;81:149–56.
36. Ruprecht K, Ferreira H, Flockerzi A, Wahl S, Sauter M, Mayer J, et al. Human endogenous retrovirus family HERV-K(HML-2) RNA transcripts are selectively packaged into retroviral particles produced by the human germ cell tumor line Tera-1 and originate mainly from a provirus on chromosome 22q11.21. J Virol. 2008;82:10008–16. https://doi.org/10.1128/JVI.01016-08.
37. Wang-Johanning F, Radvanyi L, Rycaj K, Plummer JB, Yan P, Sastry KJ, et al. Human endogenous retrovirus K triggers an antigen-specific immune response in breast cancer patients. Cancer Res. 2008;68:5869–77. https://doi.org/10.1158/0008-5472.CAN-07-6838.
38. Seifarth W, Baust C, Murr A, Skladny H, Krieg-Schneider F, Blusch J, et al. Proviral structure, chromosomal location, and expression of HERV-K-T47D, a novel human endogenous retrovirus derived from T47D particles. J Virol. 1998;72:8384–91.
39. Keydar I, Ohno T, Nayak R, Sweet R, Simoni F, Weiss F, et al. Properties of retrovirus-like particles produced by a human breast carcinoma cell line: immunological relationship with mouse mammary tumor virus proteins. Proc Natl Acad Sci U S A. 1984;81:4188–92.

40. McWilliam H, Li W, Uludag M, Squizzato S, Park YM, Buso N, et al. Analysis Tool Web Services from the EMBL-EBI. Nucleic Acids Res. 2013;41:W597–600. https://doi.org/10.1093/nar/gkt376.
41. Cartharius K, Frech K, Grote K, Klocke B, Haltmeier M, Klingenhoff A, et al. MatInspector and beyond: promoter analysis based on transcription factor binding sites. Bioinformatics. 2005;21:2933–42. https://doi.org/10.1093/bioinformatics/bti473.
42. Genomes Project C, Auton A, Brooks LD, Durbin RM, Garrison EP, Kang HM, et al. A global reference for human genetic variation. Nature. 2015;526:68–74. https://doi.org/10.1038/nature15393.
43. Lower R, Lower J, Kurth R. The viruses in all of us: characteristics and biological significance of human endogenous retrovirus sequences. Proc Natl Acad Sci U S A. 1996;93:5177–84.
44. Wang-Johanning F, Liu J, Rycaj K, Huang M, Tsai K, Rosen DG, et al. Expression of multiple human endogenous retrovirus surface envelope proteins in ovarian cancer. Int J Cancer. 2007;120:81–90. https://doi.org/10.1002/ijc.22256.
45. Fanning T, Alves G. A family of repetitive DNA sequences in Old World primates. Gene. 1997;199:279–82.
46. Shah N, Sukumar S. The Hox genes and their roles in oncogenesis. Nat Rev Cancer. 2010;10:361–71. https://doi.org/10.1038/nrc2826.
47. Tammimies K, Bieder A, Lauter G, Sugiaman-Trapman D, Torchet R, Hokkanen ME, et al. Ciliary dyslexia candidate genes DYX1C1 and DCDC2 are regulated by regulatory factor (RF) X transcription factors through X-box promoter motifs. FASEB J. 2016. https://doi.org/10.1096/fj.201500124RR.
48. Jiang S, Zhang E, Zhang R, Li X. Altered activity patterns of transcription factors induced by endoplasmic reticulum stress. BMC Biochem. 2016;17:8. https://doi.org/10.1186/s12858-016-0060-2.
49. Cook DN, Kang HS, Jetten AM. Retinoic acid-related orphan receptors (RORs): regulatory functions in immunity, development, circadian rhythm, and metabolism. Nucl Recept Res. 2015. https://doi.org/10.11131/2015/101185.
50. Morgan R, Boxall A, Harrington KJ, Simpson GR, Gillett C, Michael A, et al. Targeting the HOX/PBX dimer in breast cancer. Breast Cancer Res Treat. 2012;136:389–98. https://doi.org/10.1007/s10549-012-2259-2.
51. Burmeister T, Ebert AD, Pritze W, Loddenkemper C, Schwartz S, Thiel E. Insertional polymorphisms of endogenous HERV-K113 and HERV-K115 retroviruses in breast cancer patients and age-matched controls. AIDS Res Hum Retrovir. 2004;20:1223–9. https://doi.org/10.1089/0889222042545081.
52. Serafino A, Balestrieri E, Pierimarchi P, Matteucci C, Moroni G, Oricchio E, et al. The activation of human endogenous retrovirus K (HERV-K) is implicated in melanoma cell malignant transformation. Exp Cell Res. 2009;315:849–62. https://doi.org/10.1016/j.yexcr.2008.12.023.
53. Lemaitre C, Tsang J, Bireau C, Heidmann T, Dewannieux M. A human endogenous retrovirus-derived gene that can contribute to oncogenesis by activating the ERK pathway and inducing migration and invasion. PLoS Pathog. 2017;13:e1006451. https://doi.org/10.1371/journal.ppat.1006451.
54. Blanco-Melo D, Gifford RJ, Bieniasz PD. Co-option of an endogenous retrovirus envelope for host defense in hominid ancestors. Elife. 2017. https://doi.org/10.7554/elife.22519.
55. Samuelson LC, Wiebauer K, Gumucio DL, Meisler MH. Expression of the human amylase genes: recent origin of a salivary amylase promoter from an actin pseudogene. Nucleic Acids Res. 1988;16:8261–76.
56. Hurst TP, Magiorkinis G. Activation of the innate immune response by endogenous retroviruses. J Gen Virol. 2015;96:1207–18. https://doi.org/10.1099/jgv.0.000017.
57. Pastuzyn ED, Day CE, Kearns RB, Kyrke-Smith M, Taibi AV, McCormick J, et al. The neuronal gene arc encodes a repurposed retrotransposon gag protein that mediates intercellular RNA transfer. Cell. 2018;173:275. https://doi.org/10.1016/j.cell.2018.03.024.
58. Karolchik D, Hinrichs AS, Furey TS, Roskin KM, Sugnet CW, Haussler D, et al. The UCSC table browser data retrieval tool. Nucleic Acids Res. 2004;32:D493–6. https://doi.org/10.1093/nar/gkh103.
59. Kent WJ, Sugnet CW, Furey TS, Roskin KM, Pringle TH, Zahler AM, et al. The human genome browser at UCSC. Genome Res. 2002;2002(12):996–1006. https://doi.org/10.1101/gr.229102.
60. Tamura K, Stecher G, Peterson D, Filipski A, Kumar S. MEGA6: molecular evolutionary genetics analysis version 6.0. Mol Biol Evol. 2013;30:2725–9. https://doi.org/10.1093/molbev/mst197.

52. Serafino A, Balestrieri E, Pierimarchi P, Matteucci C, Moroni G, Oricchio E, et al. The activation of human endogenous retrovirus K (HERV-K) is implicated in melanoma cell malignant transformation. Exp Cell Res. 2009;315:849–62. https://doi.org/10.1016/j.yexcr.2008.12.023.
53. Lemaitre C, Tsang J, Bireau C, Heidmann T, Dewannieux M. A human endogenous retrovirus-derived gene that can contribute to oncogenesis by activating the ERK pathway and inducing migration and invasion. PLoS Pathog. 2017;13:e1006451. https://doi.org/10.1371/journal.ppat.1006451.
54. Blanco-Melo D, Gifford RJ, Bieniasz PD. Co-option of an endogenous retrovirus envelope for host defense in hominid ancestors. Elife. 2017. https://doi.org/10.7554/elife.22519.
55. Samuelson LC, Wiebauer K, Gumucio DL, Meisler MH. Expression of the human amylase genes: recent origin of a salivary amylase promoter from an actin pseudogene. Nucleic Acids Res. 1988;16:8261–76.
56. Hurst TP, Magiorkinis G. Activation of the innate immune response by endogenous retroviruses. J Gen Virol. 2015;96:1207–18. https://doi.org/10.1099/jgv.0.000017.
57. Pastuzyn ED, Day CE, Kearns RB, Kyrke-Smith M, Taibi AV, McCormick J, et al. The neuronal gene arc encodes a repurposed retrotransposon gag protein that mediates intercellular RNA transfer. Cell. 2018;173:275. https://doi.org/10.1016/j.cell.2018.03.024.
58. Karolchik D, Hinrichs AS, Furey TS, Roskin KM, Sugnet CW, Haussler D, et al. The UCSC table browser data retrieval tool. Nucleic Acids Res. 2004;32:D493–6. https://doi.org/10.1093/nar/gkh103.
59. Kent WJ, Sugnet CW, Furey TS, Roskin KM, Pringle TH, Zahler AM, et al. The human genome browser at UCSC. Genome Res. 2002;2002(12):996–1006. https://doi.org/10.1101/gr.229102.
60. Tamura K, Stecher G, Peterson D, Filipski A, Kumar S. MEGA6: molecular evolutionary genetics analysis version 6.0. Mol Biol Evol. 2013;30:2725–9. https://doi.org/10.1093/molbev/mst197.

10

Prediction of HIV-associated neurocognitive disorder (HAND) from three genetic features of envelope gp120 glycoprotein

Masato Ogishi[1,2*] and Hiroshi Yotsuyanagi[1]

Abstract

Background: HIV-associated neurocognitive disorder (HAND) remains an important and yet potentially underdiagnosed manifestation despite the fact that the modern combination antiretroviral therapy (cART) has achieved effective viral suppression and greatly reduced the incidence of life-threatening events. Although HIV neurotoxicity is thought to play a central role, the potential of viral genetic signature as diagnostic and/or prognostic biomarker has yet to be fully explored.

Results: Using a manually curated sequence metadataset (80 specimens, 2349 sequences), we demonstrated that only three genetic features are sufficient to predict HAND status regardless of sampling tissues; the accuracy reached 100 and 94% in the hold-out testing subdataset and the entire dataset, respectively. The three genetic features stratified HAND into four distinct clusters. Extrapolating the classification to the 1619 specimens registered in the Los Alamos HIV Sequence Database, the global HAND prevalence was estimated to be 46%, with significant regional variations (30–71%). The R package *HANDPrediction* was implemented to ensure public availability of key codes.

Conclusions: Our analysis revealed three amino acid positions in gp120 glycoprotein, providing the basis of the development of novel cART regimens specifically optimized for HAND-associated quasispecies. Moreover, the classifier can readily be translated into a diagnostic biomarker, warranting prospective validation.

Keywords: HIV-associated neurocognitive disorder (HAND), HIV envelope gp120 glycoprotein, Machine learning, Biomarker

Background

Neurocognitive impairments during the course of chronic HIV infection, called HIV-associated neurocognitive disorder (HAND), remain as an unconquered clinical entity despite the improvement of combination antiretroviral therapy (cART) over the last 20 years [1, 2]. HAND is a comprehensive concept encompassing the broad spectrum of motor, cognitive, and neuropsychiatric impairment, in which persistent HIV infection in the central nervous system (CNS) plays a fundamental role. According to the criteria proposed by the HIV Neurobehavioral Research Center (HNRC), HAND is stratified into three conditions, namely, asymptomatic neurocognitive impairment (ANI), mild neurocognitive disorder (MND), and HIV-associated dementia (HAD) [3]. In a large cohort study from the U.S., prevalence estimates of ANI, MND and HAD were inferred at 33, 12 and 2%, respectively [4]. Other cohort studies yielded similar estimates [5–7]. Despite its wide prevalence, diagnostic and therapeutic strategies are quite limited; currently, there is no molecularly defined biomarkers, and prompt initiation of cART is the only clinically available treatment, though its effectiveness on preventing the progression of neurocognitive impairment is still in hot controversy

*Correspondence: oogishi-tky@umin.ac.jp
[2] National Center for Global Health and Medicine, Tokyo, Japan
Full list of author information is available at the end of the article

[8, 9]. Indeed, as cART has become more accessible in resource-limited settings worldwide, thereby extending the expected lifespan of HIV-infected patients, the global burden of HAND is expected to be steadily on the rise. Recently, accumulating evidence suggests that persistent viral replication and ongoing diversification in the CNS compartment even in patients with undetectably suppressed viremia could lead to the emergence of neurotoxic quasispecies and thereby contribute to the progression of HAND [10, 11]. In this context, defining etiologically relevant diagnostic and/or prognostic biomarkers and optimal regimens based on those biomarkers in the era of modern cART is an inevitable step forward to improve current clinical practice.

Neurotoxic HIV viral quasispecies have been hypothesized to play an indispensable role in HAND pathogenesis. Although the mechanisms of HIV neurotoxicity has yet to be thoroughly clarified, several studies have suggested that there is a link between HAND and the neurotoxicity exerted by the orchestrated actions of several HIV proteins including trans-activating protein (*Tat*) and envelope glycoprotein (*Env*) [12]. Particularly, gp120, a fragment proteolytically cleaved from the *Env* protein, may mediate neuronal damage via direct induction of apoptosis both in rodents and primary human brain tissue culture [13–15]. On the other hand, hypervariable region 3 (V3) located at the middle of gp120 is primarily responsible for the genotypic and phenotypic diversity of HIV. A loop structure formed by V3 (V3 loop) interacts with chemokine coreceptors CCR5 and/or CXCR4, thereby determining multifaceted viral phenotypes including cell tropism [16]. Studies of CNS-derived viral isolates have indicated the links between CCR5 tropism, macrophage/microglia tropism, and the compartmentalization and persistent replication of viruses in the CNS [17–19]. Considering these insights, it is plausible to hypothesize that the gp120 glycoprotein serves as a primary, if not exclusive, determinant of both neurotropism, i.e., the ability to cross the blood–brain barrier and maintain replicative capacity in the CNS compartment, and neurotoxicity, i.e., the capability of igniting and/or fueling neurocognitive impairments. In this context, Pillai et al. [20] studied the C2V3 *env* subregion, and reported that the fifth residue of the V3 loop significantly correlated with neurocognitive deficit, although they did not explore the predictive significance of this signature. Indeed, a single amino acid signature is unlikely to be adequate to explain HIV adaptation during the course of HAND progression; thus, the combination of various signatures should be explored.

Machine learning (ML) is a highly promising technique for exploring a vastly large set of parameters to yield a potent classifier without prespecifying mathematical models. To gain optimized predictive accuracy, ML algorithms iteratively evaluate three types of error: training errors, validation errors (i.e., in-sample errors), and generalization errors (i.e., out-of-sample errors). The ultimate goal of ML-based prediction is to construct a classifier which has minimal generalization errors to unobserved real-world data. When a training dataset is provided, ML algorithms internally evaluate training errors to find the best set of parameters specific to the algorithm, and validation errors are evaluated by methods such as cross-validation (CV). When multiple algorithms with different sets of parameters are compared, the classifier with the smallest validation errors is selected. Then, the generalization errors should be evaluated with a testing dataset independent from model construction and selection.

Holman and Gabuzda applied ML-based approach to a manually collected metadataset of *env* C2V3C3 sequences derived from patients with or without HIV-associated dementia (HAD), reporting 75% accuracy for predicting HAD-associated *env* sequences [21]. Although their work provided intriguing insights into HIV neuropathogenesis, its generalizability is limited by several caveats. First, they reported the predictive accuracy via leave-one-out cross-validation. However, this corresponds to the validation error and may be a too optimistic estimation of the generalization error because of overfitting of the model against the training dataset. Rather, hold-out validation with no classifier retraining is necessary to correctly evaluate the generalization error. Second, although they only tested a simple rule-based classification algorithm, this could be outperformed by several recently implemented machine learning algorithms and an ensemble of those classifiers. Lastly, they attempted to construct a sequence-level classifier, and they empirically set a threshold at 95% of the patient's sequences for classifying the patient as having HAD. However, such empirical criteria should be carefully interpreted for potential overfitting. Moreover, since it is plausible to assume that even patients with HAD harbor non-neurotoxic quasispecies, and vice versa for patients without clinically apparent neurocognitive impairments, a patient-level set of features capturing the diversity of intrapatient quasispecies could be more predictive rather than a sequence-level set of features.

The purpose of this retrospective analysis is to propose a potential biomarker for HAND. To this end, the most predictive genetic signatures were explored by generating an ML-based HAND prediction model. A thorough literature search led to the construction of the most comprehensive metadataset to date, comprised of 2494 *env* C2V3C3 sequences from 9 studies involving 85 specimens from 43 patients. Iterative ML and stepwise feature reduction yielded three genetic features. A final ensemble classifier achieved accuracy of 100 and 94% in the hold-out testing subdataset and as a whole, respectively. Specimens from various sampling sources were classifiable using the same genetic features. Clustering analysis stratified HAND into four distinct clusters. The datasets, the main analysis workflow, and the in-house functions were made publicly available so as to maximize the reproducibility of the entire work.

Results

Construction of annotated HIV *env* sequence metadataset

A large, curated sequence dataset annotated with relevant clinical information is indispensable for ML-based prediction of the HAND status. Initially, we considered using The HIVBrainSeqDB [22] (http://hivbrainseqdb.dfci.harvard.edu/HIVSeqDB/) or The HAND Database [23] (http://www.handdatabase.org/). However, because these databases did not seem comprehensive, we decided to conduct a manual literature review. A thorough literature search resulted in the construction of a manually curated metadataset derived from 9 studies involving 40 patients, and consisting of 2494 HIV *env* C2V3C3 sequences (see "Methods" section for details), among which 2358 were unique (Additional file 1: Table 1). Sequences isolated from HAND and NonHAND cases formed several phylogenetically distinct clusters (Additional file 1: Fig. 1). Supported by this observation, we decided to further explore ML-based approach to construct a classifier predicting the HAND status from the C2V3C3 sequences.

Machine learning for predicting HAND status

The 2349 C2V3C3 sequences derived from HAND or NonHAND patients were converted into a numerical matrix using the 76 AAIndex schemes relevant to the physicochemical properties of amino acids. Patients diagnosed as either HIV-associated encephalitis (HIVE) or non-specific neuropsychiatric disorder (NPD) were excluded. Next, sequences were grouped by patient and sampling source, and representative statistics (e.g. mean and standard deviation) were calculated for each alignment position. Features with little variance, and a set of highly correlated features were excluded. In this manner, a total of 3169 patient-level predictive features were generated for 80 specimens. We performed ML with five distinct algorithms with ten different random seeds for hold-out data splitting. Stacking of the five classifiers was also attempted.

The mean and the best accuracy of the stacked classifiers trained using all features were 63 and 78%, respectively, in the hold-out validation subdataset (Fig. 1a and Additional file 1: Table 2). However, since this is extremely over-parametrized analysis, we attempted to reduce the number of features. First, algorithm-specific feature importance was calculated for each classifier, and the 20 most important ones were screened (Additional file 1: Table 3). Next, features selected by two or more distinct algorithms were selected. Finally, distributions between HAND and NonHAND cases were compared. *P* values were calculated by Welch's *t* test, and adjustment for multiple comparisons was done according to the method controlling the false discovery rate (FDR)

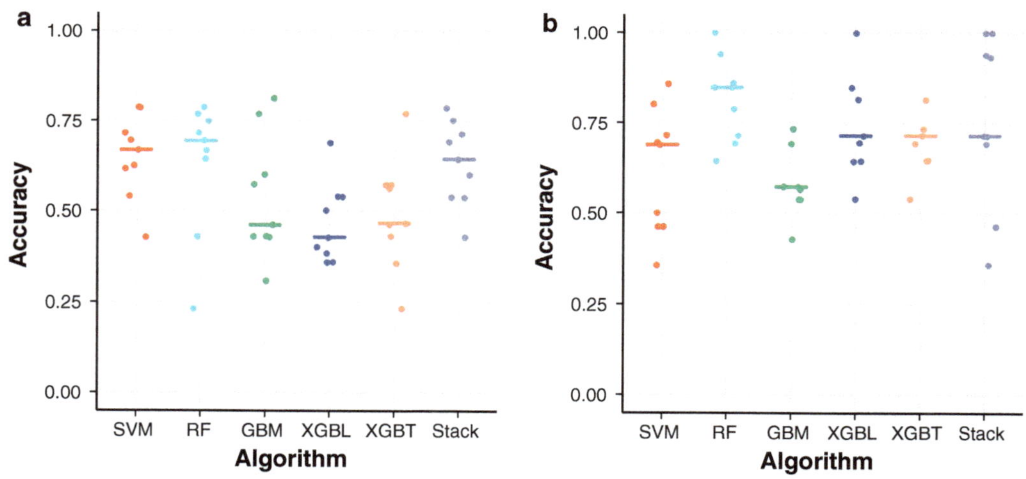

Fig. 1 Performance of ML-based classifiers predicting HAND. The in-house metadataset constructed from an extensive literature search was divided into the training and testing subdatasets. Five ML classifiers and their stacked classifier were trained using the training subdataset, and their accuracies were evaluated using the testing subdataset. Stacking algorithm was XGBT. **a** Classifiers trained using all features. **b** Classifiers trained with three minimal features. Dots represent ML attempts with different random seeds. SVM, support vector machine; RF, random forest; GBM, gradient boosting machine; XGBL, extreme gradient boosting with linear booster; XGBT, extreme gradient boosting with tree booster; Stack, the stacked classifier

originally reported by Benjamini and Hochberg [24]. Features with adjusted *P* values of lower than 0.05 were retained. In this manner, seven important features were uncovered (Additional file 1: Fig. 2). Encouraged by the observation that feature reduction did improve the predictive accuracy, we further deciphered the best set of features by means of stepwise feature reduction (Table 1 and Additional file 1: Table 4). Surprisingly, we observed the highest accuracy being 100% in the testing subdataset in two different algorithms (Fig. 1b). These classifiers used only three features. The mean and the best accuracy of the stacked classifiers were 76 and 100%, respectively (Fig. 1b and Additional file 1: Table 5). The accuracy of the best stacked classifier for the whole dataset was 94%, where the HAND status was predicted correctly in 75 out of 80 specimens. The distributions of Bayesian posterior probabilities showed that this classifier is expected to work well when the prior probability of HAND lies within the range from ~ 25% to ~ 50% (Additional file 1: Figure 3). Finally, as an external validation, all of the three specimens obtained from two patients with NPD were classified as HAND. One specimen was obtained from Patient 196, who was not diagnosed as HAD due to lack of information in the original paper despite the evidence of neuropsychiatric impairment [25]. The other two specimens were obtained from Subject 7115 at July 8th, 2002 [26]. Although the diagnosis of HAD could not be made due to confounding conditions at that moment, the same patient was diagnosed as moderate to severe HAD 2 years later. These two cases highlight the potential utility of our sequence-based HAND classifier as a diagnostic aid.

The distributions of the most important features were significantly different between the HAND and NonHAND cases (Fig. 2a). At position 291 (Pos291), the maximum value of the AAIndex DIGM050101, which represents a hydrostatic pressure asymmetry index, was shown to be the most predictive. The difference between the HAND and NonHAND cases was explained by the decrease of the frequency of 291S in the HAND group (Fig. 2b). Meanwhile, the other two important AAIndices, namely, KARP850102 and BHAR880101, are related to the flexibility of the residue. Variants enriched in the HAND cases such as 315 K/S and 340D/K/S contributed to the different feature distributions (Fig. 2).

Molecular stratification of HAND through the minimal set of genetic features

HAND is a diagnosis of exclusion, thus inherently harboring some heterogeneity. We noticed that, although the stacked classifier predicts the HAND status with high confidence, the subordinate classifiers returned considerably varied probabilities to the same cases, indicating that each of the classifiers captures distinct aspects of the triad of the genetic features. Indeed, clustering analysis revealed four distinct HAND-rich clusters (H1-H4) with characteristic genetic landscapes (Fig. 3a). Clusters enriched with NonHAND cases were combined as N. Random forest algorithm successfully constructed a classifier for these five categories, with the estimated accuracy from internal CV being 94%. A representative decision tree with the median number of nodes was shown in Fig. 3b. The tree understandably reflects the similarities between H2 and H4, and those between H1 and H3. When each amino acid variants were considered to be features, no apparent cluster-specific enrichment occurred, underscoring the effectiveness of our AAIndex-based feature generation framework (Additional file 1: Fig. 3).

To unveil the characteristics of each of the HAND clusters, we next applied the random forest classifier to the entire dataset obtained from The HAND Database (http://www.handdatabase.org/), which is a recently published, manually curated database of HIV sequences with clinical metadata [16] (Fig. 4). Collectively, 26 out of 33 (79%) HAD cases were classified into one of the HAND clusters, whereas all of the twenty cases originally annotated as not neurocognitively impaired were predicted as such (100%). Interestingly, nine out of eleven (82%) HIVE cases were predicted as N. Moreover, remaining two HIVE cases and the 6 cases with HAD and overlapping HIVE were classified as H2. These observations highlight the uniqueness of H2, potentially bridging the two distinct disease entities, namely, HAND and HIVE.

Table 1 Important AAIndices identified in this study

AAIndex ID	Description	Reference
DIGM050101	Hydrostatic pressure asymmetry index	Di Giulio [47]
KARP850102	Flexibility parameter for one rigid neighbor	Karplus and Schulz [48]
BHAR880101	Average flexibility indices	Bhaskaran and Ponnuswamy [49]

For further information regarding each AAIndex, visit the AAIndex website (http://www.genome.jp/aaindex/) [42]

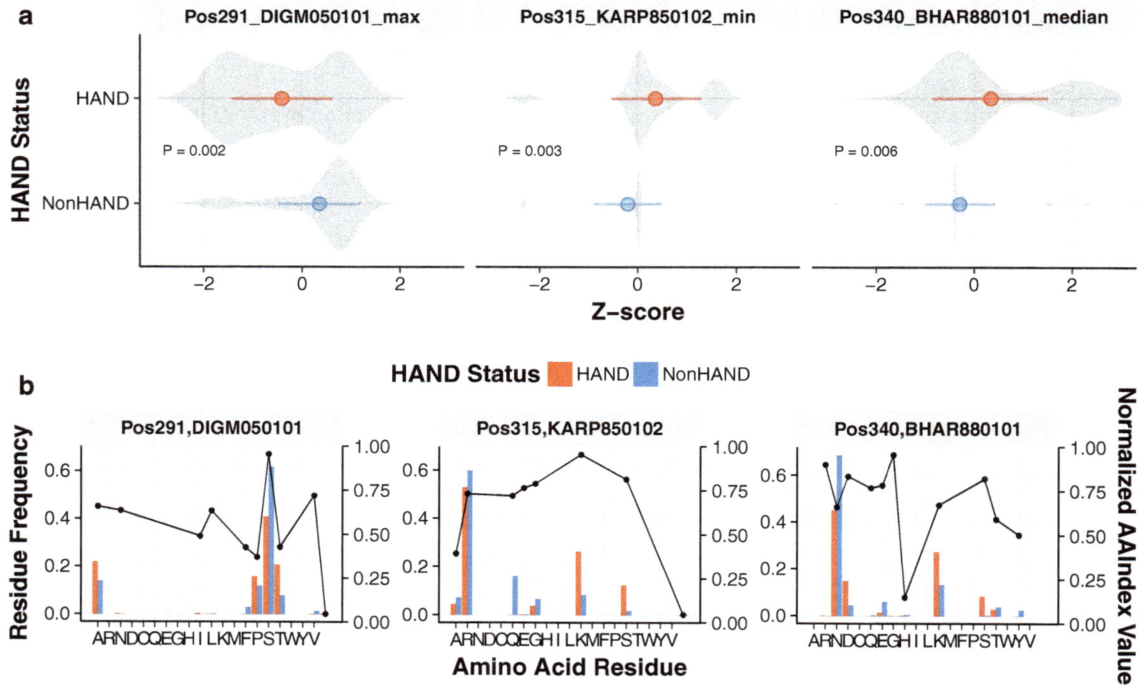

Fig. 2 The minimal set of features predictive of HAND. Model-specific feature importance was estimated using the *varImp* function implemented in the *caret* package for each of the ML algorithms except SVM. Features listed in the top 20 in two or more algorithms were selected. *P* values were calculated using Welch's *t* test and adjusted by the FDR-based method [24]. Adjusted *P* values of less than 0.05 were considered significant. In this manner, seven genetic features were retained (Additional file 1: Figure 2). Stepwise feature reduction was performed, and the minimal set of features yielding the best-performing stacked classifier was obtained. **a** Distributions of detected features among HAND and NonHAND groups. The values of each feature were converted to Z-score for visualization purposes. **b** Scaled AAIndex values and relative residue frequencies in sequence sets derived from HAND and NonHAND cases. The weights of individual sequences are normalized by the respective sequencing depths of individual patients. The alignment position numbers correspond to the positions in the HXB2 HIV-1 sequence (accession: K03455)

Estimation of the global burden of HAND

One major obstacle against the epidemiological study regarding HAND is the dearth of molecularly defined biomarkers. Currently, a careful neuropsychiatric examination is the only solid basis; in addition to this, various tests including brain CT/MRI and the cerebrospinal fluid (CSF) analysis are frequently required to exclude various mimicking diseases such as meningoencephalitis, toxoplasmosis, and primary CNS lymphoma. Biomarkers measurable from peripheral plasma could greatly reduce the burden for diagnostic procedures, and thereby facilitate epidemiological and other clinical studies particularly in resource-limited settings.

In view of this application, we retrospectively estimated the global burden of HAND from the entire sequence dataset deposited in the Los Alamos HIV Sequence Database (https://www.hiv.lanl.gov/content/sequence/HIV/mainpage.html), the largest database to date. Collectively, 46% of the cases were predicted as HAND, which was slightly higher than estimates from historical cohorts [4–7] (Fig. 5a). Among the predicted HAND cases, H1 was the most common cluster, followed by H2, H3, and H4 (Fig. 5a). Geographically dissecting, Caribbean region was the region with the largest predicted HAND burden (71%), whereas Sub-Saharan Africa was the lowest prevalence (30%). Among HAND clusters, H1 was dominant in Sub-Saharan Africa and Asia/Middle-East/Oceania (97 and 85%, respectively). In other regions, no single dominant cluster was noted. Caribbean region was characterized by the high prevalence of H3 (59%). Europe and North America regions had relatively high prevalence of H2 (54 and 38%, respectively). These regional differences in viral genetic landscape might be associated with various factors such as ethnicity and human leucocyte antigen (HLA) allele frequencies. Among the HAND clusters, H1 had a moderately higher viral load and lower CD4$^+$ T-cell count compared to H2. Meanwhile, H3 had a statistically higher CD4$^+$ T-cell count compared to H1 (Additional file 1: Figure 5). These trends were unchanged even when the *P* values were adjusted according to the FDR-based method [24].

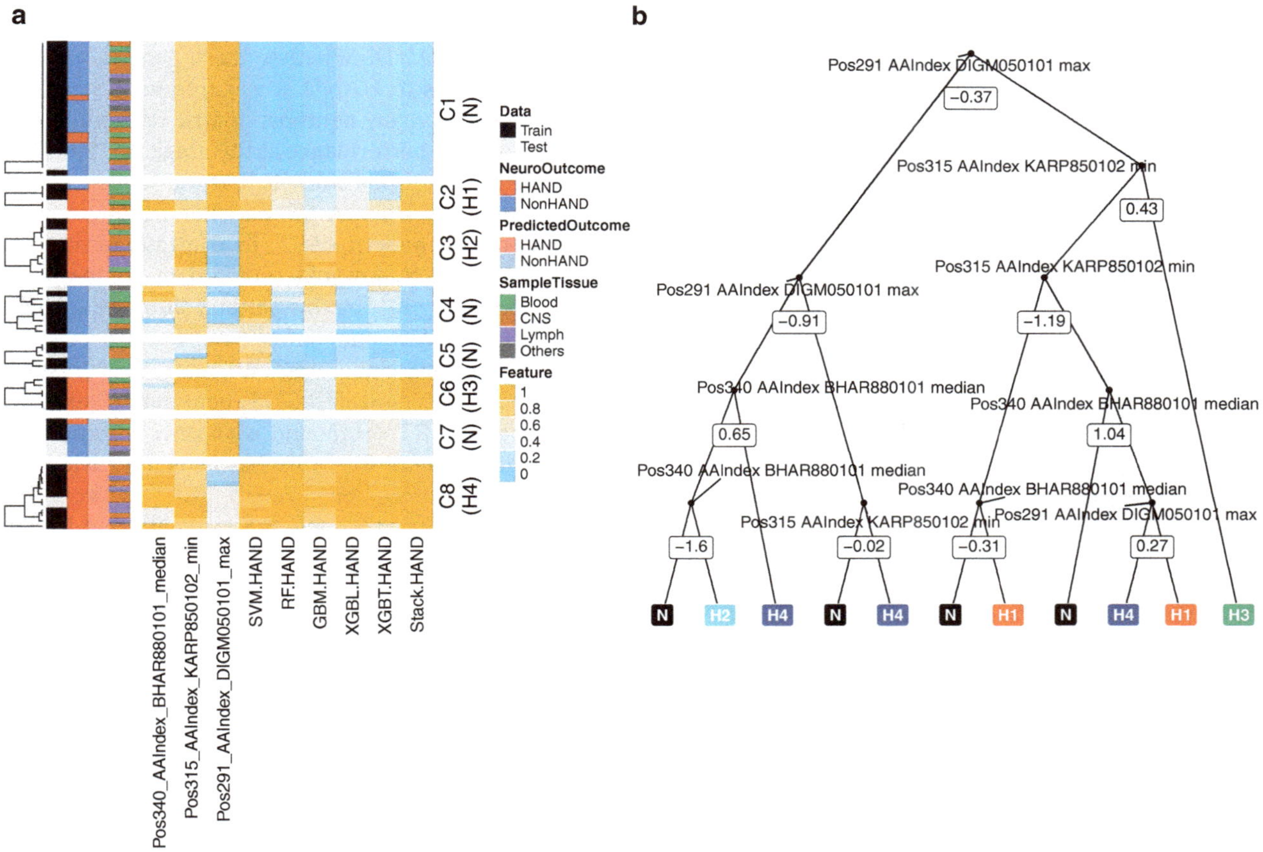

Fig. 3 Molecular stratification of HAND. **a** K-means clustering analysis. The three genetic features and the predicted probabilities from the six classifiers were used for k-means clustering analysis. Eight clusters were identified by inspecting the dendrogram manually. HAND-rich clusters, namely, C2, C3, C6, and C8, were renamed as H1, H2, H3, and H4, respectively. Other clusters were labeled as N. Heatmap was constructed using the *ComplexHeatmap* package [45]. **b** The representative decision tree for predicting the HAND clusters. The three genetic features were used to construct a multiclass RF classifier for the four HAND clusters and N. The number of nodes in the decision trees ranged from 13 to 29. The representative tree with the median number of nodes, 23, was shown. SVM, support vector machine; RF, random forest; GBM, gradient boosting machine; XGBL, extreme gradient boosting with linear booster; XGBT, extreme gradient boosting with tree booster; Stack, the stacked classifier

Data and code availability for future research

Both the datasets and the in-house functions created in this study were bundled as the R package *HANDPrediction*, and distributed on GitHub (https://github.com/masato-ogishi/HANDPrediction). To facilitate future research, the entire analysis workflow was also publicly distributed as an HTML document (Additional file 2).

Discussion

In this work, the three genetic features of the HIV *env* gene most predictive of the HAND status were identified through the construction of a highly accurate classifier via machine learning (ML). The surprisingly small number of features, three, strongly counter-argues the possibility of overfitting and supports the generalizability of the model to external datasets. The set of features stratified the 37 specimens derived from HAND cases into four clusters. The stratification process was successfully recapitulated by random forest algorithm, which enabled extrapolation of the genetic feature-based classification of HAND status. Estimation of global burden of HAND was demonstrated using the Loa Alamos HIV sequence database. The regional differences in the relative frequencies of HAND clusters probed by this retrospective analysis underscore the potential usefulness of our framework as an aid for epidemiological research, thereby warranting prospective validation.

In contrast to previous studies, neurotoxicity was stringently distinguished from neurotropism during the construction of the metadataset in this study. This is because it is inappropriate to discuss those two distinct

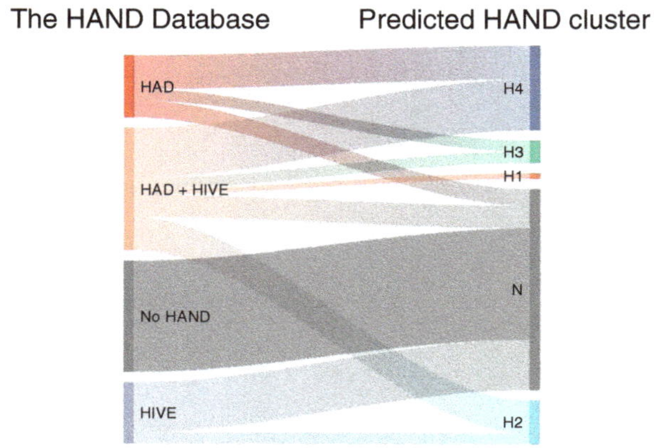

Fig. 4 HAND clusters among the cases registered in The HAND Database. A total of 1687 *env* sequences from 68 specimens were retrieved from The HAND Database [23]. The RF classifier (Fig. 3b) was applied to predict the corresponding HAND clusters. HAD, HIV-associated dementia; HIVE, HIV encephalitis

phenotypes interchangeably, since neurotoxic viral quasispecies that may trigger neurocognitive impairment could reside both inside and outside the CNS, and viral quasispecies harbored in the CNS do not necessarily exert neurotoxicity. Indeed, as shown in Fig. 3a, HAND-associated genetic signatures were shared among specimens derived from the CNS, lymphatic system, and peripheral circulation. This indicates that selection pressure outside the CNS is not a major driver for quasispecies evolution, which is consistent with a recent observational study led by Stefic et al. [27].

It is an exciting possibility that viral sequences obtained from peripheral circulation could be used as a diagnostic biomarker of HAND. Whether these genetic biomarkers provide clues to HAND at asymptomatic stage is of great interest, as many neuropsychiatric tests suffer from lower diagnostic performance at this stage [28, 29]. However, one caveat of this study is that the sequences were mainly obtained from AIDS patients without viremia suppression by modern cART. In contrast, prompt initiation of cART is the gold standard of contemporary clinical practice [30]. In this setting, immune reconstitution due to cART may affect viral quasispecies with HAND-associated signatures and alter their systemic distributions. Meanwhile, CNS penetration effectiveness score of cART compound is another consideration, since higher penetration score has been associated with lower neurocognitive impairment [31]. However, how the architecture of HIV quasispecies is affected by various cART regimens, and what roles these alterations may play in the pathogenesis of HAND, should be elucidated in future research.

Patient-level features are more informative than sequence-level features for predicting patient-level phenotypes [32]. Consistent with this viewpoint, the summary statistics representing the distribution of physicochemical properties of intrapatient viral quasispecies were used as the features on the basis of which ML was performed. One caveat of this approach is the sequence depth per patient; observed relative frequencies of each of the amino acid variants at each of the positions

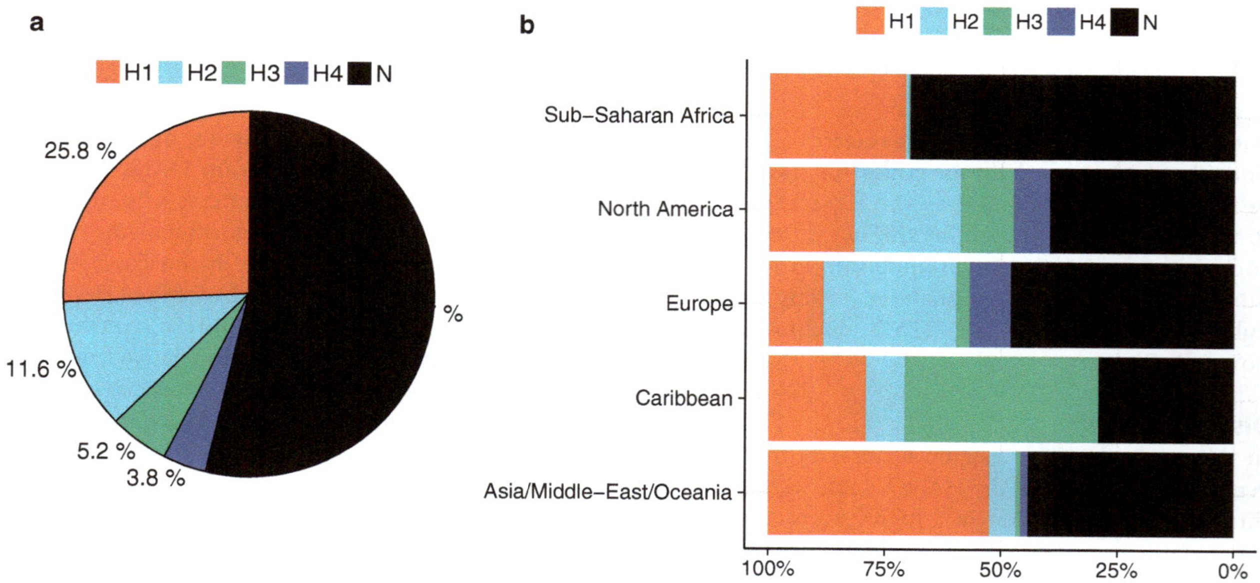

Fig. 5 Estimation of the global burden of HAND. A total of 19,800 *env* sequences from 1619 specimens were retrieved from The Los Alamos HIV Sequence Database. The RF classifier (Fig. 3b) was applied to predict the corresponding HAND clusters

may not reflect true intrapatient abundance with limited sequencing depth. Alternatively, next-generation sequencing platform could allow researchers to estimate relative abundance of variants with remarkably improved accuracy. We have previously shown that intrapatient abundances of viral quasispecies could be reliably estimated bioinformatically from short-read sequence datasets generated by the Illumina MiSeq platform [33]. This process is known as "quasispecies reconstruction". Integration of high-throughput sequencing technology and quasispecies reconstruction could enable more accurate estimations of intrapatient quasispecies abundance with augmented scalability. Such large-scale datasets could bolster the precision and accuracy of the HAND prediction framework presented in this work.

A number of gp120 variants have been associated with neurotropism and/or neurotoxicity. For example, Dunfee et al. [34] reported T283N as a neurotoxic variant causing enhanced macrophage infectivity and neuronal degeneration. Duenas-Decamp et al. [27] showed that the otherwise non-macrophage-tropic strain LN40 can be transformed into a macrophage-tropic strain by introducing 283 N substitution. However, in an already macrophage-tropic strain (B33), substitution of 283 N into 283T did not alter tropism, indicating the existence of other determinants [27]. In our analysis, three positions, namely, Pos291, Pos315, and, Pos340, were identified to be the most predictive for HAND status (Fig. 2b). Holman and Gabuzda also reported the involvement of Pos315 in HAND-predicting signature [21]. Pos315 resides in the tip of the V3-loop, and various variants such as R315K, R315T, and R315Q have been associated with reduced efficacy of neutralizing antibodies (NAbs) [35–38]. In our analysis, R315K and R315Q were enriched in the HAND and NonHAND cases, respectively (Fig. 2b). Although the other two positions, Pos291 and Pos340, were less intensively studied, S291 (enriched in NonHAND) has been associated with decreased infectivity to macrophages in R5 virus [27]. Meanwhile, compartmentalization of N340, a variant enriched in NonHAND in our analysis, to the CNS was observed in some cases [39]. Both S291 and N340 were also identified in this study (Additional file 1: Figure 4).

The current concept of HAND is heterogeneous due to its nature of being diagnosed on the basis of symptomatic criteria and by exclusion of other confounding conditions. To our knowledge, there is no attempt to date to molecularly stratify the disease entity. In this work, four HAND clusters were identified based on a clustering analysis. Particularly, H2 is interesting because it was associated with HIVE (Fig. 4). Since H2 and the closest cluster H4 were distinguished by the Pos340 feature (Fig. 3), and H4 was associated with both HAD and HAD + HIVE (Fig. 4), Pos340 might be important in separating HAND and HIVE. Moreover, geographically speaking, both H2 and H4 seemed to be enriched in Europe and North America (Fig. 5). Such geographical difference, if is the case, should be taken into consideration when interpreting various research on HAND from various nations. The biological and epidemiological relevance of those variants and clusters has yet to be elucidated, thus warranting further research.

This study has some limitations, similarly to prior studies. First, since this is a retrospective observational study, no causative link can be definitively established. Amino acid signatures detected could be relevant to the neurotoxicity of HIV, but should not be interpreted as causative of HAND. Second, although unprecedented size, the numbers of unique specimens and patients were fairly small. Although we successfully reduced the number of required genetic features down to three, the risk of overfitting to the entire dataset should not be negated. Prospective collection of the adequate size of specimens would be the only strategy to effectively resolve this concern. Third, since most of the currently available *env* sequences were derived from HAD cases, the most severe form of HAND, the utility of our analysis in predicting early-stage HAND has yet to be fully verified. Similarly to this point, the effect of cART regimens on the evolutionary trajectory of viral quasispecies should also be taken into consideration in future research. We do not argue that our analysis provides all answers; rather, we hope this work could be a starting point. Therefore, we made publicly available the datasets, the custom codes, and the entire analysis workflow for the community.

Conclusions

In this study, robust prediction of HAND status from three genetic features derived from the HIV *env* sequences was demonstrated. Furthermore, based on the combination of these three genetic features, we stratified HAND into four clusters with unique characteristics. These results could be utilized as a diagnostic aid after prospectively validation. Finally, the biological and epidemiological significance of newly discovered genetic features, potentially providing the basis of the development of novel cART regimens specifically optimized for HAND-associated quasispecies, are to be elucidated in future research.

Methods

Computational analysis

All computational analyses were conducted using R ver. 3.4.1 (https://www.r-project.org/) [40]. The latest versions of R packages were consistently used. The dataset and the scripts generated in this study are available as the

R package *HANDPrediction* on GitHub (https://github.com/masato-ogishi/HANDPrediction). The entire analysis workflow is also available as an HTML document (Additional file 2).

Assembly of the HIV *env* sequence metadataset

A thorough literature search was conducted to collect previously published studies on HIV neurotoxicity and/or neurotropism. Sequences and accompanying metadata were retrieved from the Los Alamos HIV Sequence Database (http://www.hiv.lanl.gov/content/sequence/HIV/mainpage.html/) and manually curated. Diagnoses of HIV-associated neurological conditions were retrieved from original publications for all of the cases. The subcategories of HAND (AMI, MND, and HAD) were combined as 'HAND', and the AIDS-dementia complex (ADC) was also considered 'HAND' in this study. HIVE and other NPDs were labeled as such. Cases with no neurocognitive impairments were labeled as 'NonHAND' regardless of other CNS diseases including bacterial meningitis, toxoplasmosis, and CNS lymphoma. The sample sources were categorized into one of the following categories: 'CNS', 'Blood', 'Lymph', and 'Others'.

Alignment of HIV *env* sequences

The HXB2 HIV-1 sequence (accession: K03455) was used as a reference. The *env* region was identified by mapping sequences to the reference sequence using Geneious ver 8.1.8 (www.geneious.com). The built-in Geneious mapper was used with the "Medium Sensitivity" option selected. Default parameters were used. Sequences not mapped to the reference were discarded from the metadataset. The *env* C2V3C3 regions were manually determined, clipped, and re-aligned with MAFFT [41]. The alignment was refined and translated using the HIVAlign tool with the HMM-align option selected (https://www.hiv.lanl.gov/content/sequence/VIRALIGN/viralign.html). Alignment gaps shared by the reference sequence and more than 75% of the aligned sequences were manually removed. Sequences containing stop codons inside the C2V3C3 region were discarded.

Machine learning

AAIndex matrix

AAIndex metrics (http://www.genome.jp/aaindex/) [42] were adopted as quantitative measures of biophysicochemical properties of each amino acid. A total of 531 AAIndices were retrieved from the *BioSeqClass* package available in *Bioconductor* [43]. The 76 AAIndices whose names matched with one of the following phrases were selected for machine learning: 'Hydro', 'Charge', 'Polar', 'Distribution', and 'Flexi'. A C2V3C3 sequence was converted to a numerical vector comprising a set of AAIndex values corresponding to each amino acid residue at each alignment position. For all gaps and ambiguities (i.e., two or more amino acid residues indicated), values for all AAIndices were set to zero. In this manner, all sequences were converted to a numerical matrix, which had 76 × 189 (188 alignment positions plus one gap) columns.

Hold-out validation

The metadataset was split into the training and testing subdatasets at a ratio of 4:1. Note that the metadataset was split at the patient level, not at the sequence level. Sequence-level data splitting is inappropriate because the HAND vs NonHAND status is assigned to patients, not to individual sequences, and the genetic relatedness of the sequences derived from the same patient will likely lead to biased classification.

Preprocessing

In the training subdataset, columns with zero variance and near-zero variance were removed using the *preProcess* function with the 'zv' and 'nzv' method implemented in the *caret* package [44]. Then, highly correlating columns were filtered using *preProcess* with the 'corr' method. After these filtration steps, 3169 unique features were retained. Finally, the features were centered and scaled using *preProcess* with the 'center' and 'scale' methods. All preprocessing steps were carried out with default parameter settings. In the testing phase, the same preprocessing conditions prepared in the training phase were applied.

Machine learning with different algorithms

For simplicity, binary classification was attempted, i.e., HAND vs NonHAND. The following algorithms were compared for performance: support vector machine (SVM), random forest (RF), gradient boosting machine (GBM), extreme gradient boosting with linear booster (XGBL), and extreme gradient boosting with tree booster (XGBT), all of which are implemented in the *caret* package. "Stacking" of the classifiers was done using XGBT as a supervised learning algorithm. Ten-fold repeated three-fold CV was conducted in the training phase to improve the generalizability of the classifiers. Their predictive performances, i.e., sensitivity, specificity, and overall accuracy, were estimated using the testing subdataset.

Feature importance analysis

Model-specific feature importance was estimated using the *varImp* function implemented in the *caret* package. All models except SVM tested in this study have their own feature importance measures. The 20 most important features from each of the models were combined, and features detected in two or more different models were selected. Next, the distribution of the feature values among the HAND and NonHAND groups were compared by Welch's *t* test, and *P* values were adjusted by the FDR-based method by Benjamini and Hochberg [24]. Features whose adjusted *P* values were less than 0.05 were selected. Finally, stepwise feature reduction was iteratively performed. ML was performed on the training subdataset with one of the features removed, and the accuracy in the testing subdataset. The removed feature giving the highest accuracy of the stacked classifier was removed for the next iteration.

Clustering analysis

K-means clustering was performed on the minimal set of the most important features, and the predicted probabilities by each of the classifiers. Visualization of the heatmap and dendrograms were performed using the *ComplexHeatmap* package [45]. Clusters enriched with the HAND cases were identified by manual inspection of the dendrogram. Clusters enriched with the NonHAND cases were combined and labeled as 'N'. The minimal set of the most important features was used to construct a multiclass random forest classifier classifying the HAND clusters and N using the entire dataset.

Characterization of HAND clusters using The HAND Database

The HAND Database [23] (http://database.handdatabase.org/) was used to characterize each of the HAND clusters. The entire dataset was downloaded as is. A total of 1687 *env* sequences from 68 specimens were obtained. Sequences were aligned to the HXB2 reference sequence, converted to a numerical matrix, preprocessed using the preprocessing models prepared in the training phase. For each specimen, the corresponding HAND cluster was assigned by the multiclass random forest classifier trained during the clustering analysis. The original labels of neuropathological conditions and the prediction results were linked and visualized as a Sankey plot using the *googleVis* package [46].

Estimation of the global burden of HAND using the Los Alamos HIV Sequence Database

The Los Alamos HIV Sequence Database (https://www.hiv.lanl.gov/content/sequence/HIV/mainpage.html) was used to demonstrate a retrospective estimation of the global burden of HAND. The sequences whose "culture method" were either "primary" or "uncultured" were downloaded. A total of 19800 *env* sequences from 1619 specimens were obtained. HAND status was predicted as described above.

Abbreviations

HIV: human immunodeficiency virus type 1; HAND: HIV-associated neurocognitive disorder; cART: combination antiretroviral therapy; CNS: the central nervous system; HNRC: HIV Neurobehavioral Research Center; ANI: asymptomatic neurocognitive impairment; MND: mild neurocognitive disorder; HAD: HIV-associated dementia; ML: machine learning; CV: cross-validation; HIVE: HIV-associated encephalitis; NPD: non-specific neuropsychiatric disorder; FDR: false discovery rate; CSF: cerebrospinal fluid; AIDS: acquired immunodeficiency syndrome; NAb: neutralizing antibody; ADC: AIDS-dementia complex; SVM: support vector machine; RF: random forest; GBM: gradient boosting machine; XGBL: extreme gradient boosting with linear booster; XGBT: extreme gradient boosting with tree booster; CI: confidential interval.

Authors' contributions

MO and HY designed the study; MO performed data analyses, prepared figures and tables, and drafted the manuscript; MO and HY, wrote the manuscript. Both authors read and approved the final manuscript.

Author details

[1] Division of Infectious Diseases and Applied Immunology, Research Hospital, The Institute of Medical Science, The University of Tokyo, Tokyo, Japan.
[2] National Center for Global Health and Medicine, Tokyo, Japan.

Acknowledgements

We thank Dr. Couture-Cossette for thoughtful comments.

Competing interests

The authors declare that they have no competing interests.

Funding

This research received no specific grant from any funding agency in the public, commercial, or not-for-profit sectors.

References

1. Saylor D, Dickens AM, Sacktor N, Haughey N, Slusher B, Pletnikov M, et al. HIV-associated neurocognitive disorder—pathogenesis and prospects for treatment. Nat Rev Neurol. 2016;12:234–48.
2. Clifford DB, Ances BM. HIV-associated neurocognitive disorder. Lancet Infect Dis. 2013;13:976–86.
3. Antinori A, Arendt G, Becker JT, Brew BJ, Byrd DA, Cherner M, et al. Updated research nosology for HIV-associated neurocognitive disorders. Neurology. 2007;69:1789–99.
4. Heaton RK, Clifford DB, Franklin DR, Woods SP, Ake C, Vaida F, et al. HIV-associated neurocognitive disorders persist in the era of potent antiretroviral therapy: Charter Study. Neurology. 2010;75:2087–96.
5. Chan LG, Kandiah N, Chua A. HIV-associated neurocognitive disorders (HAND) in a South Asian population—contextual application of the 2007 criteria. BMJ Open. 2012;2:e000662.
6. Yusuf AJ, Hassan A, Mamman AI, Muktar HM, Suleiman AM, Baiyewu O. Prevalence of HIV-associated neurocognitive disorder (HAND) among patients attending a tertiary health facility in Northern Nigeria. J Int Assoc Provid AIDS Care. 2014;116:1477–90.

7. Sacktor N, Skolasky RL, Seaberg E, Munro C, Becker JT, Martin E, et al. Prevalence of HIV-associated neurocognitive disorders in the Multicenter AIDS Cohort Study. Neurology. 2015;86:334–40.
8. Crum-Cianflone NF, Moore DJ, Letendre S, Roediger MP, Eberly L, Weintrob A, et al. Low prevalence of neurocognitive impairment in early diagnosed and managed HIV-infected persons. Neurology. 2013;80:371–9.
9. Oliveira MF, Chaillon A, Nakazawa M, Vargas M, Letendre SL, Strain MC, et al. Early antiretroviral therapy is associated with lower HIV DNA molecular diversity and lower inflammation in cerebrospinal fluid but does not prevent the establishment of compartmentalized HIV DNA populations. PLoS Pathog. 2017;13:e1006112.
10. Simioni S, Cavassini M, Annoni J-M, Rimbault Abraham A, Bourquin I, Schiffer V, et al. Cognitive dysfunction in HIV patients despite long-standing suppression of viremia. AIDS. 2010;24:1243–50.
11. Peluso MJ, Ferretti F, Peterson J, Lee E, Fuchs D, Boschini A, et al. Cerebrospinal fluid HIV escape associated with progressive neurologic dysfunction in patients on antiretroviral therapy with well controlled plasma viral load. AIDS. 2012;26:1765–74.
12. Toggas SM, Masliah E, Rockenstein EM, Rall GF, Abraham CR, Mucke L. Central nervous system damage produced by expression of the HIV-1 coat protein gp120 in transgenic mice. Nature. 1994;367:188–93.
13. Jana A, Pahan K. Human immunodeficiency virus type 1 gp120 induces apoptosis in human primary neurons through redox-regulated activation of neutral sphingomyelinase. Neuroscience. 2004;24:9531–40.
14. Zhang K, Rana F, Silva C, Ethier J, Wehrly K, Chesebro B, et al. Human immunodeficiency virus type 1 envelope-mediated neuronal death: uncoupling of viral replication and neurotoxicity. J Virol. 2003;77:6899–912.
15. Bachis A, Cruz MI, Mocchetti I. M-tropic HIV envelope protein gp120 exhibits a different neuropathological profile than T-tropic gp120 in rat striatum. Eur J Neurosci. 2010;32:570–8.
16. Hartley O, Klasse PJ, Sattentau QJ, Moore JP. V3: HIV's switch-hitter. AIDS Res Hum Retroviruses. 2005;21:171–89.
17. Rossi F, Querido B, Nimmagadda M, Cocklin S, Navas-Martín S, Martín-García J. The V1-V3 region of a brain-derived HIV-1 envelope glycoprotein determines macrophage tropism, low CD4 dependence, increased fusogenicity and altered sensitivity to entry inhibitors. Retrovirology. 2008;5:89.
18. Albright AV, Shieh JT, Itoh T, Lee B, Pleasure D, O'Connor MJ, et al. Microglia express CCR5, CXCR4, and CCR3, but of these, CCR5 is the principal coreceptor for human immunodeficiency virus type 1 dementia isolates. J Virol. 1999;73:205–13.
19. Gorry PR, Taylor J, Holm GH, Mehle A, Morgan T, Cayabyab M, et al. Increased CCR5 affinity and reduced CCR5/CD4 dependence of a neurovirulent primary human immunodeficiency virus type 1 isolate. J Virol. 2002;76:6277–92.
20. Pillai SK, Pond SLK, Liu Y, Good BM, Strain MC, Ellis RJ, et al. Genetic attributes of cerebrospinal fluid-derived HIV-1 env. Brain. 2006;129:1872–83.
21. Holman AG, Gabuzda D. A machine learning approach for identifying amino acid signatures in the HIV env gene predictive of dementia. PLoS ONE. 2012;7:e49538.
22. Holman AG, Mefford ME, O'Connor N, Gabuzda D. HIVBrainSeqDB: a database of annotated HIV envelope sequences from brain and other anatomical sites. AIDS Res Ther. 2010;7:43.
23. Griffin TZ, Kang W, Ma Y, Zhang M. The HAND Database: a gateway to understanding the role of HIV in HIV-associated neurocognitive disorders. BMC Med Genomics. 2015;8:70.
24. Benjamini Y, Hochberg Y. Controlling the false discovery rate: a practical and powerful approach to multiple testing. J R Stat Soc Ser B. 1995;57:289–300.
25. Shapshak P, Segal DM, Crandall KA, Fujimura RK, Zhang BT, Xin KQ, et al. Independent evolution of HIV type 1 in different brain regions. AIDS Res Hum Retroviruses. 1999;15:811–20.
26. Schnell G, Joseph S, Spudich S, Price RW, Swanstrom R. HIV-1 replication in the central nervous system occurs in two distinct cell types. PLoS Pathog. 2011;7:e1002286.
27. Duenas-Decamp MJ, Peters PJ, Burton D, Clapham PR. Determinants flanking the CD4 binding loop modulate macrophage tropism of human immunodeficiency virus type 1 R5 envelopes. J Virol. 2009;83:2575–83.
28. Bloch M, Kamminga J, Jayewardene A, Bailey M, Carberry A, Vincent T, et al. A screening strategy for HIV-associated neurocognitive disorders that accurately identifies patients requiring neurological review. Clin Infect Dis. 2016;63:687–93.
29. Krista J, Siefried BJB, Brew BJ, Siefried KJ, Draper B, Cysique LA. Is the HIV dementia scale a reliable tool for assessing HIV-related neurocognitive decline? J AIDS Clin Res. 2014;5:1–7.
30. INSIGHT START Study Group, Lundgren JD, Babiker AG, Gordin F, Emery S, Grund B, et al. Initiation of antiretroviral therapy in early asymptomatic HIV infection. N Engl J Med. 2015;373:795–807.
31. Smurzynski M, Wu K, Letendre S, Robertson K, Bosch RJ, Clifford DB, et al. Effects of central nervous system antiretroviral penetration on cognitive functioning in the ALLRT cohort. AIDS. 2011;25:357–65.
32. Vignuzzi M, Stone JK, Arnold JJ, Cameron CE, Andino R. Quasispecies diversity determines pathogenesis through cooperative interactions in a viral population. Nature. 2006;439:344–8.
33. Ogishi M, Yotsuyanagi H, Tsutsumi T, Gatanaga H, Ode H, Sugiura W, et al. Deconvoluting the composition of low-frequency hepatitis C viral quasispecies: comparison of genotypes and NS3 resistance-associated variants between HCV/HIV coinfected hemophiliacs and HCV monoinfected patients in Japan. PLoS ONE. 2015;10:e0119145.
34. Dunfee RL, Thomas ER, Gorry PR, Wang J, Taylor J, Kunstman K, et al. The HIV Env variant N283 enhances macrophage tropism and is associated with brain infection and dementia. Proc Natl Acad Sci USA. 2006;103:15160–5.
35. Pantophlet R, Wilson IA, Burton DR. Improved design of an antigen with enhanced specificity for the broadly HIV-neutralizing antibody b12. Protein Eng Des Sel. 2004;17:749–58.
36. Pantophlet R, Wrin T, Cavacini LA, Robinson JE, Burton DR. Neutralizing activity of antibodies to the V3 loop region of HIV-1 gp120 relative to their epitope fine specificity. Virology. 2008;381:251–60.
37. Pantophlet R, Aguilar-Sino RO, Wrin T, Cavacini LA, Burton DR. Analysis of the neutralization breadth of the anti-V3 antibody F425-B4e8 and re-assessment of its epitope fine specificity by scanning mutagenesis. Virology. 2007;364:441–53.
38. Kuwata T, Enomoto I, Baba M, Matsushita S. Incompatible natures of the HIV-1 envelope in resistance to the CCR5 antagonist cenicriviroc and to neutralizing antibodies. Antimicrob Agents Chemother. 2015;60:437–50.
39. Strain MC, Letendre S, Pillai SK, Russell T, Ignacio CC, Günthard HF, et al. Genetic composition of human immunodeficiency virus type 1 in cerebrospinal fluid and blood without treatment and during failing antiretroviral therapy. J Virol. 2005;79:1772–88.
40. R Core Team. R: a language and environment for statistical computing. Vienna: R Foundation for Statistical Computing; 2016.
41. Katoh K, Misawa K, Kuma K, Miyata T. MAFFT: a novel method for rapid multiple sequence alignment based on fast Fourier transform. Nucleic Acids Res. 2002;30:3059–66.
42. Kawashima S, Pokarowski P, Pokarowska M, Kolinski A, Katayama T, Kanehisa M. AAindex: amino acid index database, progress report 2008. Nucleic Acids Res. 2008;36:202–5.
43. Gentleman RC, Carey VJ, Bates DM, Bolstad B, Dettling M, Dudoit S, et al. Bioconductor: open software development for computational biology and bioinformatics. Genome Biol. 2004;5:R80.
44. Kuhn M. Building predictive models in *R* using the caret package. J Stat Softw. 2008;28:1–26.
45. Gu Z, Eils R, Schlesner M. Complex heatmaps reveal patterns and correlations in multidimensional genomic data. Bioinformatics. 2016;32:2847–9.
46. Gesmann M, de Castillo D. Using the google visualisation API with R. R J. 2011;3:40–4.
47. Di Giulio M. A comparison of proteins from *Pyrococcus furiosus* and *Pyrococcus abyssi*: barophily in the physicochemical properties of amino acids and in the genetic code. Gene. 2005;346:1–6.
48. Karplus PA, Schulz GE. Prediction of chain flexibility in proteins. Naturwissenschaften. 1985;72:212–3.
49. Bhaskaran R, Ponnuswamy PK. Positional flexibilities of amino acid residues in globular proteins. Int J Pept Protein Res. 2009;32:241–55.

11

Granulocytic myeloid-derived suppressor cells suppress virus-specific CD8$^+$ T cell responses during acute Friend retrovirus infection

Malgorzata Drabczyk-Pluta[1], Tanja Werner[1], Daniel Hoffmann[2], Qibin Leng[3], Lieping Chen[4], Ulf Dittmer[1] and Gennadiy Zelinskyy[1*]

Abstract

Background: Myeloid-derived suppressor cells (MDSCs) can suppress T cell responses in several different diseases. Previously these suppressive cells were observed to expand in HIV patients and in a mouse retrovirus model, yet their suppressive effect on virus-specific CD8$^+$ T cells in vitro and in vivo has not been characterized thus far.

Results: We used the Friend retrovirus (FV) model to demonstrate that MDSCs expand and become activated during the late phase of acute FV infection. Only the subpopulation of granulocytic MDSCs (gMDSCs) but not monocytic MDSC suppressed virus-specific CD8+ T cell proliferation and function in vitro. gMDSCs expressed arginase 1, high levels of the inhibitory ligand PD-L1 and the ATP dephosphorylating enzyme CD39 on the cell surface upon infection. All three molecules were involved in the suppressive effect of the gMDSCs in vitro. MDSC depletion experiments in FV-infected mice revealed that they restrict virus-specific CD8$^+$ T cell responses and thus affect the immune control of chronic retroviruses in vivo.

Conclusions: Our study demonstrates that MDSCs become activated and expand during the acute phase of retrovirus infection. Their suppressive activity on virus-specific CD8+ T cells may contribute to T cell dysfunction and the development of chronic infection.

Keywords: MDSC, CD8$^+$ T cells, Friend virus, PD-L1, Arginase

Background

Subpopulations of myeloid cells are part of the innate immune system and simultaneously these cells are of crucial importance for the development of adaptive immune responses. More than 30 years ago different research groups detected an accumulation of myeloid cells in growing tumors [1, 2]. Later the suppressive function and heterogeneity of these natural suppressor cells, as they were first called, was observed. It was shown that most of these cells have phenotypic and morphologic similarities to immature monocytes and granulocytes and are derived from similar myeloid precursors [3, 4]. These cells were therefore defined as myeloid-derived suppressor cells (MDSC) and divided into the subpopulations monocytic mMDSCs and granulocytic gMDSCs [4]. A hallmark of MDSCs is that they efficiently suppress effector T cell responses, including T cell proliferation and functional properties. Heterogeneity of MDSCs was observed in terms of the functional mechanisms utilized by MDSCs to suppress T lymphocytes. MDSCs can exploit the metabolism of L-arginine by arginase 1 (Arg1) [5], expression of inducible nitric oxide synthetase (iNOS) [5], and/or production of reactive oxygen species (ROS) [6] as main immunosuppressive tools. The production of inhibitory cytokines, the expression

*Correspondence: gennadiy.zelinskyy@uni-due.de
[1] Institute of Virology, University Hospital Essen, University of Duisburg-Essen, Hufelandstraße 55, 45147 Essen, Germany
Full list of author information is available at the end of the article

of ligands for inhibitory receptors on lymphocytes, and the expression of the surface enzymes CD39 (Ectonucleoside triphosphate diphosphohydrolase-1) and CD73 (Ecto-5′nucleotidase) [7, 8], regulating the metabolism of Adenosine Triphosphate were also described for subpopulations of MDSCs [9–15]. Thus MDSCs are cells with a high suppressive potential, which may simultaneously use diverse inhibitory mechanisms to inhibit T cell responses.

An activation and expansion of MDSCs was also observed during acute and chronic viral infections [16]. High numbers of MDSCs were found in patients infected with HBV [17–20] and HCV [21, 22]. Also HIV-infected patients [23, 24] have an enhanced number of mMDSCs and gMDSCs in the blood. In the murine AIDS model based on a LP-BM5 retrovirus infection of mice, Green and colleagues showed that expanded mMDSCs suppressed B cell activity and polyclonal T cell responses [25–27]. mMDSCs also influenced regulatory T cell responses in this model [28]. The data from HIV patients and from the murine LP-BM5 model demonstrate the negative influence of MDSCs on adaptive anti-retroviral immune responses. However, the suppressive effects of MDSCs on virus-specific cytotoxic $CD8^+$ T cell responses was not analyzed and the role of MDSCs for the establishment of a chronic retroviral infection remains elusive. These important questions have been addressed in the current study using the Friend virus (FV) mouse model.

FV is an oncogenic retroviral complex that can induce erythroleukemia in susceptible mice. However, resistant mouse strains, like the C57/Bl6 mice used in this study, mount a potent anti-viral immune response during the acute phase of infection that prevents the onset of leukemia [29]. Despite this efficient initial anti-viral immunity, FV escapes from T cell mediated immune control and establishes a chronic infection [30, 31]. Cytotoxic $CD8^+$ T lymphocytes (CTL) are crucial for controlling FV replication during the acute phase of infection. However, during chronic FV infection, similar to the human immunodeficiency virus (HIV) and hepatitis C virus (HCV) infections, virus-specific $CD8^+$ T cells become functionally exhausted. This exhaustion most likely contributes to the inability of the host to eliminate cells infected with the pathogen [32, 33]. Thus, our aim was to define whether MDSCs contribute to the T cell exhaustion that develops during chronic FV infection.

The current study shows the expansion of gMDSCs and mMDSCs during the late phase of acute FV infection. In contrast to the murine AIDS retrovirus model, the suppressive effect on virus-specific $CD8^+$ T cells was mainly mediated by gMDSCs. Results from in vivo MDSCs depletion experiments showed that these cells contribute to the development of chronic FV infection and hence, may be an important target for immunomodulatory therapies of acute or chronic viral infections.

Results

Myeloid-derived suppressor cells expand during FV infection

MDSCs were shown to play a role in the suppression of immune responses not only in cancer, but also in various infectious diseases such as LCMV, HCV, HBV, and HIV [34–37]. However their exact role in retroviral infections is still elusive.

In order to characterize MDSCs during FV infection, B6 mice were infected and the kinetic of the MDSC response was determined. MDSCs were characterized according to recommendations for the phenotypic definition of these cells [38–40] CD11b+ cells were gated from live, single-cell splenocytes, which were negative for CD3, CD19, NK1.1, and Ter119 lineage markers. These CD11b+ myeloid cells were divided according to their expression of Ly6C and Ly6G in monocytic (mMDSC, Ly6G-Ly6C^{high}) and granulocytic (gMDSC, Ly6G^{high} Ly6C^{low}) MDSCs (Fig. 1a). mMDSCs and gMDSCs were detectible in naïve mice, but their frequencies were below 0.3% per one million nucleated splenocytes. This number of MDSCs was stable during early FV infection until day 10. At day 14 after infection, the population of mMDSCs and gMDSCs expanded significantly. The expansion of both MDSC subpopulations peaked on day 14 post infection in the spleen (Fig. 1b) and the frequency of these cells reached more than 6000 per one million spleen cells. During the chronic phase of FV infection, the numbers of both the granulocytic and monocytic MDSCs remained slightly elevated (mean number 3200 mMDSCs and 3100 gMDSCs per one million spleen cells) in comparison to non-infected mice (mean number 2700 mMDSCs and 2500 gMDSCs per one million spleen cells) (Fig. 1b). However these differences were not significant. These data indicate that the populations of mMDSCs and gMDSCs mainly expanded at day 14 after FV infection, a time point that is concomitant with the onset of the suppression of virus-specific effector $CD8^+$ T cell responses [41, 42].

CD80 is a member of the B7 family and is a stimulatory or inhibitory molecule of T cell activation. It is a ligand for two receptors: CTLA-4 and CD28. While CD28 induces T cell activation, CTLA-4 mediates T cell suppression [43]. Activated MDSCs isolated from tumor bearing mice [44] and from cancer patients [45] show a significantly enhanced expression of CD80, defining CD80 as activation marker for MDSCs. Approximately 10% of the mMDSCs and 20% of the gMDSCs from naïve mice expressed CD80 (Fig. 1c). At day 14 post infection

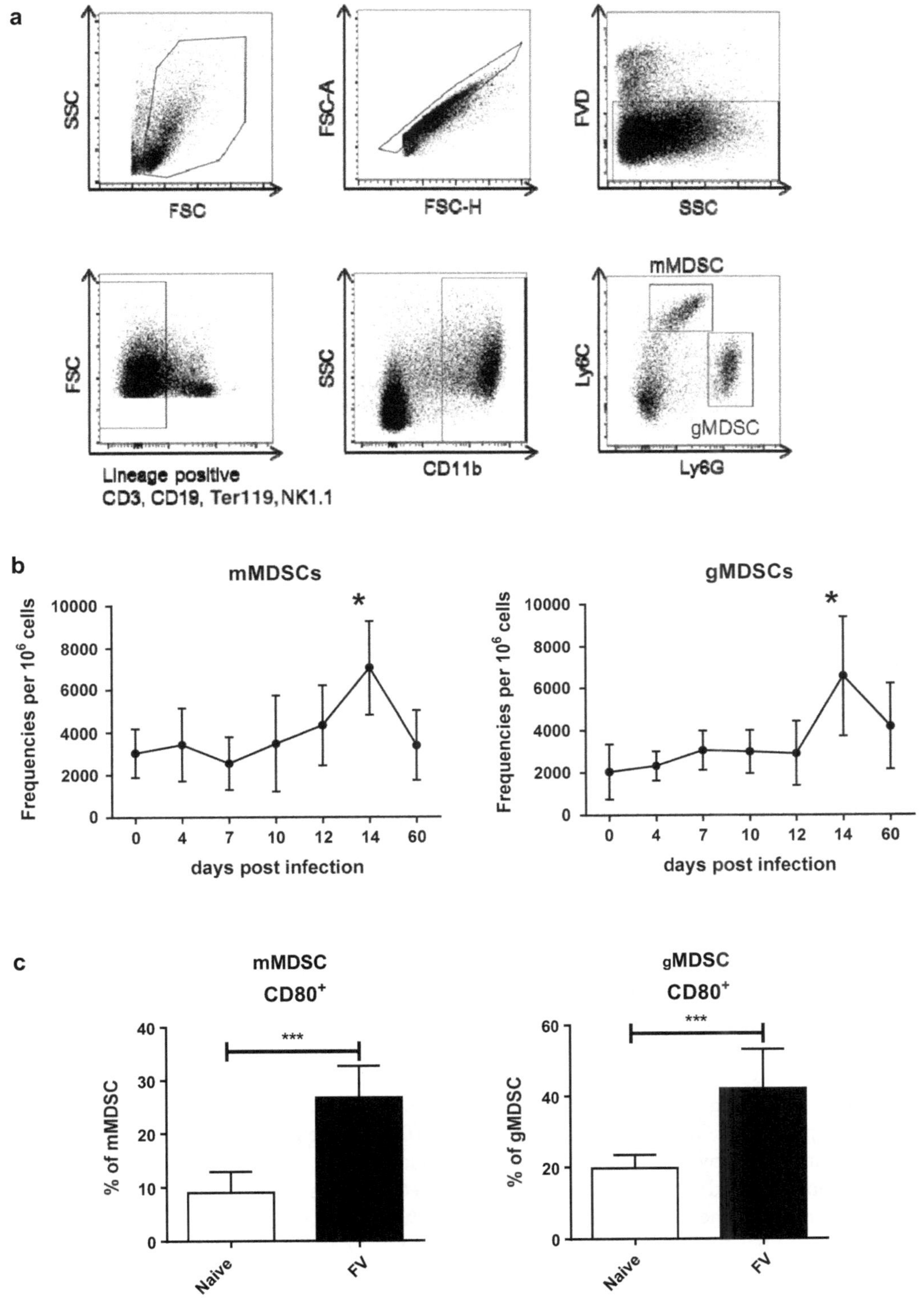

Fig. 1 MDSCs expand during acute FV infection. B6 mice were infected i.v. with 20,000 SFFU of FV or left un-infected, and MDSC numbers were measured using flow cytometry. The total spleen cells were analyzed at various time points post infection. **a** The numbers of Ly6G^{high} Ly6C^{low} and **b** Ly6G^{-} Ly6C^{high} per 1 × 10^{6} live splenocytes and the frequencies of **c** CD80 expression on the surface of Ly6G^{high} Ly6C^{low} and Ly6G^{-} Ly6C^{high} cells are displayed. At least 7 mice per group from five independent experiments were analyzed. *Bars* represent means with SD. For statistical analysis a Dunn's test with the Benjamini–Hochberg correction for multiple testing (**b**) and an unpaired t test (**c**) were performed (*<0.05; **<0.005)

a mean of 25% of the mMDSCs and 43% of the gMDSCs expressed CD80 (Fig. 1c). These data demonstrate that MDSCs expanded and became activated in the late phase of acute FV infection.

gMDSC suppress virus-specific CD8$^+$ T cell responses in vitro

MDSCs display a certain phenotype, but their main characteristic is their suppressive activity against T cell responses. We therefore analyzed whether FV-induced MDSCs can suppress FV-specific effector CD8$^+$ T cells in an in vitro model. To achieve this goal, a FV-specific CD8$^+$ T cell proliferation assay was established. Bone marrow derived dendritic cells were incubated with Violet Cell tracer labeled virus-specific CD8$^+$ T cells isolated from TCR transgenic mice, of which more than 90% of the CD8$^+$ T cells contain a TCR specific for the DbGagL FV epitope [46, 47]. The DCs were loaded with the DbGagL epitope peptide to induce a virus-specific proliferation of the CD8$^+$ T cells. mMDSCs and gMDSCs were isolated from FV-infected mice at 14 dpi, according to their expression of Ly6C and Ly6G respectively. In order to determine the suppressive effect of these subpopulations on the virus-specific CD8$^+$ T cell response, enriched mMDSCs or gMDSCs were added in a 10:1 MDSC to CD8$^+$ T cell (E:T) ratio. After 3 days of culture, CD8$^+$ T cell proliferation and effector molecule granzyme B (GzmB) expression were analyzed. At this time point an average of 90% of the CD8$^+$ T cells had undergone at least one cell division. Interestingly, this CD8$^+$ T cell proliferation was only suppressed by gMDSCs, but not by mMDSCs (Fig. 2b). To clarify the suppressive potential of gMDSCs, these cells were added to target CD8$^+$ T cells in different ratios (1:1, 2.5:1, 5:1, 10:1 gMDSCs to CD8$^+$ T cells) (Fig. 2c). An increasing reduction of CD8$^+$ T cell proliferation was observed at ratios of gMDSC to CD8+ T cells of 2.5:1 and higher. Thus, the gMDSCs mediated suppression was cell number dependent.

Additionally, the GzmB expression in activated CD8$^+$ T cells was measured. An average of 90% of all CD8$^+$ T cells in the cultures produced GzmB. This expression of GzmB was also diminished by gMDSCs, but not by mMDSCs co-incubation (Fig. 2d). At an effector target ratio of 1:10 gMDSCs reduced the percentage of GzmB$^+$ CD8$^+$ T cells to a mean of below 20% (Fig. 2d). gMDSCs from naïve mice did not suppress the proliferation or the GzmB production of FV-specific CD8$^+$ T cells, indicating that the suppressive activity of this MDSC subpopulation was induced by FV infection.

These data demonstrate that gMDSC but not mMDSCs from FV-infected mice were able to suppress virus-specific CD8$^+$ T cell responses in vitro.

Molecules involved in FV-induced T cell suppression by gMDSCs

It was previously shown that functional MDSCs express different molecules that are associated with their suppressive activity [13, 48–52]. Therefore, NOS2, Arg1, PD-L1, and CD39 expression on the suppressive gMDSCs were analyzed at the peak of their expansion after FV infection (14 dpi). Arg1 was of special interest, an enzyme converting ʟ-arginine to urea and ʟ-ornithine, followed by NOS2, an enzyme further synthetizing nitric oxide (NO) and ʟ-citrulline [14, 48, 51, 53]. Very low frequencies of gMDSCs from naïve animals expressed intracellular NOS2, which did not significantly increase after FV infection (Fig. 3a). On the contrary, the percentage of Arg1 expressing gMDSCs increased during FV infection from a mean of approximately 3% in naïve to 10% in infected mice (Fig. 3a). 30% of the naïve gMDSCs were PD-L1$^+$, which increased to about 70% after FV infection (Fig. 3a). In contrast, already 60% of the gMDSCs expressed CD39 prior to infection and this expression level did not significantly change after FV inoculation (Fig. 3a).

In order to investigate whether CD8$^+$ T cell suppression was arginase or NO dependent, the previously described proliferation assay was used. In this assay nor-NOHA, which is a pan arginase inhibitor, or L-NMMA, a pan NO inhibitor, were added to the cultures. After administration of the arginase inhibitor norNOHA, we observed an approximately 2.5 times higher percentage of proliferating CD8$^+$ T cells in the presence of gMDSCs in comparison to the untreated control (Fig. 3b). Administration of the NO inhibitor L-NMMA resulted in an almost 3 times higher percentage of proliferating CD8$^+$ T cells compared to the untreated group (Fig. 3b). The suppression of T cell responses by gMDSCs was not reversed after the control molecule D-NMMA was added (data not shown). These data suggest that ʟ-arginine metabolism is at least one mechanism of gMDSCs mediated suppression of virus-specific T cell proliferation in vitro.

To analyze the role of CD39 and PD-L1 in the T cell suppression by gMDSCs, the previously described proliferation assay was again used. For these experiments, gMDSCs were isolated from FV-infected CD39 or PD-L1 knockout mice. gMDSCs lacking PD-L1 were not able to efficiently suppress CD8$^+$ T cell proliferation in vitro (76% of all CD8$^+$ T cells proliferated, very similar to the positive control) (Fig. 3c). After incubation of gMDSCs isolated from CD39 KO mice with CD8$^+$ T cells, 64% of the CD8$^+$ T cells had undergone cell division, which was almost 2.5 higher as in the control group with gMDSCs from wild type mice (Fig. 3c). Interestingly, full suppression of T cell responses was observed when gMDSCs were isolated from CD73 (5′-nucleotidase (5′-NT)) knockout mice (data not shown). These data suggest

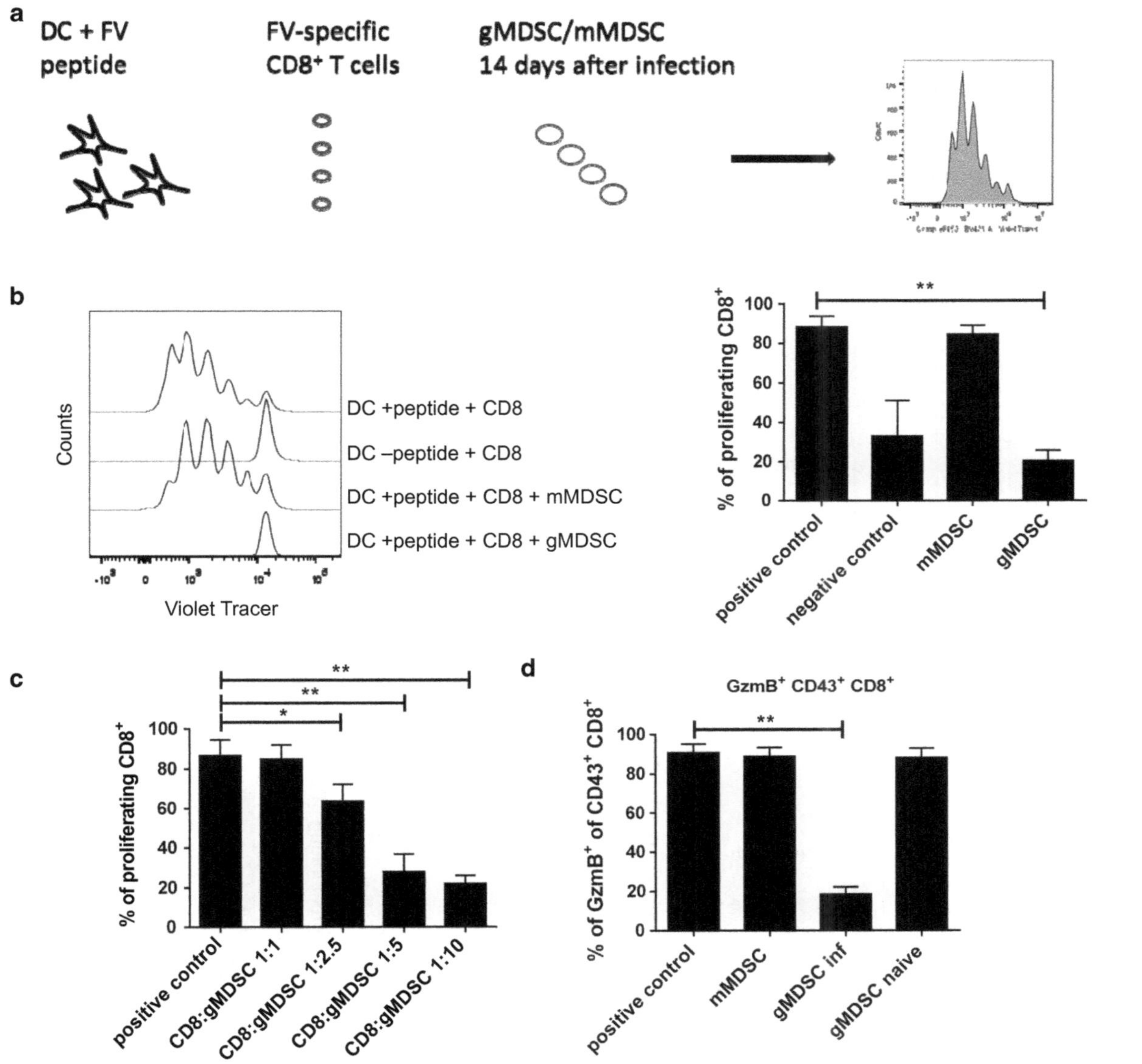

Fig. 2 Granulocytic myeloid-derived suppressor cells inhibited CD8$^+$ T cell proliferation. CD8$^+$ T cells isolated from DbGagL TCR transgene mice were incubated with dendritic cells loaded with MHC class I-restricted FV-specific CD8$^+$ T cell epitope peptide and co-incubated with either gMDSCs or mMDSCs (**a**). Representative histograms and percentages of CD8$^+$ T cells after co-incubation with or without either gMDSC or mMDSCs (in relation 1 CD8$^+$ to 10 MDSCs) from FV-infected mice are shown (**b**). CD8$^+$ T cell proliferation was measured after co-incubation with different effector ratios of gMDSCs to CD8$^+$ target cells (**c**). Frequencies of GzmB expressing CD43$^+$CD8$^+$ cells after incubation of CD8$^+$ cells with gMDSCs or mMDSCs from FV-infected mice are shown (**d**). CD8$^+$ T cells incubated with peptide loaded DC serve as a positive control, CD8$^+$ T cells incubated with non-loaded DC serve as a negative control. At least three independent experiments were analyzed. Bars represent the mean with SD. For statistical analysis, an ANOVA multiple comparison test was carried out with the group of naïve mice as a reference (*<0.05; ***<0.0005). For statistical analysis a Dunn's test with the Benjamini–Hochberg correction for multiple testing was performed (*<0.05; **<0.005)

a significant role of PD-L1 expressed and adenosine metabolism in gMDSC mediated suppression.

Selective depletion of MDSCs in FV-infected mice

After characterizing the MDSC activity in vitro it was of interest to confirm their suppressive potential in vivo during an ongoing FV infection. Different methods to eliminate, block or suppress MDSCs have been described, including antibody treatment (αGr1, αLy6G), directly acting drugs (5-Fluorouracil (5FU), Silendafil, doxorubicine) and drugs which were described to maturate MDSCs (ATRA, CpG) [54].

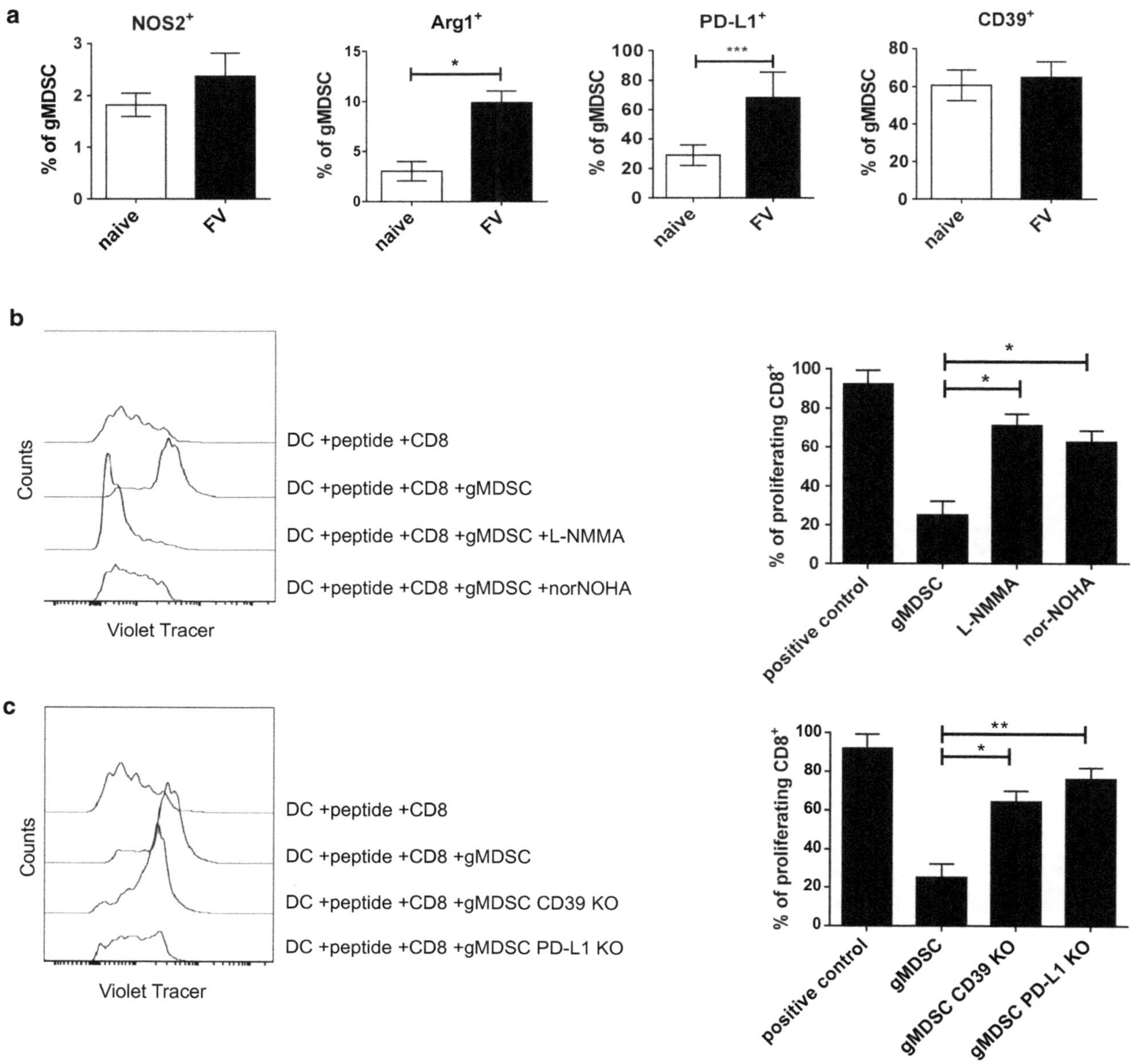

Fig. 3 Mechanisms of MDSC mediated suppression. B6 mice were infected i.v. with 20,000 SFFU of FV or left un-infected and the expression of Arg1 and NOS2 in the MDSCs was measured using flow cytometry for spleen cells at day 14 post infection. Frequencies of NOS2, Arg1, PD-L1, and CD39 expressing Ly6G^{high} Ly6C^{low} cells in the spleen are displayed (**a**). At least five mice per group from three independent experiments were analyzed. *Bars* represent the mean with SD. For statistical analysis an unpaired t test was performed (*<0.05; ***<0.0005). **b** CD8$^+$ T cells isolated from DbGagL TCR transgenic mice were incubated with dendritic cells loaded with MHC class I-restricted FV-specific CD8$^+$ T cell epitope peptide and co-incubated with gMDSCs in the ratio of 1 CD8$^+$ to 10 gMDSCs with or without the addition of L-NMMA or nor-NOHA. Representative histograms and percentages of proliferating CD8$^+$ T cell proliferation after co-incubation with or without gMDSCs in the presence or absence of L-NMMA/nor-NOHA are shown. At least three independent experiments were analyzed. *Bars* represent the mean with SD. **c** CD8$^+$ T cells isolated from DbGagL TCR transgenic mice were incubated with dendritic cells loaded with MHC class I-restricted FV-specific CD8$^+$ T cell epitope peptide and co-incubated with gMDSCs isolated from CD39 KO mice or from PD-L1 KO mice. Representative histograms and percentages of proliferating CD8$^+$ T cells after co-incubation with or without gMDSCs isolated from B6, CD39 KO and PD-L1 KO FV infected mice are shown. At least three independent experiments were analyzed. The *bars* represent the mean with SD. For statistical analysis a Dunn's test with the Benjamini–Hochberg correction for multiple testing was performed (*<0.05; **<0.005)

In order to investigate the role of MDSCs during FV infection in living mice either gMDSCs alone or both the gMDSC and mMDSC populations were depleted. By a single administration of 5FU on day 9 post FV infection (Fig. 4a), both MDSC populations were efficiently depleted in the spleen (Fig. 4b). The pyrimidine analog 5FU is a cytostatic drug with MDSCs specific cytotoxicity. Selective activity of 5FU on MDSC was shown by Vincent et al. [55], and other immune cells were not affected by the drug. We confirmed these findings by staining for all other main immune cell subsets after 5FU injection in naïve mice (data not shown). The second approach, the i.p. administration of an αLy6G antibody, selectively depleted gMDSCs (Fig. 4b). αLy6G antibody was administered four times every third day between day 5 and 13 post FV infection (Fig. 4a). First, we analyzed the $CD8^+$ T cell response in FV-infected MDSC depleted mice in the spleen. It was of interest to discover whether the population of FV-specific effector $CD8^+$ T cells was influenced by MDSCs in vivo. $CD43^+$, a sialoglycoprotein, expressed on the cell surface of T lymphocytes, is a part of the receptor-ligand complex and required for T cell activation [56, 57]. Naïve $CD8^+$ T cells do not express $CD43^+$, but $CD43^+$ becomes highly up-regulated on antigen specific effector $CD8^+$ T cells [56]. FV induced $CD8^+$ T cells express high levels of CD43 on the cell surface [58]. Interestingly, after the depletion of all MDSCs by 5FU, the frequency of $CD43^+$ $CD8^+$ T cells increased significantly in comparison to only FV-infected mice (Fig. 4c). After the gMDSC depletion using αLy6G antibodies, also significantly higher frequencies of activated $CD8^+$ T cells were observed

Next, it was investigated whether the expanded population of activated $CD8^+$ T cells was specific for the immunodominant epitope of FV. With the use of FV-D^bgagL class I tetramers it was possible to assess that the administration of 5FU did not result in increased percentages of D^bgagL-specific effector $CD8^+$ T cells. However, the depletion of gMDSCs with αLy6G antibodies significantly increased the frequency of $tetramer^+$ $CD8^+$ T cells, suggesting that mainly gMDSCs act on antigen-specific T cells.

To determine whether the expansion of $CD8^+$ T cells after MDSC depletion was driven by increased $CD8^+$ T cell proliferation, the intranuclear expression of Ki67, a nuclear protein involved in cell proliferation [59], was analyzed. After 5FU administration significantly more $CD43^+$ $CD8^+$ cells expressed Ki67 compared to the infected control group. In gMDSC depleted mice also enhanced numbers of proliferating $CD8^+$ T cells were detected. However, the difference to control mice was not statistically significant (Fig. 4c).

In order to characterize the effector functions of the expanded $CD8^+$ T cells after MDSC depletion, we stained for intracellular GzmB expression. In both groups of 5FU as well as αLy6G treated mice, frequencies of $GzmB^+$ $CD8^+$ T cells were increased in comparison to control mice, but only the difference after 5FU treatment was statistically significant (Fig. 4c). Thus, the in vivo data from MDSC depletion experiments showed that the expansion of effector $CD8^+$ T cells and their differentiation into potentially cytotoxic GzmB producing CTLs were influenced by MDSC during FV infection. In particular, gMDSCs regulated the magnitude of the CD8+ T cell response against the immunodominant CD8 epitope of FV. This effect was also seen in our in vitro suppression assays where the peptide for the FV immunodominant epitope was used.

Since $CD8^+$ T cells are necessary for controlling FV replication during acute infection [41], it was important to determine whether MDSC depletion affects viral loads. To answer this question, an infectious center assay was performed at 14 dpi. After depletion of all MDSCs with 5FU, an approximately tenfold reduction in spleen viral loads was observed. After the specific depletion of gMDSCs a more than twofold reduction was found, which was not statistically significant mainly due to the variations within the group of untreated control animals (Fig. 4e). These data suggest that MDSC-mediated suppression of $CD8^+$ T cell responses affected the elimination of FV-infected cells in vivo. They also imply that in FV infected mice both MDSC populations may contribute to the suppression of T cell responses.

Discussion

Myeloid-derived suppressor cells play an important role in regulating immune responses. Their influence may be beneficial, as they limit immunity to prevent tissue damage via T cell cytotoxicity. On the other hand, they may be suppressing immune responses against viral infections, which results in an incomplete elimination of the virus. The exact mechanisms involved in the control of viral immunity by MDSCs are not fully understood. Therefore, it is of importance to define the detailed functions of MDSCs in viral infections.

In this study (Fig. 1a) we defined MDSCs by using antibodies against $CD11b^+$, $Ly6G^+$ and $Ly6C^+$ [38]. It was also of importance to distinguish between MDSC subpopulations, since it has been shown that each subpopulation possesses distinct functions. The phenotype of MDSCs associated with the activation and with functionality of these cells was also characterized by the expression of PD-L1, CD39, CD80, as well as the functional markers Arg1 and NOS2 (Fig. 3a). The described strategy

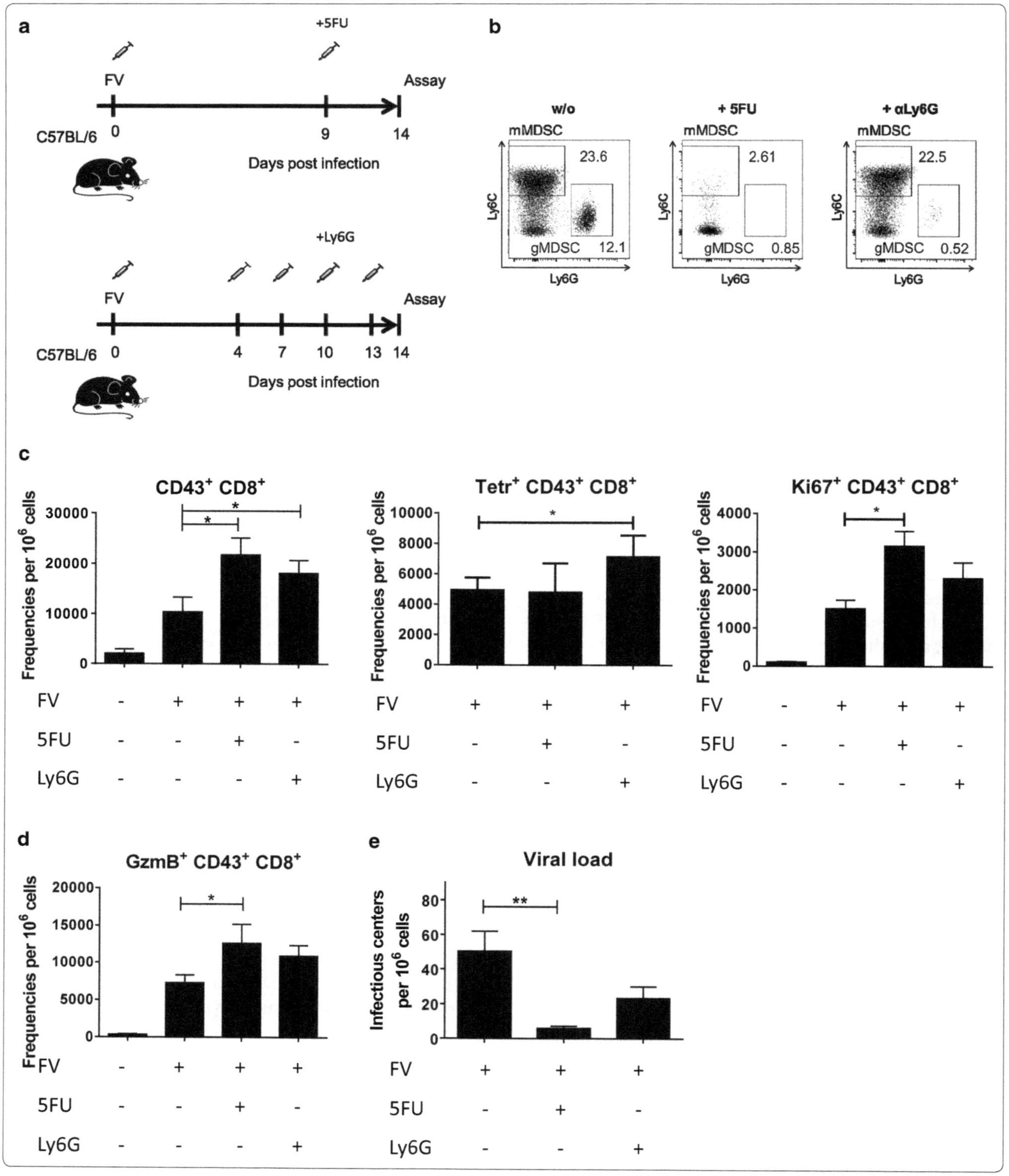
a
+5FU
FV
Assay
C57BL/6 0 9 14
Days post infection
+Ly6G
FV
Assay
C57BL/6 0 4 7 10 13 14
Days post infection
b
w/o
+ 5FU
+ αLy6G
mMDSC
23.6
gMDSC 12.1
2.61
gMDSC 0.85
22.5
gMDSC 0.52
Ly6C
Ly6G
c
CD43+ CD8+
Tetr+ CD43+ CD8+
Ki67+ CD43+ CD8+
Frequencies per 10^6 cells
FV
5FU
Ly6G
d
GzmB+ CD43+ CD8+
e
Viral load
Infectious centers per 10^6 cells

(See figure on previous page.)

Fig. 4 Depletion of MDSCs. B6 mice were infected i.v. with 20,000 SFFU of FV, and/or treated with 5FU or αLy6G. $CD8^+$ numbers were measured using flow cytometry and viral loads were estimated for spleen cells at 14 days post infection. Experimental design of the MDSC depletion by administration of 5FU or αLy6G (**a**). **b** Representative dot plots of MDSC during FV infection after administration of 5FU, αLy6G or untreated. **c** The numbers of $CD43^+$ $CD8^+$ T cells, the numbers of FV-D^bgagL class I tetramers positives $CD8^+CD43^+$, and the numbers of proliferating $Ki67^+CD43^+CD8^+$ T cells per 1×10^6 live cells in the spleen are displayed. **d** The numbers of $CD43^+$ $CD8^+$ T cells expressing GzmB per 1×10^6 live cells in the spleen are displayed. **e** The numbers of viral loads (infectious centers per 1×10^6 cells) in the spleen are displayed. At least five mice per group from three independent experiments were analyzed. Bars represent the mean with SD. For statistical analysis a Dunn's test with the Benjamini–Hochberg correction for multiple testing was performed (*<0.05; **<0.005)

allowed for a precise differentiation of the MDSC subpopulations in mice. Additionally, to correctly define the MDSC populations it was important to assess their suppressive activity as well as the biochemical and molecular markers described for MDSCs. Various studies have used different strategies for the phenotypic definition of MDSCs [60–63]. These differences make it difficult to compare results from theseinvestigations. Recently, attempts to standardize the definition of MDSC phenotypes in different species have been made [38, 64]. The current study strictly followed the state of the art recommendations ensuring that the latest standards of MDSC detection were used.

Expansion of MDSCs was previously observed in various experimental models. Recent studies suggest that MDSC may even serve as a prognostic marker for disease progression in various cancer models as well as during viral infections. Here, we demonstrate the expansion and activation of MDSC during an acute retroviral infection of mice. Elevation of MDSC numbers was also observed in blood of chronic HCV patients [21, 65, 66], as well as during human [17, 19, 20, 67] and mouse HBV infection [36, 68]. The total MDSC population as well as the gMDSC subpopulation were shown to increase in numbers during HIV infection [24, 52, 65].

In some of these studies the suppressive effect of MDSCs on T cell responses in vitro was demonstrated [48, 66, 69]. However, most studies conducted with cells from virus infected patients investigated the whole MDSCs population without distinguishing between different subpopulations [52, 65]. The effects of MDSC subpopulations on immune responses may vary significantly. Our current study demonstrates the suppressive activity of gMDSCs, but not mMDSCs, on virus-specific $CD8^+$ T cells in acute FV infection. In contrast to our findings, mMDSCs in a model of murine LP-BM5 retrovirus infection have been observed to preferentially suppress B cell responses, yet also T cell responses were downregulated [25–27]. The different results might be explained by technical differences in the T cell proliferation assays used in both studies. While we stimulated virus-specific T cells with its cognate antigen presented by DCs, a physiological way of inducing T cell proliferation and differentiation, others used a non-specific stimulation of T cells with CD3 and CD28 [25, 61]. It has already been reported that mMDSCs mainly suppress polyclonally activated T cells, as shown in tumor models [70] and infections such as LCMV, HBV, and HCV [36, 71]. On the other hand, a suppression of antigen-specific T cell responses are rather associated with gMDSCs function [72, 73]. Therefore, gMDSCs may have a significant influence on disease progression in various cancers [74], hepatic inflammation, fibrosis [75, 76], and in HIV infection [24].

An activation of MDSCs has been associated with the expression of different functional molecules. CD80, PD-L1, CD73, CD39, CD270, CD62L are just a few markers which were shown to be expressed on the cell surface of MDSCs and were linked to an activation of these cells [13, 14, 51, 77, 78]. In the current study, we analyzed the expression of CD80, as well as PD-L1 and CD39 as cell surface activation markers. An enhanced PD-1 expression on virus-specific $CD8^+$ T cells in chronically infected mice was shown to be involved in the development of $CD8^+$ T cell exhaustion. All FV-infected mice $CD8^+$ T cells also up-regulate PD-1, which plays an important role in immune evasion during acute infection [41, 79, 80]. Thus, PD-L1 expressing MDSCs might be a critical cell population inducing T cell exhaustion in chronic viral infections [34, 81]. This possibility was already discussed in various diseases, e.g. multiple melanoma [82, 83] or HIV infection [34]. Bowers et al. [34] illustrated the role of PD-1/PD-L1 signaling in T cell suppression mediated through HIV expanded gMDSC. We confirmed these results in our in vitro analysis of virus-specific $CD8^+$ T cell proliferation. gMDSCs used for this study lacked PD-L1, and were not able to suppress $CD8^+$ proliferation as efficiently as wild type gMDSCs expressing PD-L1 (Fig. 3c). However, more studies are required to determine the exact role of PD-L1 in MDSC mediated immunosuppression.

CD39 was analyzed as an additional marker associated with MDSC activation and suppression. During FV infection the expression of CD39 on MDSCs is high in both naïve and infected animals (Fig. 3a). CD39 is an essential molecule involved in an adenosine metabolism. First

CD39 metabolizes ADP/ATP to AMP, which is followed by the conversion of AMP to adenosine. This action shifts an AMP-driven proinflammatory microenvironment to anti-inflammatory conditions driven by adenosine [84]. Interestingly, CD39 expression and function were linked to immunosuppression in various diseases models [85–87]. Adenosine metabolism has a great influence on T cells, macrophages, neutrophils. Prinicipally Tregs seem to mediate suppression due to CD39 function. During HIV infection Tregs show elevated CD39 expression which has been positively correlated with a disease progression [88]. Moreover, Tregs from HIV infected patients expressed higher levels of CD39, and effector T cells from these patients showed a higher sensitivity to adenosine in vitro [87, 89]. In our work we observed a decreased MDSC suppression of $CD8^+$ T cells in the absence of CD39 (Fig. 3c). These data identify CD39 as a possible mechanism of MDSC mediated suppression. The function of CD39 on MDSCs was previously examined in an ovarian cancer model [90]. Ryzhov et al. [11] identified TGF-β as a regulator of CD39 expression on MDSCs. In colorectal cancer the expression of CD39 was also elevated on MDSC and these cells exhibited an increased inhibitory effect on T cells in comparison to cells from healthy donors [91].

The phenotypic characterization of MDSCs can be challenging, therefore their functional properties should be a focus of research. In the current work, MDSCs were functionally characterized by the expression of NOS2 and Arg1 and by specific blocking experiments in the T cell proliferation assay. Only the expression of Arg1, but not of NOS2, increased in gMDSCs after FV infection (Fig. 3a). These two important mechanisms of MDSC mediated immunosuppression are connected to ʟ-arginine metabolism. Both arginase and NO inhibitors were used in the current study, which confirmed that both pathways were partly involved in a gMDSC-mediated suppression of $CD8^+$ T cell proliferation. Although the NO inhibitor was able to partially inhibit the gMDSC-mediated immunosuppression (Fig. 3b), our previous results showed no enhanced expression of NOS2 in gMDSCs. However, it is important to note that two additional enzymes are involved in the nitric oxide metabolism: NOS1 and NOS3 [92]. The expression and activity of NOS3 was reported to play a role in gMDSC suppression using a murine model of malignant melanoma [93]. Thus, NOS3 might also be involved in the suppression by virus-induced gMDSCs, a premise which yet has to be tested.

Different attempts to study the activity and relevance of MDSCs in vivo have been presented. First, transfer experiments were performed, in which freshly isolated, lipopolisaccharid (LPS) or IFNγ induced MDSCs were transferred into recipient mice. This transfer was shown to inhibit inflammation [94] and reduce $CD8^+$ T cell responses in a melanoma model [95]. Another approach to study the function of MDSCs in vivo is a specific depletion of these cells. The most commonly used way of depletion is by administrating αLy6G or αGr1 antibodies into mice. This procedure allows the efficient depletion of all MDSCs (αGr1 antibody) or specifically of gMDSCs (αLy6G antibody).

In the current study, the MDSC function during an ongoing virus infection was investigated by using either αLy6G antibodies, which selectively deplete gMDSC, or 5FU, an anti-cancer drug known for selectively depleting all MDSCs [55]. Both αLy6G and 5FU administration resulted in the efficient depletion of MDSCs and led to an expansion of activated $CD8^+$ T cells (Fig. 4c). Although both depletion attempts have proven to be MDSC specific, it is important to note that αLy6G is also used for the depletion of neutrophils, and that 5FU appears to be cytotoxic for other cell types in higher concentration. We therefore carefully analyzed several other important immune cell populations in our depletion experiments and verified that they were not affected (data not shown). Interestingly, the depletion of all MDSCs after treatment with 5FU had a stronger effect on $CD8^+$ T cell responses than the depletion of only gMDSCs with anti-Ly6G treatment (Fig. 4c–e). Thus, mMDSCs, which were described to suppress lymphocyte responses in the murine AIDS model [25, 26], may also play role in FV infection in vivo. This data might be contradictory to the in vitro results, which showed no suppression of FV-specific T cells by mMDSCs. One possible explanation might be that gMDSCs directly affect antigen-specific $CD8^+$ T cells, whereas mMDSC only indirectly suppress CD8+ T cell responses. One possible target of mMDSCs might be $CD4^+$ T cells that then affect $CD8^+$ T cell responses. Indeed, it has been shown that at day 14 after FV infection the populations of expanded $CD4^+$ Tregs [96] and effector $CD4^+$ T cells [97] have a strong impact on the quantity and the functionality of effector $CD8^+$ T cells. It is known that MDSCs influence the expansion of Tregs [98] and $CD4^+$ T cells [99] and thus have an indirect effect on the population of effector $CD8^+$ T cells. In our in vitro assay system only DC and antigen-specific $CD8^+$ T cells, but no $CD4^+$ T cells were present. Thus, the in vitro model did not fully reflect the more complex situation in vivo. The influence of FV expanded MDSCs on different subpopulations of lymphocytes needs to be characterized in future studies.

Our data indicate in an in vivo approach that MDSC can indeed inhibit virus-specific T cell responses and interfere with virus immune control during an acute infection. This T cell suppression might be involved in the establishment or maintenance of viral chronicity.

Conclusions

Overall, this work demonstrates the important regulatory role of expanded MDSCs during the late phase of acute FV infection. The suppressive effect on $CD8^+$ T cells was predominantly mediated by gMDSCs, but mMDSCs may also contribute in vivo. Activated gMDSCs produced arginase 1 (Arg1), expressing the inhibitory ligand PD-L1 as well as the ATP dephosphorylating enzyme CD39 on their cell surface. All these effector molecules of MDSCs were involved in dampening virus-specific $CD8^+$ T cell responses in vitro. Moreover, the in vivo depletion of MDSCs resulted in augmented virus-specific cytotoxic $CD8^+$ T cell responses which correlated with a reduction in viral loads. Thus, our results demonstrate an inhibitory role of gMDSCs in FV infection and suggest these cells as a possible target for an immunomodulatory therapy of retroviral infections.

Methods

Mice

Inbred C57BL/6 (B6) mice were maintained under pathogen free conditions. Experiments were performed using B6 mice and were obtained from Charles River Laboratories. The relevant FV resistance genotype of B6 mice is H-2b/b, Fv1b/b, FV2r/r, Rfv3r/r. PD-L1 KO (B7-H1-KO) mice were originally generated by Lieping Chen [100]. CD39 KO mice on B6 background were originally produced by Simon C. Robson [101]. B6-background DbGagL TCR transgenic mice, where more than 90% of $CD8^+$ T cells contained a TCR specific for DbGagL FV epitope [46, 47]. All mice were females between 8 and 16 weeks old at day 0 of the experiments.

Virus and viral infection

The FV stock used for these experiments was a FV complex containing B-tropic Friend murine leukemia helper virus (F-MuLV) and polycythemia-inducing spleen focus forming virus, and was free of lactate dehydrogenase-elevating virus [102]. The stock was prepared as a 10% spleen cell homogenate from BALB/c mice infected 14 days previously with 3000 spleen focus-forming units of non-cloned virus stock. The virus was injected into experimental mice intravenously with 0.3 ml of PBS with 20,000 spleen focus forming units of FV.

In vivo cell depletion

C57BL/6 mice were infected with FV. For the depletion of MDSC, 10 mg/kg body mass 5-Fluorouracil (5FU) (Sigma-Aldrich) was administrated i.p. 4 days prior sacrifice. To deplete gMDSCs, 100 µg of anti-Ly6G antibody (Clone 1A8) (BioXCell) or control ratg IgG antibody (BioXCell) was administrated every third day for 3 times intraperitoneal.

Infectious center assay

Infectious centre (IC) assay was performed to determine the viral loads in infected organs. Shortly, tenfold dilutions of single-cell suspensions from the organ of interest were incubated with *Mus dunnis* cells for 3 days. After this time the cells were fixed with ethanol, stained with F-MuLV envelope-specific monoclonal antibody 720 and developed with peroxidase-conjugated goat anti-mouse antibody and aminoethylcarbazol to detect foci [103].

Cell surface and intracellular staining by flow cytometry

The antibodies used for cell surface staining were obtained from eBioscience or BioLegend, anti-Ly6G (1A8), anti-Ly6C (HK1.4), anti-Gr1 (RB6-8C5), anti-CD11b (M1/70), anti CD19 (eBio1D3), anti-CD3 (17A2), anti-CD39 (24DMS1) anti-PD-L1 (MIH5), anti-CD80 (16-10A1), anti-CD8 (53-6.7), anti-CD43 (1B11), anti-NK1.1(PK136) anti-Ter119 (Ter-119) and FC block anti-mouse CD16/CD32 [91]. Intracellular Granzym B (monoclonal anti-human granzyme B (GB11) antibody Invitrogen) staining was performed as previously described [96]. Intracellular expression of Ki67 (SolA15), Arg1 (R&D Systems) and NOS2 (CXNFT) was detected using a Foxp3 staining kit (eBioscience). Dead cells for cell-surface and intracellular staining were excluded with use of Fixable Viability Dye eFluor 780 (FVD) (eBioscience). Data were acquired on a LSR II flow cytometer (Becton–Dickinson) from 300,000 to 500,000 lymphocyte-gated events per sample. Analyses were done using FlowJo (Treestar).

Tetramer staining

To detect D^bGagL-specific $CD8^+$ T cells, splenocytes were stained with PE labelled MHC class I H2-D^b (Beckman Coulter, Marseille, France) tetramers, which are specific for FV GagL peptide [47, 104] as described previously described [96].

In vitro suppression assay

To examine the influence of MDSCs in vitro on both the proliferation and function of $CD8^+$ T cells, a standard in vitro immunosuppression assay was modified as described for the characterisation of the suppressive function of regulatory T cells. Murine bone marrow dendritic cells were generated as previously described and incubated with DbGagL peptide (5 µg/ml) in RPMI containing 10% FBS, 2 mM L-glutamine, 50 µM 2-ME and 100 U/ml each penicillin and streptomycin at for 60 min at 37 °C. FV-specific TCR Tg $CD8^+$ T cells were isolated from the spleens of DBGagL TCR Tg mice by positive selection using magnetic bead separation according to the manufacturer's instruction (Miltenyi Biotec), and then labelled with Violet Tracer dye (Life Technologies). $Ly6G^{high}$ gMDSCs and $Ly6G^{low/-m}$ MDSCs were isolated from the spleens dissected from B6, PD-L1 KO or CD39 KO mice by positive selection using magnetic bead separation according to the manufacturer's protocol. (Miltenyi Biotec) For the induction of T cell proliferation, 1×10^5 of peptide-pulsed DCs and 5×10^5 TCR Tg CD8$^+$ T cells were co-cultured on a flat-bottom 96-well plate in AIM-V (Life Technologies) containing 10% FBS. MDSCs were added to the culture simultaneously with CD8$^+$ T cells. After 72 h cells were stained for CD8, fixed, permeabilized and stained for intracellular GzmB as described above. Additionally, for experiments that examined the effect of NO and arginase, the assay described above was implemented for the characterisation of gMDSCs function in vitro with the use of inhibitors. Besides bead-isolated CD8$^+$ T cells from naïve TCRtg mice, peptide loaded DCs and Violet Tracer stained gMDSCs, an arginase inhibitor 0.5 mM nor-NOHA (NW-hydroxyl-nor-L-arginine) (Cayman Chemical) and NO inhibitor 0.5 mM L-NMMA (NG-Methyl-L-arginine acetate salt) (Sigma-Aldrich) were added at the beginning of the culture.

Statistical analysis

Statistics comparing the two groups were done using the unpaired t test (GraphPad Prism software; GraphPad Software INC., San Diego, USA). When more than two groups were compared, a Dunn test with the Benjamini–Hochberg correction for multiple testing was performed (R-package dunn.test, version 1.3.4).

Authors' contributions

GZ, UD, DH, QL, and MDP, designed the studies and wrote the manuscript; MDP and TW collected and analyzed the data; LC provided the mice. All authors read and approved the final manuscript.

Author details

[1] Institute of Virology, University Hospital Essen, University of Duisburg-Essen, Hufelandstraße 55, 45147 Essen, Germany. [2] Research Group Bioinformatics, Faculty of Biology, University of Duisburg-Essen, Essen, Germany. [3] Key Laboratory of Molecular Virology and Immunology, Institute Pasteur of Shanghai, Institutes for Biological Sciences, Chinese Academy of Sciences, Shanghai, China. [4] Department of Immunobiology, Yale School of Medicine, Yale University, New Haven, CT, USA.

Competing interests

The authors declare that they have no competing interests.

Funding

This work was supported by grants from the Deutsche Forschungsgemeinschaft (TRR60 project B8) and by the COST Action BM1404 Mye-EUNITER (http://www.mye-euniter.eu). COST is supported by the EU Framework Program Horizon 2020. The Deutsche Forschungsgemeinschaft and COST were not involved in the design of the study, in the collection, analysis, and interpretation of data or in writing the manuscript.

References

1. Bennett JA, Rao VS, Mitchell MS. Systemic bacillus Calmette–Guerin⁻ (BCG) activates natural suppressor cells. Proc Natl Acad Sci USA. 1978;75(10):5142–4.
2. Talmadge JE, Gabrilovich DI. History of myeloid-derived suppressor cells. Nat Rev Cancer. 2013;13(10):739–52.
3. Serafini P, Borrello I, Bronte V. Myeloid suppressor cells in cancer: recruitment, phenotype, properties, and mechanisms of immune suppression. Semin Cancer Biol. 2006;16(1):53–65.
4. Gabrilovich DI, Bronte V, Chen SH, Colombo MP, Ochoa A, Ostrand-Rosenberg S, et al. The terminology issue for myeloid-derived suppressor cells. Cancer Res. 2007;67(1):425 **(author reply 6)**.
5. Youn JI, Nagaraj S, Collazo M, Gabrilovich DI. Subsets of myeloid-derived suppressor cells in tumor-bearing mice. J Immunol. 2008;181(8):5791–802.
6. Kusmartsev S, Nefedova Y, Yoder D, Gabrilovich DI. Antigen-specific inhibition of CD8+ T cell response by immature myeloid cells in cancer is mediated by reactive oxygen species. J Immunol. 2004;172(2):989–99.
7. Mizumoto N, Kumamoto T, Robson SC, Sevigny J, Matsue H, Enjyoji K, et al. CD39 is the dominant Langerhans cell-associated ecto-NTPDase: modulatory roles in inflammation and immune responsiveness. Nat Med. 2002;8(4):358–65.
8. Resta R, Yamashita Y, Thompson LF. Ecto-enzyme and signaling functions of lymphocyte CD73. Immunol Rev. 1998;161:95–109.
9. Iannone R, Miele L, Maiolino P, Pinto A, Morello S. Blockade of A2b adenosine receptor reduces tumor growth and immune suppression mediated by myeloid-derived suppressor cells in a mouse model of melanoma. Neoplasia. 2013;15(12):1400–9.
10. Ryzhov S, Novitskiy SV, Goldstein AE, Biktasova A, Blackburn MR, Biaggioni I, et al. Adenosinergic regulation of the expansion and immunosuppressive activity of CD11b+Gr1+ cells. J Immunol. 2011;187(11):6120–9.
11. Ryzhov SV, Pickup MW, Chytil A, Gorska AE, Zhang Q, Owens P, et al. Role of TGF-beta signaling in generation of CD39+CD73+ myeloid cells in tumors. J Immunol. 2014;193(6):3155–64.
12. Gabrilovich DI, Nagaraj S. Myeloid-derived suppressor cells as regulators of the immune system. Nat Rev Immunol. 2009;9(3):162–74.
13. Markowitz J, Wesolowski R, Papenfuss T, Brooks TR, Carson WE 3rd. Myeloid-derived suppressor cells in breast cancer. Breast Cancer Res Treat. 2013;140(1):13–21.
14. Marvel D, Gabrilovich DI. Myeloid-derived suppressor cells in the tumor microenvironment: expect the unexpected. J Clin Investig. 2015;125(9):3356–64.
15. Ostrand-Rosenberg S, Sinha P. Myeloid-derived suppressor cells: linking inflammation and cancer. J Immunol. 2009;182(8):4499–506.
16. O'Connor MA, Rastad JL, Green WR. The role of myeloid-derived suppressor cells in viral infection. Viral Immunol. 2017;30:82–97.
17. Lu LR, Liu J, Xu Z, Zhang GL, Li DC, Lin CS. Expression and clinical significance of myeloid derived suppressor cells in chronic hepatitis B patients. Asian Pac J Cancer Prev. 2014;15(10):4367–72.
18. Pallett LJ, Gill US, Quaglia A, Sinclair LV, Jover-Cobos M, Schurich A, et al. Metabolic regulation of hepatitis B immunopathology by myeloid-derived suppressor cells. Nat Med. 2015;21(6):591–600.
19. Fang Z, Li J, Yu X, Zhang D, Ren G, Shi B, et al. Polarization of monocytic myeloid-derived suppressor cells by hepatitis B surface antigen is mediated via ERK/IL-6/STAT3 signaling feedback and restrains the activation of T cells in chronic hepatitis B virus infection. J Immunol. 2015;195(10):4873–83.
20. Huang A, Zhang B, Yan W, Wang B, Wei H, Zhang F, et al. Myeloid-derived suppressor cells regulate immune response in patients with chronic hepatitis B virus infection through PD-1-induced IL-10. J Immunol. 2014;193(11):5461–9.
21. Cai W, Qin A, Guo P, Yan D, Hu F, Yang Q, et al. Clinical significance and functional studies of myeloid-derived suppressor cells in chronic hepatitis C patients. J Clin Immunol. 2013;33(4):798–808.
22. Ning G, She L, Lu L, Liu Y, Zeng Y, Yan Y, et al. Analysis of monocytic and granulocytic myeloid-derived suppressor cells subsets in patients with hepatitis C virus infection and their clinical significance. Biomed Res Int. 2015;2015:385378.

23. Garg A, Spector SA. HIV type 1 gp120-induced expansion of myeloid derived suppressor cells is dependent on interleukin 6 and suppresses immunity. J Infect Dis. 2014;209(3):441–51.
24. Vollbrecht T, Stirner R, Tufman A, Roider J, Huber RM, Bogner JR, et al. Chronic progressive HIV-1 infection is associated with elevated levels of myeloid-derived suppressor cells. AIDS. 2012;26(12):F31–7.
25. Green KA, Cook WJ, Green WR. Myeloid-derived suppressor cells in murine retrovirus-induced AIDS inhibit T- and B-cell responses in vitro that are used to define the immunodeficiency. J Virol. 2013;87(4):2058–71.
26. Green KA, Wang L, Noelle RJ, Green WR. Selective involvement of the checkpoint regulator VISTA in suppression of B-cell, but not T-cell, responsiveness by monocytic myeloid-derived suppressor cells from mice infected with an immunodeficiency-causing retrovirus. J Virol. 2015;89(18):9693–8.
27. O'Connor MA, Fu WW, Green KA, Green WR. Subpopulations of M-MDSCs from mice infected by an immunodeficiency-causing retrovirus and their differential suppression of T- vs B-cell responses. Virology. 2015;485:263–73.
28. O'Connor MA, Vella JL, Green WR. Reciprocal relationship of T regulatory cells and monocytic myeloid-derived suppressor cells in LP-BM5 murine retrovirus-induced immunodeficiency. J Gen Virol. 2016;97(2):509–22.
29. Hasenkrug KJ, Chesebro B. Immunity to retroviral infection: the Friend virus model. Proc Natl Acad Sci USA. 1997;94(15):7811–6.
30. Zelinskyy G, Robertson SJ, Schimmer S, Messer RJ, Hasenkrug KJ, Dittmer U. CD8+ T-cell dysfunction due to cytolytic granule deficiency in persistent Friend retrovirus infection. J Virol. 2005;79(16):10619–26.
31. Akhmetzyanova I, Drabczyk M, Neff CP, Gibbert K, Dietze KK, Werner T, et al. PD-L1 expression on retrovirus-infected cells mediates immune escape from CD8+ T cell killing. PLoS Pathog. 2015;11(10):e1005224.
32. Kantzanou M, Lucas M, Barnes E, Komatsu H, Dusheiko G, Ward S, et al. Viral escape and T cell exhaustion in hepatitis C virus infection analysed using class I peptide tetramers. Immunol Lett. 2003;85(2):165–71.
33. Pantaleo G, Soudeyns H, Demarest JF, Vaccarezza M, Graziosi C, Paolucci S, et al. Evidence for rapid disappearance of initially expanded HIV-specific CD8+ T cell clones during primary HIV infection. Proc Natl Acad Sci USA. 1997;94(18):9848–53.
34. Bowers NL, Helton ES, Huijbregts RP, Goepfert PA, Heath SL, Hel Z. Immune suppression by neutrophils in HIV-1 infection: role of PD-L1/PD-1 pathway. PLoS Pathog. 2014;10(3):e1003993.
35. Tacke RS, Lee HC, Goh C, Courtney J, Polyak SJ, Rosen HR, et al. Myeloid suppressor cells induced by hepatitis C virus suppress T-cell responses through the production of reactive oxygen species. Hepatology. 2012;55(2):343–53.
36. Chen S, Akbar SM, Abe M, Hiasa Y, Onji M. Immunosuppressive functions of hepatic myeloid-derived suppressor cells of normal mice and in a murine model of chronic hepatitis B virus. Clin Exp Immunol. 2011;166(1):134–42.
37. Nagaraj S, Schrum AG, Cho HI, Celis E, Gabrilovich DI. Mechanism of T cell tolerance induced by myeloid-derived suppressor cells. J Immunol. 2010;184(6):3106–16.
38. Bronte V, Brandau S, Chen SH, Colombo MP, Frey AB, Greten TF, et al. Recommendations for myeloid-derived suppressor cell nomenclature and characterization standards. Nat Commun. 2016;7:12150.
39. Gabrilovich DI, Ostrand-Rosenberg S, Bronte V. Coordinated regulation of myeloid cells by tumours. Nat Rev Immunol. 2012;12(4):253–68.
40. Peranzoni E, Zilio S, Marigo I, Dolcetti L, Zanovello P, Mandruzzato S, et al. Myeloid-derived suppressor cell heterogeneity and subset definition. Curr Opin Immunol. 2010;22(2):238–44.
41. Zelinskyy G, Myers L, Dietze KK, Gibbert K, Roggendorf M, Liu J, et al. Virus-specific CD8+ T cells upregulate programmed death-1 expression during acute friend retrovirus infection but are highly cytotoxic and control virus replication. J Immunol. 2011;187(7):3730–7.
42. Zelinskyy G, Kraft AR, Schimmer S, Arndt T, Dittmer U. Kinetics of CD8+ effector T cell responses and induced CD4+ regulatory T cell responses during Friend retrovirus infection. Eur J Immunol. 2006;36(10):2658–70.
43. McCoy KD, Le Gros G. The role of CTLA-4 in the regulation of T cell immune responses. Immunol Cell Biol. 1999;77(1):1–10.
44. Yang R, Cai Z, Zhang Y, Yutzy WH, Roby KF, Roden RB. CD80 in immune suppression by mouse ovarian carcinoma-associated Gr-1+CD11b+ myeloid cells. Can Res. 2006;66(13):6807–15.
45. Poschke I, Mougiakakos D, Hansson J, Masucci GV, Kiessling R. Immature immunosuppressive CD14+HLA-DR-/low cells in melanoma patients are Stat3hi and overexpress CD80, CD83, and DC-sign. Can Res. 2010;70(11):4335–45.
46. Ohlen C, Kalos M, Cheng LE, Shur AC, Hong DJ, Carson BD, et al. CD8(+) T cell tolerance to a tumor-associated antigen is maintained at the level of expansion rather than effector function. J Exp Med. 2002;195(11):1407–18.
47. Chen W, Qin H, Chesebro B, Cheever MA. Identification of a gag-encoded cytotoxic T-lymphocyte epitope from FBL-3 leukemia shared by Friend, Moloney, and Rauscher murine leukemia virus-induced tumors. J Virol. 1996;70(11):7773–82.
48. Arocena AR, Onofrio LI, Pellegrini AV, Carrera Silva AE, Paroli A, Cano RC, et al. Myeloid-derived suppressor cells are key players in the resolution of inflammation during a model of acute infection. Eur J Immunol. 2014;44(1):184–94.
49. Bronte V. Myeloid-derived suppressor cells in inflammation: uncovering cell subsets with enhanced immunosuppressive functions. Eur J Immunol. 2009;39(10):2670–2.
50. De Sanctis F, Bronte V, Ugel S. Tumor-induced myeloid-derived suppressor cells. Microbiol Spectr. 2016;4(3).
51. De Sanctis F, Solito S, Ugel S, Molon B, Bronte V, Marigo I. MDSCs in cancer: conceiving new prognostic and therapeutic targets. Biochim Biophys Acta. 2016;1865(1):35–48.
52. Macatangay BJ, Landay AL, Rinaldo CR. MDSC: a new player in HIV immunopathogenesis. AIDS. 2012;26(12):1567–9.
53. Draghiciu O, Lubbers J, Nijman HW, Daemen T. Myeloid derived suppressor cells—an overview of combat strategies to increase immunotherapy efficacy. Oncoimmunology. 2015;4(1):e954829.
54. Wesolowski R, Markowitz J, Carson WE 3rd. Myeloid derived suppressor cells - a new therapeutic target in the treatment of cancer. J Immunother Cancer. 2013;1:10.
55. Vincent J, Mignot G, Chalmin F, Ladoire S, Bruchard M, Chevriaux A, et al. 5-Fluorouracil selectively kills tumor-associated myeloid-derived suppressor cells resulting in enhanced T cell-dependent antitumor immunity. Cancer Res. 2010;70(8):3052–61.
56. Matsumoto M, Atarashi K, Umemoto E, Furukawa Y, Shigeta A, Miyasaka M, et al. CD43 functions as a ligand for E-selectin on activated T cells. J Immunol. 2005;175(12):8042–50.
57. Matsumoto M, Shigeta A, Furukawa Y, Tanaka T, Miyasaka M, Hirata T. CD43 collaborates with P-selectin glycoprotein ligand-1 to mediate E-selectin-dependent T cell migration into inflamed skin. J Immunol. 2007;178(4):2499–506.
58. Zelinskyy G, Robertson SJ, Schimmer S, Messer RJ, Hasenkrug KJ, Dittmer U. CD8+ T-cell dysfunction due to cytolytic granule deficiency in persistent Friend retrovirus infection. J Virol. 2005;79(16):10619–26.
59. Scholzen T, Gerdes J. The Ki-67 protein: from the known and the unknown. J Cell Physiol. 2000;182(3):311–22.
60. Young MR, Wright MA. Myelopoiesis-associated immune suppressor cells in mice bearing metastatic Lewis lung carcinoma tumors: gamma interferon plus tumor necrosis factor alpha synergistically reduces immune suppressor and tumor growth-promoting activities of bone marrow cells and diminishes tumor recurrence and metastasis. Can Res. 1992;52(22):6335–40.
61. Kusmartsev SA, Li Y, Chen SH. Gr-1+ myeloid cells derived from tumor-bearing mice inhibit primary T cell activation induced through CD3/CD28 costimulation. J Immunol. 2000;165(2):779–85.
62. Rose S, Misharin A, Perlman H. A novel Ly6C/Ly6G-based strategy to analyze the mouse splenic myeloid compartment. Cytom A. 2012;81(4):343–50.
63. Damuzzo V, Pinton L, Desantis G, Solito S, Marigo I, Bronte V, et al. Complexity and challenges in defining myeloid-derived suppressor cells. Cytom B Clin Cytom. 2014. https://doi.org/10.1002/cytob.21206.
64. O'Connor MA, Rastad JL, Green WR. The role of myeloid-derived suppressor cells in viral infection. Viral Immunol. 2017;30(2):82–97.
65. Lei AH, Yang Q, Cai WP, Liu YF, Lan Y, Qin AP, et al. Clinical significance of myeloid-derived suppressor cells in human immunodeficiency virus-1/hepatitis C virus-coinfected patients. Scand J Immunol. 2016;83(6):438–44.

66. Zeng QL, Yang B, Sun HQ, Feng GH, Jin L, Zou ZS, et al. Myeloid-derived suppressor cells are associated with viral persistence and downregulation of TCR zeta chain expression on CD8(+) T cells in chronic hepatitis C patients. Mol Cells. 2014;37(1):66–73.
67. Pallett LJ, Gill US, Quaglia A, Sinclair LV, Jover-Cobos M, Schurich A, et al. Metabolic regulation of hepatitis B immunopathology by myeloid-derived suppressor cells. Nat Med. 2015;21(6):591–600.
68. Kong X, Sun R, Chen Y, Wei H, Tian Z. gammadeltaT cells drive myeloid-derived suppressor cell-mediated CD8+ T cell exhaustion in hepatitis B virus-induced immunotolerance. J Immunol. 2014;193(4):1645–53.
69. Arina A, Corrales L, Bronte V. Enhancing T cell therapy by overcoming the immunosuppressive tumor microenvironment. Semin Immunol. 2016;28(1):54–63.
70. Nagaraj S, Youn JI, Gabrilovich DI. Reciprocal relationship between myeloid-derived suppressor cells and T cells. J Immunol. 2013;191(1):17–23.
71. Pang X, Song H, Zhang Q, Tu Z, Niu J. Hepatitis C virus regulates the production of monocytic myeloid-derived suppressor cells from peripheral blood mononuclear cells through PI3K pathway and autocrine signaling. Clin Immunol. 2016;164:57–64.
72. Pinton L, Solito S, Damuzzo V, Francescato S, Pozzuoli A, Berizzi A, et al. Activated T cells sustain myeloid-derived suppressor cell-mediated immune suppression. Oncotarget. 2016;7(2):1168–84.
73. Solito S, Bronte V, Mandruzzato S. Antigen specificity of immune suppression by myeloid-derived suppressor cells. J Leukoc Biol. 2011;90(1):31–6.
74. Dolcetti L, Marigo I, Mantelli B, Peranzoni E, Zanovello P, Bronte V. Myeloid-derived suppressor cell role in tumor-related inflammation. Cancer Lett. 2008;267(2):216–25.
75. Hammerich L, Tacke F. Emerging roles of myeloid derived suppressor cells in hepatic inflammation and fibrosis. World J Gastrointest Pathophysiol. 2015;6(3):43–50.
76. Fletcher M, Ramirez ME, Sierra RA, Raber P, Thevenot P, Al-Khami AA, et al. L-Arginine depletion blunts antitumor T-cell responses by inducing myeloid-derived suppressor cells. Can Res. 2015;75(2):275–83.
77. Bronte V, Brandau S, Chen SH, Colombo MP, Frey AB, Greten TF, et al. Recommendations for myeloid-derived suppressor cell nomenclature and characterization standards. Nat Commun. 2016;7:12150.
78. Dolcetti L, Peranzoni E, Bronte V. Current Protocols in Immunology. In: Measurement of myeloid cell immune suppressive activity, Chap 14, Unit 14.17. Vol. 91; 2010. p. 14.17:14.17.1–14.17.25.
79. Akhmetzyanova I, Drabczyk M, Neff CP, Gibbert K, Dietze KK, Werner T, et al. PD-L1 expression on retrovirus-infected cells mediates immune escape from CD8+ T cell killing. PLoS Pathog. 2015;11(10):e1005224.
80. Barber DL, Wherry EJ, Masopust D, Zhu B, Allison JP, Sharpe AH, et al. Restoring function in exhausted CD8 T cells during chronic viral infection. Nature. 2006;439(7077):682–7.
81. Noman MZ, Desantis G, Janji B, Hasmim M, Karray S, Dessen P, et al. PD-L1 is a novel direct target of HIF-1alpha, and its blockade under hypoxia enhanced MDSC-mediated T cell activation. J Exp Med. 2014;211(5):781–90.
82. Fujimura T, Ring S, Umansky V, Mahnke K, Enk AH. Regulatory T cells stimulate B7-H1 expression in myeloid-derived suppressor cells in ret melanomas. J Invest Dermatol. 2012;132(4):1239–46.
83. Haile ST, Bosch JJ, Agu NI, Zeender AM, Somasundaram P, Srivastava MK, et al. Tumor cell programmed death ligand 1-mediated T cell suppression is overcome by coexpression of CD80. J Immunol. 2011;186(12):6822–9.
84. Antonioli L, Pacher P, Vizi ES, Hasko G. CD39 and CD73 in immunity and inflammation. Trends Mol Med. 2013;19(6):355–67.
85. Kas-Deelen AM, Bakker WW, Olinga P, Visser J, de Maar EF, van Son WJ, et al. Cytomegalovirus infection increases the expression and activity of ecto-ATPase (CD39) and ecto-5′nucleotidase (CD73) on endothelial cells. FEBS Lett. 2001;491(1–2):21–5.
86. Leal DB, Streher CA, Bertoncheli Cde M, Carli LF, Leal CA, da Silva JE, et al. HIV infection is associated with increased NTPDase activity that correlates with CD39-positive lymphocytes. Biochim Biophys Acta. 2005;1746(2):129–34.
87. Nikolova M, Carriere M, Jenabian MA, Limou S, Younas M, Kok A, et al. CD39/adenosine pathway is involved in AIDS progression. PLoS Pathog. 2011;7(7):e1002110.
88. Schulze Zur Wiesch J, Thomssen A, Hartjen P, Toth I, Lehmann C, Meyer-Olson D, et al. Comprehensive analysis of frequency and phenotype of T regulatory cells in HIV infection: CD39 expression of FoxP3+ T regulatory cells correlates with progressive disease. J Virol. 2011;85(3):1287–97.
89. Moreno-Fernandez ME, Rueda CM, Rusie LK, Chougnet CA. Regulatory T cells control HIV replication in activated T cells through a cAMP-dependent mechanism. Blood. 2011;117(20):5372–80.
90. Montalban Del Barrio I, Penski C, Schlahsa L, Stein RG, Diessner J, Wockel A, et al. Adenosine-generating ovarian cancer cells attract myeloid cells which differentiate into adenosine-generating tumor associated macrophages—a self-amplifying, CD39- and CD73-dependent mechanism for tumor immune escape. J Immunother Cancer. 2016;4:49.
91. Zhang B, Wang Z, Wu L, Zhang M, Li W, Ding J, et al. Circulating and tumor-infiltrating myeloid-derived suppressor cells in patients with colorectal carcinoma. PLoS ONE. 2013;8(2):e57114.
92. Forstermann U, Sessa WC. Nitric oxide synthases: regulation and function. Eur Heart J. 2012;33(7):829–37, 37a–d.
93. Raber PL, Thevenot P, Sierra R, Wyczechowska D, Halle D, Ramirez ME, et al. Subpopulations of myeloid-derived suppressor cells impair T cell responses through independent nitric oxide-related pathways. Int J Cancer. 2014;134(12):2853–64.
94. Greifenberg V, Ribechini E, Rossner S, Lutz MB. Myeloid-derived suppressor cell activation by combined LPS and IFN-gamma treatment impairs DC development. Eur J Immunol. 2009;39(10):2865–76.
95. Poschke I, Mao Y, Adamson L, Salazar-Onfray F, Masucci G, Kiessling R. Myeloid-derived suppressor cells impair the quality of dendritic cell vaccines. Cancer Immunol Immunother CII. 2012;61(6):827–38.
96. Zelinskyy G, Dietze KK, Husecken YP, Schimmer S, Nair S, Werner T, et al. The regulatory T-cell response during acute retroviral infection is locally defined and controls the magnitude and duration of the virus-specific cytotoxic T-cell response. Blood. 2009;114(15):3199–207.
97. Manzke N, Akhmetzyanova I, Hasenkrug KJ, Trilling M, Zelinskyy G, Dittmer U. CD4+ T cells develop antiretroviral cytotoxic activity in the absence of regulatory T cells and CD8+ T cells. J Virol. 2013;87(11):6306–13.
98. Serafini P, Mgebroff S, Noonan K, Borrello I. Myeloid-derived suppressor cells promote cross-tolerance in B-cell lymphoma by expanding regulatory T cells. Can Res. 2008;68(13):5439–49.
99. Nonaka K, Saio M, Suwa T, Frey AB, Umemura N, Imai H, et al. Skewing the Th cell phenotype toward Th1 alters the maturation of tumor-infiltrating mononuclear phagocytes. J Leukoc Biol. 2008;84(3):679–88.
100. Dong H, Zhu G, Tamada K, Flies DB, van Deursen JM, Chen L. B7-H1 determines accumulation and deletion of intrahepatic CD8(+) T lymphocytes. Immunity. 2004;20(3):327–36.
101. Enjyoji K, Sevigny J, Lin Y, Frenette PS, Christie PD, Esch JS, Imai M, Edelberg JM, Rayburn H, Lech M, Beeler DL, Csizmadia E, Wagner DD, Robson SC, Rosenberg RD. Targeted disruption of cd39/ATP diphosphohydrolase results in disordered hemostasis and thromboregulation. Nat Med 1999;5:1010–7
102. Robertson SJ, Ammann CG, Messer RJ, Carmody AB, Myers L, Dittmer U, et al. Suppression of acute anti-friend virus CD8+ T-cell responses by coinfection with lactate dehydrogenase-elevating virus. J Virol. 2008;82(1):408–18.
103. Dittmer U, Brooks DM, Hasenkrug KJ. Characterization of a live-attenuated retroviral vaccine demonstrates protection via immune mechanisms. J Virol. 1998;72(8):6554–8.
104. Schepers K, Toebes M, Sotthewes G, Vyth-Dreese FA, Dellemijn TA, Melief CJ, et al. Differential kinetics of antigen-specific CD4+ and CD8+ T cell responses in the regression of retrovirus-induced sarcomas. J Immunol. 2002;169(6):3191–9.

12

Reduction of inflammation and T cell activation after 6 months of cART initiation during acute, but not in early chronic HIV-1 infection

Hury Hellen Souza de Paula[1], Ana Cristina Garcia Ferreira[2], Diogo Gama Caetano[1], Edson Delatorre[1], Sylvia Lopes Maia Teixeira[1], Lara Esteves Coelho[2], Eduarda Grinsztejn João[2], Michelle Morata de Andrade[2], Sandra Wagner Cardoso[2], Beatriz Grinsztejn[2], Valdilea Gonçalves Veloso[2], Mariza Gonçalves Morgado[1], Monick Lindenmeyer Guimarães[1] and Fernanda Heloise Côrtes[1*]

Abstract

Objectives: To investigate the impact of early combined antiretroviral therapy (cART) on inflammation biomarkers and immune activation during acute and early chronic HIV-1 infection.

Methods: We included 12 acute (AHI), 11 early chronic (EcHI), and 18 late chronic HIV-1-infected (LcHI) individuals who were treated with cART and 18 HIV-1-uninfected (HIV-neg) individuals. Plasmatic levels of inflammation biomarkers, $CD8^+CD38^+HLA\text{-}DR^+$ T cell frequencies, CD4 T cell counts, CD4/CD8 ratio, total HIV-1 DNA and plasmatic viral load were evaluated. Mann–Whitney test, Pearson and Spearman correlation, and linear regression models were used for statistical analyses.

Results: IP-10, IL-18, and sCD163 were significantly elevated at pre-ART in the AHI and EcHI groups, showing a significant reduction after 6 months of cART in the AHI group, achieving similar levels to the HIV-neg group. For the EcHI group, the IP-10 and sCD163 levels were also significantly reduced on M6-ART; however, IP-10 levels remained higher than in the HIV-neg group, and no significant reduction of IL-18 levels was observed. The $CD8^+$ T cell activation levels were elevated in the AHI and EcHI groups at pre-ART and showed a significant reduction on M6-ART, but they were similar to levels seen for HIV-neg only after 12 months of cART. At pre-ART, IP-10 levels but not IL-18 levels were positively correlated with HIV-1 viral load in the AHI group.

Conclusions: Early initiation of cART in HIV infection can reduce systemic inflammation, but the earlier normalization of the inflammation markers was only observed when cART was initiated in the acute phase of infection. A slower dynamic of reduction was observed for $CD8^+$ T cell activation.

Keywords: HIV-1 acute infection, Inflammation, cART, Immune activation, IP-10, IL-18

*Correspondence: fheloise@ioc.fiocruz.br

[1] Laboratório de Aids e Imunologia Molecular, Instituto Oswaldo Cruz –IOC, FIOCRUZ, Av Brasil, 4365, Pavilhão Leônidas Deane, sala 401, Rio de Janeiro 21040360, Brazil

Full list of author information is available at the end of the article

Introduction

High systemic immune activation has a pivotal role in HIV-1 pathogenesis [1]. Studies have associated the elevated immune activation with $CD4^+$ T cell depletion and progression to AIDS [2–7]. During primary HIV-1 infection (PHI), viral loads can reach values higher than one million HIV-1 RNA copies/mL, and a significant decrease in $CD4^+$ T cell counts occurs [8, 9]. A high production of pro-inflammatory cytokines, called a "cytokine storm", arises in response to virus replication [10]. A controlled initial pro-inflammatory immune response may be beneficial, as observed in natural SIV hosts, which displayed significant increases in plasma cytokines and interferon-stimulated gene (ISG) expression, normalized four weeks after infection [11]. However, in the context of HIV-1 infection, the pro-inflammatory response remains elevated after the end of the acute phase, and most individuals without cART develop progressive immune dysregulation, culminating in AIDS [8].

The cART has had an enormous impact on mortality and morbidity in HIV-1 infection [12–14]. Studies in the cART era have demonstrated a shift in the causes of death and morbidities in HIV-1-infected individuals, with a proportional increase of non-AIDS-related events, such as cardiovascular, liver, and renal diseases, when compared with AIDS-related events [15–20]. Also, a higher prevalence of age-related noninfectious comorbidities was observed in HIV-1-infected individuals than in the general population [21]. Alongside this, a strong association between inflammation and AIDS and non-AIDS-events became evident [22–24]. Several biomarkers of inflammation have been associated with disease progression and mortality in HIV-1 infection [25]. Among these markers, IL-6 and CRP were associated with an increased risk of AIDS [23] or death [22, 24]. IL-6 levels, at time of seroconversion, also predicted HIV-1 disease progression [26]. The markers of monocyte activation sCD14 and sCD163 are associated with a higher risk of death [27–29]. And, during the PHI, CRP levels were significantly higher in HIV-1 infected individuals than in HIV-1 uninfected individuals [5]. Other biomarkers elevated during PHI were IP-10 and IL-18, with an association between IP-10 and a faster disease progression [30, 31]. Continued IL-18 and sCD14 elevation were also associated with clinical cART failure [32].

Although cART reduces immune activation and inflammation [5, 22], these remain higher in HIV-1-infected individuals than in HIV-1-uninfected individuals, even after long-term viral suppression [33]. Studies have shown the benefits of initiation of cART during the acute or early infection [2–4, 6, 7]. The early therapy decreases immune system damage and the establishment of a large viral reservoir [6, 7]. HIV-1 infected patients who started cART during PHI presented an earlier immune reconstitution ($CD4^+$ T cell count > 500 cells/mm^3, CD4% > 30%, and CD4/CD8 ratio > 1) than cART in chronic patients [34]. Moreover, early cART decreases T cell activation [3, 7, 35] and markers of inflammation [3, 5]. Moreover, early ART is associated with a lower risk of development of non-AIDS morbidities [36, 37], and the study with the ANRS VISCONTI cohort demonstrated the possibility of a functional cure in individuals starting ART in primary infection [38].

In the present study, we compared the impact of cART on the levels of immunological markers between primary HIV-1 infected Brazilian individuals, who initiated cART, during the acute (Fiebig I-V) and early chronic phases (Fiebig VI). We evaluated $CD4^+$ T cell counts, CD4/CD8 ratio, $CD4^+$ and $CD8^+$ T cell activation before (Pre-ART) and six (M6 ART) and 12 months (M12 ART) after cART initiation. We also evaluated plasmatic markers of inflammation and monocyte/macrophage activation at Pre-ART and M6 ART and total HIV-1 DNA at M6 and M12. We hypothesized that earlier cART would normalize levels of biomarkers associated with disease progression and death in HIV-1 infection.

Materials and methods

Study population and ethical statement

Study participants were recruited at the Instituto Nacional de Infectologia Evandro Chagas (INI), Rio de Janeiro, Brazil. The inclusion criteria include those who were over 18 years of age and with documented seroconversion within the previous 6 months. Recruitment occurred among those who sought INI or community-based out-of-health care units for HIV testing or care with HIV diagnostic tests suggesting acute HIV infection. Documentation of seroconversion could be the following: (a) a negative result for a third-generation HIV rapid test, followed by reactive HIV Ag/Ab combination assay, or a detectable HIV RNA on pooled HIV RNA testing subsequently confirmed with an individual HIV RNA test; (b) a reactive HIV serology and a documented HIV negative serology within the prior 6 months or a reactive western blot lacking p31 (pol) reactivity. The exclusion criteria were the following: lacking plasma or peripheral blood mononuclear cells (PBMC) on the baseline or M6 ART visits until March, 15th 2016 or less than 10 million cells on the PBMC samples. Of the fifty patients initially included, two did not confirm HIV infection, one abandoned the treatment, one withdrew from the study early, and 23 were excluded, as shown in Fig. 1; thus, the remaining 23 patients were included. The patients were categorized into Fiebig stages, as described in Fiebig et al. [39]. Briefly, Fiebig I was characterized by the presence of HIV-1 RNA in plasma samples (Abbott RealTime

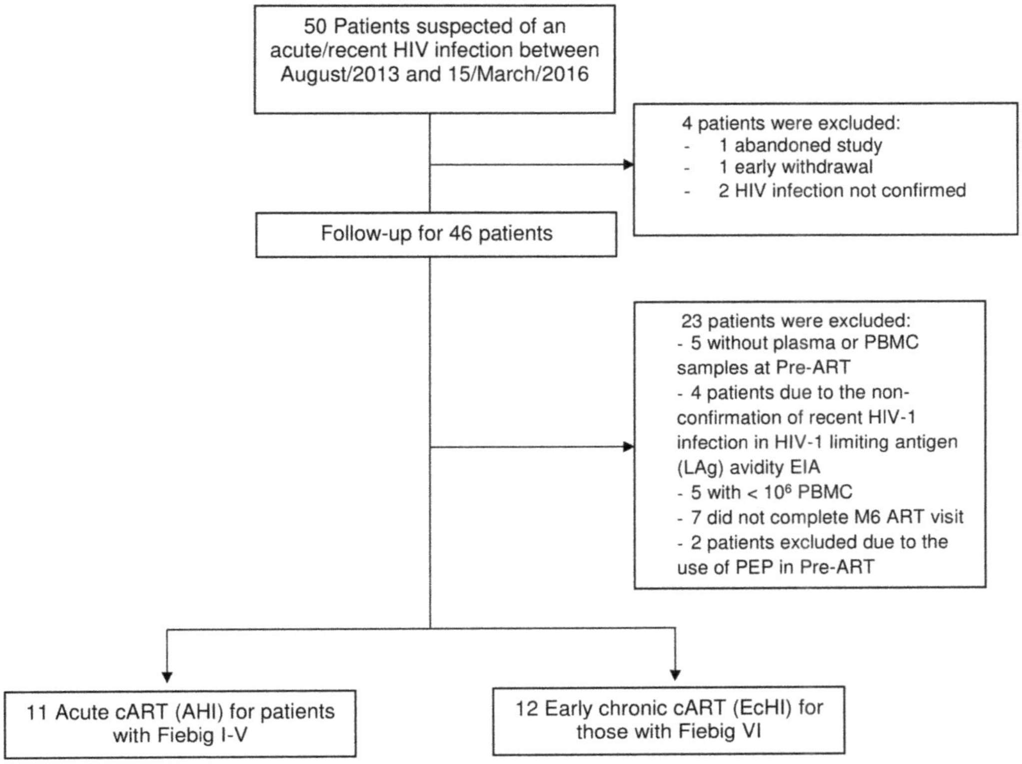

Fig. 1 Flowchart for the selection of study participants

HIV-1) together with a fourth generation ELISA negative (Enzyme Linked Fluorescent Assay BioMérieux HIV DUO Ultra Assay); Fiebig II by a HIV-1 RNA detectable in plasma and the fourth generation ELISA with antigen positive and antibody negative; Fiebig III by a HIV-1 RNA detectable in plasma and a reactive HIV-1 antibody assay (by a 3rd generation assay with detection of IgG and/or IgM anti-HIV-1 or an antibody detection in the 4th generation BioMérieux ELISA HIV DUO Ultra Assay) but a Western Blot (Western Blot Cambridge Biotech HIV-1 negative (defined by the absence of HIV-1 specifc bands); Fiebig IV reactivity profile is identical to that present in the stage III, but with undetermined pattern in Western Blot (presence HIV-1 specific bands, but do not meet the criteria for the interpretation of reactive Western Blot, which is defined by the presence of two of the following three bands: p24, gp41 or gp120/160); Fiebig V reactivity profile is identical to that verified in stage IV, but with a reactive Western Blot result (defined by the presence of two of the following three bands:p24, gp41 or gp120/160); Fiebig VI by reactivity profile is identical to that observed in the stage V, but with the complete WB reactivity pattern, including the p31 band. After, they were divided into two groups: (a) starting cART in acute HIV-1 infection (AHI) for patients with Fiebig I-V (n = 11) and (b) starting cART in early chronic HIV-1 infection (EcHI) for those with Fiebig VI (n = 12). We also included late chronic treated HIV-1 infected patients (n = 18), with more than 5 years of HIV-1 viral load suppression by cART (LcHI), and HIV-1 uninfected individuals (HIV-neg, n = 18). All patients gave informed consent, and the INI Ethical Review Board approved this study.

Sample collection and preparation

Blood samples were collected before cART initiation (Pre-ART), after six (M6 ART) and 12 months (M12 ART) of cART in EDTA and sodium heparin containing tubes. Peripheral blood mononuclear cells (PBMCs) were isolated from whole blood by Histopaque-1077 (Sigma-Aldrich, USA) density gradient centrifugation and kept in liquid nitrogen until use. Plasma aliquots were stored in a −80 °C freezer until use.

HIV-1 limiting antigen (LAg) avidity EIA

To confirm the recent HIV-1 infection status, EDTA-plasma specimens of all patients were subjected to a quantitative limiting antigen (LAg) avidity enzyme immunoassay (Sedia™ HIV-1 LAg-Avidity EIA, Sedia Biosciences Corporation, USA).

$CD4^+$ and $CD8^+$ T cell count and plasma viral load determination

Absolute $CD4^+$ and $CD8^+$ T cell counts and %CD4 were obtained using the MultiTest TruCount-kit and the MultiSet software on a FACSCalibur flow cytometer (BD Biosciences, USA). Plasma HIV-1 viral loads were measured using the Abbott RealTime HIV-1 assay (Abbott Laboratories, Germany). For two AHI patients, data of $CD4^+$ T cell counts were not available for the Pre-ART visit.

Enzyme-Linked Immunosorbent Assay (ELISA)

The plasmatic concentrations of IP-10/CXCL10, IL-18, IL-6, sCD14, sCD163, and CRP were determined in EDTA-plasma samples by ELISA according to the manufacturer's protocol at Pre-ART and M6 ART (R&D Systems, USA). For statistical analyses, points above the detection limit were replaced by the highest value in the assay. To evaluate in more detail the IP-10 and IL-18 levels in the AHI group, we also analyzed samples collected after 1, 2, and 3 months of cART initiation.

Flow cytometry

Cryopreserved PBMCs from preART, M6 ART, and M12 ART were thawed and rested overnight. Then, the PBMCs were stained with anti-CD3 APC-H7, anti-CD4 PECF594, anti-CD8 APC, anti-CD38 PerCPCy5.5 and anti-HLA-DR PE to determine the percentage of activated $CD4^+$ and $CD8^+$ T cells by flow cytometry (all antibodies were purchased from BD Biosciences). The Fixable Viability Stain 450 (FVS 450-BD Biosciences, USA) was used to exclude nonviable cells. Samples were acquired using a BD FACSAria™ IIu flow cytometer (BD Biosciences, USA), and analyses were performed with FlowJo software v.10.0.7 (Tree Star, USA). For three EcHI and one AHI patients, PBMCs from Pre-ART presented viability below the levels accepted for testing.

Quantification of HIV-1 total DNA in PBMCs

Total cellular DNA was extracted from 1×10^7 cryopreserved PBMCs from Pre-ART and M12 ART visits using the QIAamp DNA Mini Kit (Qiagen, Germany). The Generic HIV® DNA Cell Kit (Biocentric, France) was used to quantify the cell-associated HIV-1 DNA, following the manufacturer's recommendations. The lower limit of detection was 40 HIV-1 DNA copies/10^6 cells.

HIV-1 subtyping

All HIV-1 *pol* (PR/RT) sequences were submitted to REGA HIV-1 Subtyping Tool. And them pure HIV-1 subtypes and circulating recombinant forms identified were confirm by Maximum-likelihood (ML) phylogenetic trees that were reconstructed with the PhyML 3.0 program [40] using the most appropriate nucleotide substitution model selected using program jModeltest v. 3.7 [41] and the approximate likelihood-ratio test (aLRT) was used to estimate the confidence of the branching on the tree. For unique recombinant forms we performed the bootscan analysis implemented in Simplot v3.5.1 software with the following parameters: 300nt window, 20nt increments, NJ method under Kimura's two-parameter correction with 100 bootstrap replicates [42].

Statistics

The Kruskal–Wallis test and Fisher's exact test were used for the comparison of continuous variables and categorical variables, respectively, between groups. $CD4^+$ T cell absolute counts and CD4/CD8 ratio, inflammatory and activation markers were analyzed using the Mann–Whitney test to compare variables between two groups and Wilcoxon matched pairs test to compare variables between Pre-ART, M6 ART and M12 ART intragroup. Correlations were performed using the Pearson test or Spearman test according the normality of the data. Linear regression models were used to evaluate factors (age and group) associated with cytokine levels at pre-ART and M6 ART. Statistical analyses were performed using GraphPad v6.0 (Prism Software, USA) and R software.

Results

Characteristics of the study population

Fifty-nine individuals were included in this study, divided into four groups: AHI, EcHI, LcHI, and HIV-neg (Table 1). The groups with PHI (AHI and EcHI) comprised the youngest individuals (median age 30.6 [interquartile range (IQR): 27.7, 39.5] for AHI, 28.3 [IQR 26.7, 29.9] for EcHI), and the LcHI group comprised the oldest individuals (median age 44.2 [IQR 40, 49.8]). We also found a different sex distribution among the groups; the HIV-neg group had 50% women, and the EcHI and AHI groups were composed exclusively men, all men who have sex with men (MSM). The HIV-1 viral load of the AHI group was significantly higher than that of the EcHI group at Pre-ART (median of 5.9 log [IQR 4.8, 6.5] in AHI, 4 log [IQR 3.7, 4.9] in EcHI and 1.6 log [IQR 1.6–1.6] in LcHI). After 6 months of cART, most individuals with PHI had undetectable viral load, except two EcHI (VL = 94 and 96 copies/mL) and one AHI (VL = 280 copies/mL). At 12 months of cART, all PHI subjects had an undetectable viral load, except one that has suppressed on M6 and presented a blip of 460 copies/mL at M12. The $CD4^+$ T cell counts were higher in the HIV-neg group (median 831 cell/mm^3 [IQR 757.2, 1129.5]) and the LcHI group (median 853 cell/mm^3 [IQR 762, 983]) and lower in the EcHI group (median 566 cell/mm^3 [IQR 389, 666.5]) and the AHI group (median 634 cell/mm^3 [IQR 484.2, 885.5]). To EcHI and AHI individuals the HIV-1

Table 1 Demographic and clinical characteristics of study participants

	HIV-neg (n = 18)	LcHI (n = 18)	EcHI (n = 11)	AHI (n = 12)	Total (n = 59)	*P* value
Age						0.001
Median (IQR)	37.1 (31.6; 49.3)	44.2 (40;49.8)	28.3 (26.7;29.9)	30.6 (27.7;39.5)	34.5 (27.9;46.5)	
Gender (%)						< 0.001
Female	9 (50)	7 (38.9)	0 (0)	0 (0)	16 (27.1)	
Male	9 (50)	11 (61.1)	11 (100)	12 (100)	43 (72.9)	
HIV-1 viral load						< 0.001
Log_{10}, median (IQR)	NA	1.6 (1.6;1.6)	4 (3.7;4.9)	5.9 (4.8;6.5)	3.5 (1.6;4.9)	
CD4 + T cell count						0.001
Cells/mm^3, median (IQR)	831 (757.2;1129.5)	853 (762;983)	566 (389;666.5)	634 (484.2;885.5)	784 (648;909)	

Analyses with age, HIV-1 viral load and CD4$^+$ T cell count were performed using the Kruskal–Wallis

To EcHI and AHI we analyzed the pre-ART visit and for LcHI a pos-cART visit (median of 5 years after cART initiation)

Analyses with gender were performed using Fisher's exact test

LcHI late chronic HIV-1 infection; *EcHI* early chronic HIV-1 infection; *AHI* acute HIV-1 infection; *NA* not applicable; *IQR* interquartile range

subtyping was performed and the subtype B was the prevalent (Additional file 1: Table S1).

The avidity enzyme immunoassay results confirmed all AHI and EcHI individuals included in the study as recently HIV-1 infected, considering the window of 130 days established for this test.

cART initiation during acute infection can normalize CD4$^+$ T cells

We compared the CD4$^+$ T cell absolute counts and the CD4/CD8 ratio between individuals who started cART in Fiebig I-V (AHI) and Fiebig VI stages (EcHI) with HIV-neg individuals (Fig. 2). The AHI group had CD4$^+$ T cell counts (median = 838 cells/μL; IQR = 610–971) similar to the HIV-neg group (median = 831 cells/μL, IQR = 742–1227) 6 months after cART initiation, while the EcHI group, even after 12 months of cART (CD4 median = 670 cells/μL, IQR = 582–926), did not normalize CD4$^+$ T cell counts (P = 0.0139) compared with the HIV-neg group (Fig. 2a). However, when we analyzed the CD4/CD8 ratio (Fig. 2b), both groups with recent HIV-1 infection presented lower CD4/CD8 ratios than the

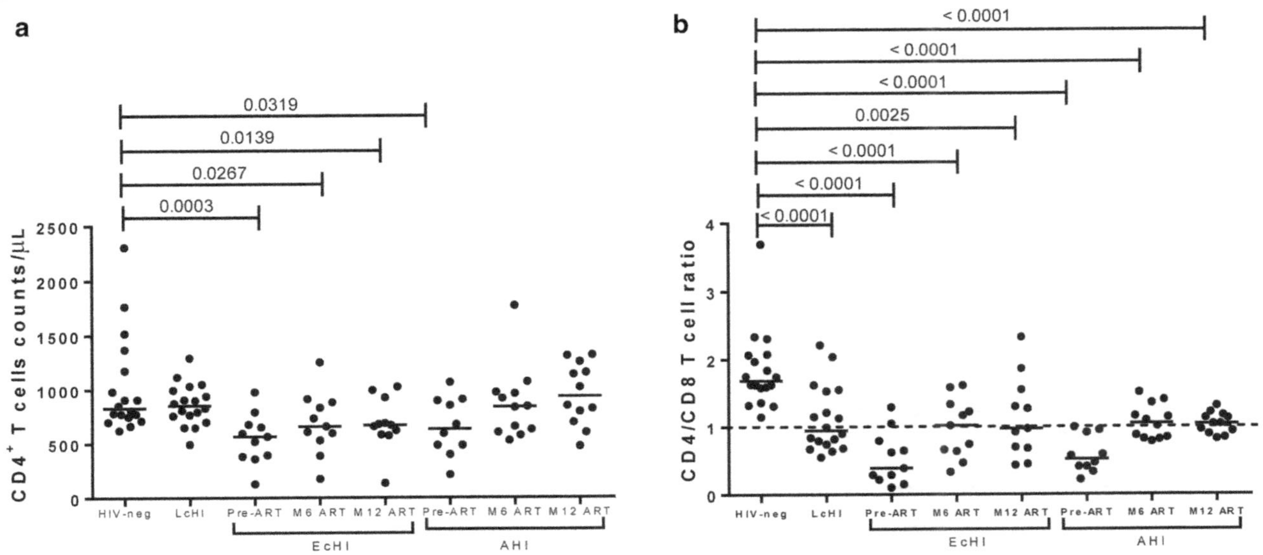

Fig. 2 CD4$^+$ T cell count and CD4/CD8 ratio on pre-ART and after six months. CD4$^+$ T cell counts (**a**) and CD4/CD8 (**b**) ratios were evaluated using the MultiTest TruCount kit. *P* values were calculated using the Mann–Whitney test. All values < 0.05 were considered statistically significant. HIV-neg, HIV-1 uninfected individuals; LcHI, late chronic treated HIV-1 infection (median of 5 years after cART initiation); EcHI, early chronic HIV-1 infection; AHI, acute HIV-1 infection; M6 ART, six months after ART start; M12 ART, 12 months after ART start

HIV-neg group after 6 and 12 months of cART. Moreover, we included a group of individuals who started cART in the late chronic phase (LcHI) and, despite a long time in cART with viral suppression and $CD4^+$ T cell counts similar to the HIV-neg group, this group also presented a lower CD4/CD8 ratio than HIV-neg.

IP-10 and IL-18 levels reduced after 6 months of cART in acute individuals

We evaluated six plasmatic markers of inflammation (IP-10/CXCL10, IL-18, IL-6, sCD14, sCD163, and CRP) before cART and 6 months after cART initiation (Fig. 3 and Additional file 1: Figure S1). These markers were chosen based in the following criteria: having a pro-inflammatory action, elevated levels in acute or chronic HIV-1 infection or being associated to AIDS progression or mortality [22, 25, 26, 28, 30, 31, 43]. IP-10 and IL-18 ($P<0.0001$ for EcHI and AHI) and sCD163 ($P=0.0049$ for EcHI and $P=0.0019$ for AHI) were higher at pre-ART in both groups of PHI individuals than in the HIV-1 uninfected group. After 6 months of cART, IP-10 levels were significantly reduced in EcHI ($P=0.0351$) and AHI ($P=0.0049$) groups when compared with the pre-ART levels. However, only the AHI group reached similar levels to those presented by the HIV-neg group. Moreover, IP-10 was higher in the LcHI group ($P<0.0001$) than the HIV-neg group (Fig. 3a). In the AHI group, cART was also able to reduce significantly ($P=0.0005$) the IL-18 levels 6 months after cART initiation compared with pre-ART levels and reached similar levels to those observed in HIV-1 negative individuals (Fig. 3b). Despite the significant reduction of sCD163 levels after 6 months of cART in the EcHI group ($P=0.0186$), but not in AHI,

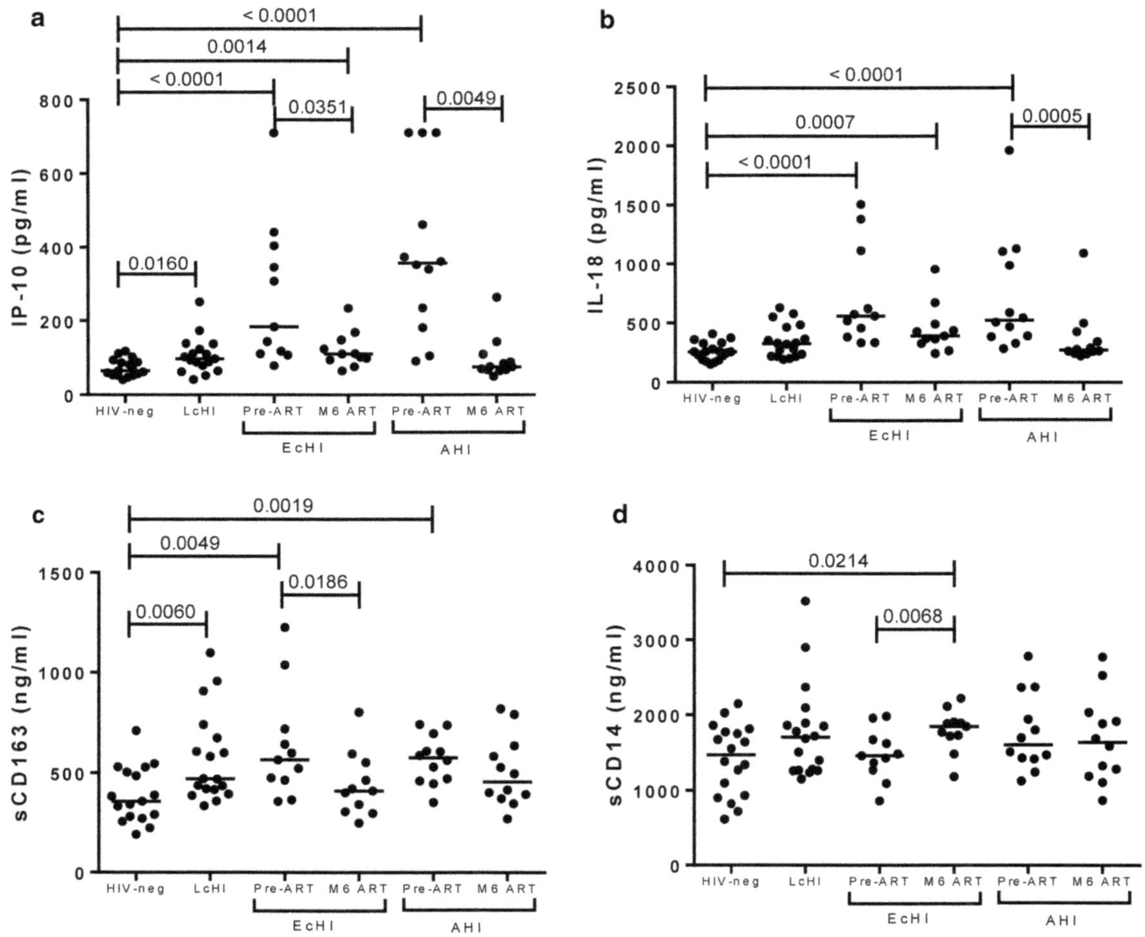

Fig. 3 Plasmatic markers of inflammation on pre-ART and after 6 months. IP-10 (**a**), IL-18 (**b**), sCD163 (**c**), and sCD14 (**d**) were assessed by ELISA. *P* values were calculated using the Mann–Whitney test. All values < 0.05 were considered statistically significant. HIV-neg, HIV-1 uninfected individuals; LcHI, late chronic treated HIV-1 infection (median of 5 years after cART initiation); EcHI, early chronic HIV-1 infection; AHI, acute HIV-1 infection; M6 ART, 6 months after ART start; and M12 ART, 12 months after ART start

compared to the pre-ART levels, both groups presented levels similar to the HIV-1-uninfected group at this visit. This marker was higher in the LcHI group than in the HIV-neg group ($P=0.0060$) (Fig. 3c). We observed higher sCD14 levels in the EcHI group after 6 months of cART than in the HIV-neg group ($P=0.0214$) and the pre-ART point of the group ($P=0.0068$) (Fig. 3d). No significant differences were observed in CRP and IL-6 among the studied groups (Additional file 1: Figure S1). We also evaluated IP-10 and IL-18 levels after 1, 2 and 3 months of cART initiation in the AHI group and observed that after 2 months, IP-10 and IL-18 reached levels that were not different from HIV-neg individuals (Additional file 1: Figure S2).

In analyses adjusted by age and group, at the pre-ART visit, the levels of IP-10 ($P<0.0001$ for EcHI and AHI), IL-18 ($P=0.0005$ for EcHI and $P<0.0001$ for AHI) and sCD163 ($P=0.0001$ for EcHI and $P=0.0056$ for AHI) remained higher than those in the HIV-1-neg group. After adjustment, we found a slightly higher level of IL-6 in the AHI group than the HIV-1-neg group at the Pre-ART visit ($P=0.0418$).

$CD8^+$ and $CD4^+$ T cell activation reach levels that are comparable with HIV-1 uninfected individuals after 12 months of cART initiation during primary infection

We observed a progressive decrease in $CD8^+$ T cell activation after cART initiation in both groups of individuals recently infected with HIV-1 (Fig. 4a). A significant decline in the frequency of activated $CD8^+$ T cells was observed comparing pre-ART and M6 ART visits in both groups. However, $CD8^+$ T cell activation was similar to the HIV-neg group only after 12 months of cART. The $CD4^+$ T cell activation, as observed to $CD8^+$ T cells, only reached similar values to the observed in HIV-1-uninfected individuals after 12 months of cART. In the AHI group, we did not observe a decrease at M6, but after 12 months after cART the levels of $CD4^+$ T cell activation were similar to the observed in HIV-neg group. In the analyses adjusted by age and group at the pre-ART visit, the frequency of activated $CD8^+$ T cells remained higher than the HIV-neg group for both the EcHI ($P=0.0264$) and AHI ($P<0.0001$) groups, data not shown.

IP-10 correlates with HIV-1 viral load at the pre-ART visit

Among the PHI individuals (AHI and EcHI groups), viral loads showed a strong positive correlation with IP-10 at pre-ART visit ($P=0.0006$ and $r=0.6610$) (Fig. 5). In the adjusted analysis, this correlation remained significant ($P=0.0019$), and we found a significant association between sCD14 and viral load ($P=0.0012$) after age adjustment (Additional file 1: Table S2). No correlation was found between IL-18 or sCD163 and viral load among the PHI individuals. We also evaluated the total HIV-1 DNA in PBMCs of AHI individuals at the pre-ART visit and found no correlation with T cell activation or markers of inflammation.

CD4/CD8 ratio correlates with inflammatory and activation markers

CD4/CD8 ratio have been pointed as a more accurate marker of immune dysfunction than absolute $CD4^+$ T cell count [44]. Negative correlations between CD4/CD8 ratio and IP-10 ($P=0.0112$ and $r=-0.5416$), IL-18

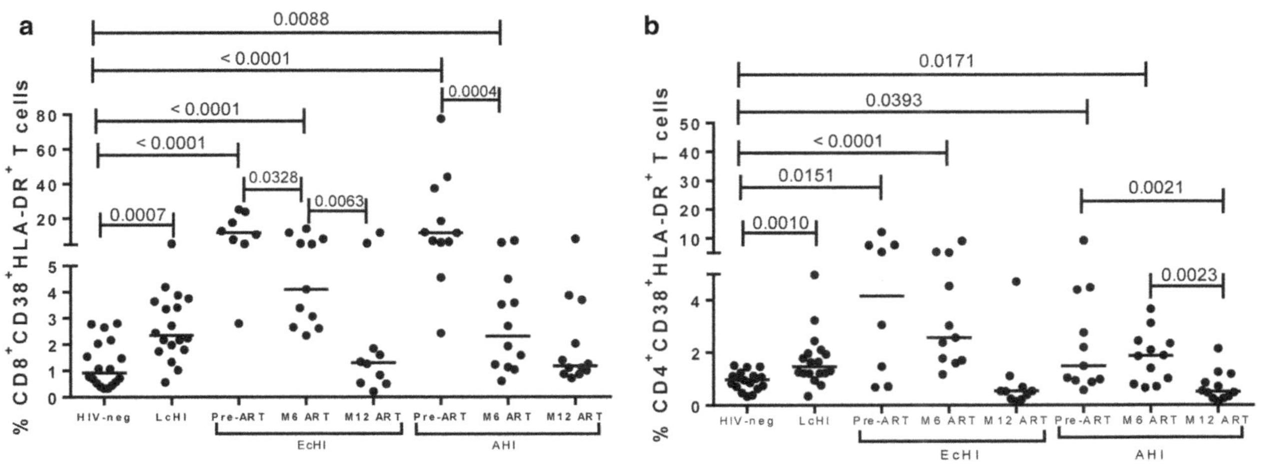

Fig. 4 T cell activation on pre-ART and 6 months after. $CD4^+$ (**a**) and $CD8^+$ (**b**) T cell activation were measured by flow cytometry. *P* values were calculated using the Mann–Whitney test. All values < 0.05 were considered statistically significant. HIV-neg, HIV-1 uninfected individuals; LcHI, late treated chronic HIV-1 infection (median of 5 years after cART initiation); EcHI, early chronic HIV-1 infection; AHI, acute HIV-1 infection; M6 ART, 6 months after ART start; and M12 ART, 12 months after ART start

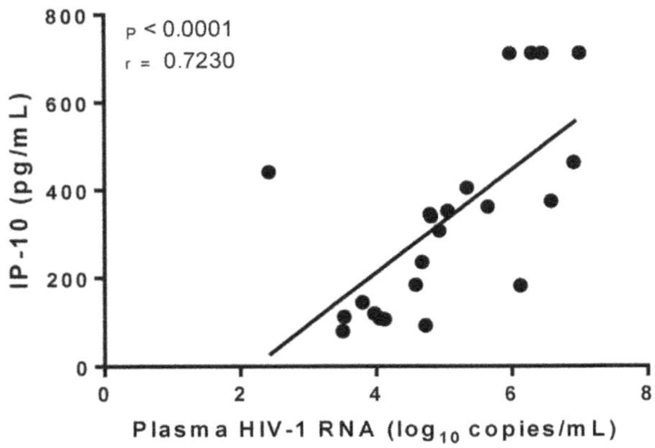

Fig. 5 Correlation between IP-10 and HIV-1 plasmatic viral load. P and r values were calculated using Pearson correlation test. All values < 0.05 were considered as statistically significant

($P=0.0068$ and $r=-0.5713$) and $CD8^+$ T cell activation ($P=0.0124$ and $r=-0.6000$) were also observed (Additional file 1: Fig S3). No correlation with sCD14, sCD163, IL-6 or CRP with CD4/CD8 ratio was observed.

Discussion

This is the first study evaluating inflammation and T cell activation in Brazilian recent HIV-1 infected individuals. In this longitudinal study, we found that cART initiation in the HIV-1 acute infection had a higher impact on inflammation reduction and $CD4^+$ T cell recovery than cART initiation in early chronic phase. IP-10, IL-18, sCD163, and $CD8^+$ T cell activation were elevated in early infected individuals from a Rio de Janeiro, Brazil when compared with the HIV-neg group. However, among the inflammatory markers, only IP-10 presented a positive correlation with the HIV-1 viral load.

There is clear evidence of the benefits of the early initiation of cART [2–7], and the WHO currently recommends the immediate cART initiation in individuals with a positive HIV-1 diagnosis, which is a recommendation followed by Brazil since 2013 [45]. The primary goals of this strategy are the following: to preserve the immune system, to avoid the early development of AIDS and non-AIDS related diseases and to reduce HIV-1 transmission. Although the $CD4^+$ T cell count is no longer a criteria to define the moment of cART initiation, it is still a valuable marker of immune reconstitution in individuals starting cART and is also routinely used in clinical follow-up in settings were viral load monitoring is poor. Here, we observed an earlier normalization of $CD4^+$ T cell counts in the group that started cART during the Fiebigs I-V acute phase (AHI). After 12 months of cART, the group that started cART only in the early chronic phase remained with lower $CD4^+$ T cell counts than the HIV-neg group. Studies have explored the CD4/CD8 ratio as a marker of persistent immune activation and inflammation, even after long-term suppressive cART [46–48]. A large study conducted with more than 400 MSM showed that 56% inverted the CD4/CD8 ratio around 3 months after seroconversion [49]. In contrast with the normalization of $CD4^+$ T cell counts in the AHI group after 6 months of cART, we observed a lower CD4/CD8 ratio in both the AHI and EcHI groups than in the HIV-neg group even at 12 months of cART. A previous study demonstrated that an altered CD4/CD8 ratio remains after 2 years of ART in individuals who started ART in acute infection, even during the Fiebig I stage [4]. In our study, the inverted CD4/CD8 ratio is mainly explained by the high $CD8^+$ T cell counts (data not shown). Once cART suppressed viral loads after 6 months of treatment, it is probably that other factors, possibly the persistent inflammation and immune activation, than only the viral loads, affect the $CD8^+$ T cell compartment. Indeed, we found negative correlations between the CD4/CD8 ratio and IP-10, IL-18, and $CD8^+$ T cell activation. Microbial translocation also correlated with $CD8^+$ T cell counts [50], and in our study, we found no higher levels of sCD14, a monocyte activation marker also used as a microbial translocation marker [51], in both AHI and EcHI groups when compared to the HIV-neg group at the pre-ART visit. An unexpected increase in sCD14 at M6 was observed in EcHI group, which could be explained by the later cART initiation in this group compared with AHI, allowing a more extensive damage in the gut mucosa. In contrast, we did not observe a higher level of this marker in LcHI group, composed by individuals with more than 5 years in cART, compared to HIV-neg group, this could indicate that a longer time of cART is necessary to decrease sCD14 to similar levels to the observed in HIV-neg individuals. A previous study [5] demonstrated that sCD14 levels remained higher in individuals treated during acute HIV infection than in HIV-uninfected individuals after 2 years, even though this group, after 12 months of treatment, presented lower sCD14 levels compared to individuals who started cART in the chronic phase.

Early cART initiation reduces inflammation and immune activation [3, 4, 6, 7]. IP-10 was one of the markers of inflammation elevated in a pre-ART visit in AHI and EcHI individuals, significantly reducing after 6 months of cART in both groups, although EcHI individuals did not achieve normal values. This marker was associated with an increased risk of HIV-1 acquisition, faster disease progression, and high plasmatic viral loads [31, 52]. IP-10 binding CXCR3 and $CD4^+CXCR3^+$ T cells are the main targets of HIV-1 infection. Thus, it is likely that a high level of IP-10 contributes to the elevated viral replication by attracting target cells that subsequently induces further IP-10 production [53]. After 2 months of treatment the levels of IP-10 in AHI group reached similar levels to the observed in HIV-1 neg-group. At this time, most individuals presented undetectable or very low (< 100 copies/mL) viral load, except for one individual (1311 copies/mL), who presented pre-ART viral load > 10 million copies/mL. We also found a high level of IL-18 at the pre-ART visit in this group. After 2 months of cART, these levels in the AHI group were similar to that observed in the HIV-neg group, following the same dynamics observed for IP-10. IL-18 is produced in response to inflammasome activation, as reviewed by Dinarello et al. 2013 [54]. The inflammasome is activated in response to several stimuli, including LPS, and HIV-1 can activate inflammasome [55, 56]. The comparison of our results of IP-10 and IL18 levels with other studies is limited by variations in assay and characteristics of study population, as age and coinfections. However, in a general way, we founded studies which showed higher [3, 6, 30, 57], lower [43], and similar [31] values of IP-10, and a similar level of IL-18 when compared with a French study [30].

Studies have demonstrated a decrease in T cell activation, mainly $CD8^+$ T cell activation, after cART initiation during the acute phase of HIV-1 infection [3, 35]. Here, we observed an earlier decrease in markers of inflammation than in T cell activation. However, we detected an earlier normalization of $CD8^+$ T cell activation after cART initiation than that reported by two previous studies, which demonstrated higher levels of this marker, even after 12 [7] or 24 months [35], than those observed in HIV-1-uninfected individuals. In both studies, the median of $CD8^+$ T cell activation was higher than that observed here, which could explain the longer time required for $CD8^+$ T cell activation normalization. Moreover, as observed to markers of inflammation, other factors, as coinfections, may also explain the differences.

The HIV-1 subtype B was the prevalent in our study groups of recent infected individuals, followed by B/C and B/F recombinants. When we compared HIV-1 subtype B versus non-B individuals, we founded no differences in inflammation or T cell activation markers. One individual from EcHI group was infected by HIV-1 subtype D, which have been associated with a faster disease progression [58]. However, this individual presented viral load and $CD4^+$ T cell count comparable with group median, and the same was observed to markers of inflammation and T cell activation.

Our study has some limitations that should be highlighted. The first limitation is the small size of our cohort. Indeed, the identification of individuals in the acute or early phase of HIV-1 infection is a major challenge. Second, our groups of recently infected individuals were composed exclusively by MSM, different from the LcHI and HIV-neg groups. Moreover, we only quantified the total HIV-1 DNA levels, with no differentiation between 2-LTR and integrated DNA and no functional assay was performed to test the size of the competent reservoir. A longer follow-up certainly will help to achieve a better understanding of the dynamics of immune activation, inflammation, and immune reconstitution after the early cART initiation.

Alltogether, our results reinforce that early cART initiation during HIV infection can reduce systemic inflammation, but the earlier normalization of the markers related to this phenomenon was only observed when cART is initiated in the acute phase. We suggest that IP-10 was the best marker to evaluate inflammation on recent HIV-1 infection, once it was elevated at pre-ART visit and showed a positive correlation with plasmatic viral load. $CD8^+$ T cell activation presented a different reduction dynamic when compared to IP-10 and IL-18, and it only reached normal levels one year after cART initiation.

Additional file

Additional file 1: Figure S1. Plasmatic markers of inflammation. CRP (**a**) and IL-6 (**b**) were measured by ELISA. *P* values were calculated using the Mann–Whitney test. All values < 0.05 were considered statistically significant. HIV-neg, HIV-1 uninfected individuals; LcHI, late chronic HIV-1 infection; EcHI, early chronic HIV-1 infection; AHI, acute HIV-1 infection; M6 ART, 6 months after cART start after cART start. **Figure S2.** Decreasing dynamics of inflammatory markers among acute treated HIV infected individuals. IP-10 (**a**) and IL-18 (**b**) were measured by ELISA at different time points after cART start. *P* value were calculated using the Mann–Whitney test. All values < 0.05 were considered statistically significant. HIV-neg, HIV-1 uninfected individuals. **Figure S3.** Correlations between markers of inflammation activation with CD4/CD8 ratio at pre-ART visit. IP-10 and CD4/CD8 ratio (**a**), IL-18 and HIV-1 CD4/CD8 ratio (**b**), CD8 activation and CD4/CD8 ratio (**c**). *P* and *r* values were calculated using the Spearman test. All values < 0.05 were considered statistically significant. **Table S1.** HIV-1 subtypes. **Table S2.** Age adjusted analyzes of the association between sCD14 levels and HIV-1 viral load at pre-ART visit.

Authors' contributions
MLG and FHC conceptualized and designed the study; FHC, MGM, and MLG contributed to the experimental design and provided intellectual input. FHC and DGC performed the sample processing, provided the clinical data, and provided intellectual input. FHC, HHSP, ED, and SLMT performed the experiments. BG, SWC, VG, EGJ, MMA and ACGF included patients and provided the clinical data. HHSP, FHC, ACGF, and LEC analyzed the data and wrote the manuscript. All authors read and approved the final manuscript.

Author details
[1] Laboratório de Aids e Imunologia Molecular, Instituto Oswaldo Cruz – IOC, FIOCRUZ, Av Brasil, 4365, Pavilhão Leônidas Deane, sala 401, Rio de Janeiro 21040360, Brazil. [2] Laboratório de Pesquisa Clínica em DST e Aids,

Instituto Nacional de Infectologia Evandro Chagas - INI, FIOCRUZ, Rio de Janeiro 25984-220, Brazil.

Acknowledgements
We thank the pasrticipants and clinicians who contributed in the study. We also thank all staff from the INI blood collection sector. We thank Mariana Gandini and the PDTIS Flow Cytometry Facility RPT08A at FIOCRUZ. Finally, we thank the Fundação de Amparo à Pesquisa do Estado do Rio de Janeiro – FAPERJ, Conselho Nacional de Desenvolvimento Científico e Tecnológico - CNPq, and Fundação Oswaldo Cruz - Fiocruz for supporting this work. E.D. is funded by a postdoctoral fellowship from the "Programa Nacional de Pós-Doutorado (PNPD)"/CAPES-Brazil.

Competing interests
The authors declare that they have no competing interests.

Funding
Supported by the Fundação de Amparo à Pesquisa do Estado do Rio de Janeiro – FAPERJ; the Conselho Nacional de Desenvolvimento Científico e Tecnológico – CNPq; and Fiocruz.

References

1. Deeks SG, Lewin SR, Havlir DV. The end of AIDS: HIV infection as a chronic disease. Lancet. 2013;382(9903):1525–33.
2. Schuetz A, Deleage C, Sereti I, Rerknimitr R, Phanuphak N, Phuang-Ngern Y, Estes JD, Sandler NG, Sukhumvittaya S, Marovich M, et al. Initiation of ART during early acute HIV infection preserves mucosal Th17 function and reverses HIV-related immune activation. PLoS Pathog. 2014;10(12):e1004543.
3. Crowell TA, Fletcher JL, Sereti I, Pinyakorn S, Dewar R, Krebs SJ, Chomchey N, Rerknimitr R, Schuetz A, Michael NL, et al. Initiation of antiretroviral therapy before detection of colonic infiltration by HIV reduces viral reservoirs, inflammation and immune activation. J Int AIDS Soc. 2016;19(1):21163.
4. Ananworanich J, Sacdalan CP, Pinyakorn S, Chomont N, de Souza M, Luekasemsuk T, Schuetz A, Krebs SJ, Dewar R, Jagodzinski L, et al. Virological and immunological characteristics of HIV-infected individuals at the earliest stage of infection. J Virus Erad. 2016;2:43–8.
5. Sereti I, Krebs SJ, Phanuphak N, Fletcher JL, Slike B, Pinyakorn S, O'Connell RJ, Rupert A, Chomont N, Valcour V, et al. Persistent, Albeit reduced, chronic inflammation in persons starting antiretroviral therapy in acute HIV infection. Clin Infect Dis. 2017;64(2):124–31.
6. Ananworanich J, Schuetz A, Vandergeeten C, Sereti I, de Souza M, Rerknimitr R, Dewar R, Marovich M, van Griensven F, Sekaly R, et al. Impact of multi-targeted antiretroviral treatment on gut T cell depletion and HIV reservoir seeding during acute HIV infection. PLoS One. 2012;7(3):e33948.
7. Jain V, Hartogensis W, Bacchetti P, Hunt PW, Hatano H, Sinclair E, Epling L, Lee TH, Busch MP, McCune JM, et al. Antiretroviral therapy initiated within 6 months of HIV infection is associated with lower T-cell activation and smaller HIV reservoir size. J Infect Dis. 2013;208(8):1202–11.
8. Pantaleo G, Fauci A. Immunopathogenesis of HIV infection. Annu Rev Microbiol. 1996;50:825–54.
9. Pastor L, Urrea V, Carrillo J, Parker E, Fuente-Soro L, Jairoce C, Mandomando I, Naniche D, Blanco J. Dynamics of CD4 and CD8 T-cell subsets and inflammatory biomarkers during early and chronic HIV infection in mozambican adults. Front Immunol. 1925;2017:8.
10. Stacey AR, Norris PJ, Qin L, Haygreen EA, Taylor E, Heitman J, Lebedeva M, DeCamp A, Li D, Grove D, et al. Induction of a striking systemic cytokine cascade prior to peak viremia in acute human immunodeficiency virus type 1 infection, in contrast to more modest and delayed responses in acute hepatitis B and C virus infections. J Virol. 2009;83(8):3719–33.
11. Jacquelin B, Mayau V, Targat B, Liovat AS, Kunkel D, Petitjean G, Dillies MA, Roques P, Butor C, Silvestri G, et al. Nonpathogenic SIV infection of African green monkeys induces a strong but rapidly controlled type I IFN response. J Clin Invest. 2009;119(12):3544–55.
12. Bhaskaran K, Hamouda O, Sannes M, Boufassa F, Johnson AM, Lambert PC, Porter K. Changes in the risk of death after HIV seroconversion compared with mortality in the general population. JAMA. 2008;300(1):51–9.
13. Mocroft A, Vella S, Benfield TL, Chiesi A, Miller V, Gargalianos P, d'Arminio Monforte A, Yust I, Bruun JN, Phillips AN, et al. Changing patterns of mortality across Europe in patients infected with HIV-1. EuroSIDA Study Group. Lancet. 1998;352(9142):1725–30.
14. Palella FJ Jr, Delaney KM, Moorman AC, Loveless MO, Fuhrer J, Satten GA, Aschman DJ, Holmberg SD. Declining morbidity and mortality among patients with advanced human immunodeficiency virus infection. HIV Outpatient Study Investigators. N Engl J Med. 1998;338(13):853–60.
15. Crum NF, Riffenburgh RH, Wegner S, Agan BK, Tasker SA, Spooner KM, Armstrong AW, Fraser S, Wallace MR. Comparisons of causes of death and mortality rates among HIV-infected persons: analysis of the pre-, early, and late HAART (highly active antiretroviral therapy) eras. J Acquir Immune Defic Syndr. 2006;41(2):194–200.
16. Lau B, Gange SJ, Moore RD. Risk of non-AIDS-related mortality may exceed risk of AIDS-related mortality among individuals enrolling into care with CD4 + counts greater than 200 cells/mm3. J Acquir Immune Defic Syndr. 2007;44(2):179–87.
17. Collaboration ATC. Causes of death in HIV-1-infected patients treated with antiretroviral therapy, 1996-2006: collaborative analysis of 13 HIV cohort studies. Clin Infect Dis. 2010;50(10):1387–96.
18. Grinsztejn B, Luz PM, Pacheco AG, Santos DV, Velasque L, Moreira RI, Guimaraes MR, Nunes EP, Lemos AS, Ribeiro SR, et al. Changing mortality profile among HIV-infected patients in Rio de Janeiro, Brazil: shifting from AIDS to non-AIDS related conditions in the HAART era. PLoS One. 2013;8(4):e59768.
19. Domingues CS, Waldman EA. Causes of death among people living with AIDS in the pre- and post-HAART Eras in the city of Sao Paulo, Brazil. PLoS One. 2014;9(12):e114661.
20. Smith CJ, Ryom L, Weber R, Morlat P, Pradier C, Reiss P, Kowalska JD, de Wit S, Law M, el Sadr W, et al. Trends in underlying causes of death in people with HIV from 1999 to 2011 (D:A:D): a multicohort collaboration. Lancet. 2014;384(9939):241–8.
21. Guaraldi G, Orlando G, Zona S, Menozzi M, Carli F, Garlassi E, Berti A, Rossi E, Roverato A, Palella F. Premature age-related comorbidities among HIV-infected persons compared with the general population. Clin Infect Dis. 2011;53(11):1120–6.
22. Kuller LH, Tracy R, Belloso W, De Wit S, Drummond F, Lane HC, Ledergerber B, Lundgren J, Neuhaus J, Nixon D, et al. Inflammatory and coagulation biomarkers and mortality in patients with HIV infection. PLoS Med. 2008;5(10):e203.
23. Boulware DR, Hullsiek KH, Puronen CE, Rupert A, Baker JV, French MA, Bohjanen PR, Novak RM, Neaton JD, Sereti I. Higher levels of CRP, D-dimer, IL-6, and hyaluronic acid before initiation of antiretroviral therapy (ART) are associated with increased risk of AIDS or death. J Infect Dis. 2011;203(11):1637–46.
24. Ledwaba L, Tavel JA, Khabo P, Maja P, Qin J, Sangweni P, Liu X, Follmann D, Metcalf JA, Orsega S, et al. Pre-ART levels of inflammation and coagulation markers are strong predictors of death in a South African cohort with advanced HIV disease. PLoS One. 2012;7(3):e24243.
25. Paiardini M, Muller-Trutwin M. HIV-associated chronic immune activation. Immunol Rev. 2013;254(1):78–101.
26. Hamlyn E, Fidler S, Stohr W, Cooper DA, Tambussi G, Schechter M, Miro JM, McClure M, Weber J, Babiker A, et al. Interleukin-6 and D-dimer levels at seroconversion as predictors of HIV-1 disease progression. Aids. 2014;28(6):869–74.
27. Knudsen TB, Ertner G, Petersen J, Moller HJ, Moestrup SK, Eugen-Olsen J, Kronborg G, Benfield T. Plasma soluble CD163 level independently predicts all-cause mortality in HIV-1-infected individuals. J Infect Dis. 2016;214(8):1198–204.
28. Sandler NG, Wand H, Roque A, Law M, Nason MC, Nixon DE, Pedersen C, Ruxrungtham K, Lewin SR, Emery S, et al. Plasma levels of soluble CD14 independently predict mortality in HIV infection. J Infect Dis. 2011;203(6):780–90.
29. Hunt PW, Sinclair E, Lecuroux C, Jacquelin B, Rodriguez B, Shive C, Clagett B, Funderburg N, Robinson J, Huang Y, Epling L, Martin JN, et al. Gut epithelial barrier dysfunction and innate immune activation predict mortality in treated HIV infection. J Infect Dis. 2014;210(8):1228–38.

30. Liovat AS, Rey-Cuille MA, Lecuroux C, Jacquelin B, Girault I, Petitjean G, Zitoun Y, Venet A, Barre-Sinoussi F, Lebon P, et al. Acute plasma biomarkers of T cell activation set-point levels and of disease progression in HIV-1 infection. PLoS One. 2012;7(10):e46143.
31. Ploquin MJ, Madec Y, Casrouge A, Huot N, Passaes C, Lécuroux C, Essat A, Boufassa F, Jacquelin B, Jochems SP, et al. Elevated basal pre-infection CXCL10 in plasma and in the small intestine after infection are associated with more rapid HIV/SIV disease onset. PLoS Pathog. 2016;12(8):e1005774.
32. Balagopal A, Gupte N, Shivakoti R, Cox AL, Yang WT, Berendes S, Mwelase N, Kanyama C, Pillay S, Samaneka W, et al. Continued elevation of interleukin-18 and interferon-gamma after initiation of antiretroviral therapy and clinical failure in a diverse multicountry human immunodeficiency virus cohort. Open Forum Infect Dis. 2016;3(3):ofw118.
33. French MA, King MS, Tschampa JM, da Silva BA, Landay AL. Serum immune activation markers are persistently increased in patients with HIV infection after 6 years of antiretroviral therapy despite suppression of viral replication and reconstitution of CD4+ T cells. J Infect Dis. 2009;200(8):1212–5.
34. Hocqueloux L, Avettand-Fenoel V, Jacquot S, Prazuck T, Legac E, Melard A, Niang M, Mille C, Le Moal G, Viard JP, et al. Long-term antiretroviral therapy initiated during primary HIV-1 infection is key to achieving both low HIV reservoirs and normal T cell counts. J Antimicrob Chemother. 2013;68(5):1169–78.
35. Vinikoor MJ, Cope A, Gay CL, Ferrari G, McGee KS, Kuruc JD, Lennox JL, Margolis DM, Hicks CB, Eron JJ. Antiretroviral therapy initiated during acute HIV infection fails to prevent persistent T-cell activation. J Acquir Immune Defic Syndr. 2013;62(5):505–8.
36. Lundgren JD, Borges AH, Neaton JD: Serious Non-AIDS Conditions in HIV: Benefit of Early ART. Curr HIV/AIDS Rep 2018.
37. Lundgren JD, Babiker AG, Gordin F, Emery S, Grund B, Sharma S, Avihingsanon A, Cooper DA, Fatkenheuer G, Llibre JM, et al. Initiation of antiretroviral therapy in early asymptomatic HIV infection. N Engl J Med. 2015;373(9):795–807.
38. Saez-Cirion A, Bacchus C, Hocqueloux L, Avettand-Fenoel V, Girault I, Lecuroux C, Potard V, Versmisse P, Melard A, Prazuck T, et al. Post-treatment HIV-1 controllers with a long-term virological remission after the interruption of early initiated antiretroviral therapy ANRS VISCONTI Study. PLoS Pathog. 2013;9(3):e1003211.
39. Fiebig E, Wright D, Rawal B, Garrett P, Schumacher R, Peddada L, Heldebrant C, Smith R, Conrad A, Kleinman S, et al. Dynamics of HIV viremia and antibody seroconversion in plasma donors: implications for diagnosis and staging of primary HIV infection. AIDS. 2003;17(13):1871–9.
40. Guindon S, Dufayard JF, Lefort V, Anisimova M, Hordijk W, Gascuel O. New algorithms and methods to estimate maximum-likelihood phylogenies: assessing the performance of PhyML 3.0. Syst Biol. 2010;59(3):307–21.
41. Darriba D, Taboada GL, Doallo R, Posada D. jModelTest 2: more models, new heuristics and parallel computing. Nat Methods. 2012;9(8):772.
42. Lole KS, Bollinger RC, Paranjape RS, Gadkari D, Kulkarni SS, Novak NG, Ingersoll R, Sheppard HW, Ray SC. Full-length human immunodeficiency virus type 1 genomes from subtype C-infected seroconverters in India, with evidence of intersubtype recombination. J Virol. 1999;73(1):152–60.
43. Burdo TH, Lo J, Abbara S, Wei J, DeLelys ME, Preffer F, Rosenberg ES, Williams KC, Grinspoon S. Soluble CD163, a novel marker of activated macrophages, is elevated and associated with noncalcified coronary plaque in HIV-infected patients. J Infect Dis. 2011;204(8):1227–36.
44. McBride JA, Striker R. Imbalance in the game of T cells: What can the CD4/CD8 T-cell ratio tell us about HIV and health? PLoS Pathog. 2017;13(11):e1006624.
45. Protocolo Clínico e diretrizes terapêuticas para o manejo da infecção pelo HIV em adultos. http://www.aids.gov.br/pcdt/protocolo-clinico.
46. Serrano-Villar S, Sainz T, Lee SA, Hunt PW, Sinclair E, Shacklett BL, Ferre AL, Hayes TL, Somsouk M, Hsue PY, et al. HIV-infected individuals with low CD4/CD8 ratio despite effective antiretroviral therapy exhibit altered T cell subsets, heightened CD8+ T cell activation, and increased risk of non-AIDS morbidity and mortality. PLoS Pathog. 2014;10(5):e1004078.
47. Sainz T, Serrano-Villar S, Diaz L, Gonzalez Tome MI, Gurbindo MD, de Jose MI, Mellado MJ, Ramos JT, Zamora J, Moreno S, et al. The CD4/CD8 ratio as a marker T-cell activation, senescence and activation/exhaustion in treated HIV-infected children and young adults. Aids. 2013;27(9):1513–6.
48. Zheng L, Taiwo B, Gandhi RT, Hunt PW, Collier AC, Flexner C, Bosch RJ. Factors associated with CD8+ T-cell activation in HIV-1-infected patients on long-term antiretroviral therapy. J Acquir Immune Defic Syndr. 2014;67(2):153–60.
49. Margolick JB, Gange SJ, Detels R, O'Gorman MR, Rinaldo CR Jr, Lai S. Impact of inversion of the CD4/CD8 ratio on the natural history of HIV-1 infection. J Acquir Immune Defic Syndr. 2006;42(5):620–6.
50. Mehraj V, Jenabian MA, Ponte R, Lebouche B, Costiniuk C, Thomas R, Baril JG, LeBlanc R, Cox J, Tremblay C, et al. The plasma levels of soluble ST2 as a marker of gut mucosal damage in early HIV infection. Aids. 2016;30(10):1617–27.
51. Hoenigl M, Perez-Santiago J, Nakazawa M, de Oliveira MF, Zhang Y, Finkelman MA, Letendre S, Smith D, Gianella S. (1−>3)-beta-d-glucan: a biomarker for microbial translocation in individuals with acute or early HIV infection? Front Immunol. 2016;7:404.
52. Kahle EM, Bolton M, Hughes JP, Donnell D, Celum C, Lingappa JR, Ronald A, Cohen CR, de Bruyn G, Fong Y, et al. Plasma cytokine levels and risk of HIV type 1 (HIV-1) transmission and acquisition: a nested case-control study among HIV-1–serodiscordant couples. J Infect Dis. 2015;211(9):1451–60.
53. Foley JF, Yu CR, Solow R, Yacobucci M, Peden KW, Farber JM. Roles for CXC chemokine ligands 10 and 11 in recruiting CD4+ T cells to HIV-1-infected monocyte-derived macrophages, dendritic cells, and lymph nodes. J Immunol. 2005;174(8):4892–900.
54. Dinarello CA, Novick D, Kim S, Kaplanski G. Interleukin-18 and IL-18 binding protein. Front Immunol. 2013;4:289.
55. Hernandez JC, Latz E, Urcuqui-Inchima S. HIV-1 induces the first signal to activate the NLRP3 inflammasome in monocyte-derived macrophages. Intervirology. 2014;57(1):36–42.
56. Chivero ET, Guo ML, Periyasamy P, Liao K, Callen SE, Buch S. HIV-1 tat primes and activates microglial NLRP3 inflammasome-mediated neuroinflammation. J Neurosci. 2017;37(13):3599–609.
57. Pastor L, Casellas A, Ruperez M, Carrillo J, Maculuve S, Jairoce C, Paredes R, Blanco J, Naniche D. Interferon-gamma-Inducible Protein 10 (IP-10) as a screening tool to optimize human immunodeficiency virus RNA monitoring in resource-limited settings. Clin Infect Dis. 2017;65(10):1670–5.
58. Leite TC, Campos DP, Coelho AB, Teixeira SL, Veloso V, Morgado MG, Guimaraes ML. Impact of HIV-1 subtypes on AIDS progression in a Brazilian cohort. AIDS Res Hum Retrovir. 2017;33(1):41–8.

13

Predominant envelope variable loop 2-specific and gp120-specific antibody-dependent cellular cytotoxicity antibody responses in acutely SIV-infected African green monkeys

Quang N. Nguyen[1†], David R. Martinez[1,2†], Jonathon E. Himes[1], R. Whitney Edwards[1,3], Qifeng Han[1], Amit Kumar[1], Riley Mangan[1], Nathan I. Nicely[1], Guanhua Xie[1], Nathan Vandergrift[1], Xiaoying Shen[1], Justin Pollara[1,3] and Sallie R. Permar[1,2,4,5*]

Abstract

Background: The initial envelope (Env)-specific antibody response in acutely HIV-1-infected individuals and simian immunodeficiency virus (SIV)-infected rhesus monkeys (RMs) is dominated by non-neutralizing antibodies targeting Env gp41. In contrast, natural primate SIV hosts, such as African green monkeys (AGMs), develop a predominant Env gp120-specific antibody response to SIV infection. However, the fine-epitope specificity and function of SIV Env-specific plasma IgG, and their potential role on autologous virus co-evolution in SIV-infected AGMs and RMs remain unclear.

Results: Unlike the dominant linear gp41-specific IgG responses in RMs, SIV-infected AGMs demonstrated a unique linear variable loop 2 (V2)-specific plasma IgG response that arose concurrently with high gp120-directed antibody-dependent cellular cytotoxicity (ADCC) activity, and SIVsab-infected cell binding responses during acute infection. Moreover, SIV variants isolated from SIV-infected AGMs exhibited high amino acid mutation frequencies within the Env V1V2 loop compared to those of RMs. Notably, the linear V2-specific IgG epitope in AGMs overlaps with an analogous region of the HIV V2 loop containing the K169 mutation epitope identified in breakthrough viruses from RV144 vaccinees.

Conclusion: Vaccine-elicited Env V2-specific IgG responses have been proposed as an immune correlate of reduced risk in HIV-1/SIV acquisition in humans and RMs. Yet the pathways to elicit these potentially-protective V2-specific IgG responses remain unclear. In this study, we demonstrate that SIV-infected AGMs, which are the natural hosts of SIV, exhibited high plasma linear V2-specific IgG binding responses that arose concurrently with SIV Env gp120-directed ADCC-mediating, and SIV-infected cell plasma IgG binding responses during acute SIV infection, which were not present in acutely SIV-infected RMs. The linear V2-specific antibody response in AGMs targets an overlapping epitope of the proposed site of vaccine-induced immune pressure defined in the moderately protective RV144 HIV-1 vaccine trial. Identifying host factors that control the early elicitation of Env V2-specific IgG and ADCC antibody responses in

*Correspondence: sallie.permar@duke.edu
†Quang N. Nguyen and David R. Martinez contributed equally to this work
[2] Department of Molecular Genetics and Microbiology, Duke University School of Medicine, Durham, NC, USA
Full list of author information is available at the end of the article

these natural SIV hosts could inform vaccination strategies aimed at rapidly inducing potentially-protective HIV-1 Env-specific responses in humans.

Keywords: SIV, Linear peptide antibody responses, Natural SIV host, African green monkey, Rhesus monkey, Antibody response, ADCC, Envelope, gp120, gp41

Background

The HIV-1 Env glycoprotein contains multiple vulnerable epitopes targeted by potent broad neutralizing antibodies (bNAbs) [1]. However, the elicitation of HIV gp120-specific bNAbs by current Env vaccination strategies is not yet feasible [2, 3]. Thus, HIV vaccine candidates currently in clinical testing focus on the elicitation of antibody specificities and functions identified as potential immune correlates of reduced infection risk in human and non-human primate vaccine efficacy studies. Immune analyses from the HVTN 505 phase IIb vaccine trial, which utilized an HIV Env gp140 protein boost immunogen and failed to show efficacy, demonstrated that the vaccine-elicited humoral responses primarily targeted HIV Env gp41 without identifiable antiviral functions [4]. Similarly, in the setting of HIV-1 infection, the initial antibody response against HIV Env is also dominated by Env gp41-specific IgG responses that are ineffective at controlling viremia [5, 6]. Interestingly, in the moderately-efficacious RV144 HIV-1 Env vaccine efficacy trial, Env-specific IgG responses targeting the variable loop 1 and 2 (V1V2) were found to be associated with reduced HIV acquisition risk [7]. The following sieve analysis of the breakthrough virus variants localized the site of vaccine-induced immune pressure to two amino acid residue positions within V2 loop [8–10]. Moreover, the V2 epitope that was associated with immune escape in RV144 vaccinees spans the region capable of engaging the gut-homing integrin receptor $\alpha4\beta7$, which has been implicated in the trafficking of immune cells to the gut associated lymphoid tissue, and the enhancement of cell-to-cell HIV transmission [11–13]. Furthermore, the HIV Env V2-specific IgG responses in RV144 vaccinees mediated ADCC activity [14, 15]. Notably, V2-specific IgG responses were also associated with a reduced risk of SIVmac251 acquisition in RMs that received a similar vaccine regimen to that in the RV144 HIV-1 trial [16]. While these types of V2-specific IgG responses are a major clinical endpoint of HIV vaccination, our understanding of factors that control the elicitation of V2-specific IgG responses and ADCC function by existing vaccination strategies remains limited [3, 17].

Previous studies have investigated the proportion of SIV Env-specific IgG responses in the setting of infection in natural and non-natural SIV hosts. Antibody responses in acute SIV infection of RMs—a non-natural SIV host species and model of AIDS pathogenesis—predominantly target the SIV Env gp41 region [18–20]. Moreover, in SIV infected RMs, Env-specific IgG autologous virus neutralizing responses do not arise until approximately 1 year post-infection [18]. In contrast, AGMs—which are thought to have co-evolved with SIV for at least 30,000 years—sustain a non-pathogenic SIV infection, and do not exhibit a predominant gp41-specific IgG response over the course of infection [18, 19, 21]. SIV Env-specific IgG responses in chronically SIV-infected AGMs more frequently target gp120 epitopes compared to SIV-infected RMs [20]. While previous studies reported that B cell depletion in SIV-infected AGMs has no appreciable effect in disease progression outcome [22], it has also been shown that SIV-associated B cell dysfunction is associated with pathogenic SIV infection, and not in non-pathogenic SIV infection in natural SIV hosts [19, 23]. Moreover, we previously demonstrated that SIV Env gp120-specific IgG monoclonal antibodies (mAbs) isolated from chronically SIV-infected AGMs mediated robust virus capture activity, and ADCC—an antibody function that has been associated with delayed progression to AIDS in RMs [20, 24]. Moreover, SIV Env gp120-specific mAbs in chronically SIV-infected AGMs exhibited higher binding levels against SIV-infected $CD4^+$ target cells compared to gp41-specific antibodies [20]. In this study, we further explore the kinetics of ADCC and SIV-infected cell binding responses, the fine epitope-specificity of the early Env-specific IgG responses in AGMs and RMs, and their potential role in autologous virus evolution at key vulnerable Env sites. A deeper understanding of the fine-specificity, kinetics, and antiviral function of gp120-specific IgG responses in AGMs may help guide future HIV vaccination strategies aimed at eliciting potentially-protective Env-specific IgG responses in humans.

Results

Kinetics of SIV Env-specific plasma IgG binding response in SIV-infected AGMs and RMs

We previously showed that SIV-infected AGMs have a predominant gp120-specific antibody response compared to SIV-infected RMs [18]. In contrast, SIV-infected RMs have a more focused gp140-specific antibody response compared to SIV-infected AGMs [18]. We set out to further investigate the early acute kinetics of the gp120-specific IgG responses in AGMs and RMs by examining plasma

IgG binding against the autologous SIVsab92018ivTF/SIVmac251 Env gp120 proteins in both species. Prior to SIV infection, AGM and RM plasma IgG exhibited relatively low binding against SIV Env proteins (Fig. 1). By 3 wpi, AGMs had higher magnitude SIV Env-specific IgG binding responses to the autologous SIV gp120 compared to RMs (logED$_{50}$ median in AGMs vs. RMs: 3.0 vs. 2.3, FDR $p=0.032$) (Fig. 1). By 15 weeks post-infection (wpi) and 1 year post-infection (1 ypi), there was no difference in plasma gp120-specific binding in SIVsab92018ivTF and SIVmac251-infected AGMs and RMs (Fig. 1).

Linear SIV Env peptide-specific plasma IgG binding responses in SIV-infected AGMs and RMs

To map the fine-epitope specificity of SIV Env-specific IgG responses in SIV-infected AGMs and RMs, we measured plasma IgG responses against a linear overlapping peptide library spanning the entire species-specific SIVsab92018WT/SIVmac239 Env gp160 for each species. SIV Env linear peptide-specific plasma IgG binding responses were undetectable prior to infection (Fig. 2a). By 15 wpi, both species demonstrated strong plasma antibody responses against peptides analogous to the HIV gp120-gp41 fusion domain, gp41 immunodominant region, and the N-terminal region of gp41 cytoplasmic tail (Fig. 2a) [5]. Notably, by 15 wpi, RM plasma demonstrated high IgG binding against peptides of the variable loop 1 (V1) and variable loop 3 (V3) regions as well as binding to a large number of peptides within the gp41 subunit, including those of the membrane-proximal external region (MPER), which remained high binding responses at 1 ypi (Fig. 2a). In addition to an appreciable linear V3-specific IgG response by 15 wpi (Fig. 2a), 3 of 6 AGMs (AGMs 90, 93, 94) exhibited strong linear V2-specific IgG response that was markedly undetectable in all RMs (Fig. 2b). By 1 ypi, all AGMs had a high plasma IgG binding response against the linear V2 epitopes, yet this response remained undetectable in RMs (Fig. 2b). To more closely examine the kinetics of V2-specific IgG response in AGMs, we assessed AGM plasma IgG binding to the overlapping peptide library spanning SIVsab Env gp160 at earlier time-points during acute SIV infection. No appreciable plasma IgG binding to linear V2 peptides was detected at 3 wpi (Fig. 2c). Interestingly, in 3 of 6 SIV-infected AGMs (AGMs 90, 93, 94) plasma IgG binding responses against 3 overlapping linear V2 peptides appeared by 6 wpi (Fig. 2c).

Autologous and heterologous V1V2- and linear V2-specific plasma IgG binding responses in SIV-infected AGMs and RMs

Understanding the development of the predominant V2-specific IgG responses in AGMs is of interest, as it remains unclear how to elicit HIV V1V2-specific IgG responses that correlated with reduced HIV acquisition risk in the moderately protective RV144 vaccine trial [7–10]. To account for the few existing amino acid (AA) differences in the overlapping SIVsab92018WT and SIVmac239 V2 peptide libraries tested (peptides 40–42, and 42–44, respectively) compared to the AGM and RM challenge viruses, we measured AGM and RM plasma IgG binding against linear peptides that cover the same V2 region and directly match the AA sequence of the SIVsab92018ivTF (SFAMAGYRRDVKKNYSTVWYDQE) and SIVmac251 (KFNMTGLKRDKTKEYNETWYSTD) challenge viruses. SIV-infected AGMs exhibited high plasma IgG binding against the linear V2 SIVsab92018ivTF peptide at 15 wpi (median binding OD, 1.35 [range,

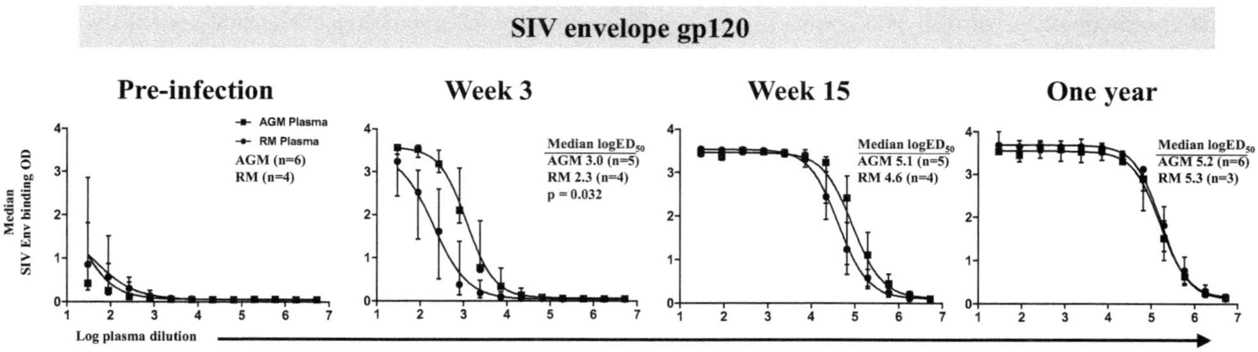

Fig. 1 Early development of SIV Env gp120 IgG responses in SIV-infected AGMs compared to those in SIV-infected RMs. Recombinant SIVsab92018ivTF or SIVmac251 envelope (Env) gp120 were tested by ELISA using plasma samples from SIVsab92018ivTF-infected AGMs and SIVmac251-infected RMs throughout infection. Shown are the median OD binding curves with ranges for all monkeys against Env specificity and the corresponding median log half-maximal effective plasma dilution (logED$_{50}$) at each time-point. Reported false discovery rate (FDR) p value by the Wilcoxon-Mann–Whitney test indicates significant difference in the logED$_{50}$ values for each Env specificity between AGM and RM plasma binding at a given time-point

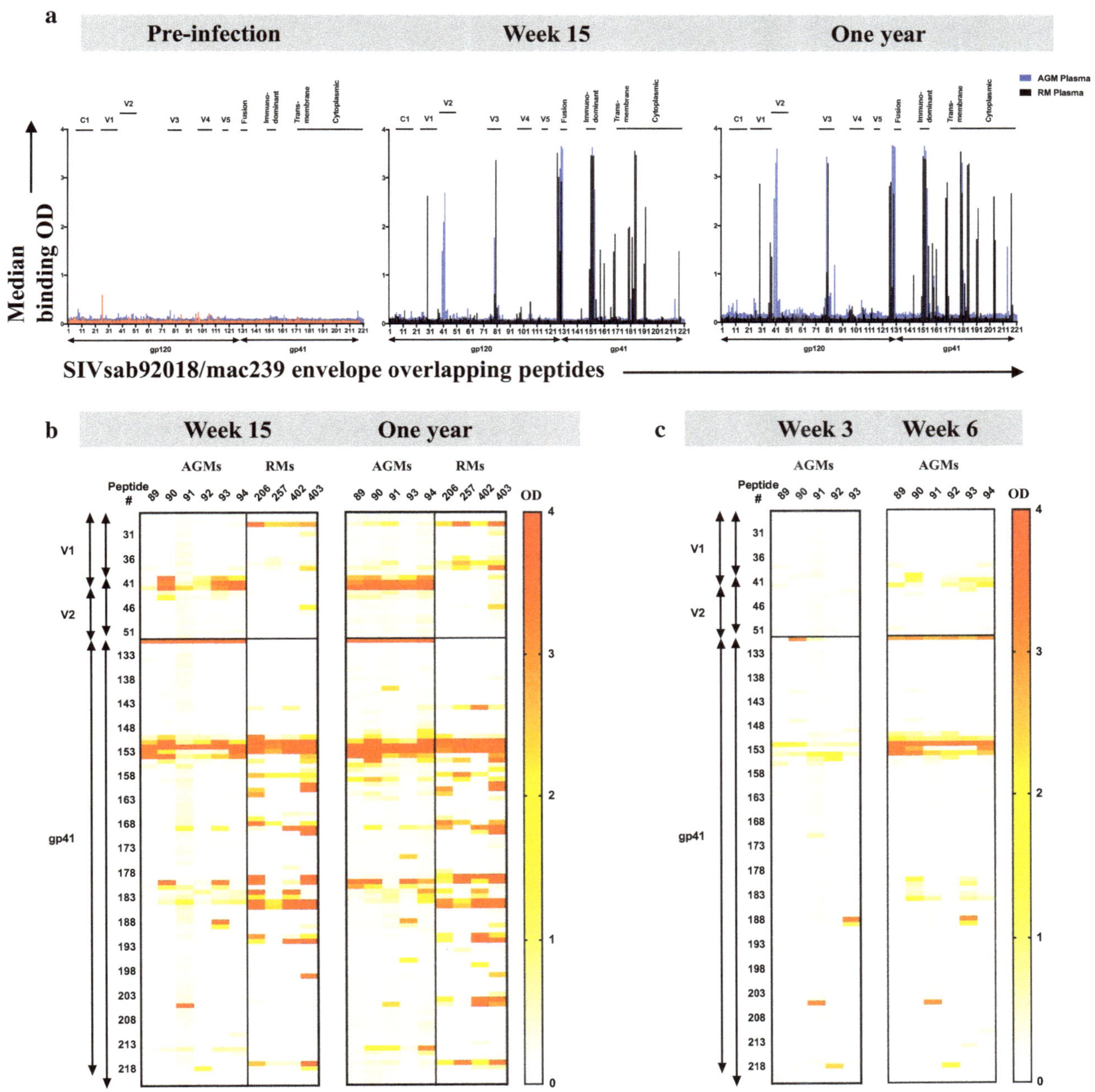

Fig. 2 Distinct pattern of SIV Env linear peptide-specific plasma IgG responses in AGMs and RMs. Plasma samples from SIVsab92018ivTF-infected AGMs and SIVmac251-infected RMs during acute and chronic infection were tested against species-specific overlapping linear peptides spanning SIV envelope (Env) gp160 by ELISA. **a** Each peak represents the median plasma binding levels of all monkeys against each peptide across Env gp160. Each block represents the plasma binding level of **b** AGMs and RMs at week 15 and 1 year against each peptide within the V1, V2, and gp41 regions, or **c** AGMs at week 3 and week 6. Linear V2-specific plasma IgG response emerged at week 6 in AGMs. At week 15, AGM plasma showed a unique binding to a cluster of peptides in the V2 and V3 loops that increased in magnitude through chronic infection in addition to binding clusters nearby the gp120/gp41 fusion domain and immunodominant region of Env gp41

0.34–2.54]) and 1 ypi (median binding OD, 2.76 [range, 2.22–3.11]) (Fig. 3a), whereas RM plasma remained non-reactive against the linear V2 SIVmac251 peptide (median binding OD, 0.04 [range, 0.02–0.11] at 1 ypi) (Fig. 3b).

To examine the cross-reactivity of the predominant V2- and V1V2-specific IgG responses in SIV-infected AGMs, we assessed binding responses to both autologous and heterologous SIVmac and HIV-1 linear V2 peptides and

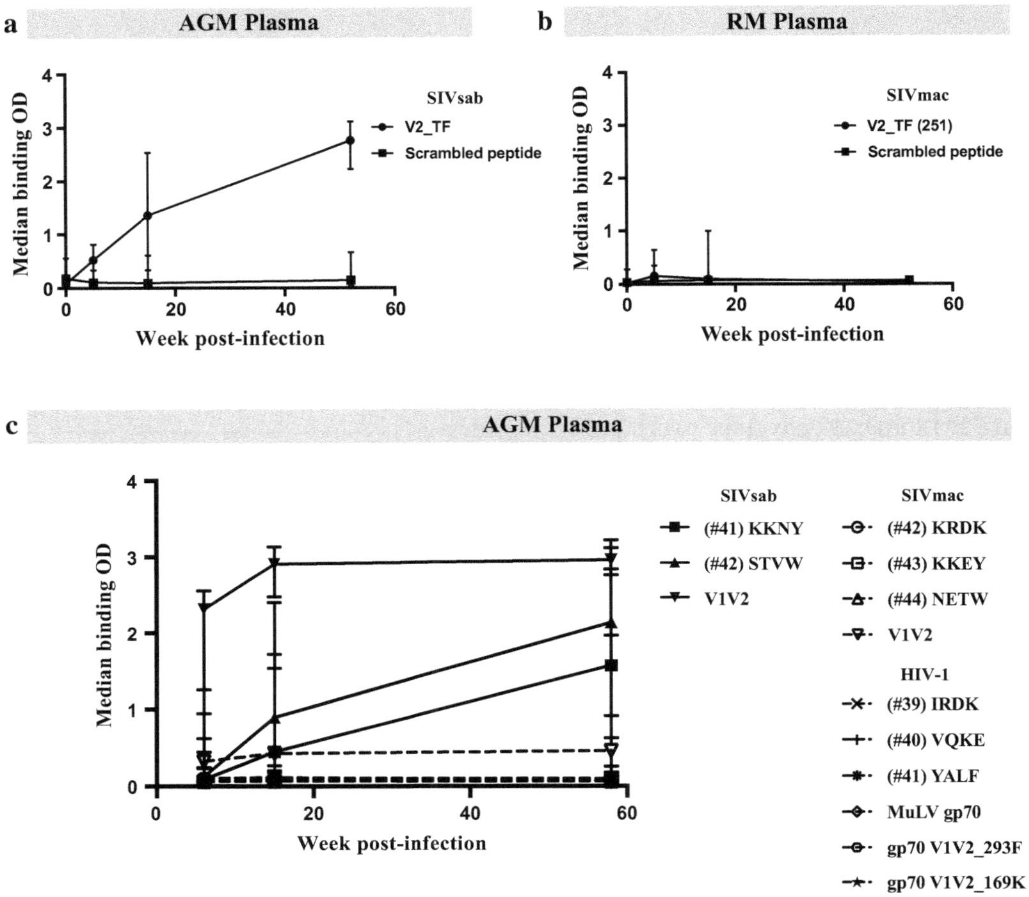

Fig. 3 Autologous and heterologous V1V2- and linear V2-specific plasma IgG binding responses in SIV-infected AGMs. Autologous SIVsab92018ivTF and SIVmac251 linear V2 peptides were tested by ELISA using plasma samples from **a** SIVsab92018ivTF-infected AGMs (n = 5 for 15 wpi, and n = 6 for all other time-points), and **b** SIVmac251-infected RMs (n = 3 for 5 wpi, and n = 4 for all other time-points) throughout infection. **c** Heterologous plasma linear V2-specific IgG binding in 6 SIVsab92018ivTF-infected AGMs was assessed against SIVsab92018WT, and SIVmac239 and HIV-1 Consensus subtype B (ConB) overlapping linear V2 peptides with (#) indicate peptide number, and the last four amino acids from each peptide. Cross-reactivity of AGM V1V2-specific plasma IgG binding was also tested using SIVsab92018ivTF, and SIVmac251 and HIV-1 ConB V1V2 proteins by ELISA. Shown are the median OD binding curves with range for all monkeys against Env specificity at each time-point

V1V2 proteins. The V2-specific IgG response in AGMs was not cross-reactive with heterologous V2 linear SIV and HIV peptides or V1V2 proteins (Fig. 3c). However, SIV-infected AGM plasma demonstrated strong IgG binding to the autologous SIVsab Env V1V2 protein as early as 6 wpi (median binding OD, 2.3 [range, 0.6–2.6]) and this response increased in magnitude (median binding OD, 2.9 [range, 2.5–3.1]) by 15 wpi, and (median binding OD, 3.0 [range, 2.8–3.2]) by 1 ypi.

SIV Env mutation frequency of autologous circulating viruses isolated from plasma of SIV-infected AGMs and RMs

To explore the potential role of the predominant gp120- and V1V2-specific IgG binding responses in AGMs and gp41-specific IgG responses in RMs on autologous viral evolution, we examined the Env gp160 AA sequence variability of single genome Env variants isolated from plasma of SIVsab92018ivTF-infected AGMs at 22 wpi and 1 ypi, and SIVmac251-infected RMs at 18 wpi and 1 ypi. Plasma SIV Env variants from both species demonstrated appreciable mutation frequency in the region of gp120 V4 loop at 1 ypi (Fig. 4a). Yet, the Env variants isolated from SIV-infected RMs exhibited high variability in Env gp41 region, compared to Env variants from SIV-infected AGMs (Fig. 4a). By 1 ypi, Env variants in RMs also showed appreciable sequence variability in a region near the V3 and V5 loops (Fig. 4a). In contrast, plasma Env variants isolated from AGMs showed high sequence variability within C1 region that increased in frequency throughout infection compared to those of SIV-infected RMs (Fig. 4a). Moreover, plasma SIVsab Env variants in AGMs accumulated higher mutation frequencies within

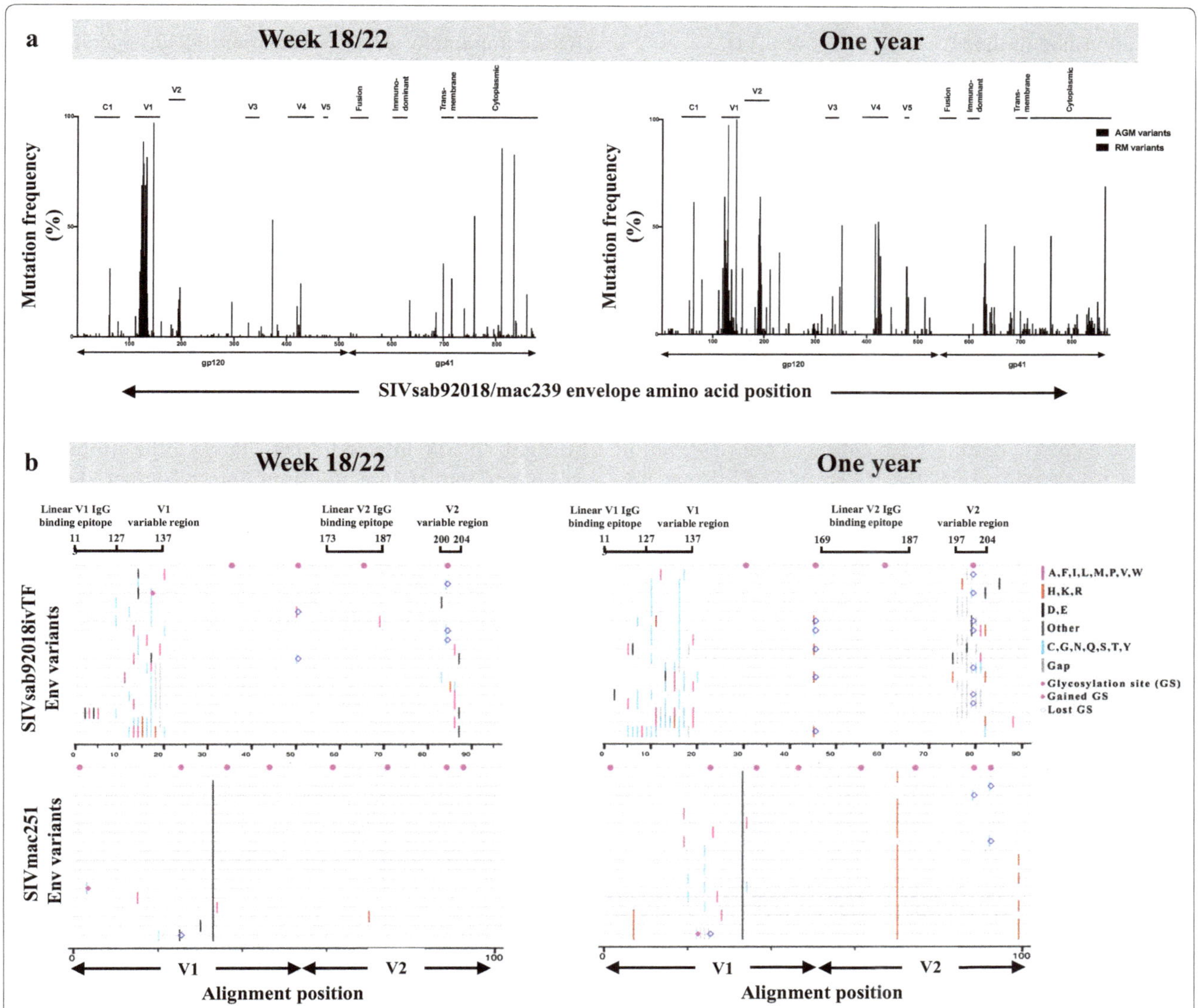

Fig. 4 High mutation frequency in V1 and V2 loops of SIV Env variants in SIV-infected AGMs. **a** Mutation frequency in the protein sequence of plasma SIV Env variants isolated from SIVsab92018ivTF-infected AGMs and SIVmac251-infected RMs was quantified as the percent of Env variants showing a different amino acid compared to that of the challenge viral Env from the total number of variants. **b** Representative protein sequence alignments within V1V2 region of plasma SIV Env variants (depicted by each line) with mismatches highlighted relative to the respective wild-type challenge env (the top line). Hanging bars with SIVmac239 Env sequence number indicate relative positions of linear SIVmac V1 or SIVsab V2 IgG epitopes, and the two regions having high protein sequence variability in V1 and V2 loops. Purple circles indicate potential N-linked glycosylation sites (GS). Purple diamonds or blue empty diamonds indicate gained or lost potential N-linked GS in Env variants compared to corresponding challenge viral Env sequences, respectively

V1 (mutation frequency mean in AGM vs. RM variants, 5.4 vs. 0.22% at 22 or 18 wpi, and 6.7 vs. 3.5% at 1 ypi, respectively), and V2 loops (mutation frequency mean in AGM vs. RM variants, 1.8 vs. 0.086% at 22 or 18 wpi, and 5.9 vs. 3.2% at 1 ypi, respectively) compared to RM Env variants over the course of infection (Fig. 4b). Particularly, AGM plasma SIVsab Env variants exhibited high sequence variability at AA residue positions 127–137 within the V1 loop, and 200–204 within the V2 loop at 22 wpi and at 197–204 within the V2 loop at 1 ypi (SIVmac239 Env number sequence) (Fig. 4b). The high mutation frequency regions within SIVsab92018ivTF V2 are both near the epitope of the high magnitude linear V2-specific responses and within the vicinity of an analogous region of the HIV V2 loop where the K169 mutation hotspot was identified in breakthrough viruses isolated from HIV RV144 vaccinees (Fig. 4b) [9]. This SIVsab V2 loop site of high sequence variability is also near the analogous putative gut-homing receptor α4β7-binding motif identified in the HIV-1 Env (Fig. 4b) [11].

As N-linked glycans on Env can evolve in response to the autologous virus-specific IgG responses, we also assessed the N-linked glycosylation pattern in SIV plasma Env variants. AGM SIVsab Env plasma variants exhibited a relatively high loss of predicted N-linked GS motifs within the V2 variable region compared to RM SIVmac Env variants (percent of variants exhibiting a GS loss in AGMs vs. RMs: 35% at 22 wpi and 69% at 1 ypi vs. 0% at 18 wpi and 16% at 1 ypi, respectively) at equivalent AA positions (SIVmac239 Env position 202) (Fig. 4b). SIVsab Env variants also showed a 7 and 31% loss of predicted GS motifs (SIVmac239 Env position 167) at 22 wpi and 1 ypi, respectively, within the V1 loop. A loss in GS motifs within the corresponding V1 region was not observed in SIVmac Env variants (Fig. 4b). However, RM SIVmac Env variants isolated at 1 ypi had 30 and 16% loss of predicted GS motifs within the V5 loop that were not observed in AGM SIVsab Env variants from 1 ypi (SIVmac239 Env positions 476 and 479, respectively).

ADCC kinetics of plasma IgG responses in SIV-infected AGMs and RMs

ADCC-mediating IgG responses and IgG targeting linear V2 epitopes were associated with a decreased risk of HIV-1 infection in the RV144 vaccine trial [25]. To explore the ADCC function of Env gp120-predominant antibody responses and their relationship to the early development of linear V2-specific IgG responses in AGMs, we evaluated the kinetics of plasma Env-specific ADCC responses in AGMs compared to those of RMs. ADCC activity was not detected in AGMs or RMs prior to SIV infection (Fig. 5a, b). As early as 6 wpi in AGMs and 5 wpi in RMs, high ADCC activity was detected against SIV Env gp140-coated cells in both species (ADCC titer as reciprocal plasma dilution median in AGMs vs. RMs: 52,225 [range, 36,931–90,286] vs. 61,826 [range, 54,503–64,725], raw $p=0.71$, FDR $p=0.95$) (Fig. 5a). Gp140-specific IgG-mediated ADCC activity increased in both species by 15 wpi in AGMs and 17 wpi in RMs (ADCC titer median in AGMs vs. RMs: 89,851 [range, 64,474–317,680] vs. 409,600 [range, 409,600–409,600], raw $p=0.024$, FDR $p=0.063$) (Fig. 5a). Interestingly, SIV-infected AGMs demonstrated high magnitude ADCC activity against SIV Env gp120-coated cells at 6 wpi, and this response was remarkably undetectable in acutely SIV-infected RMs (ADCC titer median in AGMs: 949.3 [range, 319–3746], raw $p=0.024$, FDR $p=0.063$) (Fig. 5b). Importantly, the more rapid development of gp120-specific ADCC-mediating response in AGMs arose concurrently with the gp120 and V2 loop-specific antibody responses in AGMs, which are distinct from that in SIV-infected RMs. Yet, both AGMs and RMs had comparable ADCC-mediating gp120-specific IgG responses by 15 or 17 wpi (ADCC titer median in AGMs vs. RMs: 11,169 [range, 4867–25,125] vs. 13,120 [range, 6036–23,722], raw $p=0.90$, FDR $p=1$) (Fig. 5b).

Plasma IgG binding to SIV-infected CD4$^+$ T cells

To examine the infected cell binding activity of the early acute SIV Env-specific IgG response in AGMs and RMs, we measured the binding of Env-specific IgG to CD4$^+$ T cells infected with SIVsab92018ivTF and SIVmac251 infectious molecular clones (IMC). For all experiments, we obtained high levels of infection with > 55% (range, 56–82% p24/p27+) of cells being infected for both viruses. Env-specific IgG plasma responses in AGMs recognized SIV-infected CD4$^+$ target cells more effectively and earlier in acute time points compared to RM plasma (median % FITC positivity in AGMs vs. RMs, 40.8% [range, 29.3–55.6%] vs. 1.0% [range, 0.1–1.1%] at week 6 or 5, raw $p=0.0095$, FDR $p=0.057$; 72.5% [range, 58.6–78.3%] vs. 23.2% [range, 17.3–24.6%] at week 15 or 17, raw $p=0.0095$, FDR $p=0.057$, respectively) (Fig. 5c).

Limited contribution of linear V2 epitopes to early ADCC and autologous SIV neutralizing responses in SIV-infected AGMs

Given the early development of ADCC and neutralizing activity that arose concurrently with the linear V2-specific plasma IgG binding response in AGMs during acute SIV infection, we sought to define potential contribution of the linear V2-specific IgG response to early gp120-directed functional responses that are unique to AGMs. With a SIVsab linear V2 peptide-coated IgG depletion column of AGM plasma, we achieved an enriched 76 ± 4.4% (mean and SD) depletion of linear V2-specific antibodies compared to 53 ± 8.4% (mean and SD) depletion with a scrambled peptide-coated column. We then compared the gp120-specific ADCC and autologous neutralizing function of AGM V2 and scrambled peptide-depleted plasma at 6 and 15 wpi, respectively [18]. Despite a reduction in ADCC activity compared to that of non-depleted plasma (median ADCC titers for non-depleted vs. V2-depleted response, 949.3 [range, 318.8–3746] vs. 314.4 [range, 100–576.3], raw $p=0.031$, FDR $p=0.063$; non-depleted vs. scrambled peptide-depleted response, 294.6 [range, 100–679.5], raw $p=0.031$, FDR $p=0.063$), we observed similar ADCC antibody titers in linear V2 peptide and scrambled peptide-depleted plasma at 6 wpi (raw $p=1$, FDR $p=1$) (Fig. 5d). Similarly, we observed an equivalent decrease in plasma SIVsab neutralizing activity following the linear V2 and scrambled peptide-depletion of AGM plasma at 15 wpi (median ID_{50} for non-depleted vs. V2-depleted activity,

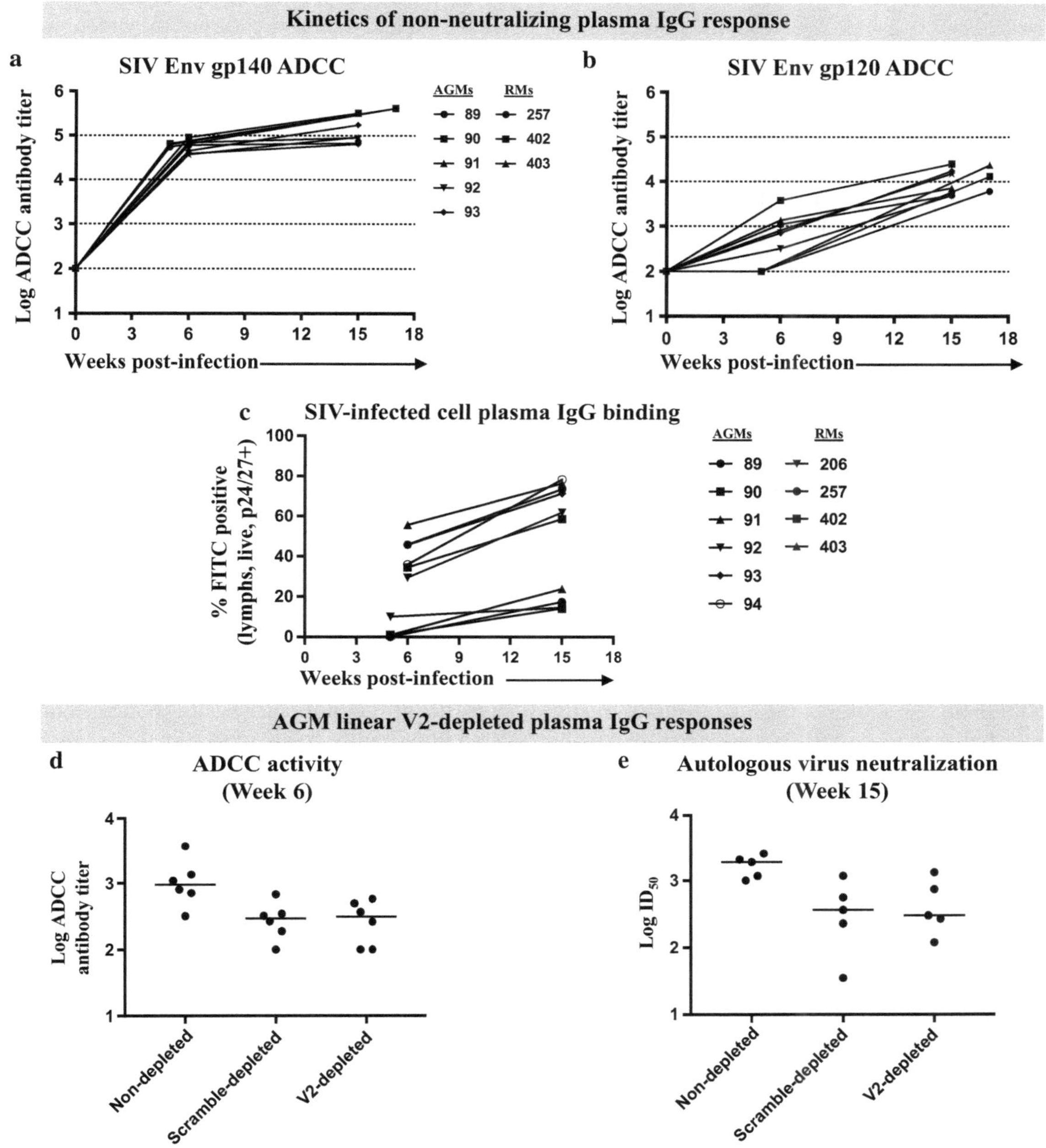

Fig. 5 Early development of SIV Env gp120-directed plasma IgG ADCC and SIV-infected cell binding in AGMs. ADCC activity against the autologous challenge SIV envelope (Env) **a** gp140 and **b** gp120-coated cells of plasma IgG from SIVsab92018ivTF-infected AGMs and SIVmac251-infected RMs prior to infection, at week 6 (AGM) or 5 (RM), and 15 (AGM) or 17 (RM) was quantified as ADCC antibody titers. **c** Binding to SIV-infected cells was measured as % FITC positivity after week 0 background subtraction. **d** Plasma gp120-specific ADCC responses (median shown) after depletion with a linear V2 peptide or scrambled peptide in SIV-infected AGMs at week 6. **e** Neutralizing responses (median shown) in SIVsab92018ivTF-infected AGMs against the autologous challenge virus at week 15 after depletion with a linear V2 peptide and scrambled peptide

1933 [range, 1007–2592] vs. 303 [range, 120–1340], raw $p=0.13$, FDR $p=0.19$; non-depleted vs. scrambled peptide-depleted activity, 367 [range, 35–1184], raw $p=0.063$, FDR $p=0.11$; scrambled peptide-depleted vs. V2-depleted activity, raw $p=1$, FDR $p=1$) (Fig. 5e). AGM and RM plasma IgG binding to scrambled peptide was negligible (median binding OD, 0.21 [range, 0.15–0.25], and 0.08 [range, 0.04–0.12], respectively).

Discussion

Vaccine-elicited V2-specific IgG responses that mediate ADCC activity have been proposed as an immune correlate of reduced HIV-1 infection risk in the moderately protective RV144 vaccine efficacy trial [3, 17, 26]. However, factors controlling the elicitation of V2-specific IgG responses capable of mediating ADCC are not well established. We previously reported that natural SIV hosts, AGMs, have a SIV Env gp120-biased antibody response to SIV infection, and rapidly develop autologous neutralizing antibodies, which are unique from both SIV-infected RMs and HIV-infected human non-natural hosts [18–20]. In this study, we identified that the distinct plasma linear V2-specific IgG responses arose at 6 wpi in 50% of all AGMs, which were markedly undetectable in all RMs. Moreover, acutely SIV-infected AGMs mediated a more rapid development of gp120-directed ADCC activity, and SIV-infected cell binding response compared to SIV-infected RMs. In addition, single genome Env variant analyses demonstrated an increase in AA mutation frequency, and a high loss of predicted N-linked GS within the V1 and V2 loops of SIVsab Env variants over time. Interestingly, the target of linear V2-specific plasma IgG binding responses in AGMs localized to an equivalent HIV V2 loop epitope near residues that were associated with immune escape from the potentially-protective V2-specific IgG responses in RV144 vaccinees [8, 9]. This SIVsab linear V2 epitope is also in close vicinity of the putative binding motif of the gut-homing receptor α4β7 in HIV-1 Env, underlining the potential importance of the V2 loop region for virus infectivity and escape from humoral responses in the natural SIV hosts.

In this study, we found that SIV-infected AGMs exhibited a predominant SIV Env gp120-directed response very early during acute SIV infection compared to the RM counterparts. To further examine the gp120/gp41 antibody differential antibody responses in these species, we measured their SIV Env linear peptide-specific IgG binding responses against overlapping SIVsab92018WT and SIVmac239 peptide libraries spanning Env gp160, respectively. Interestingly, we observed higher binding responses against linear V2 peptides in AGMs but not in RMs. Given that a few AA differences occur in these overlapping peptide libraries compared to the SIVsab92018ivTF and SIVmac251 challenge viruses, we also tested AGM and RM plasma IgG binding against linear V2 peptides matching the challenge viruses. AGMs showed higher magnitude linear V2-specific IgG binding against the linear V2 SIVsab92018ivTF peptide compared to RM plasma binding to linear V2 SIVmac251 peptide, suggesting that the observed binding differences against linear V2 peptides in AGMs and RMs are not due differences in the AA sequences between the SIV Env overlapping peptide library and the SIV challenge viruses. While the mechanism of the predominant, early plasma linear V2-specific IgG response in AGMs remains unclear, it is possible that AGMs mediate a higher frequency of SIV linear-epitope specific responses given their lower frequency of SIV immune-complex trapping by follicular dendritic cells in germinal centers [27, 28].

We observed that V1 and V2 loops of SIVsab variants had higher mutation frequencies compared to SIVmac variants isolated from RMs. This high Env mutation frequency in the V1V2 region has also been reported in SIVsab isolates from naturally SIV-infected AGMs [29, 30]. Therefore, it is plausible that the early development of predominant gp120-specific and high magnitude V2-specific IgG responses in AGMs may influence the unique Env mutation pattern observed within the V1V2 loop of autologous SIVsab circulating viruses [25, 31, 32]. It is also worth noting that the hypervariable loops of HIV-1 have long been suspected to influence conformational state changes in Env [33]. In addition, mutations within V1 or V2 loop could influence the conformation of nearby epitopes via glycan shielding [31, 32, 34, 35].

Comparisons in SIV and HIV Env-specific antibody responses are admittedly limited by a lack of SIV Env trimer crystal structures [32, 36–38]. The structure of a core SIV gp120 monomer has been determined and differed significantly in the orientation of its inner and outer domains compared to HIV gp120 monomers, suggesting a differing functional SIV Env trimer arrangement ([39]). Nevertheless, SIV V1V2 has been recognized to correspond to HIV-1 V1V2, and the SIV and HIV V2 loops share certain structural motifs, specifically the [ED] [VLI] motif and a lysine-rich motif that appears as K[KV] QK in HIV and KK (alternately, K[KT]K) in SIV [29, 36, 37, 40]. In fact, an RV144 vaccine regimen utilizing SIV gp120 as the immunogen elicited V2 loop-specific IgG responses that were associated with protection against SIVmac251 in RMs [16]. Furthermore, V2-specific antibody responses were inversely correlated with peak and set-point viral loads in SIV-infected RMs [41]. Thus, the V2 loop in HIV and SIV may share immunogenic and functional characteristics that are capable of eliciting potentially-protective IgG responses in non-human primates.

Vaccine-elicited V2-specific IgG responses from the RV144 vaccine efficacy trial may have mediated protection through an ADCC mechanism [7, 9, 15]. In addition to the previously reported early kinetics of autologous virus neutralizing responses in AGMs [18], we observed an early SIV Env gp120-specific ADCC response in AGMs compared to RMs. Moreover, AGM plasma IgG showed higher SIV-infected cell binding activity compared to RM plasma, suggesting that AGM

SIV Env-specific IgG responses may better recognize epitopes exposed on infected cells. Similar functional immune profiles using AGM sera have been observed in previous studies of naturally SIV-infected AGMs [42, 43]. These results suggest that AGM SIV Env-reactive plasma IgG can mediate robust non-neutralizing antiviral functions and increased infected cell binding activity during early acute SIV infection compared to RM plasma antibodies. Intriguingly, we observed similar ADCC activity in V2 and scrambled peptide-depleted AGM plasma, suggesting that other sites within SIVsab gp120 may mediate the early acute ADCC responses. However, it should be noted that there was non-specific IgG depletion in the scrambled peptide depletion experiments [44]. Furthermore, vaccine-elicited antibody responses in the RV144 efficacy trial were mapped to both conformational and linear V2 epitopes, suggesting that V2-specific IgG responses with anti-viral activity may not only target linear V2 epitopes [9, 14]. Interestingly, it has been shown that conformational epitope recognition of SIV Env-specific antibody responses differs greatly between nonpathogenic and pathogenic SIV infections in natural SIV hosts [23]. Thus, ADCC-mediating IgG responses in humans and AGMs may target both linear V2 and conformational gp120 epitopes within HIV and SIV Env.

Conclusion

We have demonstrated that acutely SIV-infected AGMs develop high magnitude plasma linear V2-specific IgG responses that arise concurrently with ADCC-mediating and SIV-infected cellular binding responses targeting gp120 epitopes. Yet, these early plasma linear V2-specific and gp120-specific IgG responses with ADCC activity were undetectable in RMs. Moreover, the antibody responses targeting either the linear V2 or other non-linear gp120 variable loop epitopes that are generated more quickly in AGMs than those in RMs could contribute to the observed faster accumulation of diversity of autologous circulating viruses within the V1V2 loop in SIV-infected AGMs. Our findings support the potential use of AGMs as an alternative non-human primate model for HIV-1 vaccine development, particularly for the evaluation of immunogens that target high magnitude neutralizing and non-neutralizing gp120-specific IgG responses. A greater understanding of host immune factors that lead to the unique functional gp120-specific IgG responses in these natural SIV hosts may provide valuable insights for future HIV vaccine immunogen design.

Methods

Nonhuman primates and sample collection

Six female AGMs (*Chlorocebus sabaeus*) and four female Indian RMs (*Macaca mulatta*) between four and 11 years of age were intravenously inoculated with, infectious molecular clones of SIVsab92018ivTF, or SIVmac251.30 virus isolate, respectively [19, 20, 45]. All four RMs maintained high-level viremia compared to those of AGMs over the course of infection [18]. Blood was collected in EDTA tubes at 0, 2, 3, 5, 6, 15, 17, and 18 (RMs) or 22 (AGMs) weeks post-infection (wpi), and again at 1 year post-infection (ypi) in both species, and plasma was then isolated. Animals were housed and maintained according to the *Guide for the Care and Use of Laboratory Animals* [46].

HIV/SIV Env protein and peptides

To assess the fine-specificity of SIV Env-specific plasma IgG binding responses in AGMs and RMs, we tested plasma samples for IgG binding to recombinant SIVsab92018ivTF Env gp120, [47] SIVmac251 Env gp120, [47] HIV-1 murine leukemia virus (MuLV) gp70_His6/Mon (293F), gp70 V1V2 CaseA subtype B (293F), gp70 V1V2 CaseA2 subtype B (169 K), and [48] SIVsab92018ivTF or SIVmac251 V1V2 proteins [18]. To examine Env linear peptide IgG binding responses, we used the SIVagm Sabaeus 92018 (Cat #10451), SIVmac239 (Cat #6883, AAA47637.1), and HIV-1 ConB (Cat #9480) Env peptide sets obtained through the NIH NIAID AIDS Reagent Program, Division of AIDS in addition to SIVsab92018ivTF V2 peptide (SFAMAGYRRDVKKNYSTVWYDQE) and SIVmac251 V2 peptide (KFNMTGLKRDKTKEYNETWYSTD) (CPC Scientific, Sunnyvale, CA). SIVmac239 Env peptide regions were annotated based on the HIV-2/SIV protein annotations from HIV Sequence Compendium 2016 [49]. To annotate the regions for SIVsab92018 Env overlapping peptide, its sequence was aligned with SIVmac239 (M33262) using the HIVAlign tool (https://www.hiv.lanl.gov/content/sequence/VIRALIGN/viralign.html). An additional table file details the information of both SIVsab92018 and SIVmac239 Env overlapping peptides for each Env region (see Additional file 1).

HIV/SIV Env-specific IgG binding enzyme-linked immunosorbent assays (ELISAs)

To assess SIV Env-specific plasma IgG binding kinetics, plasma samples from SIVsab92018ivTF-infected AGMs and SIVmac251-infected RMs were tested for IgG binding to recombinant SIVsab92018ivTF or SIVmac251 Env gp120 proteins at the following time-points: 0, 3, 15 wpi, and 1 ypi according to sample availability. To measure linear SIV Env peptide-specific plasma IgG binding responses, plasma samples from SIV-infected AGMs and RMs from 0, 3, 6, 15 wpi, and 1 ypi were tested for IgG binding to SIVsab92018WT and SIVmac239 overlapping linear peptide libraries, respectively. To confirm similar IgG responses to the autologous linear V2 peptides, plasma samples from SIV-infected AGMs and RMs at 5, 15 wpi, and 1 ypi were tested against SIVsab92018ivTF and SIVmac251 linear V2 peptides, respectively. To examine the magnitude and cross-reactivity of the heterologous linear V2 and V1V2 protein-specific plasma IgG binding, plasma samples from SIV-infected AGMs at 6, 15 wpi, and 1 ypi were serially diluted from a 1:100 starting dilution and tested against SIVmac239 and HIV-1 Consensus Subtype B (ConB) overlapping linear V2 peptides, and SIVsab92018ivTF, SIVmac251, and HIV-1 V1V2 proteins. The SIV Env/peptide-specific IgG binding ELISAs were carried out with Env proteins or peptides diluted to 3 μg/mL in 0.1 M sodium bicarbonate, and coated on plates for 1 h. Plasma samples were diluted at 1:100. Plates were washed and blocked with SuperBlock (4% whey protein, 15% goat serum, and 0.5% Tween 20 diluted in 1X phosphate-buffered saline (PBS)) at room temperature for 1 h. Plasma from chronically SIV-infected AGMs and RMs were used as positive controls. Influenza hemagglutinin-specific mAb CH65 [50] and an anti-respiratory syncytial virus antibody, Pavilizumab (Medimmune, Inc, Quakertown, PA), were used as negative controls. Scrambled peptides and species-specific SIV Env gp140 proteins were used as additional negative and positive controls, respectively. Horseradish peroxidase (HRP)-conjugated polyclonal goat anti-monkey IgG (gamma chain) antibody (Rockland Immunochemicals, Gilbertsville, PA) was added and incubated for 1 h at 1:10,000 dilution. Plates were washed and developed with SureBlue reserve TMB substrate for 5 min, and the reaction was stopped by adding an equal volume of TMB stop solution (KPL, Gaithersburg, MD). Optical densities (ODs) were measured at 450 nm using a Spectramax Plus spectrophotometer (Molecular Devices, Sunnyvale, CA). A line of best fit calculated by a 4-parameter logistic curve was used to interpolate the log half-maximal effective plasma dilution (log ED_{50}) using GraphPad Prism 7 (Graph Pad, La Jolla, CA).

Peptide depletion and SIV neutralization assays

To deplete linear V2-specific IgG responses, CNBr-activated Sepharose 4B columns (GE Healthcare, Pittsburgh, PA) were coated with 2 mg of SIVsab92018WT V2 peptide (SFAMAGYRRDVKKNYSTVWDDQE) (CPC Scientific, Sunnyvale, CA), and a scrambled peptide of similar length according to manufacturer's guidelines. AGM plasma samples were run through the columns, and the flow-through was run for a total of 10 times. Depletion efficiency of linear V2 peptide-specific antibodies was quantified as the percent depletion, which equals 1- ((OD of peptide-depleted V2 binding/OD of non-depleted V2 binding) * 100). Non-depleted, linear V2 peptide-depleted, and scrambled peptide-depleted plasma samples from SIVsab92018ivTF-infected AGMs at 15 wpi were tested against SIVsab92018ivTF infectious molecular clone (IMC) in the TZM-bl cell neutralization assay [18]. In brief, samples were serially diluted in duplicate, and incubated with virus (50,000 relative light unit [RLU] equivalents) for 1 h at 37 °C. Cells were added at a density of 10,000 cells per well, and incubated for 48 h at 37 °C + 5% CO_2. A reduction in luciferase activity was quantified by the Brightglow Luciferase detection system (Promega, Madison, WI). The sample dilution at which a 50% RLU reduction was observed was defined as the inhibitory dilution (ID_{50}).

ADCC assay

We used the GranToxiLux (GTL) ADCC assay to measure the ADCC activity of whole plasma samples, linear V2 peptide-depleted plasma IgG, and/or scrambled peptide-depleted plasma IgG isolated from SIVsab92018ivTF-infected AGMs and SIVmac251-infected RMs at 0, 6 (AGMs) or 5 (RMs), and 15 (AGMs) or 17 (RMs) wpi as described previously [51]. The CEM.NKR_{CCR5} target cells [52] were coated with recombinant SIVsab92018ivTF/SIVmac251 gp120, or avi-tagged gp140 Env representing the same virus isolates. Cryopreserved human peripheral blood mononuclear cells from an HIV-seronegative donor with the 158F/F polymorphic variant of Fcγ receptor 3A served as effector cells [53]. Plasma samples were tested using 4-fold serial dilutions starting at 1:100 dilution. Data were analyzed using FlowJo 9.8.2 (Tree Star Inc., Ashland, OR). The % Granzyme B (GzB) activity was defined as the percentage of cells positive for proteolytically active GzB out of the total viable target cell population. Final results were calculated after subtracting the background % GzB activity observed in wells containing effector and target cells in the absence of plasma or IgG samples. Plasma ADCC antibody titers were determined by interpolating the plasma dilutions that intersect the positive cut-off (8% GzB activity) using Graph Pad Prism 7 (Graph Pad, La Jolla CA).

Plasma IgG binding to Env on the surface of SIV-infected $CD4^+$ cells

Indirect surface staining was used to evaluate the ability of SIV Env-specific plasma IgG from AGMs and RMs to bind the surface of CEM.NKR_{CCR5} $CD4^+$ T cells [54] infected with SIVsab92018ivTF or SIVmac251. Infections with replication competent infectious molecular clone virus (IMC) were performed using DEAE-Dextran as described previously [51, 55]. At 48 h postinfection, the infected CEM.NKR_{CCR5} cells were incubated with a 1:100 dilution of plasma for 2 h at 37 °C and then stained with a vital dye (Live/Dead Fixable Aqua Dead Cell Stain; Invitrogen) to exclude nonviable cells from subsequent analyses. Primary Ab binding was detected by secondary labeling with fluorescein isothiocyanate (FITC)-conjugated goat anti-rhesus IgG (SouthernBiotech Inc., Birmingham, AL), and SIV-infected cells were identified by staining for intracellular expression of p24/27 (KC57-RD1; Beckman Coulter) using standard methods. Binding was quantified as % of live, p24/27 positive, FITC positive cells after subtracting the background binding of week 0 samples. Mock-infected cells were used to establish negative gates.

Single genome amplification (SGA) and sequencing of plasma SIV env variants

SIV env variants from SIVsab92018ivTF-infected AGM plasma samples at 22 and 52 wpi (n = 4, 71 variants, and n = 3, 39 variants, respectively), [56] and SIVmac251-infected RM plasma samples at 18 and 50 wpi (n = 4, 107 variants; and n = 3, 67 variants, respectively) were amplified and sequenced using limiting dilution PCR [56]. For this study, SGAs isolated from AGMs and RMs were aligned against sequences of the corresponding wild-type SIVsab92018 and SIVmac251 using Seaview [57]. For SIVsab92018WT Env sequence (871 amino acid (AA)), the regions include constant region 1 (C1) (54 AA), V1 (42 AA), V2 (45 AA), and gp41 (345 AA). For SIVmac251 Env sequence (876 AA), the regions include C1 (49 AA), V1 (58 AA), V2 (43 AA), and gp41 (348 AA). Mutation frequency of Env variants at each time-point was calculated as the percent of variants with different AA compared to respective SIVsab/mac virus Env sequence at each residue within a defined Env region. Mutations in Env variants were graphically represented in relationship to their respective challenge Env using Highlighter tool (https://www.hiv.lanl.gov/content/sequence/HIGHLIGHT/highlighter_top.html). Analysis of N-linked glycosylation sites (GS) (Nx[ST] pattern) of plasma Env variants were performed using N-GlycoSite tool (https://www.hiv.lanl.gov/content/sequence/GLYCOSITE/glycosite.html).

Statistical analysis

Wilcoxon-Mann–Whitney was used to compare the Env protein-specific plasma IgG binding $\log EC_{50}$ values between AGMs and RMs (Fig. 1). The exact Wilcoxon Rank Sum test was used to compare the gp140 and gp120 ADCC titers in AGMs and RMs (Fig. 5a, b), and the % infected cells in AGMs and RMs (Fig. 5c). The Wilcoxon Signed-Rank test to compare the non-depleted and V2/scrambled peptide depleted ADCC activity (Fig. 5d), and neutralizing activity (Fig. 5e). All analyses were adjusted for multiple comparisons by false discovery rate (FDR) p value correction.

Authors' contributions

QNN, DRM, SRP designed, analyzed, interpreted the data, and wrote the manuscript; JEH and QNN performed and analyzed peptide binding ELISA; DRM, RM and QH performed and analyzed protein binding ELISA; DRM performed and analyzed depletion and neutralization data; RWE and JP performed and analyzed ADCC and virus binding data; QNN and AK analyzed virus envelope variant analysis; NIN and XS contributed important insights for the interpretation and discussion of the results. All authors read and approved the final manuscript.

Author details

[1] Duke Human Vaccine Institute, Duke University School of Medicine, Durham, NC, USA. [2] Department of Molecular Genetics and Microbiology, Duke University School of Medicine, Durham, NC, USA. [3] Department of Surgery, Duke University School of Medicine, Durham, NC, USA. [4] Department of Pediatrics, Duke University School of Medicine, Durham, NC, USA. [5] Department of Immunology, Duke University School of Medicine, Durham, NC, USA.

Acknowledgements

We thank Thomas Jeffries Jr. and Robert Parks for their technical expertise; Kevin Saunders and Melissa Cooper for providing HIV-1 V1V2 proteins; Josh Eudailey for processing stool bacterial lysates; and Meng Chen for statistical advice. The funders had no role in study design, data collection and interpretation, or the decision to submit the work for publication.

Competing interests

The authors declare that they have no competing interests.

Consent for publication

All authors consented to the publication of this manuscript. The funding agencies had no role in the decision to publish this manuscript. The content of this manuscript is the sole view of the authors and does not necessarily reflect the view of the funding sources.

Funding

Support for this work was provided by an R01 grant to Sallie R. Permar from the National Institutes of Health (NIH grant 5R01AI106380-04). David R. Martinez is supported by an ASM Robert D. Watkins Graduate Research Fellowship, an NIH NIAID Ruth L. Kirschstein National Research Service Award F31 F31AI127303, and a Burroughs Wellcome Graduate Diversity Enrichment Award.

References

1. Kwong PD, Mascola JR. Human antibodies that neutralize HIV-1: identification, structures, and B cell ontogenies. Immunity. 2012;37:412–25.
2. Haynes BF, Mascola JR. The quest for an antibody-based HIV vaccine. Immunol Rev. 2017;275:5–10.
3. Pollara J, Easterhoff D, Fouda GG. Lessons learned from human HIV vaccine trials. Curr Opin HIV AIDS. 2017;12:216–21.
4. Williams WB, Liao H-X, Moody MA, Kepler TB, Alam SM, Gao F, Wiehe K, Trama AM, Jones K, Zhang R, et al. Diversion of HIV-1 vaccine-induced immunity by gp41-microbiota cross-reactive antibodies. Science. 2015;349:aab1253.
5. Tomaras GD, Yates NL, Liu P, Qin L, Fouda GG, Chavez LL, Decamp AC, Parks RJ, Ashley VC, Lucas JT, et al. Initial B-cell responses to transmitted human immunodeficiency virus type 1: virion-binding immunoglobulin

5. Tomaras GD, Yates NL, Liu P, Qin L, Fouda GG, Chavez LL, Decamp AC, Parks RJ, Ashley VC, Lucas JT, et al. Initial B-cell responses to transmitted human immunodeficiency virus type 1: virion-binding immunoglobulin M (IgM) and IgG antibodies followed by plasma anti-gp41 antibodies with ineffective control of initial viremia. J Virol. 2008;82:12449–63.
6. Liao HX, Chen X, Munshaw S, Zhang R, Marshall DJ, Vandergrift N, Whitesides JF, Lu X, Yu JS, Hwang KK, et al. Initial antibodies binding to HIV-1 gp41 in acutely infected subjects are polyreactive and highly mutated. J Exp Med. 2011;208:2237–49.
7. Haynes BF, Gilbert PB, McElrath MJ, Zolla-Pazner S, Tomaras GD, Alam SM, Evans DT, Montefiori DC, Karnasuta C, Sutthent R, et al. Immune-correlates analysis of an HIV-1 vaccine efficacy trial. N Engl J Med. 2012;366:1275–86.
8. Rolland M, Edlefsen PT, Larsen BB, Tovanabutra S, Sanders-Buell E, Hertz T, deCamp AC, Carrico C, Menis S, Magaret CA, et al. Increased HIV-1 vaccine efficacy against viruses with genetic signatures in Env V2. Nature. 2012;490:417–20.
9. Karasavvas N, Billings E, Rao M, Williams C, Zolla-Pazner S, Bailer RT, Koup RA, Madnote S, Arworn D, Shen X, et al. The thai phase III HIV type 1 vaccine trial (RV144) regimen induces antibodies that target conserved regions within the V2 loop of gp120. AIDS Res Hum Retrovir. 2012;28:1444–57.
10. Zolla-Pazner S, deCamp AC, Cardozo T, Karasavvas N, Gottardo R, Williams C, Morris DE, Tomaras G, Rao M, Billings E, et al. Analysis of V2 antibody responses induced in vaccinees in the ALVAC/AIDSVAX HIV-1 vaccine efficacy trial. PLoS ONE. 2013;8:e53629.
11. Tassaneetrithep B, Tivon D, Swetnam J, Karasavvas N, Michael NL, Kim JH, Marovich M, Cardozo T. Cryptic determinant of α4β7 binding in the V2 loop of HIV-1 gp120. PLoS ONE. 2014;9:e108446.
12. Ansari AA, Reimann KA, Mayne AE, Takahashi Y, Stephenson ST, Wang R, Wang X, Li J, Price AA, Little DM, et al. Blocking of α4β7 gut-homing integrin during acute infection leads to decreased plasma and gastrointestinal tissue viral loads in simian immunodeficiency virus-infected rhesus macaques. J Immunol. 2011;186:1044–59.
13. Peachman KK, Karasavvas N, Chenine AL, McLinden R, Rerks-Ngarm S, Jaranit K, Nitayaphan S, Pitisuttithum P, Tovanabutra S, Zolla-Pazner S, et al. Identification of new regions in HIV-1 gp120 variable 2 and 3 loops that bind to α4β7 integrin receptor. PLoS ONE. 2015;10:e0143895.
14. Bonsignori M, Pollara J, Moody MA, Alpert MD, Chen X, Hwang KK, Gilbert PB, Huang Y, Gurley TC, Kozink DM, et al. Antibody-dependent cellular cytotoxicity-mediating antibodies from an HIV-1 vaccine efficacy trial target multiple epitopes and preferentially use the VH1 gene family. J Virol. 2012;86:11521–32.
15. Pollara J, Bonsignori M, Moody MA, Liu P, Alam SM, Hwang KK, Gurley TC, Kozink DM, Armand LC, Marshall DJ, et al. HIV-1 vaccine-induced C1 and V2 Env-specific antibodies synergize for increased antiviral activities. J Virol. 2014;88:7715–26.
16. Vaccari M, Gordon SN, Fourati S, Schifanella L, Liyanage NP, Cameron M, Keele BF, Shen X, Tomaras GD, Billings E, et al. Adjuvant-dependent innate and adaptive immune signatures of risk of SIVmac251 acquisition. Nat Med. 2016;22:762–70.
17. Kim JH, Excler JL, Michael NL. Lessons from the RV144 Thai phase III HIV-1 vaccine trial and the search for correlates of protection. Annu Rev Med. 2015;66:423–37.
18. Amos JD, Himes JE, Armand L, Gurley TC, Martinez DR, Colvin L, Beck K, Overman RG, Liao HX, Moody MA, Permar SR. Rapid development of gp120-focused neutralizing B Cell responses during acute simian immunodeficiency virus infection of African green monkeys. J Virol. 2015;89:9485–98.
19. Amos JD, Wilks AB, Fouda GG, Smith SD, Colvin L, Mahlokozera T, Ho C, Beck K, Overman RG, DeMarco CT, et al. Lack of B cell dysfunction is associated with functional, gp120-dominant antibody responses in breast milk of simian immunodeficiency virus-infected African green monkeys. J Virol. 2013;87:11121–34.
20. Zhang R, Martinez DR, Nguyen QN, Pollara J, Arifin T, Stolarchuk C, Foulger A, Amos JD, Parks R, Himes JE, et al. Envelope-specific B-cell populations in African green monkeys chronically infected with simian immunodeficiency virus. Nat Commun. 2016;7:12131.
21. Chahroudi A, Bosinger SE, Vanderford TH, Paiardini M, Silvestri G. Natural SIV hosts: showing AIDS the door. Science. 2012;335:1188–93.
22. Gaufin T, Pattison M, Gautam R, Stoulig C, Dufour J, MacFarland J, Mandell D, Tatum C, Marx MH, Ribeiro RM, et al. Effect of B-cell depletion on viral replication and clinical outcome of simian immunodeficiency virus infection in a natural host. J Virol. 2009;83:10347–57.
23. Brocca-Cofano E, Kuhrt D, Siewe B, Xu C, Haret-Richter GS, Craigo J, Labranche C, Montefiori DC, Landay A, Apetrei C, Pandrea I. Pathogenic correlates of simian immunodeficiency virus-associated B cell dysfunction. J Virol. 2017;91:e01051-17.
24. Banks ND, Kinsey N, Clements J, Hildreth JE. Sustained antibody-dependent cell-mediated cytotoxicity (ADCC) in SIV-infected macaques correlates with delayed progression to AIDS. AIDS Res Hum Retrovir. 2002;18:1197–205.
25. Liao HX, Bonsignori M, Alam SM, McLellan JS, Tomaras GD, Moody MA, Kozink DM, Hwang KK, Chen X, Tsao CY, et al. Vaccine induction of antibodies against a structurally heterogeneous site of immune pressure within HIV-1 envelope protein variable regions 1 and 2. Immunity. 2013;38:176–86.
26. Hsu DC, O'Connell RJ. Progress in HIV vaccine development. Hum Vaccin Immunother. 2017;13:1018–30.
27. Brenchley JM, Vinton C, Tabb B, Hao XP, Connick E, Paiardini M, Lifson JD, Silvestri G, Estes JD. Differential infection patterns of CD4+ T cells and lymphoid tissue viral burden distinguish progressive and nonprogressive lentiviral infections. Blood. 2012;120:4172–81.
28. Beer B, Scherer J, zur Megede J, Norley S, Baier M, Kurth R. Lack of dichotomy between virus load of peripheral blood and lymph nodes during long-term simian immunodeficiency virus infection of African green monkeys. Virology. 1996;219:367–75.
29. Baier M, Dittmar MT, Cichutek K, Kurth R. Development of vivo of genetic variability of simian immunodeficiency virus. Proc Natl Acad Sci USA. 1991;88:8126–30.
30. Müller MC, Saksena NK, Nerrienet E, Chappey C, Hervé VM, Durand JP, Legal-Campodonico P, Lang MC, Digoutte JP, Georges AJ. Simian immunodeficiency viruses from central and western Africa: evidence for a new species-specific lentivirus in tantalus monkeys. J Virol. 1993;67:1227–35.
31. Shen G, Upadhyay C, Zhang J, Pan R, Zolla-Pazner S, Kong XP, Hioe CE. Rationally targeted mutations at the V1V2 domain of the HIV-1 envelope to augment virus neutralization by anti-V1V2 monoclonal antibodies. PLoS ONE. 2015;10:e0141233.
32. Jiang X, Totrov M, Li W, Sampson JM, Williams C, Lu H, Wu X, Lu S, Wang S, Zolla-Pazner S, Kong XP. Rationally designed immunogens targeting HIV-1 gp120 V1V2 induce distinct conformation-specific antibody responses in rabbits. J Virol. 2016;90:11007–19.
33. Moscoso CG, Xing L, Hui J, Hu J, Kalkhoran MB, Yenigun OM, Sun Y, Paavolainen L, Martin L, Vahlne A, et al. Trimeric HIV Env provides epitope occlusion mediated by hypervariable loops. Sci Rep. 2014;4:7025.
34. Ly A, Stamatatos L. V2 loop glycosylation of the human immunodeficiency virus type 1 SF162 envelope facilitates interaction of this protein with CD4 and CCR5 receptors and protects the virus from neutralization by anti-V3 loop and anti-CD4 binding site antibodies. J Virol. 2000;74:6769–76.
35. Losman B, Bolmstedt A, Schønning K, Björndal A, Westin C, Fenyö EM, Olofsson S. Protection of neutralization epitopes in the V3 loop of oligomeric human immunodeficiency virus type 1 glycoprotein 120 by N-linked oligosaccharides in the V1 region. AIDS Res Hum Retrovir. 2001;17:1067–76.
36. Zanetti G, Briggs JA, Grünewald K, Sattentau QJ, Fuller SD. Cryo-electron tomographic structure of an immunodeficiency virus envelope complex in situ. PLoS Pathog. 2006;2:e83.
37. Hu G, Liu J, Taylor KA, Roux KH. Structural comparison of HIV-1 envelope spikes with and without the V1/V2 loop. J Virol. 2011;85:2741–50.
38. Bohl C, Bowder D, Thompson J, Abrahamyan L, Gonzalez-Ramirez S, Mao Y, Sodroski J, Wood C, Xiang SH. A twin-cysteine motif in the V2 region of gp120 is associated with SIV envelope trimer stabilization. PLoS ONE. 2013;8:e69406.
39. Chen B, Vogan EM, Gong H, Skehel JJ, Wiley DC, Harrison SC. Structure of an unliganded simian immunodeficiency virus gp120 core. Nature. 2005;433:834–41.
40. Cole KS, Alvarez M, Elliott DH, Lam H, Martin E, Chau T, Micken K, Rowles JL, Clements JE, Murphey-Corb M, et al. Characterization of neutralization epitopes of simian immunodeficiency virus (SIV) recognized by rhesus

monoclonal antibodies derived from monkeys infected with an attenuated SIV strain. Virology. 2001;290:59–73.
41. Guo J, Zuo T, Cheng L, Wu X, Tang J, Sun C, Feng L, Chen L, Zhang L, Chen Z. Simian immunodeficiency virus infection evades vaccine-elicited antibody responses to V2 region. J Acquir Immune Defic Syndr. 2015;68:502–10.
42. Lozano Reina JM, Favre D, Kasakow Z, Mayau V, Nugeyre MT, Ka T, Faye A, Miller CJ, Scott-Algara D, McCune JM, et al. Gag p27-specific B- and T-cell responses in Simian immunodeficiency virus SIVagm-infected African green monkeys. J Virol. 2009;83:2770–7.
43. Norley SG, Kraus G, Ennen J, Bonilla J, König H, Kurth R. Immunological studies of the basis for the apathogenicity of simian immunodeficiency virus from African green monkeys. Proc Natl Acad Sci USA. 1990;87:9067–71.
44. Martinez DR, Vandergrift N, Douglas AO, McGuire E, Bainbridge J, Nicely NI, Montefiori DC, Tomaras GD, Fouda GG, Permar SR. Maternal binding and neutralizing IgG responses targeting the C terminal region of the V3 loop are predictive of reduced peripartum HIV-1 transmission risk. J Virol. 2017;91:e02422-16.
45. Permar SR, Kang HH, Carville A, Mansfield KG, Gelman RS, Rao SS, Whitney JB, Letvin NL. Potent simian immunodeficiency virus-specific cellular immune responses in the breast milk of simian immunodeficiency virus-infected, lactating rhesus monkeys. J Immunol. 2008;181:3643–50.
46. National Research Council. Guide for the care and use of laboratory animals. Washington, DC: National Academy Press; 1996.
47. Gnanadurai CW, Pandrea I, Parrish NF, Kraus MH, Learn GH, Salazar MG, Sauermann U, Töpfer K, Gautam R, Münch J, et al. Genetic identity and biological phenotype of a transmitted/founder virus representative of nonpathogenic simian immunodeficiency virus infection in African green monkeys. J Virol. 2010;84:12245–54.
48. Yates NL, Liao HX, Fong Y, deCamp A, Vandergrift NA, Williams WT, Alam SM, Ferrari G, Yang ZY, Seaton KE, et al. Vaccine-induced Env V1–V2 IgG3 correlates with lower HIV-1 infection risk and declines soon after vaccination. Sci Transl Med. 2014;6:228ra239.
49. Foley B, Leitner T, Apetrei C, Hahn B, Mizrachi I, Mullins J, Rambaut A, Wolinsky S, Korber B. HIV sequence compendium 2016. LA-UR-16-25625 edition. Los Alamos: Los Alamos National Laboratory, Theoretical Biology and Biophysics; 2016. p. 2016.
50. Whittle JR, Zhang R, Khurana S, King LR, Manischewitz J, Golding H, Dormitzer PR, Haynes BF, Walter EB, Moody MA, et al. Broadly neutralizing human antibody that recognizes the receptor-binding pocket of influenza virus hemagglutinin. Proc Natl Acad Sci USA. 2011;108:14216–21.
51. Pollara J, Hart L, Brewer F, Pickeral J, Packard BZ, Hoxie JA, Komoriya A, Ochsenbauer C, Kappes JC, Roederer M, et al. High-throughput quantitative analysis of HIV-1 and SIV-specific ADCC-mediating antibody responses. Cytometry A. 2011;79:603–12.
52. Trkola A, Matthews J, Gordon C, Ketas T, Moore JP. A cell line-based neutralization assay for primary human immunodeficiency virus type 1 isolates that use either the CCR5 or the CXCR4 coreceptor. J Virol. 1999;73:8966–74.
53. Koene HR, Kleijer M, Algra J, Roos D, von dem Borne AE, de Haas M. Fc gammaRIIIa-158V/F polymorphism influences the binding of IgG by natural killer cell Fc gammaRIIIa, independently of the Fc gammaRIIIa-48L/R/H phenotype. Blood. 1997;90:1109–14.
54. Ferrari G, Pollara J, Kozink D, Harms T, Drinker M, Freel S, Moody MA, Alam SM, Tomaras GD, Ochsenbauer C, et al. An HIV-1 gp120 envelope human monoclonal antibody that recognizes a C1 conformational epitope mediates potent antibody-dependent cellular cytotoxicity (ADCC) activity and defines a common ADCC epitope in human HIV-1 serum. J Virol. 2011;85:7029–36.
55. Edmonds TG, Ding H, Yuan X, Wei Q, Smith KS, Conway JA, Wieczorek L, Brown B, Polonis V, West JT, et al. Replication competent molecular clones of HIV-1 expressing Renilla luciferase facilitate the analysis of antibody inhibition in PBMC. Virology. 2010;408:1–13.
56. Ho C, Wu S, Amos JD, Colvin L, Smith SD, Wilks AB, Demarco CT, Brinkley C, Denny TN, Schmitz JE, et al. Transient compartmentalization of simian immunodeficiency virus variants in the breast milk of african green monkeys. J Virol. 2013;87:11292–9.
57. Gouy M, Guindon S, Gascuel O. SeaView version 4: a multiplatform graphical user interface for sequence alignment and phylogenetic tree building. Mol Biol Evol. 2010;27:221–4.

14

Computational analysis of envelope glycoproteins from diverse geographical isolates of bovine leukemia virus identifies highly conserved peptide motifs

Aneta Pluta[1*], Lorraine M. Albritton[2], Marzena Rola-Łuszczak[1] and Jacek Kuźmak[1]

Abstract

Background: Bovine leukemia virus (BLV) is a deltaretrovirus infecting bovine B cells and causing enzootic bovine leucosis. The SU or surface subunit, gp51, of its envelope glycoprotein is involved in receptor recognition and virion attachment. It contains the major neutralizing and CD4+ and CD8+ T cell epitopes found in naturally infected animals. In this study, we aimed to determine global variation and conservation within gp51 in the context of developing an effective global BLV vaccine.

Results: A total of 256 sequences extracted from the NCBI database and collected in different parts of the world, were studied to identify conserved segments along the *env* gene sequences that encode the gp51 protein. Using the MEME server and the conserved DNA Region module for analysis within DnaSP, we identified six conserved segments, referred to as A–F, and five semi-conserved segments, referred to as G–K. The amino acid conservation ranged from 98.8 to 99.8% in conserved segments A to F, while segments G to K had 89.6–95.2% conserved amino acid sequence. Selection analysis of individual segments revealed that residues of conserved segments had undergone purifying selection, whereas, particular residues in the semi-conserved segments are currently undergoing positive selection, specifically at amino acid positions 48 in segment K, 74 in segment G, 82 in segment I, 133 and 142 in segment J, and residue 291 in segment H. Each of the codons for these six residues contain the most highly variable nucleotides within their respective semi-conserved segments.

Conclusions: The data described here show that the consensus amino acid sequence constitutes a strong candidate from which a global vaccine can be derived for use in countries where eradication by culling is not economically feasible. The most conserved segments overlap with amino acids in known immunodeterminants, specifically in epitopes D–D′, E-E′, CD8+ T-cell epitopes, neutralizing domain 1 and CD4+ T-cell epitopes. Two of the segments reported here represent unique segments that do not overlap with previously identified antigenic determinants. We propose that evidence of positive selection in some residues of the semi-conserved segments suggests that their variation is involved in viral strategy to escape immune surveillance of the host.

Keywords: BLV, Deltaretrovirus, Env, Genetic variability, MEME, Conserved segments, Antigenic determinants, Positive selection

*Correspondence: aneta.pluta@piwet.pulawy.pl
[1] OIE Reference Laboratory for EBL, Department of Biochemistry, National Veterinary Research Institute, Pulawy, Poland
Full list of author information is available at the end of the article

Background

Bovine leukemia virus (BLV), which belongs to the *Deltaretrovirus* genus of the *Retroviridae* family, is the causative agent of Enzootic Bovine Leukosis (EBL) [1, 2]. The majority of BLV-infected cattle remain asymptomatic throughout their lives. However, about 5–10% of infected animals develop malignant tumors and 30% of infected individuals develop B-cell lymphocytosis characterized by extensive proliferation of CD5+ cells [3].

Similar to other retroviruses, the genome of BLV contains three major genes, *gag*, *pol* and *env*. The first two genes yield the virion structural proteins and the viral enzymes, including reverse transcriptase. The *env* gene encodes the envelope glycoprotein precursor (gp72), the viral membrane protein against which neutralizing antibodies would be expected to act. During synthesis, gp72 is cleaved into two associated components, a surface glycoprotein subunit of 51 kDa size (SU, gp51), implicated in receptor recognition and virion attachment, and a transmembrane glycoprotein subunit (TM, gp30) that induces the fusion of viral and cellular membranes necessary for virus to penetrate into host cell cytoplasm [4, 5].

The genetic variation in field isolates of BLV from naturally infected livestock represents a low mutation rate which appears to be less than that of other genera in *Retroviridae*. The genetic variability of BLV strains is typically analyzed by comparing the proviral sequences present in naturally infected cattle to the reference sequences of the FLK/BLV strain isolated from a BLV-infected foetal lamb kidney cell line (Accession no. M35242.1, USA) or to the field strain (Accession no. K02120, Japan). While the in vivo mutation rates measured on the *env* genes of BLV and lentiviruses are comparable at 0.009 and 0.0085 nucleotide changes per year, respectively, 81% of the nucleotide changes made by lentivirus reverse transcriptase (RT) are nonsynonomous whereas 50% of the BLV RT errors change the amino acid [6–9]. In vitro, the BLV RT is 10-fold more faithful than other retroviral polymerases and errs on average at an estimated one out of 20,800 nucleotides for an error rate of 4.8×10^{-6} nucleotides compared to $2.5–5.9 \times 10^{-4}$ for purified HIV-1 RT and 3.4×10^{-5} measured during single cycle HIV-1 infection, 5.9×10^{-5} for avian myeloblastosis RT, 3.3×10^{-5} for Moloney murine leukemia virus RT, and 1.2×10^{-5} nucleotides for avian spleen necrosis virus RT [9–12].

Genetic studies have determined that the level of variation in proviral nucleotide and amino acid sequences is very low among infected cattle from geographically different parts of the world [13–27]. Phylogenetic analysis of the *env* gene of BLV isolates from these different geographical locations identified ten genotypes, G1–G10 [16]. Even among these ten genotypes the variation is low and only twenty amino acids among the 515 residues in gp72 differ among all the genotypes combined.

The lower non-synonomous mutation rate in BLV infection and very low interstrain variation in the envelope glycoprotein were predicted to be advantageous in developing a BLV vaccine [6]. A number of B-cell and T-cell determinants have been identified on gp51. The antigenic sites are the G, H, and F conformational epitopes, the A, B-B′, D–D′ and E-E′ linear epitopes and three neutralization domains ND1, ND2 and ND3 involved in virus neutralization and syncytium inhibition [28–31]. Five T-cell epitopes (CD4+ T-cell epitope, CD8+ T-cell epitope, gp51N5, gp51N11 and gp51N12) located within gp51 have been shown to be immunologic targets of cytotoxic T lymphocytes (CTL) and involved in induction of proliferative responses in BLV infected cattle [31–33].

The variation of these promising B-cell and T-cell determinants across global isolates has not yet been determined. Here we apply computational analysis of gp51 nucleotide and deduced amino acid sequences from the 256 BLV *env* gene sequences in the NCBI database to investigate global variation and conservation within SU. The results revealed the presence of highly conserved segments as well as several semi-conserved segments. Together these analyses define a hierarchy of conservation among the sequences of the B-cell and T-cell determinants that will be useful in developing an effective global BLV vaccine.

Results

Identification and selection of segments using MEME and DnaSP analyses

To analyze BLV isolates collected in different parts of the world we mapped conserved segments along the *env* gene sequence that encodes the gp51 protein. Two hundred and fifty-six sequences encoding residues 1–267 of the 268 amino acids in the mature gp51 surface subunit of BLV isolates representative for all known genotypes (G1–G10) were extracted from the NCBI nucleotide database and used for this study. Sequences were aligned and the alignment submitted to Multiple Expectation Maximization for Motif Elicitation (MEME) to identify sequence patterns or segments that occur repeatedly in the aligned group of 256 sequences analyzed. The MEME server generated twenty blocks of nucleotide sequence that occur once in the gp51 coding sequence and have E-values consistent with a very low probability that the MEME motif (term used interchangeably with a segment) is randomly present in the 256 BLV *env* genes. Eleven blocks were selected for further analysis based on considerations of four criteria: (1) low probability that the repeated occurrence of the sequence segment was random within the aligned coding sequences of gp51; (2) degree of amino acid conservation; (3) degree of nucleotide conservation; and (4) the genetic distance exhibited within a MEME motif. The eleven MEME motifs range in length from six to nineteen amino acids and have an average length just under twelve amino acids.

Sequence Logos were generated to illustrate the consensus amino acid and nucleotide sequences of each MEME motif obtained using all BLV genotypes in the public databases (Fig. 1). The first six blocks identified by MEME analysis (referred to here as A–F) represent MEME motifs with E-values from 2.4e−1098 to 4.9e−1070. Blocks 16–20 (referred to here as G–K) showed MEME motifs with E-values from 4.4e−1020 to 4.6e−455. Blocks 7–15 were not further analyzed because they had intermediate characteristics. The amino acid conservation ranged from 98.8 to 99.8% in MEME motifs A to F. The remaining MEME motifs G to K had 89.6–95.2% conserved amino acid sequence.

The alignment was separately interrogated using the Conserved DNA Region module of DnaSP to identify segments of the nucleotide sequence conservation with the threshold for nucleotide conservation set at 80%. This analysis identified six statistically significant stable segments, with p value < 0.05 and nucleotide sequence conservation greater than or equal to 0.8 (80%). These segments were located exactly within the six blocks of segments A–F identified as nonrandom using MEME analysis. Segments G–J in MEME blocks 16–20 fell below the 80% nucleotide conservation threshold and their p values were not calculated (Fig. 1).

To estimate genetic distances for each of these MEME blocks p distance nucleotide substitution model in the

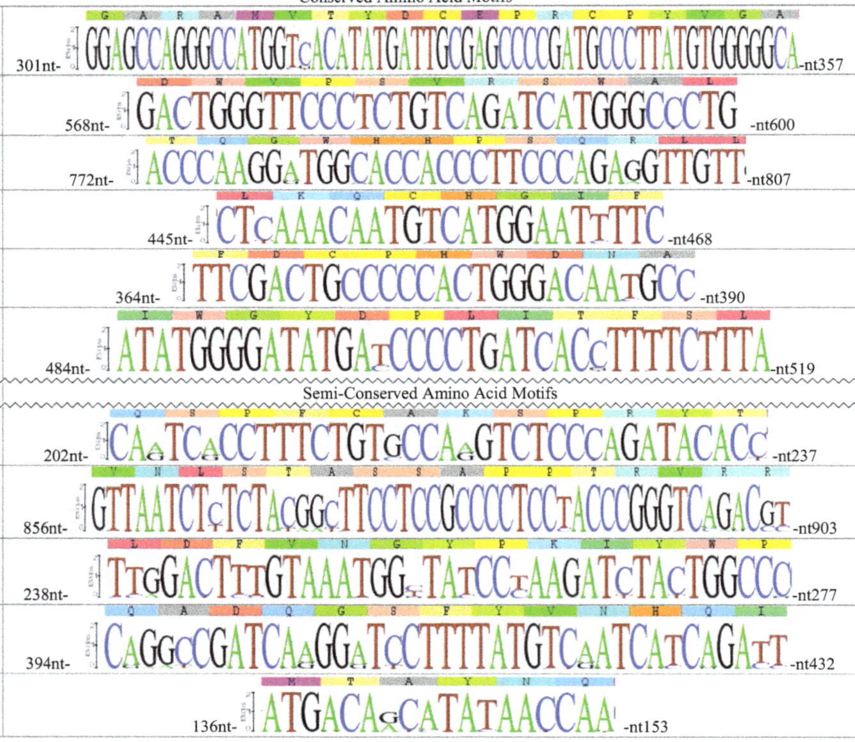

	MEME Motif Logos	MEME E-value	DnaSP Conservation	DnaSP p-value	Genetic distance nt (%)
	Conserved Amino Acid Motifs				
A	301nt- … -nt357	2.4e-1098	0.806	0.001	1.0
B	568nt- … -nt600	3.2e-1095	0.905	0.041	0.6
C	772nt- … -nt807	2.4e-1094	0.923	0.011	1.6
D	445nt- … -nt468	2.2e-1094	0.800	0.045	1.7
E	364nt- … -nt390	6.0e-1086	0.844	0.044	1.7
F	484nt- … -nt519	4.9e-1070	0.880	0.032	1.7
	Semi-Conserved Amino Acid Motifs				
G	202nt- … -nt237	4.4e-1020	b/t	n/a	6.0
H	856nt- … -nt903	8.2e-1016	b/t	n/a	5.8
I	238nt- … -nt277	8.2e-1009	b/t	n/a	7.0
J	394nt- … -nt432	2.3e-997	b/t	n/a	7.8
K	136nt- … -nt153	4.6e-455	b/t	n/a	6.7

Fig. 1 Sequence Logos of the predicted segments. WebLogo was used as described in Materials and Methods to generate a quantitative graphical representation of the 256 BLV nucleotide sequences in the multiple alignment from which MEME analysis identified conserved segments found in the gp51-encoding, 804 bp fragment of the *env* gene. The height of individual nucleotides within each Logo is proportional to the percentage of the 256 isolates in which that nucleotide occurred in that position. Nucleotides shown with maximum height were present at that position in all the isolates, whereas nucleotides with reduced height occurred in some but not all isolates. Nucleotides are numbered based on their position in the coding sequence of the envelope protein beginning with the "A" nucleotide of the start codon. Software parameters where chosen as followed: distribution of MEME motif occurrences (zero or one per sequence), number of different motifs (20), minimum motif width (15) and maximum motif width (60). DnaSP conservation values below the 80% or 0.800 threshold of conservation set for the analysis are indicated as b/t (below threshold) and their p values indicated as n/a (Not applicable). The genetic distance in each MEME block was calculated showing that MEME blocks A–F are conserved and MEME blocks G–K are highly variable. %, resulting values of the mean genetic distances were multiplied by 100. Detailed data about nucleotides less frequently observed at specific positions of the selected segments are shown in Additional file 2: Table S2

Molecular Evolutionary Genetics Analysis software (MEGA) version 5.2.2. was used. The genetic distance relates to divergence between any two strands of aligned DNA sequences, and it is expressed as the number of nucleotide differences per site. In this study we calculated the overall mean, which is the arithmetic mean of all individual pairwise distances between the sequences under study. The six conserved segments had genetic distances from 0.6 to 1.7%, while distance values from 5.8 to 7.8% characterized segments G–K (Fig. 1). Based on these findings taken together, we grouped segments A–F together as conserved in amino acid and nucleotide sequence and segments G–K as semi-conserved in amino acid sequence.

Evidence of purifying selection in segments A–F and positive selection in segments G–J

Genetic diversity within a coding sequence from a population of related sequences can be assessed by comparing the rates of synonymous and non-synonymous substitutions per site. An excess of non-synonymous mutations suggests selection for change, or positive selection, which is often the result of an antagonistic interaction between a virus and its host. This high genetic variation confers a fitness advantage to the pathogen in its attempt to evade host defenses. In contrast, an excess of synonymous mutations suggests that there is selection against change, or purifying selection, meaning the sequence is likely to be beneficial to the host. The negatively selected sites point to functional constraint, and could be used in vaccine design. The high degree of amino acid and nucleotide sequence conservation of segments A–F among the 256 BLV sequences suggested that these portions of the *env* gene are very stable during replication and might be undergoing negative or purifying selection associated with predominantly synonymous substitutions. Whereas, the nucleotide variation of segments G–J suggested the possibility that they contained sequences that are less stable and under positive selection. To examine this possibility, the frequency of non-synonymous (dN) and synonymous (dS) nucleotide substitutions per site within each segment was calculated and the ratio dN/dS was estimated.

Selection analysis revealed that the rates of synonymous nucleotide substitutions per site (dS) were higher than the non-synonymous rates (dN) within all analyzed segments showing a high probability of purifying selection. These differences were statistically significant as judged by Z-test (Z-test: $p < 0.0001$ for A-J segments and $p < 0.05$ for K segment). The dN/dS values were substantially less than 1 for all segments, also in agreement with the existence of strong purifying selection pressure. Nevertheless, dN/dS values noted for semi-conserved segments G–K were substantially higher than those noted for conserved segments A–F, suggesting the existence of putative sites undergoing positive but not purifying selection.

To test this possibility, sliding window analysis was performed and the rate of DNA polymorphism and divergence per codon was calculated. A standard indication of selective pressure is the ratio of dN/dS, where dN/dS $\sim$ 1 signifies neutral evolution or a balance in evolutionary pressure on that sequence, dN/dS < 1 indicates negative or purifying selection and a ratio > 1 indicates positive selection pressure [5]. The following segments with putative positive selection sites within particular semi-conserved segments were identified: 213–228 nt, 861–879 nt, 238–252 nt, 394–402 nt, 414–432 nt and 138–150 nt for segments: G–K, respectively, based on the ratio of synonymous sites (dS) to nonsynonymous sites (dN) within each codon (Fig. 2). While the majority of codons located in these segments had elevated dN/dS ratios that were nonetheless < 1, six codons in segments G–K had dN/dS ratios > 1 identifying them as the major sites where positive selection occurs. These were codons 74 in segment G, 291 in segment H, 82 in segment I, 133 and 141 in segment J and 48 in segment K (Table 1).

Refinement of the gp51 sequence segment map

In the process of evaluating the sequence segments (MEME motifs), we placed them on the consensus amino acid sequence generated in the initial MEME analysis. It was then apparent that two pairs of conserved MEME motifs, e.g., A and E, and D and F, and one pair of semi-conserved MEME motifs, e.g., G and I, were located very close to each other in the primary sequence. Any of these pairs might actually represent a larger segment that was split up because the maximum MEME motif width was set at 60 nucleotides or 20 amino acid codons in our initial MEME analysis. To determine if this was the case, the 256 gp51-encoding sequences were submitted a second time to the MEME server using a minimum MEME motif width of 60 nucleotides and a maximum width of 99 nucleotides (20–33 amino acid codons). Using these parameters, the output should cover the 268 codons of gp51 almost twice over (minimum length of 20 codons $\times$ 25 motif block output = 500 codons). Since all of the segments discovered in the initial analyses are shorter than the minimum 20 amino acid width in the new analysis, we expected that the original A-K MEME motifs would be included in the blocks from this new analysis only if they are part of longer conserved regions.

Six blocks were identified by MEME using the wider motif parameter and named MEME motifs I–VI (Table 2). E-values ranged from 6.7e−2042 to 7.5e−1028 and amino acid conservation ranged from 94.4 to 99.6%. Genetic distances ranged from 1.2 to 4.6%, however,

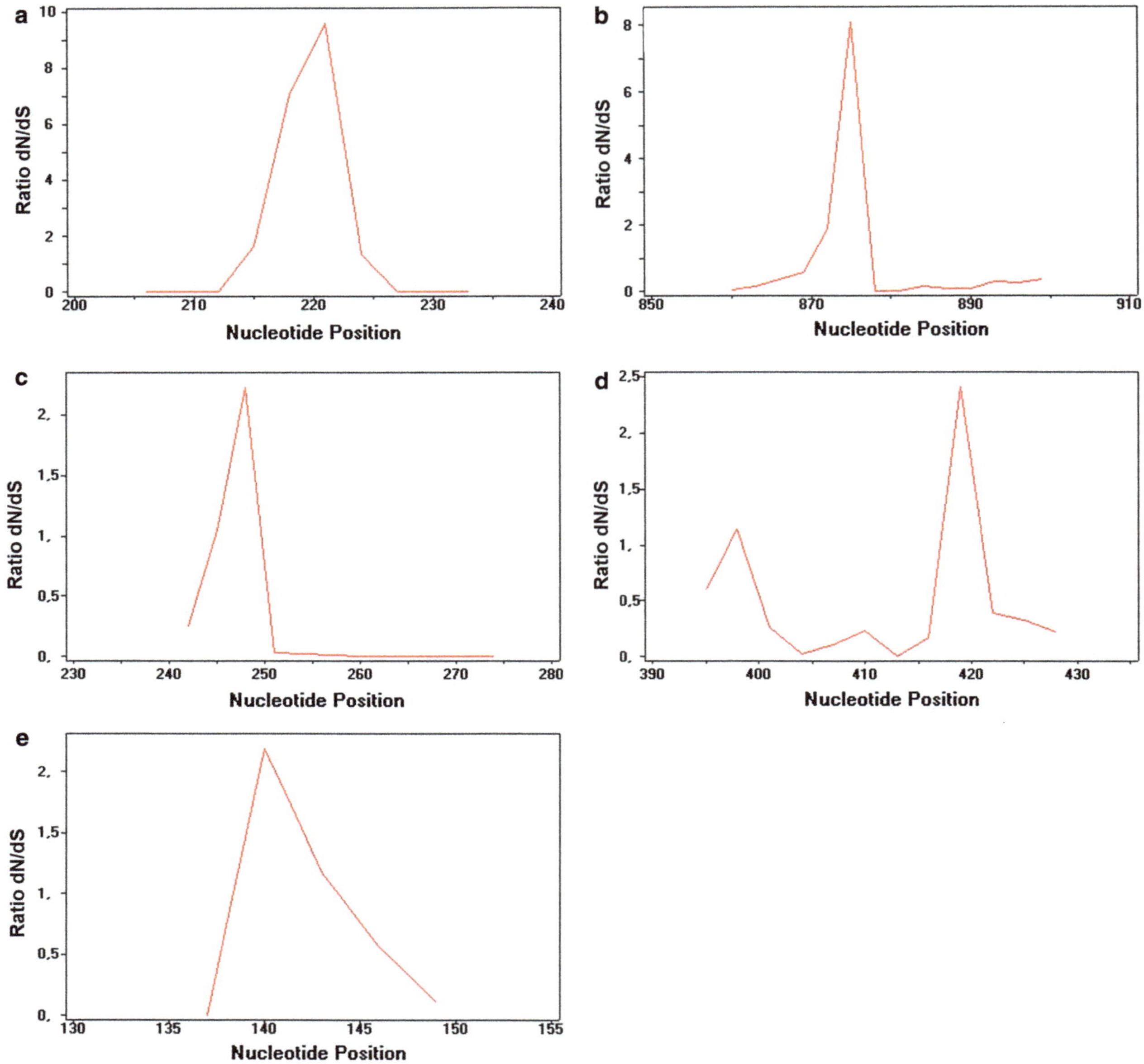

Fig. 2 Sliding window analysis based on the rate of DNA polymorphism and divergence in synonymous sites and non-synonymous sites. Values shown on the y-axis are the ratio of the rate of non-synonymous substitutions dN to the rate of synonymous substitutions dS at each nucleotide position calculated using the following sliding window options: window length: 9, step size: 3. The segments analyzed were: **a** 202–237 (segment G), **b** 856–903 (segment H), **c** 238–277 (segment I), **d** 394–432 (segment J), **e** 136–153 (segment K). Nucleotide positions were numbered beginning with the adenosine residue in the first position of the ATG start codon of the gp51 coding sequence

only MEME motifs I, IV and V had p values < 0.05 for nucleotide conservation indicating that these were statistically significant stable nucleotide segments. Eight short MEME motifs (A–G and I) are part of five of the longer MEME motifs (I–III and V–VI). Three of the longer MEME motifs contain adjacent shorter MEME motifs found in the initial analysis. Among the initial eleven MEME motifs discovered, H, J and K were not found to be part of longer MEME motifs. Long segments discovered in the second MEME and DnaSP analyses of the BLV gp51 *env* sequences are highly conserved at the amino acid level but less well conserved at the nucleotide level (Table 2).

Relationship of segments to known and predicted functional domains of gp51

To determine possible association of segments with functional domains within gp51 glycoprotein, the segment

Table 1 Computational analyses indicate that conserved MEME motifs G–K of gp51 are under positive selection

MEME motift[a]	dS[b]	dN[c]	Z-test	dN/dS ratio	Sites of positive selection[d]	p values[e]
A	0.03450	0.00629	< 0.0001	0.182	0	NA
B	0.03541	0.00253	< 0.0001	0.071	0	NA
C	0.08121	0.01129	< 0.0001	0.139	0	NA
D	0.09827	0.00000	< 0.0001	0.000	0	NA
E	0.09267	0.00133	< 0.0001	0.014	0	NA
F	0.09420	0.00000	< 0.0001	0.000	0	NA
G	0.11690	0.02867	< 0.0001	0.245	Lys^{74}, Arg^{74}	< 0.05
H	0.09437	0.04350	< 0.0001	0.461	Ala^{291}, Val^{291}, Thr^{291}	< 0.05
I	0.24183	0.01934	< 0.0001	0.080	Phe^{82}, Ser^{82}, Leu^{82}	< 0.05
J	0.14623	0.05945	< 0.0001	0.381	Val^{133}, Thr^{133}, Pro^{133}, Asp^{133}, Ala^{133}, Asp^{141}, Asn^{141}	< 0.05
K	0.10056	0.05619	< 0.050	0.559	Thr^{48}, Tyr^{48}, Val^{48}, Ile^{48}	< 0.05

[a] MEME motifs A–F are highly conserved at the amino and nucleotide sequence levels, while MEME motifs G–K are conserved at the amino acid but not the nucleotide level

[b] dS, rate of synonymous substitutions

[c] dN, rate of non-synonymous substitutions

[d] Positions of positive selection sites are the amino acid numbers beginning from the N-terminus of the precursor

[e] *p*-values were calculated for the positive selection sites; NA, not applicable

Table 2 Long conserved MEME motifs found in BLV gp51 *env* sequences

Block	MEME Motif	MEME E-value	Amino acid conservation (%)	Genetic distance (%)	Nucleotide conservation	p value
1.	I	6.7e−2042	98.0	1.2	0.813	0.007
2.	II	8.4e−2032	99.5	2.3	0.800	0.063
3.	III	9.7e−2032	99.1	3	0.800	0.056
4.	IV	1.8e−2099	99.6	1.3	0.807	0.037
5.	V	5.4e−2042	97.3	2.5	0.818	0.029
6.	VI	7.5e−1028	94.4	4.6	b/t	n/a

b/t, value is below threshold of 80% conservation or 0.80

n/a, not applicable

%, resulting values of mean genetic distances were multiplied by 100

sequences were aligned with the nucleotide and amino acid MEME consensus sequence (Fig. 3). Short segment K is located near the N-terminus of gp51. While this segment encompasses a single residue of conformational epitope G, represented by Tyr-48, it is otherwise unique in its lack of association with known antigenic determinants and function. The other segments overlap with previously identified functional and immune recognition sequences. Conserved segments A–F are located within the central and the C-terminal part of gp51 protein between amino acids 105–280, overlapping mainly linear and T-cell epitopes. In addition, segment A overlaps with the C-terminal portion of the B cell epitope (gp51p16-19) described by Bai et al. in 2015 [33].

While the three-dimensional structure for a portion of TM proteins from BLV (gp30) and the human deltaretrovirus HTLV-1 (gp21) have been solved, no structure is available for gp51 or the SU (gp46) from HTLV-1. However, the envelope glycoproteins of both retroviruses belong to the cd09948:Ebola_RSV-like_HR1-HR2 and pfam00429:TLV_coat ENV polyprotein (coat polyprotein) families of glycoproteins which include the envelope glycoproteins of HTLV-1, BLV, the avian retroviruses Rous sarcoma virus and Avian Leukosis Virus subgroup J, Friend Murine Leukemia Virus (Fr57 MLV) and Feline leukemia virus type B (FeLV-B) and Ebola virus [34, 35]. Members of these families contain related functional domains of similar sequence and there is evidence that they use a shared mechanism for attachment and membrane fusion during virus entry based on isomerization of a disulphide bond between the attachment and fusion subunits [35]. Structures from three members of the families have been solved, e.g., that of the established receptor binding domains of gammaretroviruses Fr57 MLV

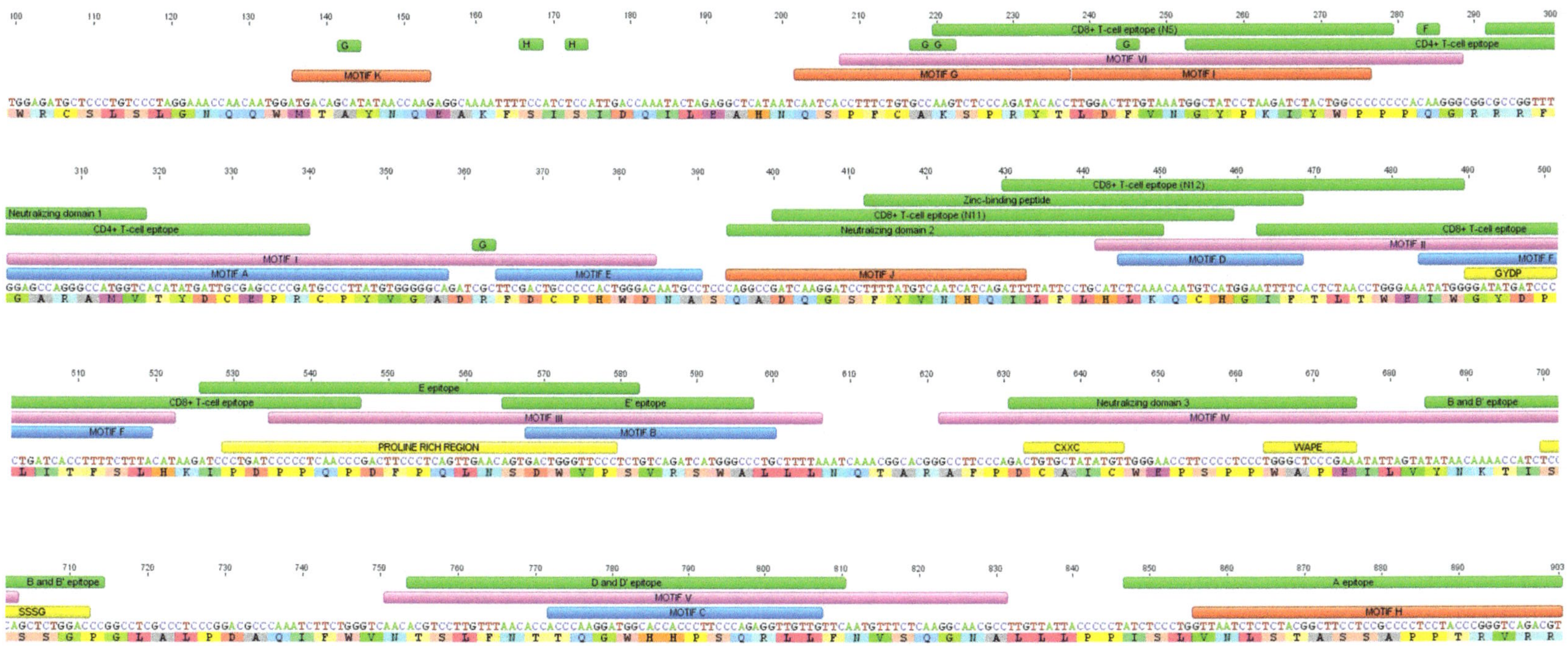

Fig. 3 Distribution of segments along the surface glycoprotein gp51. The labeled green bars in the upper part of figure refer to the coding sequences of antigenic determinants and to the Zinc-binding peptide. MEME motifs obtained by the initial MEME analysis with a motif width range set at 5–20 amino acids are indicated by blue (A–F) and red (G–K) bars above the consensus sequence. MEME motifs I–VI defined in the second MEME analysis using a motif width range of 20–33 amino acids are indicated by purple bars. Epitopes B, B′, E, and E′ (linear), G, H and F (conformational), ND1, 2, and 3 (neutralization domains), and B-cell and T-cell determinants are indicated by green bars. Previously reported conserved functional sequence segments and the putative proline rich region are indicated by yellow bars above the sequence. The sequence of mature gp51 is shown. Nucleotides are numbered based on their position in the coding sequence of the envelope protein beginning with the "A" nucleotide of the start codon

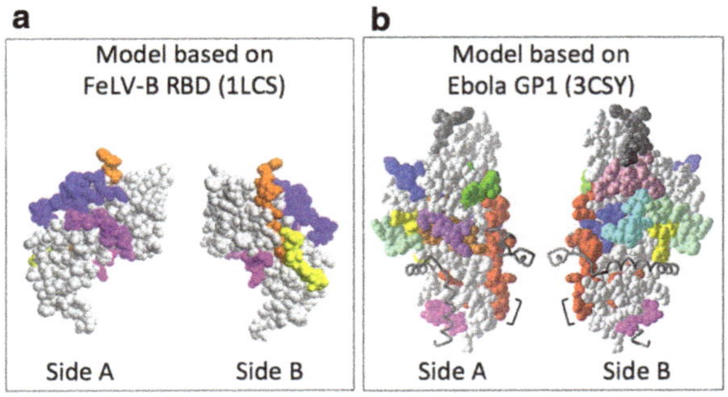

Fig. 4 Representation of the conserved and semi-conserved segments on 3 D glycoprotein structure models of gp51. Homology models are shown as space-filled images of opposite sides arbitrarily named side A and side B. Each is based on alignment of residues 1–156 of mature gp51 with the analogous sequences of FeLV-B RBD (accession number 1LCS), or alignment of residues 1–298 with the amino acid sequence of Ebola Zaire GP1 (accession number 3CSY). The model of the gp51 RBD contain conserved segments A (blue), E (violet), D (yellow), and F (orange), and semi-conserved segments K (magenta), G (cyan), I (light green), and J (light blue). The model of the entire gp51 based on Ebola GP1 contains these segments plus segments included in residues 156–298, specifically conserved segments B (green), IV (red), and C (pink), and semi-conserved segment H (gray)

and FeLV-B, and of the ectodomain of the mature envelope glycoprotein from Ebola virus.

Using homology modeling based on the gammaretroviral receptor binding domain (RBD) structures, we placed the conserved segments within residues 1- 156 of mature gp51 in three-dimensional structural models. A previously published BLV model based on Fr57 MLV structure was supported by natural variants in BLV envelope protein, but was recognized as flawed because there were several small gaps in the main peptide chain of the BLV RBD that the model could not resolve [34]. The subsequent publication of the FeLV-B RBD structure (1LCS) provided a second template structure upon which the BLV RBD could be modeled. Figure 4a shows a model based on the FeLV-B structure which is reported here for the first time. This new model resolves the gaps in the BLV RBD peptide chain that were the weakness of the previous model. The predicted locations of surface-exposed residues in the conserved and semi-conserved segments identified in the analyses reported here are shown in colors by motif segment.

While highly speculative, a model based on the alignment of conserved domains previously identified in the cd09948:Ebola_RSV-like_HR1-HR2 and pfam00429:TLV_coat ENV polyprotein (coat polyprotein) families [35] was also generated. Since the Ebola glycoprotein GP structure is the only structure that includes more than the RBD for a member of these families, its use as the model template afforded an opportunity to model the six segments which are carboxy-terminal to residue 156 of mature gp51 and lie outside the alignments with the RBD of FeLV-B and Fr57 MLV (Fig. 4b). It also could provide insight into segments that are surface-exposed in the RBD model but which are buried in gp51 by residues 197–267 of gp51. This model predicts that the majority of residues within segments A–F are exposed on the surface of the gp51 glycoprotein, including the following residues: 106-VTYDCEPRCPY-116 of segment A, 194-PSVRS-197 of segment B, 256-TQGWHHPSQRLL-269 of segment C, as well 164-GYDP-167 of segment F. Conversely, conserved segments D and E as well as residues 198-WALLLNQ-204 in segment B are predicted to be buried in a beta-strand within the core of the structure in the model shown in Fig. 4b.

Discussion

Since the discovery of BLV as the causative agent in EBL in 1969 [36], efforts at global elimination of BLV have centered on testing and removal of infected animals from herds as the preferred method of eradication of the disease. This strategy has achieved eradication in some countries particularly those with a low initial prevalence of BLV-infected cattle. However, EBL remains endemic in locations where the high prevalence rate is an economic barrier to the culling strategy and large numbers of cattle exhibit the most common clinical manifestation of BLV infection, persistent lymphocytosis. These locations include North and South America, Middle Eastern and Southeast Asian nations and the Caribbean region (OIE World Animal Health Organization from the years 2015 and 2016). The genetic selection of cattle resistant to BLV infection might be beneficial for minimizing economic losses by reducing disease severity but not for eradication because it currently appears that complete resistance to BLV cannot be obtained by selection of cattle with BLV-resistant alleles of the DRB3.2 gene [37, 38].

A vaccine that protects against BLV transmission would provide an economical alternative method of eradication in countries with high prevalence rates, but the prospect for a preventive or therapeutic anti-deltaretrovirus vaccine are uncertain [39, 40]. One way to make an anti-BLV vaccine would be a peptide-based strategy consisting of highly conserved immunogenic peptide constructs capable of eliciting broadly neutralizing antibody responses [40–42].

The initial goals of our study were to determine a consensus gp51 amino acid sequence for BLV isolates from endemic locations all over the world, identify which, if any, of the known BLV antigenic determinants within glycoprotein gp51 are conserved globally within and between all known genotypes, and to gain insights into the evolution of BLV gp51 during natural in vivo infection particularly with respect to known neutralizing domains and antibody and T-cell epitopes. Toward this objective, all primary isolate sequences that contain at least the coding sequence of residue 1–297 of the 298 residues in mature gp51 were mined from the NCBI public database and submitted to computational analysis from which a consensus amino acid sequence was obtained (Additional file 1: Table S1).

Based on the results reported here, we draw five conclusions. First, we conclude that the consensus amino acid sequence derived from MEME analysis represents the primary consensus sequence of global BLV gp51. The set of 256 proviruses used to generate the consensus sequence represents nucleotide sequences isolated from infected cattle in twenty-one countries and includes isolates from all of the ten known BLV genotypes.

Second, we propose to classify the new segments exhibiting DNA conservation greater than or equal to 80% and genetic distances ranging less than or equal to 3.0% as conserved, and those showing less than 80% DNA conservation and a somewhat higher degree of variability greater than or equal to 4.6% as semi-conserved. Based on these criteria, gp 51 contains six highly conserved segments and five semi-conserved segments of five to twenty residues in length that are distributed throughout gp51. Two pairs of conserved segments (A and E, and D and F) were found to be part of larger conserved segments I and II, respectively, when the analysis parameters were set for segments of longer length (20–33 amino acids) than in our initial analyses. Similarly, semi-conserved segments G and I were found to be part of a larger semi-conserved segment VI.

Third, we conclude that among the segments reported here, only the conserved ones exhibited evidence of purifying selection of their sequences, e.g., dN/dS ratios of substantially less than 1 ($\leq$ 0.182). These relatively low levels of variability reflect a general feature of BLV genomes, that is, its characteristically high degree of conservation among different strains. Low variability among strains is also characteristic of other viruses belonging to the genus *Deltaretrovirus* [6, 12]. It was recently shown that the homology of the *env* gene was 94.5–97.7% amongst BLV isolates representative of the ten genotypes [16]. This level of variability is much lower than that reported for conserved regions identified in studies by others of human cytomegalovirus or herpes simplex virus type 1 [43, 44].

We favor that the purifying selection during evolution of the conserved segments suggests that these segments may contain critical core structure sequences, or sequences that interface with the TM gp30, or codons for the residues that interact with the BLV entry receptor. The homology models predict that part of the core structure of the RBD is formed by conserved segments D and F. In the models these segments form adjacent beta-strands previously proposed to face inward toward the trimer interface [35]. The presence of the previously recognized small motif CXXC in large conserved segment IV identifies this segment as containing residues in the gp51:gp30 heterodimer interface. In native envelope glycoprotein on virions, one of the cysteine residues in CXXC motif of gp51 is covalently bound to a cysteine in the conserved CX_6CC motif of gp30 and makes an essential contribution to the association of gp51:gp30 heterodimers [45, 46]. In addition, the C-terminal domain of HTLV-1 surface subunit has been reported to be involved in transducing the envelope activation signal upon receptor binding [47, 48], suggesting that conserved segment C in the analogous domain of gp51 may also participate in activation of membrane fusion.

With respect to residues that interact with the BLV entry receptor, neither the receptor nor the gp51 residues with which it interacts are known. However the location of the BLV RBD within the amino-terminal 156 residues of the mature BLV gp51 subunit is supported by homology to the established RBD of the HTLV-1 and the gammaretroviruses [35, 49, 50] and the positions of the proline rich region and the conserved CXXC motif in gp51 [46]. Assuming that all strains of BLV use a common entry receptor, it is most likely that the receptor binding site lies within a highly conserved segment. Four conserved (A, D, E and F) and four semi-conserved segments (G, I, K and J) lie within the first 156 residues (Fig. 3).

We favor that the binding site for the receptor includes conserved segments A and E for the following reasons. First, the receptor binding site is likely to be within or adjacent to a neutralizing domain. There are two neutralizing domains within residues 1–156, ND1 and ND2. Two segments overlap with these domains: segment A overlaps with the C-terminal two-thirds of ND1 and segment J overlaps with the N-terminal two-thirds of ND2. Second, the receptor binding site is likely to be located on the analogous surface to that of the established sites on the Fr57 MLV and Fe-LV-B RBDs. The homology models of gp51 based on these two RBD structures predict that segment A lies on the analogous surface whereas all but the first three residues of segment J are predicted to be buried within the core of the model. Lastly, while segment E is not included in a known neutralizing domain, the homology models predict that it lies beside segment A and that together they form a convex surface comparable in size and predicted location to the known receptor binding sites of Fr57 MLV and FeLV-B. In agreement with this last prediction by the models, analysis for segments of 20–33 amino acid length identified segments A and E as belonging to a single long conserved segment I.

Fourth, we conclude that amino acid positions 48 in segment K, 74 in segment G, 82 in segment I, 133 and 142 in segment J, and residue 291 in segment H are undergoing positive selection. Each of the codons for these six particular residues contains the most highly variable nucleotides within their respective semi-conserved segments. We favor that the genetic evidence of positive selection in these residues is consistent with viral evolution to escape immune surveillance as judged by the location of each of the six residues within known epitopes and neutralizing domains. Residues 48, 74 and 82 comprise three of the four amino acids in conformational epitope G, residues 133 and 142 are located in ND2 (which overlaps with the Zinc binding peptide), and residue 291 is in epitope A.

It has been postulated that some divergent residues within conserved segments may be under positive selection [32, 51]. Zhao and coworkers investigated possible selection over the entire glycoprotein sequence from forty-four unique isolates and found evidence of negative selection in the coding sequences for most of the known functional domains. However, those authors found evidence of positive selection only in the immunostimulatory conformational epitope G and linear epitopes D, D′ and concluded that these segments and not others were under selection for immune escape. Using a different approach to study selection in natural BLV infection, that is, analysis of individual conserved and semi-conserved segments instead of analysis of the whole glycoprotein, we confirmed that residues in conformation epitope G are under positive selection but found no evidence of positive selection in segment C which comprises a large portion of linear epitopes D, D′. In contrast to their finding of no evidence for positive selection of residues within other segments of gp51, we found evidence that ND2 (which includes segment J) and epitope A at the C-terminal end of gp51 (which includes segment H) are undergoing positive selection. We believe that the difference between our reports attributes to the segment-based analysis reported here being a more sensitive method for identifying rare residues of BLV that are under selection but in otherwise highly conserved segments.

Lastly, we conclude that the consensus amino acid sequence constitutes a strong candidate from which a universal vaccine can be derived for use in countries where eradication by culling is not economically feasible. Two of the MEME motifs represent unique segments that are not contained within previously identified immune determinants. Based on their location relative to that of known functional and structural domains in other retroviral SU, we propose that one (segment K) of these newly identified segments is involved in the folding of native gp51 and the other (segment E) is a strong candidate for part of the receptor binding sequence. Notably, none of the conserved or semi-conserved segments include residues in conformational epitope H. Taken together the results suggest that globally dispersed variations in segments E and K and in epitope H sequences make them poor candidates for use in a peptide vaccine strategy. In contrast, highly conserved segment A, which is another strong candidate for the receptor binding site, contains previously identified neutralizing and T- cell determinants that suggest it is a good candidate for use in a peptide vaccine, one capable perhaps of eliciting neutralizing antibody responses.

All of the B-cell and T-cell determinants and neutralizing domains, and all but one (H) of the antigenic epitopes contain and overlap with the new segments reported here. The new semi-conserved segments overlap with conformation epitopes G and F, linear epitopes A and B-B′, neutralizing domains 2 and 3, CD4+ T-cell epitopes N5 and N11. Of the three amino acid differences between the consensus sequence and that of strain FLK, two residues (74 and 82) are in the CD8+ T-cell epitope N5 and in conformational epitope G and the third is in the D, D′ epitope. This low level of difference is in line with the unusually high fidelity of the BLV polymerase and high level of conservation seen across global isolates. The most conserved segments encompass amino acids in epitopes D–D′, E-E′, two CD8+ T-cell epitopes, neutralizing domain 1 and two CD4+ T-cell epitopes, suggesting that their consensus sequence has limited global variation and are prime candidates for development of an globally effective peptide vaccine.

Methods

Sequence motif discovery

MEME version 4.10.0 [52], a hidden Markov model based software, and DnaSP version 5.10 software [53] were used to identify segments that occur in a group of 256 sequences extracted from the NCBI nucleotide database and represent all known BLV *env* sequences. Seventy-one proviruses were from three East Asian countries (Myanmar, South Korea and Thailand), sixty-three were from six South American countries (Argentina, Boliva, Brazil, Peru, Paraguay and Uruguay), thirty-nine from Russia, thirty-four from Japan, twenty-six from four European countries (Belgium, Italy, Moldova and Poland), eighteen from three North American sites (Caribbean, Costa Rico and the United States), and five proviruses from Australia, Iran and the Ukraine combined (Additional file 1: Table S1). Sequences derived from PCR-amplification of proviruses using the nested primer sets related to those introduced by Fechner et al. in 1997 were excluded from the analysis because they comprise fragments of < 500 bp and encode only the central half of gp51 [54]. The coding sequence for the 33 amino acid signal peptide was also excluded from analysis because the majority of the sequences in the database that encode the entire mature gp51 are derived from an 804 bp restriction fragment that does not contain signal peptide coding sequence.

The following steps were performed during the MEME and DnaSP analysis. First, a multiple sequence alignment was generated using ClustalW implemented in Geneious Pro 5.5.9 Software (Biomatters Ltd). Second, the aligned sequences were submitted to the MEME server (http://meme-suite.org/tools/meme). The MEME analysis was run in zoops (zero or one per sequence) mode with the maximum number of motifs set at 20. The optimal motif width range was set between 15 and 60 nucleotides (*i.e.*, 5 and 20 amino acids in length) to reflect the length of the synthetic peptides commonly used in epitope mapping as well as the typical number of amino acid (aa) residues identified in antigenic determinants. The E-values are the estimated log likelihood ratio of a sequence pattern calculated by MEME analysis as a measure of how different each nucleotide in the segment (motif) is from an n-order Markov model matrix of the same width but randomized sequence. A segment with an E-value below $e^{-9.2}$ has an estimated < 0.0001 probability that its sequence occurs randomly within the aligned input sequences. Third, the alignment was separately imported into the Conserved DNA Region module implemented in DnaSP Ver. 5.10.01 software and the minimum conservation threshold set at 80%. Lastly, the combined analyses were used to identify conserved segments of interest in the *env* genes.

Weblogo was then used to generate a quantitative graphical representation of the segments and sequence alignments. To estimate the level of nucleotide sequence divergence for conserved and variable segments, maximum composite likelihood model analysis (MEGA 5.2.2) was conducted as described [55]. Synonymous and non-synonymous sites were identified using the DnaSP 5.0. Statistical analyses identifying synonymous and non-synonymous sites was performed using STATISTICA ver. 10 (StatSoft, Dell Software, USA). Specifically, the number of synonymous substitutions per synonymous site (dS), the number of non-synonymous substitutions per non-synonymous site (dN) and their variances were calculated and applied to a selection Z-test to evaluate whether the segments were under positive or purifying selection. For comparison of two paired, nonparametric groups of data points, dN versus dS, a Wilcoxon matched pairs test was used to calculate the *p* values. Based on the results of the first analyses, the 256 member sequence alignment was submitted to the MEME server again using zoops mode with the maximum number of motifs set at 25 and the optimal motif width range set between 60 and 99 nucleotides (i.e., 20 and 33 amino acids in length) and the motif block output was subjected to further analysis as described above.

Molecular modeling

The consensus amino acid sequence for residues 3–156 comprising the putative RBD of mature SU sequence derived in the MEME analysis was aligned with the analogous extracellular amino acid sequence of the two gammaretroviral SU whose crystal structure has been solved, e.g., residues 9–236 of the mature SU from Fr57 MLV isolate 57 [56] and residues 4–207 of the mature SU from FeLV-B virus [57]. Each alignment was submitted for molecular modeling [58] using the 2.0 Å structure of the Fr57 MLV RBD (1AOL) and the 2.5 Å structure of FeLV-B RBD (1LCS) as templates. Noting that the BLV envelope glycoprotein is a member of the HTLV-1 superfamily of envelope glycoproteins [35], residues downstream of BLV gp51 residue 156 were modeled based on the known structure of the only member of the HTLV-1 superfamily for which structure of the analogous residues downstream of the RBD is available, e.g., GP of the Zaire strain of Ebola virus. Alignments of the consensus BLV gp51 sequence and that of Ebola virus GP were submitted for molecular modeling [58] using the 3.4 Å structure of the pre-fusion GP1-GP2 trimer (3CSY) [59] and the 3.3 Å structure of the host-primed GP1 and GP2 complexed with the virus binding domain of its host cell fusion receptor Niemann-Pick C1 (5HJ3) [60]. Visualization of the predicted structures and figure preparation were performed using the MolSoft ICM Browser. Conserved and semi-conserved segments were visualized on structure models of the envelope protein.

Additional files

Additional file 1: Table S1. Accession numbers and country or region of origin of the sequences analyzed in the study. §—direct submission to GenBank.

Additional file 2: Table S2. The conserved and semi-conserved segments in BLV envelope glycoprotein. A white color reflect segments highly conserved at the amino and nucleotide sequence levels, dark yellow color reflect segments conserved at the amino acid level. # The sequences are from MEME output and they reflect the nucleotides that are most frequent at these positions. For example, having a T in the output sequence does not mean that all gp51 *env* segments of BLVs will have a T nucleotide at that position. A [] indicates more than one nucleotide is frequent at this position. A "-" horizontal dashes between nucleotides denote the next codon. *The amino acid residues that are most frequent at these positions.

Authors' contributions

AP designed and performed bioinformatic and statistical analyses. LMA performed molecular modeling. AP, LMA analyzed and interpreted the data and the results. AP, LMA wrote the manuscript. All authors helped edit the paper. All authors read and approved the final manuscript.

Author details

[1] OIE Reference Laboratory for EBL, Department of Biochemistry, National Veterinary Research Institute, Pulawy, Poland. [2] Department of Microbiology, Immunology and Biochemistry, College of Medicine, The University of Tennessee Health Science Center, Memphis, TN, USA.

Acknowledgements

The authors would like to acknowledge the funding of this work: this research was funded by KNOW (Leading National Research Centre) Scientific Consortium "Healthy Animal- Safe Food", decision of Ministry of Science and Higher Education of Poland No. 05-1/KNOW2/2015.

Competing interests

The authors declare that they have no competing interests

Funding

Not applicable

References

1. King AMQ, Adams J, Carstens EB, Lefkowitz EJ. Virus Taxonomy: classification and nomenclature of viruses: ninth report of the ICTV. Amsterdam: Elsevier/Academic Press; 2012. p. 1344.
2. Kettmann R, Burny A, Callebaut I, Droogmans L, Mammerickx M, Willems L, Portetelle D. Bovine leukemia virus. The retroviridae. New York: Plenum Press; 1994. p. 39–81.
3. Ghysdael J, Bruck C, Kettmann R, Burny A. Bovine Leukemia virus. Curr Top Microbiol Immunol. 1984;112:1–19.
4. Temin HM. Origin and general nature of retroviruses. In: Levy J, editor. The retroviridae. New York: Plenum Press; 1992. p. 1–18.
5. Zhao X, McGirr KM, Buehring GC. Potential evolutionary influences on overlapping reading frames in the bovine leukemia virus pXBL region. Genomics. 2007;89:502–11.
6. Willems L, Thienpont E, Kerkhofs P, Burny A, Mammerickx M, Kettmann R. Bovine leukemia virus an animal model for the study of intrastrain variability. J Virol. 1993;67:1086–9.
7. Burns DP, Desrosiers RC. Selection of genetic variants of simian immunodeficiency virus in persistently infected rhesus monkeys. J Virol. 1991;65(4):1843–54.
8. Overbaugh J, Rudensey LM, Papenhausen MD, Benveniste RE, Morton WR. Variation in simian immunodeficiency virus env is confined to V1 and V4 during progression to simian AIDS. J Virol. 1991;65(12):7025–31.
9. Johnson PR, Hamm TE, Goldstein S, Kitov S, Hirsch VM. The genetic fate of molecularly cloned simian immunodeficiency virus in experimentally infected macaques. Virology. 1991;185(1):217–28.
10. Preston BD, Poiesz BJ, Loeb LA. Fidelity of HIV-1 reverse transcriptase. Science. 1988;242(4882):1168–71.
11. Roberts JD, Bebenek K, Kunkel TA. The accuracy of reverse transcriptase from HIV-1. Science. 1988;242(4882):1171–3.
12. Mansky LM, Temin HM. Lower mutation rate of bovine leukemia virus relative to that of spleen necrosis virus. J Virol. 1994;68:494–9.
13. Camargos MF, Pereda A, Stancek D, Rocha MA, dos Reis JK, Greiser-Wilke I, Leite RC. Molecular characterization of the env gene from Brazilian field isolates of Bovine leukemia virus. Virus Genes. 2007;34(3):343–50.
14. Rodriguez SM, Golemba MD, Campos RH, Trono K, Jones LR. Bovine leukemia virus can be classified into seven genotypes: evidence for the existence of two novel clades. J Gen Virol. 2009;90:2788–97.
15. Hemmatzadeh F. Sequencing and phylogenetic analysis of gp51 gene of bovine leukaemia virus in Iranian isolates. Vet Res Commun. 2007;31(6):783–9.
16. Polat M, Takeshima SN, Hosomichi K, Kim J, Miyasaka T, Yamada K, Arainga M, Murakami T, Matsumoto Y, Barra Diaz V, Panei CJ, González ET, Kanemaki M, Onuma M, Giovambattista G, Aida Y. A new genotype of bovine leukemia virus in South America identified by NGS-based whole genome sequencing and molecular evolutionary genetic analysis. Retrovirology. 2016;13(1):4.
17. Moratorio G, Obal G, Dubra A, Correa A, Bianchi S, Buschiazzo A, Cristina J, Pritsch O. Phylogenetic analysis of bovine leukemia viruses isolated in South America reveals diversification in seven distinct genotypes. Arch Virol. 2010;155(4):481–9.
18. Dube S, Abbott L, Dube DK, Dolcini G, Gutierrez S, Ceriani C, Juliarena M, Ferrer J, Perzova R, Poiesz BJ. The complete genomic sequence of an in vivo low replicating BLV strain. Virol J. 2009;6:120.
19. Lee E, Kim EJ, Ratthanophart J, Vitoonpong R, Kim BH, Cho IS, Song JY, Lee KK, Shin YK. Molecular epidemiological and serological studies of bovine leukemia virus (BLV) infection in Thailand cattle. Infect Genet Evol. 2016;41:245–54.
20. Yang Y, Kelly PJ, Bai J, Zhang R, Wang Ch. First molecular characterization of bovine leukemia virus infections in the Caribbean. PLoS ONE. 2016;11(12):e0168379.
21. Polat M, Ohno A, Takeshima SN, Kim J, Kikuya M, Matsumoto Y, Mingala CN, Onuma M, Aida Y. Detection and molecular characterization of bovine leukemia virus in Philippine cattle. Arch Virol. 2015. https://doi.org/10.1007/s00705-014-2280-3.
22. Matsumura K, Inoue E, Osawa Y, Okazaki K. Molecular epidemiology of bovine leukemia virus associated with enzootic bovine leukosis in Japan. Virus Res. 2011. https://doi.org/10.1016/j.virusres.2010.11.005.
23. Yang Y, Fan W, Mao Y, Yang Z, Lu G, Zhang R, Zhang H, Szeto C, Wang C. Bovine leukemia virus infection in cattle of China: association with reduced milk production and increased somatic cell score. J Dairy Sci. 2016. https://doi.org/10.3168/jds.2015-10580.
24. Rola-Łuszczak M, Pluta A, Olech M, Donnik I, Petropavlovskiy M, Gerilovych A, Vinogradova I, Choudhury B, Kuźmak J. The molecular characterization of bovine leukaemia virus isolates from Eastern Europe and Siberia and its impact on phylogeny. PLoS ONE. 2013. https://doi.org/10.1371/journal.pone.0058705.
25. Balić D, Lojkić I, Periškić M, Bedeković T, Jungić A, Lemo N, Roić B, Cač Z, Barbić L, Madić J. Identification of a new genotype of bovine leukemia virus. Arch Virol. 2012. https://doi.org/10.1007/s00705-012-1300-4.
26. Pluta A, Rola-Łuszczak M, Kubiś P, Balov S, Moskalik R, Choudhury B, Kuźmak J. Molecular characterization of bovine leukemia virus from Moldovan dairy cattle. Arch Virol. 2017. https://doi.org/10.1007/s00705-017-3241-4.
27. Lee AJ, Kim EJ, Joung HK, Kim BH, Song JY, Cho IS, Lee KK, Yeun-Kyung Shin YK. Sequencing and phylogenetic analysis of the gp51 gene from Korean bovine leukemia virus isolates. Virol J. 2015. https://doi.org/10.1186/s12985-015-0286-4.
28. Bruck C, Mathot S, Portetelle D, Berte C, Herion P, Burny A. Monoclonal antibodies define eight independent antigenic regions on the bovine leukemia virus (BLV) envelope glycoprotein gp51. Virology. 1982;122(2):342–52.
29. Portetelle D, Dandoy C, Burny A, Zavada J, Siakkou H, Gras-Masse H, Drobecq H. Tartar A Synthetic peptides approach to identification of epitopes on bovine leukemia virus envelope glycoprotein gp51. Virology. 1989. https://doi.org/10.1016/0042-6822(89)90038-x.
30. Callebaut I, Burny A, Krchnák V, Gras-Masse H, Wathelet B, Portetelle D. Use of synthetic peptides to map sequential epitopes recognized by monoclonal antibodies on the bovine leukemia virus external glycoprotein. Virology. 1991;185(1):48–55.
31. Callebaut I, Voneche V, Mager A, Fumiere O, Krchnak V, Merza M, Zavada J, Mammerickx M, Burny A, Portetelle D. Mapping of B-neutralizing and T-helper cell epitopes on the bovine leukemia virus external glycoprotein gp51. J Virol. 1993;67:5321–7.
32. Zhao X, Buehring GC. Natural genetic variations in bovine leukemia virus envelope gene: possible effects of selection and escape. Virology. 2007;366:150–65.
33. Bai L, Takeshima SN, Isogai E, Kohara J, Aida Y. Novel CD8(+) cytotoxic T cell epitopes in bovine leukemia virus with cattle. Vaccine. 2015. https://doi.org/10.1016/j.vaccine.2015.10.128.
34. National Center for Biotechnology Information, U.S. National Library of Medicine 8600 Rockville 2017. https://www.ncbi.nlm.nih.gov/Structure/cdd/cddsrv.cgi?uid=306850 and https://www.ncbi.nlm.nih.gov/Structure/cdd/cddsrv.cgi?hslf=1&uid=cd09948&#seqhrch. Accessed 1 July 2017.
35. Johnston ER, Albritton LM, Radke K. Envelope proteins containing single amino acid substitutions support a structural model of the receptor-

binding domain of bovine leukemia virus surface protein. J Virol. 2002;76(21):10861–72.
36. Miller JM, Miller LD, Olson C, Gillette KG. Virus-like particles in phytohemagglutinin-stimulated lymphocyte cultures with reference to bovine lymphosarcoma. J Natl Cancer Inst. 1969;43(6):1297–305.
37. Esteban EN, Poli M, Poiesz B, Ceriani C, Dube S, Gutierrez S, Dolcini G, Gagliardi R, Perez S, Lützelschwab C, Feldman L, Juliarena MA. Bovine leukemia virus (BLV), proposed control and eradication programs by marker assisted breeding of genetically resistant cattle. In: Rechi LJ, editor. Animal genetics. Hauppauge, NY: Nova Science Publishers Inc; 2009. p. 1–24.
38. Juliarena MA, Lendez PA, Gutierrez SE, Forletti A, Rensetti DE, Ceriani MC. Partial molecular characterization of different proviral strains of bovine leukemia virus. Arch Virol. 2013;158:63–70.
39. Altanerova V, Holicova D, Kucerova L, Altaner C, Lairmore MD, Boris-Lawrie K. Long-term infection with retroviral structural gene vector provides protection against bovine leukemia virus disease in rabbits. Virology. 2004;329(2):434–9.
40. Gutiérrez G, Rodríguez SM, Brogniez A, Gillet N, Golime R, Burny A, Jaworski JP, Alvarez I, Vagnoni L, Trono K, Willems L. Vaccination against δ-retroviruses: the bovine leukemia virus paradigm. Viruses. 2014. https://doi.org/10.3390/v6062416.
41. Sugimoto M, Ohishi K, Ikawa Y. Role of cell-mediated immunity in bovine leukemia virus (BLV) infection in ruminants: its implication for the vaccination strategy against retroviruses. Ther Immunol. 1994;1(5):297–301.
42. Ohishi K, Kabeya H, Amanuma H, Onuma M. Peptide-based bovine leukemia virus (BLV) vaccine that induces BLV-Env specific Th-1 type immunity. Leukemia. 1997;11(Suppl 3):223–6.
43. Woon HG, Scott GM, Yiu KL, Miles DH, Rawlinson WD. Identification of putative functional motifs in viral proteins essential for human cytomegalovirus DNA replication. Virus Genes. 2008. https://doi.org/10.1007/s11262-008-0255-8.
44. Kolb AW, Schmidt TR, Dyer DW, Brandt CR. Sequence variation in the herpes simplex virus U(S)1 ocular virulence determinant. Invest Ophthalmol Vis Sci. 2011. https://doi.org/10.1167/iovs.10-7032.
45. Gallaher WR, Ball JM, Garry RF, Martin-Amedee AM, Montelaro RC. A general model for the surface glycoproteins of HIV and other retroviruses. AIDS Res Hum Retroviruses. 1995;11(2):191–202.
46. Johnston ER, Radke K. The SU and TM envelope protein subunits of bovine leukemia virus are linked by disulfide bonds, both in cells and in virions. J Virol. 2000;74(6):2930–5.
47. Wu SR, Sjöberg M, Wallin M, Lindqvist B, Ekström M, Hebert H, Koeck PJ, Garoff H. Turning of the receptor-binding domains opens up the murine leukaemia virus Env for membrane fusion. EMBO J. 2008;27:2799–808.
48. Kuo CW, Mirsaliotis A, Brighty DW. Antibodies to the envelope glycoprotein of human T cell leukemia virus type 1 robustly activate cell-mediated cytotoxic responses and directly neutralize viral infectivity at multiple steps of the entry process. J Immunol. 2011. https://doi.org/10.4049/jimmunol.1100070.
49. Battini JL, Danos O, Heard JM. Receptor-binding domain of murine leukemia virus envelope glycoproteins. J Virol. 1995;69(2):713–9.
50. Kim FJ, Manel N, Garrido EN, Valle C, Sitbon M, Battini JL. HTLV-1 and -2 envelope SU subdomains and critical determinants in receptor binding. Retrovirology. 2004;2(1):41.
51. Monti G, Schrijver R, Beier D. Genetic diversity and spread of Bovine leukaemia virus isolates in Argentine dairy cattle. Arch Virol. 2005;150(3):443–58.
52. Bailey TL, Williams N, Misleh C, Li WW. MEME: discovering and analyzing DNA and protein sequence motifs. Institute of Molecular Bioscience, The University of Queensland, St Lucia. Nucleic Acids Res. 2006. https://doi.org/10.1093/nar/gkl198.
53. Librado P, Rozas J. DnaSP v5: a software for comprehensive analysis of DNA polymorphism data. Bioinformatics. 2009;25:1451–2.
54. Fechner H, Blankenstein P, Looman AC, Elwert J, Geue L, Albrecht C, Kurg A, Beier D, Marquardt O, Ebner D. Provirus variants of the bovine leukemia virus and their relation to the serological status of naturally infected cattle. Virology. 1997;237(2):261–9.
55. Tamura K, Peterson D, Peterson N, Stecher G, Nei M, Kumar S. MEGA5: molecular evolutionary genetics analysis using maximum likelihood, evolutionary distance, and maximum parsimony methods. Mol Biol Evol. 2011;28:2731–9.
56. Fass D, Davey RA, Hamson CA, Kim PS, Cunningham JM, Berger JM. Structure of a murine leukemia virus receptor-binding glycoprotein at 2.0 angstrom resolution. Science. 1997;277(5332):1662–6.
57. Barnett AL, Wensel DL, Li W, Fass D, Cunningham JM. Structure and mechanism of a coreceptor for infection by a pathogenic feline retrovirus. J Virol. 2003;77(4):2717–29.
58. Arnold K, Bordoli L, Kopp J, Schwede T. The SWISS-MODEL Workspace: a web-based environment for protein structure homology modelling. Bioinformatics. 2006;22:195–201.
59. Lee JE, Fusco ML, Hessell AJ, Oswald WB, Burton DR, Saphire EO. Structure of the Ebola virus glycoprotein bound to an antibody from a human survivor. Nature. 2008. https://doi.org/10.1038/nature07082.
60. Bornholdt ZA, Ndungo E, Fusco ML, Bale S, Flyak AI, Crowe JE Jr, Chandran K, Saphire EO. Host-primed Ebola virus GP exposes a hydrophobic NPC1 receptor-binding pocket, revealing a target for broadly neutralizing antibodies. MBio. 2016. https://doi.org/10.1128/mbio.02154-15.

Permissions

All chapters in this book were first published in RETROVIROLOGY, by BioMed Central; hereby published with permission under the Creative Commons Attribution License or equivalent. Every chapter published in this book has been scrutinized by our experts. Their significance has been extensively debated. The topics covered herein carry significant findings which will fuel the growth of the discipline. They may even be implemented as practical applications or may be referred to as a beginning point for another development.

The contributors of this book come from diverse backgrounds, making this book a truly international effort. This book will bring forth new frontiers with its revolutionizing research information and detailed analysis of the nascent developments around the world.

We would like to thank all the contributing authors for lending their expertise to make the book truly unique. They have played a crucial role in the development of this book. Without their invaluable contributions this book wouldn't have been possible. They have made vital efforts to compile up to date information on the varied aspects of this subject to make this book a valuable addition to the collection of many professionals and students.

This book was conceptualized with the vision of imparting up-to-date information and advanced data in this field. To ensure the same, a matchless editorial board was set up. Every individual on the board went through rigorous rounds of assessment to prove their worth. After which they invested a large part of their time researching and compiling the most relevant data for our readers.

The editorial board has been involved in producing this book since its inception. They have spent rigorous hours researching and exploring the diverse topics which have resulted in the successful publishing of this book. They have passed on their knowledge of decades through this book. To expedite this challenging task, the publisher supported the team at every step. A small team of assistant editors was also appointed to further simplify the editing procedure and attain best results for the readers.

Apart from the editorial board, the designing team has also invested a significant amount of their time in understanding the subject and creating the most relevant covers. They scrutinized every image to scout for the most suitable representation of the subject and create an appropriate cover for the book.

The publishing team has been an ardent support to the editorial, designing and production team. Their endless efforts to recruit the best for this project, has resulted in the accomplishment of this book. They are a veteran in the field of academics and their pool of knowledge is as vast as their experience in printing. Their expertise and guidance has proved useful at every step. Their uncompromising quality standards have made this book an exceptional effort. Their encouragement from time to time has been an inspiration for everyone.

The publisher and the editorial board hope that this book will prove to be a valuable piece of knowledge for researchers, students, practitioners and scholars across the globe.

List of Contributors

Julien Matysiak, Paul Lesbats and Eric Mauro
Fundamental Microbiology and Pathogenicity Laboratory, UMR 5234 CNRS, University of Bordeaux, SFR TransBioMed, 146 rue Léo Saignat, 33076 Bordeaux Cedex, France

Jean-William Dupuy
Centre Génomique fonctionnelle Bordeaux, Plateforme Proteome, Université de Bordeaux, Bordeaux, France

Angelica P. Lopez and Manuel Llano
Department of Biological Sciences, University of Texas at El Paso, El Paso, TX, USA

Marc Ruff
Département de Biologie Structurale Intégrative, UDS, U596 INSERM, UMR7104 CNRS, IGBMC(Institut de Génétique et de Biologie Moléculaire et Cellulaire), Illkirch-Graffenstaden, France

Olivier Delelis
LBPA, UMR8113, CNRS, ENS-Cachan, Cachan, France
Viral DNA Integration and Chromatin Dynamics Network (DyNAVir), Paris, France

Marc Lavigne
Department of Virology, UMR 3569, CNRS, Institut Pasteur, Paris, France
Institut Cochin-INSERM U1016-CNRS UMR8104, Université Paris Descartes, Paris, France
Viral DNA Integration and Chromatin Dynamics Network (DyNAVir), Paris, France

Delphine Lapaillerie, Mohamed Salah Benleulmi, Christina Calmels, Marie-Line Andreola and Vincent Parissi
Fundamental Microbiology and Pathogenicity Laboratory, UMR 5234 CNRS, University of Bordeaux, SFR TransBioMed, 146 rue Léo Saignat, 33076 Bordeaux Cedex, France
International Associated Laboratory (LIA) of Microbiology and Immunology, CNRS/University de Bordeaux/Heinrich Pette Institute-Leibniz Institute for Experimental Virology, Bordeaux, France
Viral DNA Integration and Chromatin Dynamics Network (DyNAVir), Paris, France

Steven J. Smith and Stephen H. Hughes
HIV Dynamics and Replication Program, National Cancer Institute-Frederick, National Institutes of Health, Frederick, MD, USA

Xue Zhi Zhao and Terrence R. Burke Jr.
Chemical Biology Laboratory, National Cancer Institute-Frederick, National Institutes of Health, Frederick, MD, USA

Kenta Iijima and Yukihito Ishizaka
Department of Intractable Diseases, National Center for Global Health and Medicine, 1-21-1 Toyama, Shinjuku-ku, Tokyo 162-8655, Japan

Junya Kobayashi
Department of Genome Repair Dynamics, Radiation Biology Center, Kyoto University, Yoshidakonoe-cho, Sakyo-ku, Kyoto 606-8501, Japan

Carin K. Ingemarsdotter, Jingwei Zeng and Ziqi Long
Department of Medicine, University of Cambridge, Addenbrookes Hospital, Level 5, Hills Rd, Cambridge CB2 0QQ, UK

Andrew M. L. Lever
Department of Medicine, University of Cambridge, Addenbrookes Hospital, Box 157, Level 5, Hills Rd, Cambridge CB2 0QQ, UK
Department of Medicine, National University of Singapore, Singapore, Singapore

Julia C. Kenyon
Department of Medicine, University of Cambridge, Addenbrookes Hospital, Box 157, Level 5, Hills Rd, Cambridge CB2 0QQ, UK
Department of Microbiology and Immunology, National University of Singapore, Singapore, Singapore
Homerton College, University of Cambridge, Cambridge, UK

Yoshio Koyanagi
Laboratory of Systems Virology, Institute for Frontier Life and Medical Sciences,
Kyoto University, Kyoto, Japan

Yoriyuki Konno and Shumpei Nagaoka
Laboratory of Systems Virology, Institute for Frontier Life and Medical Sciences, Kyoto University, Kyoto, Japan
Graduate School of Biostudies, Kyoto University, Kyoto, Japan

Izumi Kimura
Laboratory of Systems Virology, Institute for Frontier Life and Medical Sciences, Kyoto University, Kyoto, Japan
Graduate School of Pharmaceutical Sciences, Kyoto University, Kyoto, Japan

Keisuke Yamamoto
Laboratory of Systems Virology, Institute for Frontier Life and Medical Sciences, Kyoto University, Kyoto, Japan
Graduate School of Medicine, Kyoto University, Kyoto, Japan

Yumiko Kagawa
Laboratory of Systems Virology, Institute for Frontier Life and Medical Sciences, Kyoto University, Kyoto, Japan
Faculty of Medicine, Kyoto University, Kyoto, Japan

Ryuichi Kumata
Laboratory of Systems Virology, Institute for Frontier Life and Medical Sciences, Kyoto University, Kyoto, Japan
Faculty of Science, Kyoto University, Kyoto, Japan

Hirofumi Aso
Laboratory of Systems Virology, Institute for Frontier Life and Medical Sciences, Kyoto University, Kyoto, Japan
Graduate School of Pharmaceutical Sciences, Kyoto University, Kyoto, Japan
Faculty of Pharmaceutical Sciences, Kyoto University, Kyoto, Japan

Mahoko Takahashi Ueda
Micro/Nano Technology Center, Tokai University, Kanagawa, Japan

So Nakagawa
Micro/Nano Technology Center, Tokai University, Kanagawa, Japan
Department of Molecular Life Science, Tokai University School of Medicine, Tokai University, Kanagawa, Japan

Tomoko Kobayashi
Department of Animal Science, Faculty of Agriculture, Tokyo University of Agriculture, Kanagawa, Japan

Kei Sato
Laboratory of Systems Virology, Institute for Frontier Life and Medical Sciences, Kyoto University, Kyoto, Japan
CREST, Japan Science and Technology Agency, Saitama, Japan
Division of Systems Virology, Department of Infectious Disease Control, International Research Center for Infectious Diseases, Institute of Medical Science, The University of Tokyo, 4-6-1 Shirokanedai, Minato-ku, Tokyo 1088639, Japan

Eugene L. Asahchop, Manmeet K. Mamik, Wing F. Chan and William G. Branton
Department of Medicine (Neurology), University of Alberta, Edmonton, AB, Canada

Christopher Power
Department of Medicine (Neurology), University of Alberta, Edmonton, AB, Canada
Department of Psychiatry, University of Alberta, Edmonton, AB, Canada
Department of Medicine, University of Calgary, Calgary, AB, Canada

Glen B. Baker
Department of Psychiatry, University of Alberta, Edmonton, AB, Canada

Lothar Resch
Department of Pathology, University of Calgary, Calgary, AB, Canada

M. John Gill
Department of Medicine, University of Calgary, Calgary, AB, Canada

Elie Haddad
CHU Sainte-Justine, Montréal, Canada
Department of Pediatrics, Université de Montréal, Montréal, Canada

Eric A. Cohen
Department of Microbiology, Infectiology and Immunology, Université de Montréal, Montréal, Canada
Montreal Clinical Research Institute, Montréal, Canada

Oussama Meziane
Montreal Clinical Research Institute, Montréal, Canada

Jean V. Guimond
CIUSSS du Centre-Sud-de-l'ile-Montréal, CLSC des Faubourgs, Montréal, QC, Canada

Mark A. Wainberg
McGill University AIDS Centre, Lady Davis Institute for Medical Research, Jewish General Hospital, Montréal, QC, Canada

Meijuan Niu and Abdelkrim Temzi
HIV-1 RNA Trafficking Laboratory, Lady Davis Institute at the Jewish General Hospital, Montreal, QC H3T 1E2, Canada

Shringar Rao
HIV-1 RNA Trafficking Laboratory, Lady Davis Institute at the Jewish General Hospital, Montreal, QC H3T 1E2, Canada
Department of Microbiology and Immunology, McGill University, Montreal, QC H3A 2B4, Canada

Andrew J. Mouland
HIV-1 RNA Trafficking Laboratory, Lady Davis Institute at the Jewish General Hospital, Montreal, QC H3T 1E2, Canada
Department of Microbiology and Immunology, McGill University, Montreal, QC H3A 2B4, Canada
Department of Medicine, McGill University, Montreal, QC H3A 0G4, Canada

Raquel Amorim
HIV-1 RNA Trafficking Laboratory, Lady Davis Institute at the Jewish General Hospital, Montreal, QC H3T 1E2, Canada
Department of Medicine, McGill University, Montreal, QC H3A 0G4, Canada

Fu-Hsien Yu and Chin-Tien Wang
Department of Medical Research, Taipei Veterans General Hospital and Institute of Clinical Medicine, National Yang-Ming University School of Medicine, 201, Sec. 2, Shih-Pai Road, Taipei 11217, Taiwan

John M. Coffin and Zachary H. Williams
Department of Molecular Biology and Microbiology, Tufts University School of Medicine, Boston, MA, USA

Charlotte Kuperwasser
Department of Developmental, Chemical, and Molecular Biology, Tufts University School of Medicine, Boston, MA, USA
Raymond and Beverly Sackler Convergence Laboratory, Tufts University School of Medicine, Boston, MA, USA

Meagan Montesion
Department of Molecular Biology and Microbiology, Tufts University School of Medicine, Boston, MA, USA
Foundation Medicine, Inc., Cambridge, MA, USA

Ravi P. Subramanian
Department of Molecular Biology and Microbiology, Tufts University School of Medicine, Boston, MA, USA
Excerpta Medica, New York, NY, USA

Hiroshi Yotsuyanagi
Division of Infectious Diseases and Applied Immunology, Research Hospital, The Institute of Medical Science, The University of Tokyo, Tokyo, Japan

Masato Ogishi
Division of Infectious Diseases and Applied Immunology, Research Hospital, The Institute of Medical Science, The University of Tokyo, Tokyo, Japan
National Center for Global Health and Medicine, Tokyo, Japan

Malgorzata Drabczyk-Pluta, Tanja Werner, Ulf Dittmer and Gennadiy Zelinskyy
Institute of Virology, University Hospital Essen, University of Duisburg-Essen, Hufelandstraße 55, 45147 Essen, Germany

Daniel Hoffmann
Research Group Bioinformatics, Faculty of Biology, University of Duisburg-Essen, Essen, Germany

Qibin Leng
Key Laboratory of Molecular Virology and Immunology, Institute Pasteur of Shanghai, Institutes for Biological Sciences, Chinese Academy of Sciences, Shanghai, China

Lieping Chen
Department of Immunobiology, Yale School of Medicine, Yale University, New Haven, CT, USA

Hury Hellen Souza de Paula, Diogo Gama Caetano, Edson Delatorre, Sylvia Lopes Maia Teixeira, Mariza Gonçalves Morgado, Monick Lindenmeyer Guimarães and Fernanda Heloise Côrtes
Laboratório de Aids e Imunologia Molecular, Instituto Oswaldo Cruz - IOC, FIOCRUZ, Av Brasil, 4365, Pavilhão Leônidas Deane, sala 401, Rio de Janeiro 21040360, Brazil

Ana Cristina Garcia Ferreira, Lara Esteves Coelho, Eduarda Grinsztejn João, Michelle Morata de Andrade, Sandra Wagner Cardoso, Beatriz Grinsztejn and Valdilea Gonçalves Veloso
Laboratório de Pesquisa Clínica em DST e Aids, Instituto Nacional de Infectologia Evandro Chagas - INI, FIOCRUZ, Rio de Janeiro 25984-220, Brazil

Quang N. Nguyen, Jonathon E. Himes, Qifeng Han, Amit Kumar, Riley Mangan, Nathan I. Nicely, Guanhua Xie, Nathan Vandergrift and Xiaoying Shen
Duke Human Vaccine Institute, Duke University School of Medicine, Durham, NC, USA

David R. Martinez
Duke Human Vaccine Institute, Duke University School of Medicine, Durham, NC, USA
Department of Molecular Genetics and Microbiology, Duke University School of Medicine, Durham, NC, USA

R. Whitney Edwards and Justin Pollara
Duke Human Vaccine Institute, Duke University School of Medicine, Durham, NC, USA
Department of Surgery, Duke University School of Medicine, Durham, NC, USA

Sallie R. Permar
Duke Human Vaccine Institute, Duke University School of Medicine, Durham, NC, USA
Department of Molecular Genetics and Microbiology, Duke University School of Medicine, Durham, NC, USA
Department of Pediatrics, Duke University School of Medicine, Durham, NC, USA
Department of Immunology, Duke University School of Medicine, Durham, NC, USA

Aneta Pluta, Marzena Rola-Łuszczak and Jacek Kuźmak
OIE Reference Laboratory for EBL, Department of Biochemistry, National Veterinary Research Institute, Pulawy, Poland

Lorraine M. Albritton
Department of Microbiology, Immunology and Biochemistry, College of Medicine, The University of Tennessee Health Science Center, Memphis, TN, USA

Index

CPSIA information can be obtained
at www.ICGtesting.com
Printed in the USA
LVHW060859030122
707669LV00002B/248

9 781639 272648